Food Microbiology
AN INTRODUCTION

THIRD EDITION

THIRD EDITION

Food Microbiology

AN INTRODUCTION

Thomas J. Montville
Department of Food Science
School of Environmental and Biological Sciences
Rutgers, the State University of New Jersey
New Brunswick, New Jersey

Karl R. Matthews
Department of Food Science
School of Environmental and Biological Sciences
Rutgers, the State University of New Jersey
New Brunswick, New Jersey

Kalmia E. Kniel
Department of Animal and Food Sciences
University of Delaware
Newark, Delaware

Washington, DC

Library of Congress Cataloging-in-Publication Data

Montville, Thomas J.

Food microbiology : an introduction / Thomas J. Montville and Karl R. Matthews ; Kalmia E. Kniel.—3rd ed.

p. cm.

Includes bibliographical references and index.

ISBN 978-1-55581-636-0 (hardcover)—ISBN 978-1-55581-720-6 (e-book)

1. Food—Microbiology. I. Matthews, Karl R. II. Kniel, Kalmia E. III. Title.

QR115.M625 2012

664.001'579—dc23

2012000682

10 9 8 7 6 5 4 3 2 1

Printed in the United States of America

Address editorial correspondence to ASM Press, 1752 N St. NW, Washington, DC 20036-2904, USA

E-mail: books@asmusa.org

Send orders to ASM Press, P.O. Box 605, Herndon, VA 20172, USA

Phone: (800) 546-2416 or (703) 661-1593

Fax: (703) 661-1501

Online: estore.asm.org

doi:10.1128/9781555817206

Cover and interior design: Susan Brown Schmidler

Cover photos

Background: Electron micrograph of a biofilm (imaged by A. E. H. Shearer). See Fig. 2.3 legend for details. Top row, left to right: *Aspergillus* ear rot in corn before harvest (Fig. 22.5, courtesy of Karl R. Matthews); cheeses (source: iStockphoto); cantaloupe (photo by Scott Bauer, U.S. Department of Agriculture); peanuts (source: Wikimedia Commons [user: "Flyingdream"]); lettuce (source: iStockphoto). Bottom row: (Left) Phase-contrast photomicrograph of *Bacillus anthracis* spores in their mother cells, taken through a green filter at ×400 magnification (courtesy of Thomas J. Montville); (right) low-temperature electron micrograph of a cluster of *E. coli* bacteria (photo by Eric Erbe, U.S. Department of Agriculture).

We dedicate this book
to our teachers,
to our colleagues,
and
to our students
who have taught us so much.

Contents

SECTION II

Foodborne Pathogenic Bacteria 105

Preface

FOOD MICROBIOLOGY is a dynamic area of study that reaches into every home and supports a multibillion-dollar food industry. This book provides a taste of its complexity and challenge. The third edition has been thoroughly updated and revised to reflect our newest understanding of the field.

The safety of food requires more than mere memorization of microbiological minutiae. It calls for critical thinking, innovative approaches, and healthy skepticism. We have tried to foster these skills so that today's students will be able to solve tomorrow's problems. Students and instructors have asked us to provide answers to the homework problems that have appeared in previous editions. We have done so for many of the problems, although frequently there is no single "correct answer."

We would have never attempted to write a textbook on such a complex topic as food microbiology from scratch. Fortunately, ASM Press had published an advanced food microbiology reference book for researchers, graduate students, and professors. The "big book," *Food Microbiology: Fundamentals and Frontiers,* was written by an army of subject area experts who presumed that the reader had a working knowledge of microbiology, biochemistry, and genetics. The success of the first two editions of that book gave us the courage (and the resource) to write a food microbiology textbook for undergraduates. *Food Microbiology: An Introduction* is the child of the "big book."

Academic integrity, and with it the need to acknowledge the work of others, is a big issue on most campuses. Some of the chapters in this book are entirely the work of the authors. Other chapters have taken the work of subject authors from *Food Microbiology: Fundamentals and Frontiers* and rewritten or edited it to various degrees so that it is accessible to undergraduates. In some cases, this meant adding foundational material; in others, it meant deleting details that only an expert needs to know. A few chapters needed only minimal editing. Most chapters in this book are quite different from those originally written for the "big book." We acknowledge the subject experts whose chapters in *Food Microbiology: Fundamentals and Frontiers* provided the foundation for our writing. They are Gary R. Acuff,

John W. Austin, J. Stan Bailey, Dane Bernard, Larry R. Beuchat, Gregory A. Bohach, Robert E. Brackett, Robert L. Buchanan, Herbert J. Buckenhüskes, Lloyd B. Bullerman, Iain Campbell, Michael L. Chikindas, Dean O. Cliver, Jean-Yves D'Aoust, P. Michael Davidson, James S. Dickson, Michael P. Doyle, Józef Farkas, Peter Feng, Graham H. Fleet, Joseph F. Frank, H. Ray Gamble, Per Einar Granum, Paul A. Hartman, Eugene G. Hayunga, Craig W. Hedberg, Ailsa D. Hocking, Lynn M. Jablonski, Timothy C. Jackson, Eric A. Johnson, Mark E. Johnson, James B. Kaper, Jimmy T. Keeton, Charles W. Kim, Sylvia M. Kirov, Todd R. Klaenhammer, Keith A. Lampel, Alex S. Lopez, Douglas L. Marshall, Anthony T. Maurelli, John Maurer, Bruce A. McClane, Jianghong Meng, Kenneth B. Miller, Irving Nachamkin, James D. Oliver, Ynes R. Ortega, Merle D. Pierson, John I. Pitt, Steven C. Ricke, Roy M. Robins-Browne, Peter Setlow, L. Michele Smoot, James L. Steele, Bala Swaminathan, Sterling S. Thompson, Richard C. Whiting, Karen Winkowski, Irene Zabala Díaz, Tong Zhao, and Shaohua Zhao. In all of our chapters, we have tried to write in a style, at a level, and in language appropriate for undergraduates. To enhance our book's utility as a textbook, we have added case studies, crossword puzzles, chapter summaries, questions for critical thought, a glossary, and even a few cartoons.

This book is divided into four sections. Students should be aware that the third edition contains a substantial amount of material not found in previous editions. Since different instructors' courses present the material in different order, the revised chapters are more self-contained, and pathogens are presented in alphabetical order. The first section of the book covers the foundational material, describing how bacteria grow in food, how the food affects their growth, control of microbial growth, spores, detection, and microbiological criteria. Instructors may choose to use the other three sections in virtually any order. The foodborne pathogenic bacteria are covered in section II. Section III contains chapters on beneficial microbes, spoilage organisms, and pathogens that are not bacteria. Lactic acid bacteria and yeast fermentations are covered separately. Molds are covered both as spoilage organisms and as potential toxin producers. Since viruses may cause more than half of all foodborne illnesses, treatment of viruses has been expanded to include explanations of lytic and temperate phages, the importance of bacteriophage infection prevention to the dairy industry, and the recent adoption of phages for pathogen control. Prions are not bacteria, molds, or viruses; in fact, they are not microbes at all. However, they are a major biological concern to the public and food safety experts, and they are covered in the same chapter as viruses. We have added a chapter on parasites, which are important sources of disease in many part of the world. Section IV covers the chemical, biological, and physical methods of controlling foodborne microbes and closes by examining industrial and regulatory strategies for ensuring food safety.

The reader should be grateful to the students who reviewed each chapter for level and depth of coverage, writing style, and "what an undergraduate could be expected to know," in addition to grammar and usage. Hanna Clune, Danielle Voss, and Jennifer Merle have taken on this role for this edition.

We thank Eleanor Riemer, our acquisitions editor, and Ken April, our production editor, at ASM Press. This edition would not have happened without Eleanor's enthusiasm and encouragement. Ken's attention to detail was critical for us "to get it right."

We hope that *Food Microbiology: An Introduction* makes the subject come alive and encourages you to explore careers in food microbiology. Remember, there will always be people who have to eat and there will always be microbes. Food microbiologists have great long-term job security.

Thomas J. Montville
Karl R. Matthews
Kalmia E. Kniel

About the Authors

THOMAS J. MONTVILLE is Professor II (distinguished) of Food Microbiology at Rutgers, the State University of New Jersey, where he received his B.S. in 1975. Dr. Montville received his Ph.D. from the Massachusetts Institute of Technology (MIT) and then worked at the U.S. Department of Agriculture (USDA) as a research microbiologist before returning to Rutgers as a professor. He has published over 100 research papers on *Clostridium botulinum*, *Listeria monocytogenes*, antimicrobial peptides, and, more recently, *Bacillus anthracis* spores. Dr. Montville was a member of the Food and Drug Administration's Food Advisory Committee, the Institute of Food Technologists' expert panel on antimicrobial resistance, and many grant review panels. Dr. Montville is a fellow of the American Academy of Microbiology and a fellow of the Institute of Food Technologists.

Author's Statement

My desire to know how things work drew me to science. When I was a child, my relatives saved their broken appliances so that I could take them apart and see how they worked. My attraction to science was fed by microscopes; the ability to see bacteria sucked me into the field of microbiology. Rods, cocci, spores, motile, tumbling, germinating before my very eyes!

Careers in science can take strange turns. My undergraduate goal was to get a good job where I didn't have to work the night shift, but my professors badgered me to attend graduate school, and I was admitted to MIT. My research there, on the dental bacterium *Streptococcus mutans*, had nothing to do with my career in food microbiology, but it taught me about science on a grand scale. The USDA was a great place to start a research career, but an opening at Rutgers allowed me to reclaim my laboratory station from undergraduate Applied Microbiology as part of my research laboratory. It's possible, and even desirable, for scientists to have a life outside the lab. I am a serious bicyclist, having ridden from California to Maine. (See pedalingprof.blogspot.com when you are too tired to study.) I've skied the Rockies with my son and completed the Philadelphia Marathon with my daughter. But at the end of an athletic event it's always good to *sit* in front of a microscope.

KARL R. MATTHEWS is Professor of Microbial Food Safety at Rutgers University. He received a Ph.D. from the University of Kentucky in 1988. Dr. Matthews has earned an international reputation for his work on the interaction of foodborne pathogens with fresh produce. This includes demonstrating the internal localization of bacteria during growth of leafy greens. He further showed that the internalization process is a passive event by demonstrating the internalization of fluorescent polystyrene beads. Dr. Matthews has also been active in research on antimicrobial resistance of foodborne bacteria, specifically on intrinsic mechanisms of resistance and transfer of resistance genes among bacteria in food.

Author's Statement

My interest in microbiology was sparked one summer when I was working on a dairy farm. I regularly drank raw milk, but one time after doing so I became extremely ill (I won't go into the messy details). I became intrigued by microorganisms associated with milk and the disease bovine mastitis. These beginnings led me to an exciting career in food microbiology, where every day seems to bring a new challenge to be addressed.

KALMIA E. KNIEL is Associate Professor of Food Parasitology and Virology in the Department of Animal and Food Sciences at the University of Delaware. She received her Ph.D. from Virginia Tech in food science and technology. Her doctoral work focused on protozoan parasites. After that, she was a postdoctoral microbiologist at the USDA Agricultural Research Service's Animal Parasitic Diseases Laboratory. She is now nationally recognized as a leading expert in transmission of viruses, protozoa, and bacteria in the preharvest environment. Dr. Kniel has been active in researching the mechanisms behind the survival and inactivation of norovirus, hepatitis A virus, and other enteric viruses prevalent in our water and foods. She is an active advocate for teaching food safety at all levels and has been involved with elementary and secondary education. At the University of Delaware, she teaches courses on foodborne outbreak investigations and the basics of food science and food safety.

Author's Statement

I received my first microscope when I was 10, and I was hooked. My children now use it to look at plant cells and pond life, and what a kick it is seeing them light up looking into those little lenses. Serving as a teaching assistant for a pathogenic bacteriology laboratory course changed my life. I relished working with the students, challenging them with fecal unknowns and mock sputum samples. The best part was seeing their reaction as they identified an unknown bacterium and observed growth on their petri plates. It's that level of excitement from students of all ages that I love. Working with students in my laboratory and in the classroom is an honor. We all share a great curiosity for science. I believe that food microbiology is the greatest science, as it includes basic scientific inquiries with an applied twist. I have been fortunate to work with a myriad of amazing people, and with them at my side I look forward to the challenges of every day.

SECTION

I

Basics of Food Microbiology

1

The Trajectory of Food Microbiology

LEARNING OBJECTIVES

The information in this chapter will help the student:

- increase awareness of the antiquity of microbial life and the newness of food microbiology as a scientific field
- appreciate how fundamental discoveries in microbiology still influence the practice of food microbiology
- understand the origins of food microbiology and thus anticipate its forward path

INTRODUCTION

A former president of the American Society for Microbiology (Box 1.1) defined *microbiology* as an artificial subdiscipline of biology based on size. This suggests that basic biological principles hold true, and are often discovered, in the field of microbiology. *Food microbiology* is a further subdivision of microbiology. It studies microbes that grow in food and how food environments influence microbes. In some ways, food microbiology has changed radically in the last 20 years. The number of recognized foodborne pathogens has doubled. "Safety through end product testing" has given way to the "safety by design" provided by Hazard Analysis and Critical Control Points (HACCP). Genetic and immunological probes have replaced biochemical tests and reduced testing time from days to minutes. In other ways, food microbiology is still near the beginning. Louis Pasteur would find his pipettes in a modern laboratory. Julius Richard Petri would find his plates (albeit plastic rather than glass). Hans Christian Gram would find all the reagents required for his stain. Food microbiologists still study only the microbes that we can see under the microscope and grow on agar media in petri dishes. Experts suggest that only 1% of all the bacteria in the biosphere can be detected by cultural methods.

This chapter's discussion of microbes per se sets the stage for a historical review of food microbiology. The bulk of it concerns bacteria. Most of this book deals with bacteria. Less is known about viruses and prions, so little that they are covered later in a single chapter. This chapter ends with some thoughts about future developments in the field.

WHO'S ON FIRST?

Let there be no doubt about it: the microbes were here first. It is a microbial world (Fig. 1.1). If the earth came into being at 12:01 a.m. of a 24-h day, microbes would arrive at dawn and remain the only living things until

doi:10.1128/9781555817206.ch01

BOX 1.1

Preparing for the future

Membership in professional societies is a great way to advance professionally, even as a student (who benefits from reduced membership fees and is eligible for a variety of scholarships). Professional societies provide continuing-education and employment services to their members, provide expertise to those making laws and public policies, have annual meetings for presentation of the latest science, and publish books and journals. The three main societies for food microbiologists are listed below. Applications for membership can be obtained from their websites.

The American Society for Microbiology (ASM) (http://www.asm.org) is the largest life science society in the world. It has 27 divisions covering all facets of microbiology from microbial pathogenesis to immunology to antimicrobial agents to food microbiology, and it publishes 12 scholarly journals. ASM Press is the publisher of this book.

The Institute of Food Technologists (IFT) (http://www.ift.org) devotes itself to all areas of food science by discipline (food microbiology, food chemistry, and food engineering), as well as by commodity (cereals, fruits and vegetables, and seafood) and by processing (refrigerated foods). The IFT is a nonprofit scientific society with 22,000 members (as of 2006) working in food science, food technology, and related professions in industry, academia, and government. The IFT publishes four journals, sponsors a variety of short courses, and contributes to public policy and opinion at national, state, and local levels.

The International Association for Food Protection (IAFP) (http://www.foodprotection.org) is the only professional society devoted exclusively to food safety microbiology. IAFP is dedicated to the education and service of its members, as well as industry personnel. Members keep informed of the latest scientific, technical, and practical developments in food safety and sanitation. IAFP publishes two scientific journals, *Food Protection Trends* and *Journal of Food Protection*.

well after dusk. Around 9 p.m., larger animals would emerge, and a few seconds before midnight, humans (Fig. 1.2) would appear. The microbes were here first, they cohabit the planet with us, and they will be here after humans are gone. Life is not sterile. Microbes can never be (nor should they be) conquered, once and for all. The food microbiologist can only create foods that microbes do not "like," manipulate the growth of microbes that are in food, kill them, or exclude them by physical barriers.

Bacteria live in airless bogs, thermal vents, boiling geysers, us, and foods. We are lucky that they are here, for microbes form the foundation of the biosphere. We could not exist without microbes, but they would do just fine without us. Photosynthetic bacteria fix carbon into usable forms and make much of our oxygen. *Rhizobium* bacteria fix air's elemental nitrogen into ammonia that can be used for a variety of life processes. Degradative enzymes allow ruminants to digest cellulose. Microbes recycle the dead into basic components that can be used again and again. Microbes in our intestines aid in digestion, produce vitamins, and prevent colonization by pathogens. For the most part, microbes are our friends.

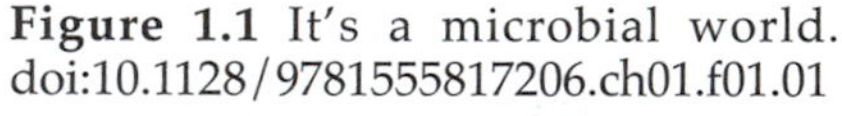

Figure 1.1 It's a microbial world. doi:10.1128/9781555817206.ch01.f01.01

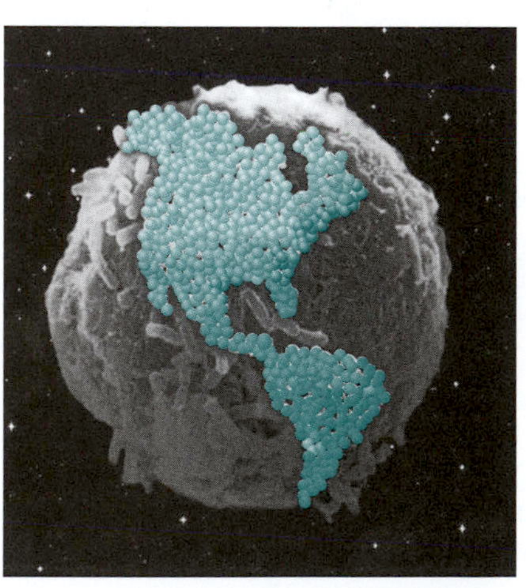

FOOD MICROBIOLOGY, PAST AND PRESENT

From the dawn of civilization until about 10,000 years ago, humans were hunter-gatherers. Humans were lucky to have enough. There was neither surplus nor a settled place to store it. Preservation was not an issue. With the shift to agricultural societies, storage, spoilage, and preservation became important challenges. The first preservation methods were undoubtedly accidental. Sun-dried, salted, or frozen foods did not spoil. In the classic "turning lemons into lemonade" style, early humans learned that "spoiled" milk could be acceptable or even desirable if viewed as "fermented." Fermenting food became an organized activity around 4000 B.C.E.

Figure 1.2 Evolution as seen through the eyes of Calvin and Hobbes. *Calvin and Hobbes* © Bill Watterson. Distributed by Universal Press Syndicate. Reprinted with permission. All rights reserved.

(Table 1.1). Breweries and bakeries sprung up long before the idea of yeast was conceived.

Humans remained ignorant of microbes for thousands of years. In 1665, Robert Hooke published *Micrographia,* the first illustrated book on microscopy that detailed the structure of *Mucor,* a microscopic fungus. In 1676, Antonie van Leeuwenhoek (Fig. 1.3) used a crude microscope (Fig. 1.4) of Hooke's design to see small living things in pond water. *Micro*biology (Fig. 1.5) was born!

Nonetheless, it took another 200 years to prove that microbes exist and cause fermentative processes. In the mid-1700s, Lazzaro Spallanzani showed that boiled meat placed in a sealed container did not spoil. Advocates of spontaneous generation, however, argued that air is needed for life and that the air was sealed out. It took another 100 years for Louis Pasteur's elegant "swan-necked flask" experiment (see Box 27.1) to replicate Spallanzani's experiment in a way that allowed access to air but not microbes. Napoleon, needing to feed his troops as they traveled across Europe, offered a prize to anyone who could preserve food. Nicolas Appert (Fig. 1.6) won this prize when he discovered that foods would not spoil if they were heated in sealed containers, i.e., were canned. Thus, canning was invented without any knowledge of microbiology. Indeed, it was not until the 1900s that mathematical bases for canning processes were developed.

Robert Koch was a giant of microbiology. In the late 1800s, he established criteria for proving that a bacterium caused a disease. To satisfy "Koch's postulates," one had to (i) isolate the suspected bacterium in pure culture from the diseased animal, (ii) expose a healthy animal to the bacterium and make it sick, and (iii) reisolate the bacterium from the newly infected animal.

This approach was so brilliant that it is still used today to prove that a disease has a microbial origin. Microbiology remains based on the study of single bacterial species in pure culture. Unfortunately, most microbes cannot be cultured and exist in nature as community members, not in pure culture. Two other innovations from Koch's laboratory are still with us. Julius Richard Petri, an assistant in Koch's laboratory, invented the

Figure 1.3 Antonie van Leeuwenhoek.
doi:10.1128/9781555817206.ch01.f01.03

Table 1.1 Significant events in the history of food microbiology[a]

Decade	Event
~4000 B.C.E.	Fermentation of food becomes an organized activity.
1670 C.E.	Hooke and van Leeuwenhoek observe microscopic fungi and bacteria; microbiology is born.
1760	Spallanzani's experiments with boiled beef strike a blow against spontaneous generation.
1800	Nicolas Appert invents the canning process. Amazingly, this is still a mainstay of food processing 200 years later.
1810	Peter Durand patents the tin can, making Appert's life much easier.
1850	Appert and Raymond Chevallier-Appert are issued a patent for steam sterilization (retort). The use of steam under pressure increases process temperatures, decreases process times, and radically improves the quality of canned food.
	Louis Pasteur demonstrates that living organisms cause lactic and alcoholic fermentations.
1860	Pasteur disproves spontaneous generation. Life can come only from other life.
	Joseph Lister develops the concept of antiseptic practice. Persuading surgeons to wash their hands saves thousands of lives.
1880	Robert Koch postulates bacteria as causative agents of disease. To this day, Koch's postulates remain the "gold standard" for proving that bacteria cause disease.
	Hans Christian Gram invents the Gram stain.
	Julius Richard Petri invents the petri dish. Petri worked in Koch's laboratory, where the usefulness of agar was also discovered.
	A. A. Gartner isolates *Salmonella enterica* serovar Enteritidis from a food poisoning outbreak. A century later, salmonellae are still the leading cause of death among foodborne microbes.
1890	Pasteurization of milk begins in the United States.
1900	The Food and Drug Act is passed in response to Upton Sinclair's exposé of the meat industry in *The Jungle*.
1920	The U.S. Public Health Commission publishes methods to prevent botulism.
	Alexander Fleming discovers antibiotics. Less than 100 years later, multiply antibiotic-resistant pathogens threaten to cause possible epidemics.
	Clarence Birdseye introduces frozen foods to the retail marketplace.
1930	G. M. Dack confirms that *Staphylococcus aureus* makes toxin.
	Home refrigerators are widely introduced.
	The Food, Drug, and Cosmetic Act strengthens regulation over foods.
	Viruses are discovered.
1940	The first freeze-dried food is developed.
	The supermarket replaces assorted shops as the place to buy food.
1950	James Watson and Francis Crick discover the structure of DNA. Rosalind Franklin plays a large but uncredited role.
	Research on food irradiation begins.
1960	C. Duncan and D. Strong demonstrate that perfringens food poisoning is caused by a toxin.
	The role of fungal toxins is discovered when "turkey X" disease breaks out in poultry.
1970	S. Cohen discovers genetic recombination in bacteria.
	Larry McKay reports the presence of plasmids in gram-positive bacteria.
	Good manufacturing practices are introduced for low-acid foods.
	Invention of monoclonal antibodies lays the foundation for mass-produced antibody-based tests.
1980	The first recognized outbreak of listeriosis occurs.
	Escherichia coli O157:H7 is first recognized as a pathogen.
	The first genetic probe for detection of *Salmonella* is developed.
	Polymerase chain reaction is invented.
	Prions are discovered.
1990	Irradiation is approved for pathogen control in meat and poultry.
	The mad cow disease crisis hits the United Kingdom.
	HACCP is required by the U.S. Department of Agriculture.
2000	Irradiation is approved for shell eggs.
	Regulatory concerns about bioterroristic contamination of food lead to new laws on facility registration, product traceability, and prenotification of imports.
	The first report of mad cow disease in the United States is made.
2012	**You are here.**

[a]Compiled from Brock (1999), Hartman (2001), and other sources.

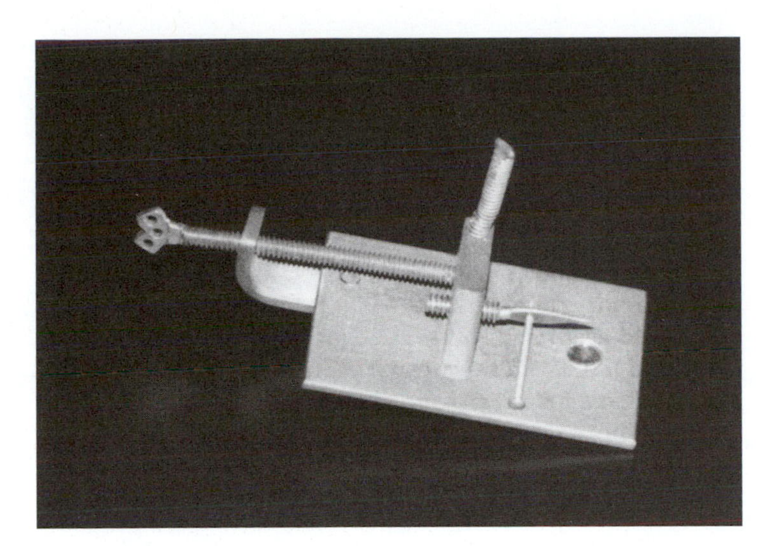

Figure 1.4 Replica of van Leeuwenhoek microscope. doi:10.1128/9781555817206.ch01.f01.04

Authors' note

I never appreciated van Leeuwenhoek's skill as a microscopist until I tried to use this replica microscope.

petri dish as a minor modification of Koch's plating method. And workers in the Koch laboratory were often frustrated by the gelatin used in the plating method. Gelatin would not solidify when the temperature was too warm and was dissolved by certain organisms (that produced enzymes that degrade gelatin). Walter Hesse, a physician who had joined the Koch

Figure 1.5 The invention of the microscope made laboratories much more interesting. *In the Bleachers* © Steve Moore. Reprinted with permission of Universal Press Syndicate. All rights reserved.

Figure 1.6 Stamp honoring Nicolas Appert. doi:10.1128/9781555817206.ch01.f01.06

laboratory to study infectious bacteria in the air, shared this frustration with his wife, Fanny. Frau Hesse had been using agar to thicken jams and jellies for years and suggested that they try agar in the laboratory. You know the rest of the story.

The first half of the 20th century was marked by the discovery of the "traditional" foodborne pathogens. *Salmonella* species came from warm-blooded animals. *Clostridium botulinum* became a problem with improperly canned foods. *Staphylococcus aureus* became associated with poor hygiene, *Bacillus cereus* contaminated starchy foods, and other food-organism combinations became associated with certain illnesses. Viruses were first crystallized and associated with disease in the 1930s and are a major cause of foodborne illness. However, relatively little is known about them and they are "understudied" relative to bacteria. In the middle of the 20th century, biology made a great leap forward: James Watson and Francis Crick discovered the structure of DNA (Fig. 1.7). This gave birth to the era of molecular genetics. This has led to a better understanding of how bacteria cause illness, revolutionary genetic and antibody-based detection methods for bacteria, genetic "fingerprints" as epidemiological tools, and the ability to genetically alter fermentation organisms to improve their industrial characteristics. The tools of molecular biology are becoming increasingly important to food microbiology.

Figure 1.7 The discovery of DNA's double helix marked a quantum leap in the history of microbiology. doi:10.1128/9781555817206.ch01.f01.07

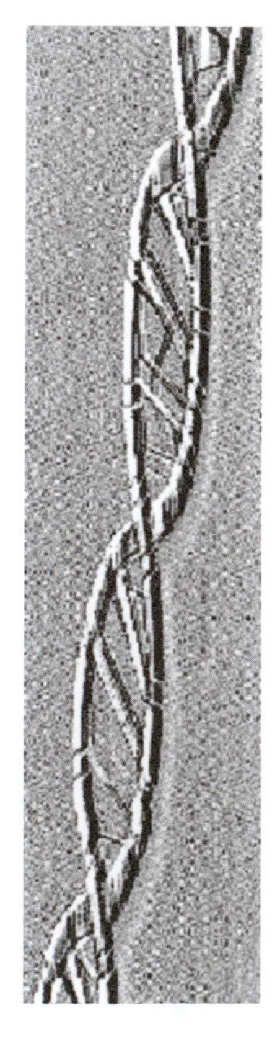

The movement away from massive end product testing to safety by design also started in this era. "Good manufacturing practices" provide manufacturers with procedures that should yield safe products. HACCP codified this into a safety assurance system. Food irradiation is approved as a "kill" step in the processing of raw poultry and meat. Risk assessment provides the basis for more sophisticated regulatory approaches based on the end result (fewer sick people) rather than dictating processes that give fewer microbes.

TO THE FUTURE AND BEYOND

My mentor once told me, "If you stare into the crystal ball too long, you end up with shards of glass in your face." It is hard to predict the future. However, it is safe to say that there will be greater demand for more, safer foods, that innovations in medical microbiology will continue to migrate into food microbiology, and that entirely new concepts will alter the course of food microbiology.

About one-third of the world's food supply is lost to spoilage. A global population that will double in your lifetime cannot afford this loss. If everyone is to eat, that level of spoilage must be reduced. Hunger hovers over humanity and may come to dwarf safety as an issue. Fermentation as a low-energy "appropriate technology" may become more important as a preservation method. Microbes themselves may become a food source. (Yeast was touted as "single-cell protein" in the 1970s.) Probiotic bacteria may help people maintain health through their diet. Perhaps, just as pesticides have been cloned into plants to kill insects, antimicrobials will be cloned into food to make them kill bacteria.

While 30% of the world's population is worried about getting *enough* food, the third of the world's population that worries about "overnutrition" demands *safer* food. Zero risk is impossible, and the cost of safety increases exponentially as one approaches zero. How safe is "safe enough"? When should a society shift resources from increased safety to increased availability? The requirements of international trade mandate the harmonization of safety standards. Globalization of the food supply creates new challenges to food safety. Ideally, First and Third World countries should have similar sanitary standards. But how can this be achieved in countries where indoor plumbing and rudimentary sanitation are scarce and refrigeration is almost nonexistent?

Scientific innovations usually appear first in medical technology and have an impact on the food system relatively later. This can be seen through the adaptation of rapid and automated methods, food-processing facilities increasingly adapting the methods of pharmaceutical houses, and increased emphasis on molecular understanding. Look at what is "hot" in the biomedical area today, and you will see the food microbiology of tomorrow.

This chapter highlights the scientific limits of microbiology. They are starting to fall. The causes of half of the cases of foodborne illness are unknown. Could they be caused by viruses? (It is estimated that viruses might be responsible for as much as 60% of foodborne illnesses.) Could they be due to microbes that are there but nonculturable? Will new molecular techniques for studying these bacteria allow us to understand and control them? The first 200 years of microbiology has been devoted to the study of bacteria as if they were noninteracting ball bearings. We now know that they "talk" to each other using chemical signals and sensors. This quorum sensing helps microbial communities decide when they will take action (for example, to mount an infectious attack). If bacteria talk to each other, what are they saying? If we knew, could we prevent spoilage and disease by scrambling the signal, drowning the signal out with other biochemical noise, or inactivating the receivers? What about the 99% of bacteria that remain undiscovered? What is in their gene pool? How might we use those genes and their products to make food better and safer? What are the roles of prions and viruses in foodborne disease? How do we participate in the ethical debate that surrounds genetic engineering, food irradiation, and other new technologies?

The most exciting era of food microbiology may lie in its future. It is the period that you will live in. You may create the next epoch in food microbiology. To prepare for this, gain a foundational understanding of the field today, learn how to think critically and creatively, and learn to love these creatures that are so small but so smart.

An Undergraduate's Research Experience

Rebecca Montville, The Center School, New York, New York

A lucky chance to do undergraduate research changed my life. I became one of the nation's experts on the microbiology of hand washing, traveled nationally and internationally to present my research, and published several papers in international journals.

How did this happen? As a college freshman, I wasn't sure about a major or where my college career should lead me. My introductory classes in math and science both bored and alienated me. To make some extra money, I started doing clerical work for Professor Don Schaffner. This turned out to be auspicious. One fateful day, he asked me if I would be interested in working on a quantitative risk assessment (QRA) (see box in chapter 29) of hand washing. Hand washing always interested me, so I happily accepted the offer. For my summer job, I worked to become an expert on the literature of previous studies. Then I could convert those data into mathematical equations. These equations were used to generate computer models that predicted what would happen under various hand washing conditions.

When the hand washing QRA was published, it was time for a new project. From my extensive reading I knew there were limited data on the effectiveness of glove use in the food service industry. I pitched it as a research problem to Dr. Schaffner. He supported my idea. The research was performed in a traditional research laboratory and was very different than the computer modeling I had done on the QRA. It felt like I was starting over from scratch. Fortunately, my labmates patiently taught me everything I needed to know. My research led to a number of extraordinary experiences. Dr. Schaffner invited me to Washington, DC, when he presented our hand washing QRA to the Food and Drug Administration. I was able to travel and present my research at conferences in Chicago, Dallas, Minneapolis, and Detroit. After a few years in the lab, I became the expert and got the chance to mentor other undergraduates and high school students. I enjoyed passing on all that my labmates had taught me.

What did I learn from all of this? I learned how to do literature searches, read technical papers, and use highly technical software. I learned how to critique the methods and analysis of others' research and, perhaps most importantly, how scientists communicate with each other. I learned more doing this research than I did in all of my classes combined. I learned how much I really loved research science. This helped define my college career. I combed the course catalogs and found a major that encompassed much of what I had done and learned: biomathematics. Now that it had a focus, coursework became more interesting. As an added bonus, all of these research experiences look really great on my résumé. After graduation, I went to work in a virology lab at Yale University. Now I teach middle school math and science and try to excite a future generation of scientists.

Do I have a word of wisdom for you? Whenever someone tells me that they are considering a career in science, I offer this advice: find someone who is researching a topic that interests you and ask if they will take you on as an intern. Asking that question changed my life.

Summary

- Microbes were the first living things on earth and remain the foundation of the biosphere.
- The fundamental biological discoveries of the 1800s, such as the disproof of spontaneous generation, Koch's postulates, and pure culture of microbes, continue to influence the field.
- Molecular biology and genetics have begun to have an impact on food microbiology and will shape its future.
- Food microbiology is a relatively young field of study.

Suggested reading

Brock, T. D. 1999. *Milestones in Microbiology: 1546 to 1940*. ASM Press, Washington, DC.

Hartman, P. 2001. The evolution of food microbiology, p. 3–12. *In* M. P. Doyle, L. R. Beuchat, and T. J. Montville (ed.), *Food Microbiology: Fundamentals and Frontiers*, 2nd ed. ASM Press, Washington, DC.

Needham, C., M. Hoagland, K. McPherson, and B. Dodson. 2000. *Intimate Strangers: Unseen Life on Earth*. ASM Press, Washington, DC.

Questions for critical thought

1. What would have happened if there had been spores in Pasteur's swan-necked flask?
2. Can you think of another experiment that would disprove spontaneous generation? How would food safety be affected if there were spontaneous generation?
3. Extend Table 1.1 50 years into the future. Enter what you think might happen.
4. Discuss three historical discoveries in microbiology that still influence food microbiology.
5. It is said that if Pasteur walked into a food microbiology laboratory today, he could be comfortably at work in about 15 min. Is this true? Is it good?
6. If 99% of bacteria are nonculturable, how do we know that they are there?
7. Looking back at Table 1.1, what historical development do you think has had the greatest impact? Why?

2

Microbial Growth, Survival, and Death in Foods

LEARNING OBJECTIVES

The information in this chapter will enable the student to:

- distinguish among different methods of culturing foodborne microbes and choose the correct one for a given application
- recognize how intrinsic and extrinsic factors are used in controlling microbial growth
- use the water activity concept to distinguish between "bulk water" and water that is unavailable for microbial growth
- understand the effect of water activity on microbial growth and microbial ecology
- quantitatively predict the size of microbial populations and their rate of growth using equations that describe the kinetics of microbial growth
- qualitatively describe how altering the variables in the equations for the kinetics of microbial growth alters growth patterns
- relate biochemical pathways to energy generation and metabolic products of foodborne bacteria

INTRODUCTION

Never say that you are *just* a food microbiologist. Food microbiology is a specialty area in microbiology that is regulated by many government agencies (Box 2.1). Just as a heart surgeon is more than a mere surgeon, a food microbiologist is a microbiologist with a specialty. Food microbiologists must understand microbes, and in addition to that they must understand complex food systems. Finally, they must integrate both to solve microbiological problems in complex food ecosystems. Even the question "How many bacteria are in that food?" has no simple answer. The first part of this chapter examines the question "How do we detect and quantify bacteria in food?" We present the concept of culturability and ask if it really reflects *viability*. Even though injured or otherwise nonculturable cells cannot form colonies in petri dishes, they do exist and can cause illness.

The second part of the chapter presents the importance of food intrinsic (inherent to) and extrinsic (external) factors for bacterial growth. Ecology teaches that the *interaction* of factors determines which organisms grow in an environment. The use of multiple environmental factors (i.e., pH, salt concentration, temperature, etc.) to inhibit microbial growth is called hurdle technology. Hurdle technology is an increasingly important preservation strategy.

doi:10.1128/9781555817206.ch02

BOX 2.1

Agencies involved in food safety

The U.S. Department of Agriculture (USDA) regulates meat and poultry through its Food Safety and Inspection Service (FSIS). All meat and poultry must be processed under continuous USDA inspection. The USDA does not have the authority to force a recall, but it can pressure food processors to initiate a "voluntary" recall. The USDA's Agricultural Marketing Service is responsible for the regulation of organic food, while the Agricultural Research Service provides the science that the regulations are based on and that is required to promote U.S. agriculture.

The Food and Drug Administration (FDA) regulates all foods except meat, poultry, and alcohol (which is regulated by the Bureau of Alcohol, Firearms, Tobacco and Explosives). The FDA periodically inspects food-manufacturing facilities, must approve all thermal processes for low-acid canned foods, has the authority to issue mandatory recalls, and can quarantine foods under its authority for administrative detention. The FDA also regulates drugs and medical devices. The food side of the agency resides in the Center for Food Safety and Applied Nutrition (CFSAN).

The Centers for Disease Control and Prevention (CDC) is the part of the Department of Health and Human Services that provides epidemiological expertise to investigate outbreaks of foodborne illness and tracks the incidence of foodborne illness over time. It oversees PulseNet, a "DNA fingerprinting clearinghouse" that can link geographically dispersed cases to a common outbreak. The CDC has no regulatory authority.

The Environmental Protection Agency (EPA) has regulatory authority over insecticides, fungicides, and herbicides that are applied to crops.

Consider this: a "bargain pot pie with chicken" containing <3% chicken would be regulated by the FDA, but if the company started an "upscale chicken pot pie" with 30% chicken, it would be regulated by the USDA. If the upscale pot pie were so popular that the company added an upscale vegetarian pot pie to its product lineup, that would be regulated by the FDA. Similarly, when benzoic acid is added to yogurt or cottage cheese to inhibit microbial growth, it is regulated by the FDA. When the same compound is used in the field to prevent fungi from growing on grains, the EPA has regulatory authority.

The kinetics of microbial growth are covered in the third part of this chapter. All four phases of the microbial growth curve are important to food microbiologists. If the lag phase can be extended beyond the normal shelf life, the food is microbially safe. If the growth rate can be slowed so that the population is too small to cause illness, the product is safe. Increasing the rate of the death phase can help ensure safety, as in the case of cheddaring cheese made from unpasteurized milk. Quantifying microbial growth rates and predicting microbial population sizes are important skills for students of food microbiology.

The fourth part of this chapter introduces the physiology and metabolism of foodborne microbes. We deliberately cover it last lest students panic at the sight of the word "biochemistry." This book assumes no background in biochemistry, but the biochemistry associated with foodborne microbes can be very important. The biochemical pathways used by lactic acid bacteria determine the characteristics of fermented foods. Spoilage is also caused by biochemicals. Lactic acid makes milk sour; carbon dioxide causes cans to swell. The ability of bacteria to grow under adverse conditions is governed by their ability to generate different amounts of energy (in the form of adenosine 5′-triphosphate [ATP]) from different biochemical pathways. Bacteria grown in the presence of oxygen (aerobically) produce more energy (ATP) than those grown in the absence of oxygen (anaerobically) (Box 2.2). Thus, organisms such as *Staphylococcus aureus* can withstand hostile environments, such as relatively high salt concentrations, better under aerobic conditions than under anaerobic conditions.

BOX 2.2

Mechanisms of energy generation

Anaerobes cannot liberate all of the energy in a substrate. They typically make one or two ATP molecules by burning 1 mol of glucose. This process is called *substrate-level phosphorylation*. Strict anaerobes are killed by air.

Aerobes can burn glucose completely in the presence of oxygen. They generate 34 ATP molecules from 1 mol of glucose through a process called *respiration*. Strict aerobes cannot grow in the absence of oxygen.

Facultative anaerobes can generate energy by both substrate-level phosphorylation and respiration. They can grow in a wider variety of environments than strict aerobes or strict anaerobes.

Chemical structure of ATP. doi:10.1128/9781555817206.ch02.fBox2.2

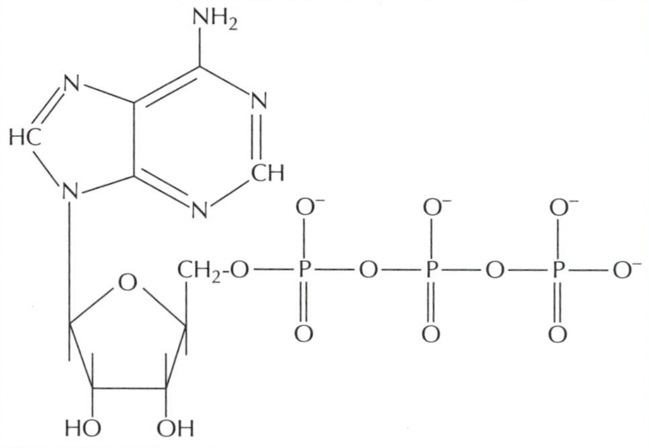

FOOD ECOSYSTEMS, HOMEOSTASIS, AND HURDLE TECHNOLOGY

Foods as Ecosystems

Foods are ecosystems composed of the environment and the organisms that live in it. The food environment is composed of intrinsic factors (inherent to the food) (i.e., pH, water activity, and nutrients) and extrinsic factors (external to it) (i.e., temperature, gaseous environment, and the presence of other bacteria). Intrinsic and extrinsic factors can be manipulated to preserve food. When applied to microbiology, ecology can be defined as "the study of the *interactions* between the chemical, physical, and structural aspects of a niche and the composition of its specific microbial population" (International Commission on Microbiological Specifications for Foods, 1980). "Interactions" emphasizes the dynamic complexity of food ecosystems.

Foods can be heterogeneous on a scale of micrometers. Heterogeneity and gradients of pH, oxygen, nutrients, etc., are key ecological factors in foods. Foods may contain multiple microenvironments. This is well illustrated by the food poisoning outbreaks associated with "aerobic" foods caused by a "strict anaerobe" like *Clostridium botulinum*. Growth of *C. botulinum* in potatoes, sautéed onions, and coleslaw has caused botulism. The food's oxygen is driven out during cooking and diffuses back in so slowly that most of the product remains anaerobic.

CLASSICAL MICROBIOLOGY AND ITS LIMITATIONS

Limitations of Detection and Enumeration Methods

All methods based on the plate count have the same limitations. The plate count assumes that every cell forms one colony and that every colony originates from only one cell. The ability of a given cell to grow to the size of a macroscopic colony depends on many things. These include the cell's physiological state, the growth medium used (and the presence or absence of selective agents), the incubation temperature, etc.

Figure 2.1 illustrates the concepts of plate counting, qualitative detection, and most-probable-number (MPN) methodologies. Consider a 50-ml sample of milk that contains 10^6 CFU of anaerobic bacteria per ml, 10^5 aerobic bacteria per ml, a significant number of *Staphylococcus aureus* organisms (a pathogen), some *Escherichia coli* organisms, and a few *Salmonella* organisms. ("CFU" stands for "colony-forming unit," which ideally would be a single cell but in reality could be a chain of 10 or a clump of 100 cells.) Since the number of bacteria is either too large to count directly or too small to find easily, samples have to be diluted or enriched.

Authors' note

Singular and plural: learn these now and you will start to sound like a microbiologist. "Media" is the plural of "medium." "Bacteria" is the plural of "bacterium." "Data" is the plural for "datum," although in practice "data" is used for both the singular and plural.

Plate Counts

Plate counts are used to quantify bacterial populations of >250 CFU/ml (for liquids) or >2,500 CFU/ml (for solids, which must be diluted 1:10 into a liquid to be pipettable). The food sample is first homogenized 1:10 (weight:volume) in a buffer to give a 10-fold dilution. This is then diluted through a series of 10-fold dilutions to give 10-, 100-, 1,000-, 10,000-fold, etc., dilutions and spread onto an agar medium. After aerobic incubation at 35°C, the average number of colonies per plate (on plates having between 25 and 250 colonies) is counted and multiplied by the dilution factor to give the number of bacteria in the original sample. This sounds simple, right? Wrong. At one time, this procedure was referred to as the total plate count. However, it counts only those bacteria that can grow on that agar medium at 35°C in the presence of air. The larger population of anaerobes is not detected because they cannot grow in the presence of air. Bacteria that grow only below 15°C might be a good predictor of the refrigerated milk's shelf life. They are not detected either. *Staphylococcus aureus*, an opportunistic pathogen (i.e., one which under some conditions can make you sick), would likely go undetected and be lost in the crowd of harmless bacteria. Variations of the standard aerobic plate count consider these issues. Plate counts can be determined at different temperatures or under different atmospheres to increase their specificity for certain types of bacteria.

Selective, or Differential, Media

The concept of selective media is simple. Some ingredient (such as an unusual sugar) that favors the growth of the targeted organism is included in the medium. The medium might also contain ingredients (such as antibiotics, salt, surfactants, etc.) that inhibit the growth of competing bacteria. Table 2.1 gives some examples of selective media. In the case of Baird-Parker agar (see Fig. 16.2), lithium chloride, glycine, and tellurite allow *S. aureus* to grow while suppressing a larger number of other organisms. However, if injured (see below), target organisms may be killed by the selective agent. In addition, the selective media are not entirely selective. *Lactobacillus* de Man-Rogosa-Sharpe (MRS) medium is selective for lactobacilli but also

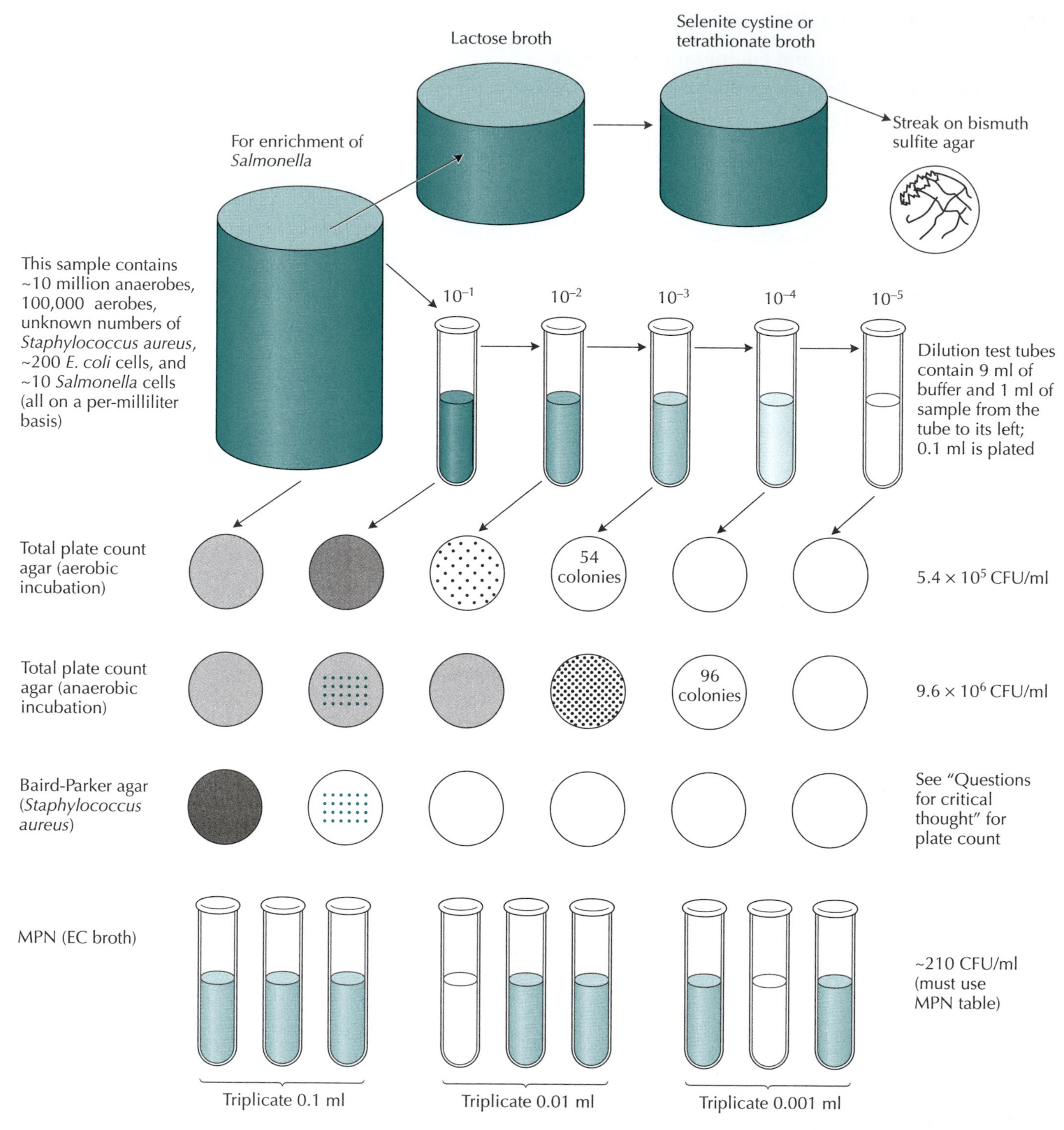

Figure 2.1 Methods of enumerating bacteria. The sample (which represents a 10^0 dilution) is diluted through a series of 10-fold dilutions before enumeration. To enumerate the bacteria, 0.1 ml from each dilution tube is plated on an agar plate of appropriate medium. After 24- to 48-h incubations, the colonies are counted and the number is multiplied by the dilution factor to give the microbial load of the original sample. Gray scale indicates culture density. If the number of cells present is expected to be low, the MPN technique is used. In this case, the sample is diluted into triplicate series of test tubes containing the media, and the pattern of positive tubes is used to determine the MPN from an MPN table (Table 2.3). doi:10.1128/9781555817206.ch02.f02.01

Table 2.1 Some selective media used in food microbiology

Organism(s)	Medium	Selective agent(s)
Staphylococcus aureus	Baird-Parker agar[a]	Lithium chloride, glycine, and tellurite
Listeria species	LPM[b]	Lithium chloride, phenylethanol, antibiotic
Lactic acid bacteria	de Man-Rogosa-Sharpe	Low pH, surfactant
Campylobacter species	Abeyta-Hunt-Bark agar	Microaerobic incubation, antibiotics

[a]The biggest honor for a microbiologist is to have an organism named after oneself. Not far behind, and much easier to obtain, is to name a medium after oneself.

[b]LPM, lithium chloride-phenylethanol-moxalactam.

supports the growth of *Listeria monocytogenes*. Thus, the identity of suspected pathogens isolated on selective media must be confirmed by biochemical or genetic methods.

Differential media make a colony of the target organism look different and stand out from the background of other organisms. For example, *S. aureus* colonies are black with a precipitation zone on Baird-Parker agar. *Listeria monocytogenes* colonies have a blue sheen on McBride's agar and turn Fraser broth black.

The CFU-per-gram values derived from plate counts are also influenced by intrinsic factors. Table 2.2 illustrates these points by providing *D* values (time to kill 90% of the population) at 55°C for *L. monocytogenes* organisms with different thermal histories (heat shocked for 10 min at 80°C or not heat shocked) on selective (McBride's) or nonselective (TSAY) medium under an aerobic or anaerobic atmosphere. Cells that have been heat shocked and recovered on TSAY medium are about four times more heat resistant (as evidenced by the larger *D* values) than cells which have not been heat shocked and recovered on McBride's medium under aerobic conditions. How might this affect processing times and temperatures?

Most-Probable-Number Methods

MPN methods are used to *estimate* the MPN of organisms in a sample when only low levels (<30 CFU/ml) are present (Fig. 2.1). For example, *E. coli* and coliform bacteria might be present in milk at very low levels. Selective agents are added to broth medium. One milliliter of sample is transferred to triplicate test tubes of media. At least three 10-fold serial dilutions are also inoculated. The samples are incubated and the pattern of turbid tubes with microbial growth is recorded. This pattern is used to derive the MPN

Table 2.2 Influence of thermal history and enumeration protocols on experimentally determined *D* values at 55°C for *L. monocytogenes*

	D_{55} value (min)			
	TSAY medium		McBride's medium	
Atmosphere	With heat shock	Without heat shock	With heat shock	Without heat shock
Aerobic	18.7	8.8	9.5	6.6
Anaerobic	26.4	12.0	No growth	No growth

of bacteria from a statistical chart (Table 2.3). The lower sensitivity limit (for example) for a three-tube MPN is <3.6 cells/g. Put differently, 3.6 cells/g is the lowest concentration that can accurately be quantified by a three-tube MPN. Using 10 tubes at each dilution instead of 3 can lower the detection level to <0.96 cells/g, but the 10-tube MPN is rarely cost-effective and seldom used. The numerical value obtained from an MPN determination is much less precise than that obtained from a plate count.

However, MPN methodology is often used in the rapid analysis of environmental and food samples. Mechanized rapid methods detect specific bacterial deoxyribonucleic acid (DNA) using polymerase chain reaction (PCR) or specific proteins in enzyme immunoassays. These testing systems are usually designed to detect one specific genus of bacteria at a time, although this is changing. Using the MPN strategy combined with a simple sample preparation allows faster screening methods (Box 2.3), which are needed by large food companies to check the safety of their products. Traditional direct plating methods may be used to confirm positive MPN results.

Enrichment Techniques

There is a "zero tolerance" (actually 0 cells/25 g) for *Salmonella*, *Listeria monocytogenes*, and *E. coli* O157:H7 in ready-to-eat foods. The good news is that this makes it unnecessary to count these pathogens. The bad news

Table 2.3 MPN table[a]

No. of positive tubes			MPN/g	95% Confidence interval limits		No. of positive tubes			MPN/g	95% Confidence interval limits	
0.10	0.01	0.001		Low	High	0.10	0.01	0.001		Low	High
0	0	0	<3.0		9.5	2	2	0	21	4.5	42
0	0	1	3.0	0.15	9.6	2	2	1	28	8.7	94
0	1	0	3.0	0.15	11	2	2	2	35	8.7	94
0	1	1	6.1	1.2	18	2	3	0	29	8.7	94
0	2	0	6.2	1.2	18	2	3	1	36	8.7	94
0	3	0	9.4	3.6	38	3	0	0	23	4.6	94
1	0	0	3.6	0.1	18	3	0	1	38	8.7	110
1	0	1	7.2	1.3	18	3	0	2	64	17	180
1	0	2	11	3.6	38	3	1	0	43	9	180
1	1	0	7.4	1.3	20	3	1	1	75	17	200
1	1	1	11	3.6	38	3	1	2	120	37	420
1	2	0	11	3.6	42	3	1	3	160	40	420
1	2	1	15	4.5	42	3	2	0	93	18	420
1	3	0	16	4.5	42	3	2	1	150	37	420
2	0	0	9.2	1.4	38	3	2	2	210	40	430
2	0	1	14	3.6	42	3	2	3	290	90	1,000
2	0	2	20	4.5	42	3	3	0	240	42	1,000
2	1	0	15	3.7	42	3	3	1	460	90	2,000
2	1	1	20	4.5	42	3	3	2	1,100	180	4,100
2	1	2	27	8.7	94	3	3	3	>1,100	420	

[a]For three tubes with 0.1-, 0.01-, and 0.001-g inocula. Source: http://www.fda.gov/Food/ScienceResearch/LaboratoryMethods/BacteriologicalAnalyticalManualBAM/ucm1096546.htm.

BOX 2.3

Testing programs

Testing programs are increasingly important to food companies of all sizes. Ingredients are tested before being used, and finished products are tested before being made available to consumers. The first part of any testing program is the sampling, which is not as easy as it sounds. Sampling protocols must be statistically valid for the specific type of food being tested. Then samples are assessed for the presence of microbial pathogens or toxins. Tests must also be validated for the specific types of foods being tested. This is a problem when an outbreak of foodborne illness occurs with a new or unusual food; it takes time to get a test validated to detect a specific pathogen in a specific foodstuff. This happened in 2009 when an outbreak of gastrointestinal illness was caused by *E. coli* O157:H7 in raw cookie dough. Up until that point, *E. coli* O157:H7 had never been isolated from this food and would not have been suspected. Product testing and testing programs called food safety audits play an important role under the Food Safety Modernization Act (passed by Congress in December 2010 and signed by President Barack Obama in January 2011). It is important to keep in mind that testing must be used in conjunction with appropriate training measures and good sanitation.

is that one must be able to find one or more *Salmonella* cells hiding among a million other bacteria. This is done by coupling preenrichment media with selective enrichment, followed by plating on selective media and biochemical confirmation or genetic authentication (Fig. 2.1). The preenrichment in a nonselective medium allows injured pathogens to repair, recover, and regain resistance to selective agents. The enrichment step uses selective agents to promote pathogen growth to high levels while inhibiting growth of the competing organisms. The pathogen can then be isolated on a selective agar and identified using other methods.

PHYSIOLOGICAL STATES OF BACTERIA

Introduction

Not all cells are created equal. Some injured cells lose their ability to grow on selective media. Other cells, though viable, cannot grow on any medium at all. A biochemical "awareness" allows some bacteria to act collaboratively and transmit environmental information to regulate gene expression. Bacteria can even exist in structured communities called biofilms. In fact, in reality, outside of a laboratory, bacterial communities are highly dynamic and often involve all four of the previous conditions.

Injury

Injured cells and cells that are viable but nonculturable (VNC) pose additional problems to food microbiologists. Injury is defined as the inability of cells exposed to sublethal stress to grow on selective media, while retaining culturability on nonselective media. Cells that are VNC cannot be cultured on *any* medium. Cells in either state cannot be detected by colony formation. However, they can make people sick and can be recovered by standard methods once they recover from the injury.

Sublethal levels of heat, radiation, acid, or sanitizers may injure rather than kill cells. Injured cells are less resistant to selective agents or have increased nutritional requirements. The degree of injury is influenced by time, temperature, selective-agent concentration, experimental methodology, and other factors. For example, a sanitizer test may appear to kill listeria cells. However, these cells might be recovered using listeria repair broth. Media for the recovery of injured cells are often specific for a given bacterium or stress condition.

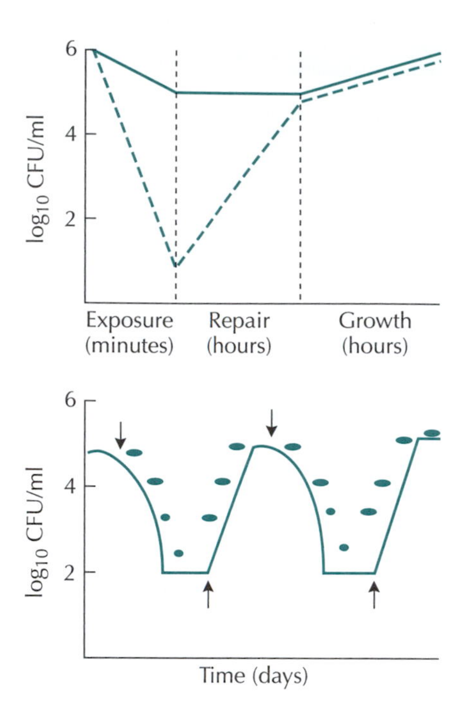

Figure 2.2 Data indicative of injury and repair **(top)** or VNC **(bottom)**. The top panel shows bacteria plated on selective (dashed line) or nonselective (solid line) medium during exposure to some stressor. The decrease in CFU on a nonselective medium represents the true lethality, while the difference between the values obtained on each medium is defined as injury. During repair, resistance to selective agents is regained and the value obtained on the selective medium approaches that of the nonselective medium. The lower panel shows the decrease in viability during stress (↓) when the bacteria are cultured on a nonselective medium. The dark shapes represent cellular morphology as cells enter the VNC state. Note that when the stress is removed (↑), the cells become culturable and regain their normal morphology. doi: 10.1128/9781555817206.ch02.fo2.02

Data that illustrate injury and the VNC state are given in Fig. 2.2. In the case of injury, cells subjected to a mild stress are plated on a rich nonselective medium and on a selective medium. The difference between the populations of cells able to form colonies on each medium represents the number of injured cells. If 1×10^7 CFU of a population/ml can grow on the selective medium and 1×10^4 CFU/ml can grow on the nonselective medium, then 9.9×10^6 CFU/ml are injured. (Students who have difficulty working with logs can solve this as 10,000,000 − 10,000 = 9,990,000.)

Cell injury is a threat to food safety for several reasons. (i) If injured cells are classified as dead during heat resistance determinations, the effect of heating will be overestimated and the resulting heat process will be ineffective. (ii) Injured cells that escape detection at the time of postprocessing sampling may repair before the food is eaten and then cause illness. (iii) The "selective agents" may be common food ingredients such as salt or organic acids or even suboptimal temperature. These would prevent injured cells from growing and result in an underestimation of the microbial levels.

Cells injured by heating, freezing, and detergents usually leak intracellular constituents from damaged membranes. Membrane integrity is reestablished during repair. Osmoprotectants can prevent or minimize freeze injury in *L. monocytogenes*. Oxygen toxicity also causes injury. Recovery of injured cells is often enhanced by adding peroxide-detoxifying agents such as catalase or pyruvate to the recovery medium or by excluding oxygen through the use of anaerobic incubation conditions or adding Oxyrase (which enzymatically reduces oxygen) to the recovery medium.

During repair, cells recover from injury. Repair requires ribonucleic acid (RNA) and protein synthesis and often lengthens the lag phase. Environmental factors influence the extent and rate of repair. The microbial stability of cured luncheon meat is due to the extended lag period required for the repair of spoilage organisms. The extent and rate of repair are influenced by environmental factors. For example, *L. monocytogenes* injured at 55°C for 20 min starts to repair immediately at 37°C and has completely recovered by 9 h. But repair at 4°C is delayed for 8 to 10 days and full recovery requires 16 to 19 days.

Viable but Nonculturable

In contrast to injured cells which can form colonies on nonselective media, cells that are VNC cannot be cultured on *any* medium. For example, *Salmonella*, *Campylobacter*, *Escherichia*, *Shigella*, and *Vibrio* species can be viable but not be culturable by normal methods. The differentiation of vegetative cells into a dormant VNC state is a survival strategy for many nonsporulating bacteria. During the transition to the VNC state, rod-shaped cells shrink and become small spherical bodies. These are totally different from bacterial spores. It takes from 2 days to several weeks for an entire population of vegetative cells to become VNC.

Although they cannot be cultured, the viability of VNC cells is demonstrated through cytological methods. The structural integrity of the bacterial cytoplasmic membrane can be determined by the permeability of cells to fluorescent nucleic acid stains. Bacteria with intact cell membranes stain fluorescent green, whereas bacteria with damaged membranes stain fluorescent red. Iodonitrotetrazolium violet can also identify VNC cells. Respiring cells reduce iodonitrotetrazolium violet to form an insoluble compound detectable by microscopic observation. Additional methods for detecting VNC cells are being developed as understanding of bacteria

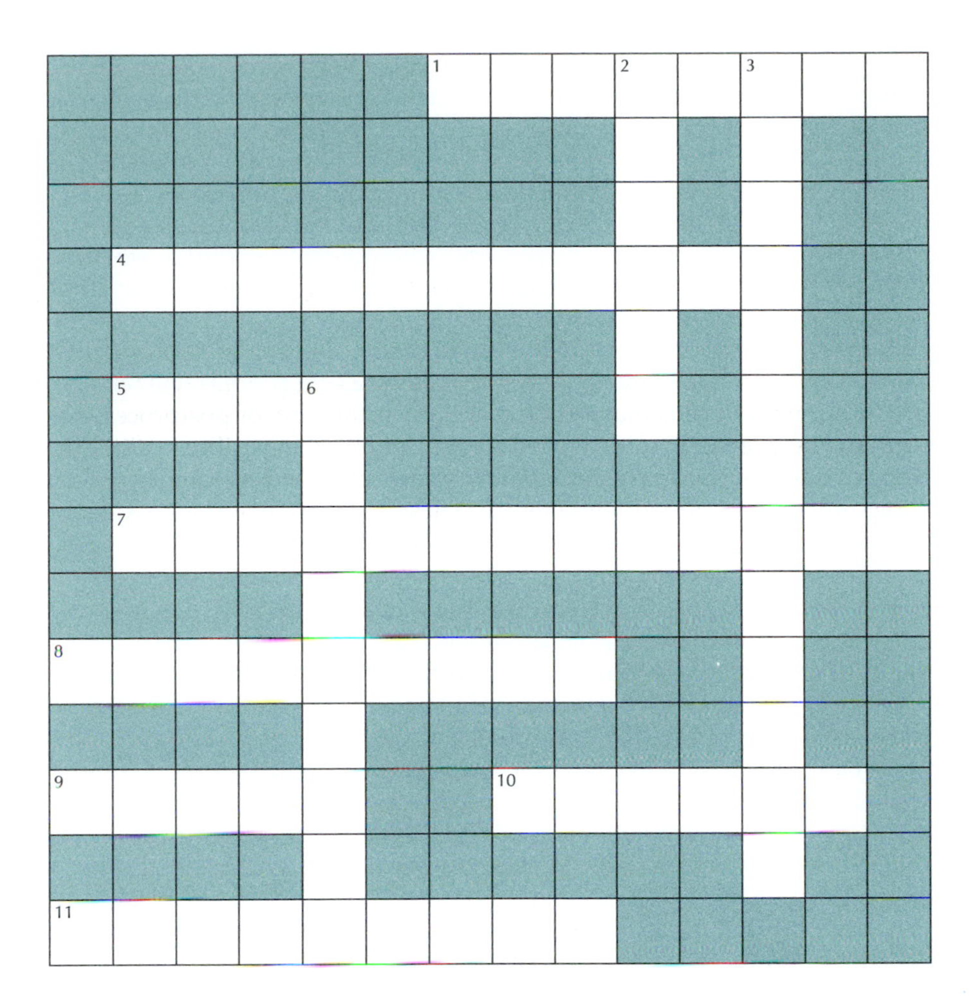

Across

1. MPN methods are used to ________ low numbers of bacteria
4. Arguably the most important extrinsic factor
5. This helps determine the safety of the foods held at abuse temperature
7. One way to determine viability
8. A type of factor inherent to a food that affects the growth of bacteria
9. Organisms that grow at low water activity and low pH
10. A technology that uses more than one inhibitor to stop microbial growth
11. A type of medium that allows only specific bacteria to grow

Down

2. A sublethal level of stress may ________ rather than kill cells
3. When cells grow in the absence of air, they grow ________
6. A type of factor that is not part of a food

at the molecular level and techniques for genetic manipulation advance. The detection of specific RNA by reverse transcriptase PCR is one such method. Alternatively, reporter genes such as green fluorescent protein-tagged and bioluminescence-tagged genes can identify cells that are synthesizing proteins, even though they cannot be cultured.

Nonculturable *Salmonella enterica* serovar Enteritidis populations (<10 CFU/ml) starved at 7°C have been quantified as 10^4 viable cells per ml using nonculture methods. Experimental data in Fig. 2.2 illustrate a

Vibrio population that appears to have died off (i.e., gone through a 6-log reduction in CFU per milliliter) even though $>10^5$ cells per ml are quantified as viable by nonculture methods.

Because the VNC state is most often induced by nutrient limitation in aquatic environments, it might appear irrelevant to the nutrient-rich milieu of food. However, the VNC state can be induced by changes in salt concentration, exposure to hypochlorite, and shifts in temperature. Foodborne pathogens in nutritionally rich media can become VNC when shifted to refrigeration temperatures. When *Vibrio vulnificus* populations (10^5 viable cells) are shifted to refrigeration temperatures the entire population becomes nonculturable (<0.04 CFU/ml) but maintains lethality in mice, in which the cells can be resuscitated. VNC campylobacters are resuscitated when inoculated into fertilized chicken eggs at 37°C for 48 h. *Escherichia coli* and *Salmonella enterica* serovar Typhimurium can enter a VNC state following chlorination of wastewater. Although nonculturable, they may still present a public health hazard. Temperature changes can induce the VNC state. When starved at 4 or 30°C for more than a month, *Vibrio harveyi* becomes VNC at 4°C but remains culturable at 30°C. In contrast, *E. coli* enters the VNC state at 30°C but dies at 4°C.

Resuscitation of VNC cells is demonstrated by an increase in culturability that is not accompanied by an increase in the total cell numbers. The return to culturability can be induced by temperature shifts or gradual return of nutrients. The same population of bacteria can go through multiple cycles of the VNC and culturable states in the absence of growth. Inhibitors of transcription and translation inhibit resuscitation. The addition of catalase or sodium pyruvate to media restores culturability of VNC *E. coli* O157 and *Vibrio parahaemolyticus*. This suggests that the transfer of cells to nutrient-rich media initiates rapid production of superoxide and free radicals. Catalase hydrolyzes H_2O_2. Sodium pyruvate degrades H_2O_2 through decarboxylation of the α-keto acid to form acetic acid and CO_2.

VNC cells are found in marine, soil, and gastrointestinal environments. Indeed, as many as 99% of bacteria in the biosphere may be nonculturable. Increased awareness of the VNC state may lead to a reexamination of our reliance on cultural methods to monitor the viability of microbes in foods. The mechanisms of VNC state formation, the mechanisms by which VNC cells resuscitate, and the mechanisms that regulate resuscitation are largely unknown. The relationship between viability and culturability needs to be better understood. Does the VNC state reflect a true state of bacteria, or simply our inability to culture them?

Quorum Sensing and Signal Transduction

Introduction

It turns out that most bacteria do not exist as solitary isolated beings. They communicate with each other. There is an explosion of information on cellular communication among foodborne microbes. Quorum sensing and signal transduction both regulate genes that would be superfluous to isolated cells but advantageous to large populations. Cellular communication occurs by two different mechanisms: quorum sensing and signal transduction. In quorum sensing systems, autoinducers diffuse into the cell. When the autoinducer reaches some threshold concentration it evokes a physiological response. Two-component signal transduction systems are comprised of a signal-binding sensor that spans the membrane and a protein that responds to the signal.

Quorum Sensing

Bacteria use quorum sensing to determine if their population is big enough to justify some action. The term quorum sensing is derived from human legislative bodies where a quorum (i.e., a certain number of participants) is required before action can be taken. In microbial quorum sensing, cells produce a signal compound that diffuses into the environment. When the bacterial population is low, the extracellular concentration of the signal molecule remains low and the signal molecule continues to diffuse away from the cell. However, when many cells produce the signal molecule, its extracellular concentration increases and it diffuses back into the cell. When the signal compound diffuses back into the cell, it binds to an intracellular regulator protein that affects transcription of a regulon(s) to elicit a cellular response. The gene for the signal compound is on the same regulon and hence is autoinduced.

In gram-negative bacteria, *N*-acyl homoserine lactones (AHLs; also abbreviated in the literature as HSLs) generally act as signaling molecules. These are referred to as autoinducer 1 and are synthesized by AHL synthase, encoded by the *luxI* gene. At high concentrations, the autoinducers bind to and activate a transcriptional activator, which, in turn, induces target gene expression.

Examples of cell signaling in foodborne microbes. It is tempting to speculate that quorum sensing has a role in spoilage. However, there are not many data to support this. Many studies have used a bioluminescence response in *Vibrio harveyi* as evidence of quorum sensing in a foodborne organism. However, these studies do not prove much unless the autoinducing compound has been isolated and the regulated phenotype identified in that microbe. *Campylobacter*, *Salmonella*, *E. coli* O157:H7, *Enterobacteriaceae*, *Pseudomonas*, *Aeromonas*, *Shewanella*, and *Photobacterium* produce a bioluminescence response in *Vibrio harveyi*. Such signals have been detected in broth, chicken soup, milk, bean sprouts, vacuum-packed beef, fish fillets, and turkey. There are many phenotypes for which a role for cell signaling has been established at a genetic level.

Molecules in foods may mimic or alter quorum sensing signaling systems of spoilage and pathogenic bacteria in food. Probiotic bacteria have health-promoting effects. How probiotic bacteria such as *Lactobacillus acidophilus* produce these effects is not entirely understood. Probiotic bacteria prevent infection with *E. coli* O157:H7 in mouse models. Research suggests that probiotics produce small molecules that interfere with *E. coli* O157:H7 quorum sensing systems. The yeast *Saccharomyces cerevisiae*, commonly used in making bread and beer, exhibits quorum sensing behavior. The communication molecules are aromatic alcohols. The transition between the solitary yeast form and the filamentous form is regulated by quorum sensing.

A more rigorous study casts doubt on the linkage of quorum sensing and spoilage. Meat inoculated with wild-type strains or AHL synthase knockout mutants spoiled at the same rate. Furthermore, addition of quorum sensing inhibitors did not influence spoilage. This indicates that quorum sensing does not regulate spoilage in vacuum-packed meat.

There are four criteria for quorum sensing in a given organism. (i) The production of the signal compound is specific to an event. (ii) The signal accumulates extracellularly. (iii) A unique response is generated

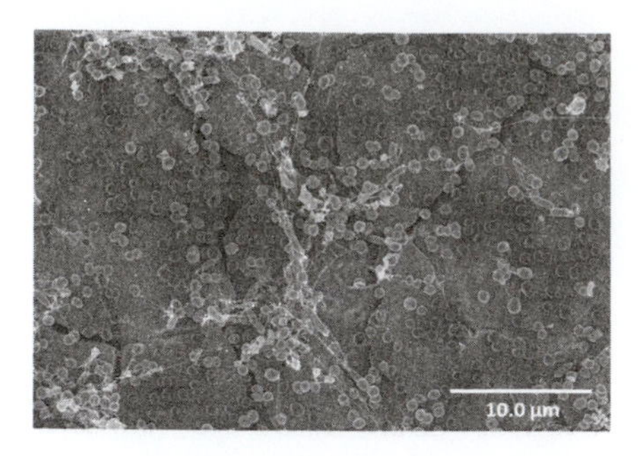

Figure 2.3 Scanning electron microscope image of a biofilm developing on stainless steel. The biofilm contains *Staphylococcus epidermidis*, *E. coli*, and *Pseudomonas fluorescens*. The size is indicated by the bar. Imaged by A. E. H. Shearer, University of Delaware. doi: 10/1128/9781555817206.ch02.f02.03

when the signal compound reaches a threshold concentration and binds to a specific receptor. (iv) The response goes beyond metabolism or detoxification of the signal compound. Rarely are these criteria met.

Signal Transduction

Two-component signal transduction systems consist of a membrane-spanning histidine kinase receptor and an intracellular response regulator. The signal molecule acts on the extracellular side of the protein kinase. (Unlike in quorum sensing, the trigger molecule does not diffuse into the cell.) The kinase transduces (i.e., transmits) the signal across the membrane through a conformational change. This increases the kinase activity on the intracellular side of the cell membrane. The increased kinase activity phosphorylates a response regulator protein. The phosphorylated response regulator can modulate gene expression, enzymatic activity, flagellar rotation, or other phenotypes.

Biofilms

A biofilm is an aggregation of cells, often of many species, into complex structures on solid surfaces. Biofilm formation is a multistep process. Biofilms can form on any surface. A developing biofilm is shown in the electron micrograph in Fig. 2.3. First, the solid surface is conditioned so that cells can be absorbed. Biopolymer formation follows rapidly and anchors the cells. The synthesis of this matrix may be upregulated by quorum sensing when the local concentration of cells increases due to their adsorption. This is followed by the formation of microcolonies having defined boundaries. These allow fluid channels to run through the biomatrix. This may be considered a primitive circulatory system. These circulatory systems require higher-level organization, quorum sensing, or some kind of cell-to-cell communication to prevent undifferentiated growth. This would clog the channels which bring nutrients and remove wastes. Finally, cells can detach from the biofilm to initiate new biofilms or move to other environments (such as from a mixer into a food). A definite role for quorum sensing in biofilm formation has been proved in some bacteria. But the evidence in foodborne pathogens is conflicting.

Cells in biofilms are more resistant to heat, chemicals, and sanitizers than are planktonic (free individual) cells. The lethality of a combination of sodium hypochlorite and heat to *Listeria monocytogenes* is approximately 100 times lower for biofilms than for free cells. Increased chemical resistance is attributed to the very slow growth of cells in biofilms and not the diffusional barrier created by the biofilm. Indeed, cells in the nutrient-depleted interior of the microcolony may be in the VNC state. Biofilms pose special challenges to the food industry. The foodborne pathogens *E. coli* O157:H7, *L. monocytogenes*, *Yersinia enterocolitica*, and *Campylobacter jejuni* form biofilms on food surfaces and food contact equipment, leading to serious safety issues. Only proper cleaning can ensure that the cells in the nascent biofilm can be reached by sanitizers before they become recalcitrant in fully developed biofilms. Trisodium phosphate is effective for *E. coli* O157:H7, *C. jejuni*, and *Salmonella* Typhimurium cells in a biofilm. Other methods for control of biofilms include superhigh magnetic fields, ultrasound treatment, proteolytic and glycolytic enzymes, and high-voltage pulsed electric fields. The design of equipment with smooth highly polished surfaces also impedes biofilm formation by making the initial adsorption step more difficult.

There continues to be a great need for research that will yield a better understanding of biofilms. Planktonic cells are easy to study, and pure culture is the foundation of microbiology as we know it. However, in most natural environments bacteria reproduce on surfaces rather than in liquids. To a large degree, food microbiologists study microbes in their domesticated (i.e., planktonic) setting rather than their natural attached state. More researchers are addressing bacterial biofilms on food samples and in food service situations.

FACTORS THAT INFLUENCE MICROBIAL GROWTH

Intrinsic Factors

Characteristics inherent to the food itself are called intrinsic factors. These include pH, water activity (a_w), oxidation-reduction potential, naturally occurring compounds that influence microbial growth, and compounds added as preservatives. Most of these factors are covered separately in the chapters on physical and chemical methods of food preservation. Because of their importance, we give special attention to pH, a_w, and temperature (an external factor).

pH

The pH value of a food is a log scale measurement of its acidity. pH is defined by the relationship $pH = -\log [H^+]$. Because each unit on the pH scale represents a 10-fold difference, a food with a pH of 6 is 10 times more acidic than one with a pH of 7 and pH 5.0 is 100 times more acidic. That a low pH number represents a high acidity invariably causes "misspeaking" even among professionals. Try to avoid this error. The lower the pH, the more energy the cell has to spend on maintaining its intracellular pH near neutrality and the less energy it has to grow, produce toxins, etc. Table 2.4 lists typical pH values for several foods. Note that foods with lower pH values generally require milder processing than foods having higher pH values. Most foodborne pathogens cannot grow below pH 4.4. They may persist at lower pH values or die off very slowly. Spoilage organisms such as lactic acid bacteria and fungi can grow in acid products, but there are rarely microbial problems in foods with pH values of <3.0. Environmental extremophiles such as *Thiobacillus thiooxidans* and *Sulfolobus acidocaldarius* grow at pH values of <1.0! Fortunately, these organisms are not associated with foods. The optimum pH for most foodborne microbes is near neutrality. The upper limit for growth is pH 8 to 9, but no foods are this alkaline.

Authors' note

"pH" is a noun, not a verb. Your instructors will cringe if they hear you say, "I pHed the sample."

Table 2.4 Typical pH values of foods

pH	Food(s)
>7.0	Egg albumen
7.0–6.5	Milk, ham, bacon, poultry, fish
6.5–5.3	Raw beef, vegetables, vacuum-packed meat, melons
5.3–4.5	Cottage cheese, fermented vegetables, fermented meats (e.g., summer sausage), many sauces and soups
<4.5	Tomatoes, fruits and fruit juices, yogurt, pickles, sauerkraut

It's easy to make a place for bacteria to grow.
Reprinted with permission from CartoonStock.

In addition to the energy expense required to maintain intracellular neutrality, pH influences gene expression. The expression of genes governing proton transport, amino acid degradation, adaptation to acidic or basic conditions, and even virulence can be regulated by the external pH. Cells sense changes in acidity through several mechanisms.

1. Organic acids enter the cell only in the protonated (undissociated) form. Charged compounds cannot cross the membrane. So the uncharged acid enters the cell and then dissociates, releasing H^+ (the proton) in the cell cytoplasm. The cell may sense the accumulation of the resulting anions, which cannot pass back across the cell membrane. Additionally, they may sense the released protons, which acidify the cytoplasm.
2. The change in pH results in a change in the transmembrane proton gradient (i.e., the difference in pH between the inside and outside of the cell) which can also serve as a sensor to start or stop energy-dependent reactions.
3. The changes in acidity inside the cell may result in the protonation or deprotonation of amino acids in proteins. This may alter the secondary or tertiary structure of the proteins, changing their function and signaling the cell about the change in pH.

Cells must maintain their intracellular pH (pH_i) above some critical pH_i at which intracellular proteins become denatured. *Salmonella* serovar

Typhimurium has three progressively more stringent mechanisms to maintain pH_i that supports life. These three mechanisms are the homeostatic response, the acid tolerance response, and the synthesis of acid shock proteins.

At an external pH (pH_o) of >6.0, salmonella cells adjust their pH_i through the homeostatic response. The homeostatic response maintains pH_i by increasing the activity of proton pumps so that they expel more protons from the cytoplasm. The homeostatic mechanism is "always on" and functions in the presence of protein synthesis inhibitors.

The acid tolerance response (ATR) is triggered by a pH_o of 5.5 to 6.0. This mechanism is sensitive to protein synthesis inhibitors; there are at least 18 ATR-induced proteins. The ATR appears to involve the membrane-bound ATPase proton pump and maintains a pH_i of >5.0 at pH_o values as low as 4.0. The loss of ATPase activity caused by gene disruption mutations or metabolic inhibitors abolishes the ATR but not the pH homeostatic mechanism.

The ATR may confer cross-protection to other environmental stressors. Acid adaptation increases heat and freeze-thaw resistance of *E. coli* O157:H7. The exposure of *Salmonella* serovar Typhimurium cells to pH 5.8 for a few cell doublings renders the cells less sensitive to sodium chloride and heat. The rate of survival of acid-adapted *L. monocytogenes* exposed to nisin is approximately 10-fold greater than that of nonadapted cells. Acid-adapted *L. monocytogenes* also has increased resistance against heat shock, osmotic stress, and alcohol stress. Acid adaptation of *E. coli* O157:H7 enhances thermotolerance.

The synthesis of acid shock proteins is the third way that cells regulate pH_i. The synthesis of these proteins is triggered by a pH_o from 3.0 to 5.0. They constitute a set of regulatory proteins distinct from the ATR proteins.

Water Activity

Water is a major factor in controlling microbial growth and chemical reactions in food. This inhibition is caused by the *availability* of the water in the food rather than the amount of water. Water can be chemically bound to food molecules or immobilized by capillary action in the food microstructure. The water is bound by ionic interactions, hydrogen bonding, capillary action, etc., with constituents of the food. You may have experienced the difficulty of removing water from a capillary tube. The water bound by capillary action is not available for chemical reactions or microbial growth. The amount of unbound or available water determines if microbes can grow.

a_w is the measure of available water or, strictly speaking, the energy of water in foods. a_w is defined as the ratio of the vapor pressure of water in a food, P, to the vapor pressure of pure water, P_0, at the same temperature:

$$a_w = \frac{P}{P_0}$$

The movement of water vapor from a food to the air depends on the moisture content, the food composition, the temperature, and the humidity. At constant temperature, water in the food equilibrates with water vapor in the air. This is the food's equilibrium moisture content. Think of potato chips at a picnic on a humid day; they gain moisture and become soggy. Similarly, bread left out in dry winter air loses moisture and becomes stale. At the equilibrium moisture content, the food neither gains nor loses water to the air. The relative humidity of the air surrounding the

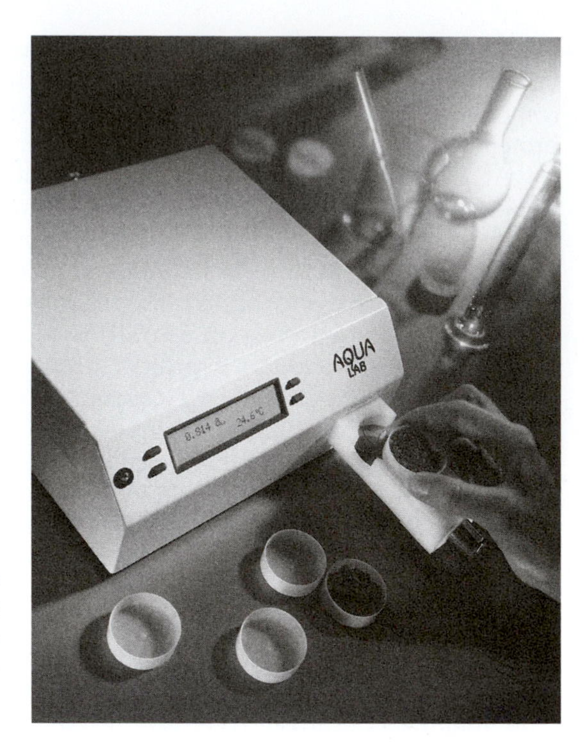

Figure 2.4 The AquaLab Series 3 is a meter that uses the dew point to measure a_w. Courtesy of Decagon Devices. doi:10.1128/9781555817206.ch02.f02.04

food is thus the equilibrium relative humidity (ERH). So, a_w can also be defined as follows:

$$a_w = \text{ERH}\ (\%)/100$$

The relative humidity relationship is used by instruments that use dew point to measure a_w (Fig. 2.4). The food is put in a small cup and the water in the food is allowed to equilibrate (i.e., to its ERH) with the headspace water vapor. The dew point (the temperature at which the vapor in the air condenses to visible water) is a primary measurement of vapor pressure. So, the vapor pressure can be measured by cooling the chamber until condensation is detected optically. The a_w of the sample is then the ratio of the condensation vapor pressure at dew point temperature to the saturation vapor pressure at the sample temperature. These measurements can be made to an a_w of 0.001 in less than 5 min.

The third way to express a_w is on a mole fraction basis as a_w = moles of solvent/(moles of solvent + moles of solute). In symbolic language, this is expressed as follows:

$$a_w = n_1/(n_1 + n_2)$$

where n_1 is the moles of solvent (usually water) and n_2 is the moles of solute. (Can you see why a_w values must be less than 1?) Since there are 55.5 mol of water in 1 liter (1,000 g/18 gmol^{-1}), the equation for calculating a_w becomes

$$a_w = 55.5\ /(55.5 + n_2)$$

The theoretical amounts of sugar and salt required to achieve a given a_w are shown in Table 2.5. While it appears that half as much sodium chloride is required as sucrose (on a molar basis), 1 mol of sodium chloride

Table 2.5 Theoretical amounts of salt and sugar required to achieve various a_ws

a_w	Salt		Sugar	
	Molality[a]	%	Molality	%
0.99	0.3	1.7	0.6	17.0
0.98	0.5	2.8	1.1	27.3
0.96	1.1	6.0	2.3	44.0
0.94	1.0	9.5	3.6	55.3
0.92	2.4	12.3	4.9	62.6
0.90	3.1	15.3		
0.88	3.7	17.8		
0.86	4.5	20.8		
0.84	5.3	23.7		
0.80	6.9	28.8		
0.75	9.2	34.9		
0.70	11.9	41.0		
0.65	15.0	46.7		
0.60	18.5	52.0		

[a]Do you remember the difference between molality and molarity?

disassociates into 1 mol of sodium and 1 mol of chloride to give 2 mol of solute.

These equations can be used to calculate the a_w of a known solute concentration or to determine how much solute should be added to obtain a given a_w, as shown below.

Example 1
What will be the a_w of a solution containing 10% (weight/weight) glucose?

1. Assume 1 liter of solution.
2. Calculate the molal concentration of glucose.
   a. 100 g/900 g of water
   b. For the molal concentration, this equals 111 g of glucose/1,000 g of water.
   c. 111 g/180 $gmol^{-1}$ (the formula weight for glucose) = 0.61 mol
3. Calculate the a_w.

$$a_w = 55.5/(55.5 + n_2) = 55.5/(55.5 + 0.61) = 55.5/56.1 = 0.989$$

Example 2
What percent glucose would it take to reach an a_w of 0.90?

1. Assume 1 liter of solution.
2. Calculate the molal concentration of glucose.
   a. $a_w = 55.5 / (55.5 + n_2)$
   b. $0.90 = 55.5/(55.5 + n_2)$
   c. $0.90(55.5) + 0.90(n_2) = 55.5$
   d. $49.95 + 0.90(n_2) = 55.5$
   e. $0.90(n_2) = 5.5$
   f. $n_2 = 6.1$ molal

Figure 2.5 When water turns from solid to liquid, it becomes more available for microbial growth. *Lucky Cow* © 2004 Mark Pett. Distributed by Universal Press Syndicate. Reprinted with permission.

3. Convert molal concentration to percent.
   a. 6.1 molal solution = 6.1 mol × 180 g/mol = 1,098 g of glucose
   b. 1,098 g/(1,098 g of glucose + 1,000 g of water) = 52%

Foods at the same a_w may have different moisture contents due to their chemical compositions and water binding capacities (see Table 25.2).

Dehydration preserves food by removing available water, i.e., reducing a_w. Hot air removes water by evaporation. Freeze-drying removes water by sublimation (the conversion of ice to vapor without passing through the liquid stage) after freezing. These processes reduce the a_w of the food to levels that inhibit growth. Freezing reduces a_w, and when the food thaws, the a_w increases (Fig. 2.5).

The a_ws of various foods are listed in Table 2.6. High-moisture foods such as fruits, vegetables, meats, and fish have a_ws of ≥0.98. Intermediate-moisture foods (e.g., jams and sausages) have a_w levels of 0.7 to 0.85. Additional preservative factors (e.g., reduced pH, preservatives, and pasteurization) are required for their microbiological stability.

Various microbes have different a_w requirements. Decreasing the a_w increases the lag phase of growth, decreases the growth rate, and decreases the

Table 2.6 Typical a_ws of various foods[a]

Foods	a_w
Fresh, raw fruits, vegetables, meat, and fish	>0.98
Cooked meat, bread	0.95–0.98
Cured meat products, cheeses	0.91–0.95
Sausages, syrups	0.87–0.91
Rice, beans, peas	0.80–0.87
Jams, marmalades	0.75–0.80
Candies	0.65–0.75
Dried fruits	0.60–0.65
Dehydrated vermicelli, spices, milk powder	0.20–0.60

[a]Reprinted from J. Farkas, p. 567–591, *in* M. P. Doyle, L. R. Beuchat, and T. J. Montville (ed.), *Food Microbiology: Fundamentals and Frontiers*, 2nd ed. (ASM Press, Washington, DC, 2001).

Table 2.7 Minimal a_ws required for growth of foodborne microbes at 25°C[a]

Group of microorganisms	Minimal a_w required
Most bacteria	0.91–0.88
Most yeasts	0.88
Regular molds	0.80
Halophilic bacteria	0.75
Xerotolerant molds	0.71
Xerophilic molds and osmophilic yeasts	0.62–0.60

[a]Reprinted from J. Farkas, p. 567–591, *in* M. P. Doyle, L. R. Beuchat, and T. J. Montville (ed.), *Food Microbiology: Fundamentals and Frontiers*, 2nd ed. (ASM Press, Washington, DC, 2001).

number of cells at stationary phase. Foodborne microbes are grouped by their minimal a_w requirements in Table 2.7. Gram-negative species usually require the highest a_w. Gram-negative bacteria such as *Pseudomonas* spp. and most members of the family *Enterobacteriaceae* usually grow only above a_ws of 0.96 and 0.93, respectively. Gram-positive non-spore-forming bacteria are less sensitive to reduced a_w. Many *Lactobacillaceae* have a minimum a_w near 0.94. Some *Micrococcaceae* grow below an a_w of 0.90. Staphylococci are unique among foodborne pathogens because they can grow at a minimum a_w of about 0.86. However, they do not make toxins below an a_w of 0.93. Most spore-forming bacteria do not grow below an a_w of 0.93. Spore germination and outgrowth of *Bacillus cereus* are prevented at a_ws of 0.97 to 0.93. The minimum a_w for *Clostridium perfringens* spore germination and growth is between 0.97 and 0.95.

Several yeast species grow at a_w levels lower than those of bacteria. Salt-tolerant species such as *Debaryomyces hansenii*, *Hansenula anomala*, and *Candida pseudotropicalis* grow well on cured meats and pickles at NaCl concentrations of up to 11% (a_w = 0.93). Some xerotolerant species (such as *Zygosaccharomyces rouxii*, *Zygosaccharomyces bailii*, and *Zygosaccharomyces bisporus*) grow and spoil foods such as jams, honey, and syrups having a high sugar content (and correspondingly low a_w).

Molds generally grow at a_ws lower than do foodborne bacteria. The most common xerotolerant molds belong to the genus *Eurotium*. Their minimal a_w for growth is 0.71 to 0.77, while the optimal a_w is 0.96. True xerophilic molds such as *Monascus* (*Xeromyces*) *bisporus* do not grow at a_ws of >0.97 to 0.99. The relationship of a_w to mold growth and toxin formation is complex. Under marginal environmental conditions of low pH, low a_w, or low temperature, molds may be able to grow but not make toxins.

The varied a_w growth limits of bacteria and fungi reflect the mechanisms that help them grow at low a_ws. Bacteria protect themselves from osmotic stress by accumulating compatible solutes intracellularly. Compatible solutes equilibrate the cells' intracellular a_w to that of the environment but do not interfere with cellular metabolism. Some bacteria accumulate K^+ ions and amino acids, such as proline. Halotolerant and xerotolerant fungi concentrate polyols such as glycerol, erythritol, and arabitol.

Extrinsic Factors

Extrinsic factors are *external* to the food. Temperature and gas composition are the main extrinsic factors influencing microbial growth. The influence of temperature on microbial growth and physiology is huge. While the influence of temperature on growth rate is obvious and covered in some detail, the influence of temperature on gene expression is equally important. Cells grown at refrigeration temperature are not just "slower" than those grown at room temperature; they express different genes and are physiologically different. Later chapters provide details about how temperature regulates the expression of genes governing traits ranging from motility to virulence.

Cells need many metabolic capabilities to grow in the cold. Cells maintain membrane fluidity at low temperatures through a process called homeoviscous adaptation. As temperature decreases, cells synthesize increasing amounts of mono- and diunsaturated fatty acids. The "kinks" caused by the double bonds prevent the tight packing of the fatty acids which would cause them to solidify. The membrane's physical state can regulate gene expression, particularly those genes that respond to temperature. The accumulation of compatible solutes at low temperatures is similar to their accumulation during low a_w. Foods are a rich source of compatible solutes for foodborne bacteria. Compatible solutes that can be used directly by bacteria include peptides, amino acids, betaine, sugars, taurine, and carnitine. Cold shock proteins also contribute to growth at low temperatures. Cold shock proteins appear to function as RNA chaperones, minimizing the folding of messenger RNA (mRNA). This helps the translation process.

Temperature can also regulate the expression of virulence. The expression of 16 proteins on seven operons on the *Y. enterocolitica* virulence plasmid is high at 37°C, weak at 22°C, and undetectable at 4°C. Similarly, the gene(s) required for *Shigella* virulence is expressed at 37°C but not at 30°C. The expression of genes required for *L. monocytogenes* virulence is also regulated by temperature. Cells grown at 4, 25, and 37°C all synthesize internalin, a protein required for penetration of the host cell. Cells grown at 37°C, but not at 25 or 4°C, can lyse blood cells. Temperature influences the expression of the *Vibrio cholerae toxT* and *toxR* genes, which are essential for cholera toxin production. Esp proteins are required for signal transduction events leading to the formation of the attaching and effacing lesions linked to virulence. In enterohemorrhagic *E. coli*, temperature modulates transcription of the *esp* genes; synthesis of Esp proteins is enhanced when bacteria are grown at 37°C.

The growth temperature can influence a cell's thermal sensitivity. *L. monocytogenes* cells preheated at 48°C have increased thermal resistance. Holding listeria cells at 48°C for 2 h in sausages increases their *D* values at 64°C 2.4-fold. This thermotolerance is maintained when cells are held for 24 h at 4°C. Subjecting *E. coli* O157:H7 to sublethal heating at 46°C increases their *D* value at 60°C 1.5-fold.

Shock proteins synthesized in response to one stressor may provide cross-protection against other stressors. *Bacillus subtilis* exposed to mild heat stress can survive lethal temperatures and exposure to toxic concentrations of NaCl. Heat-adapted (50°C for 45 min) *Listeria* is also more resistant to acid shock. Similarly, sublethal heat treatment of *E. coli* O157:H7 cells increases their tolerance to acidic conditions.

Temperature control is a key factor for food safety. The old adage "Keep hot foods hot and cold foods cold" (Fig. 2.6) has much validity. The general temperature guideline is captured in the "40-140" rule. Foods should be held below 40°F or above 140°F, since these are the temperature ranges for microbial growth. As a corollary to the 40-140 rule, foods should move through the zone of temperature abuse as rapidly as possible to prevent microbial growth. Note that slow cooling causes more cases of foodborne disease than improper heating. The cooling time is proportional to the square of the

Figure 2.6 One of the major rules of food safety is "Keep hot foods hot and cold foods cold." *Baldo* © 2004 Baldo Partnership. Distributed by Universal Press Syndicate. Reprinted with permission. All rights reserved.

shortest dimension of the food sample. That is, doubling the diameter of a pot increases the cooling time fourfold. An 8-gallon container of sauce placed in a 40°F refrigerator can take 25 h to cool from 105 to 57°F! The solution to the problem of slow cooling is to use shallow trays rather than deep pots.

A more sophisticated model, developed by Frank Bryan, combines time with temperature. As you'll see later in this chapter, time and temperature are in the exponential portion of the growth equation (equation 1; note that the influence of temperature is incorporated into the growth rate, μ). Figure 2.7 associates the degree of danger in a given time with specific temperatures. Foods held at the heart of the danger zone, ~35°C (imagine a picnic on a hot summer day), become dangerous in a few hours. On a cool autumn day, it might take a few days to reach the same degree of hazard. Figure 2.8 further parses these into zones of risk and hazard. Students should understand the difference between hazard and risk. Hazard is the potential to cause harm or be a source of damage. Risk is the probability that the hazard will lead to injury. One might consider ground beef hazardous due to the potential presence of pathogenic bacteria; if it is cooked to well done, there is no risk. Eating rare hamburger has a higher risk. In some ways, the practice of food microbiology is one of risk management, that is, reducing the level of some hazard to an acceptable probability.

A rule of thumb in chemistry suggests that reaction rates (or specific growth rates, μ) double with every 10°C increase in temperature. This simplifying assumption is valid for bacterial growth rates only over a limited temperature range (Fig. 2.9). Above the optimal growth temperature, the growth rates decrease rapidly. Below the optimum, growth rates also decrease but more gradually. Bacteria can be classified as psychrophiles, psychrotrophs, mesophiles, and thermophiles according to how temperature influences their growth.

Both psychrophiles and psychrotrophs grow, albeit slowly, at 0°C. True psychrophiles have optimum growth rates at 15°C and cannot grow above 25°C. Psychrotrophs, such as *Listeria monocytogenes* and *Clostridium botulinum* type E, have an optimum growth temperature of ~25°C and cannot grow at >40°C. Because these foodborne pathogens, and even some mesophilic *Staphylococcus aureus* organisms, can grow at <10°C, conventional refrigeration cannot ensure food safety.

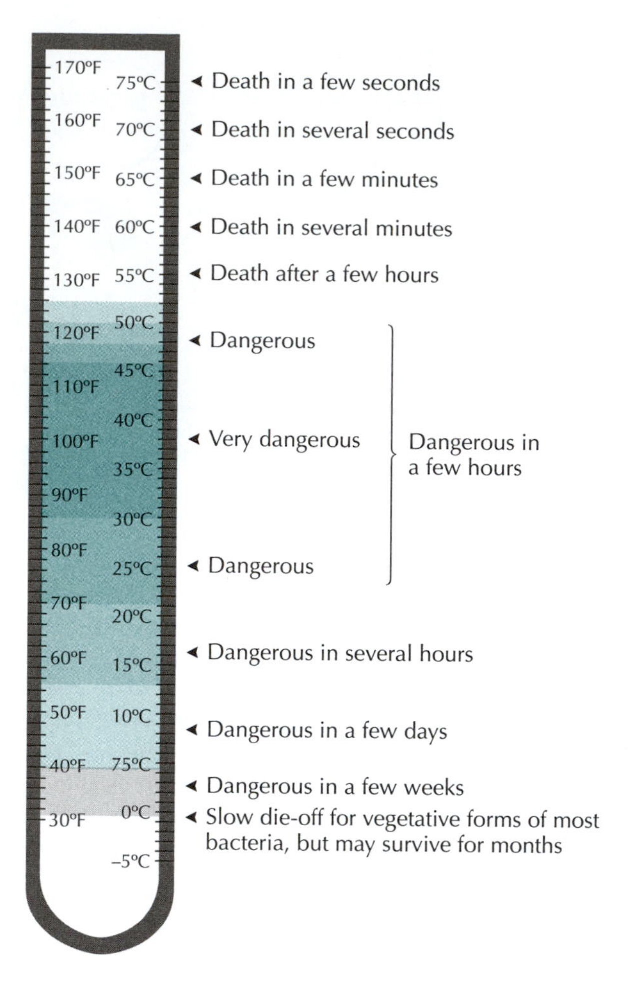

Figure 2.7 Effects of time and temperature on bacteria. Redrawn from an illustration by Frank Bryan, with permission of *Food Safety Magazine*. doi:10.1128/9781555817206.ch02.f02.07

Homeostasis and Hurdle Technology

Consumer demands for minimally processed fresh foods have decreased the use of intrinsic factors such as acidity and salt as the sole means of inhibiting microbes. Many foods now use multiple-hurdle technology to inhibit microbial growth. Instead of setting one environmental parameter to the extreme limit for growth, hurdle technology "deoptimizes" several factors. For example, limiting the amount of available water to an a_w of <0.85 *or* a limiting pH of 4.6 prevents the growth of foodborne pathogens. Hurdle technology might obtain similar inhibition at pH 5.2 *and* an a_w of 0.92. Table 2.8 shows how different combinations of pH and a_w determine the refrigeration requirements of different foods. Combinations of pH and a_w are also part of the federal canning regulations. Foods with a pH of <4.6 *or* those with a pH of >4.6 *but* an a_w of <0.85 are considered high-acid foods and can be processed in open kettles of boiling water. Foods with pH of >4.6 *and* an a_w of >0.85 are considered low-acid foods and must be processed at higher temperatures in retorts (industrial-size pressure cookers) under pressure using processes approved by the Food and Drug Administration (FDA).

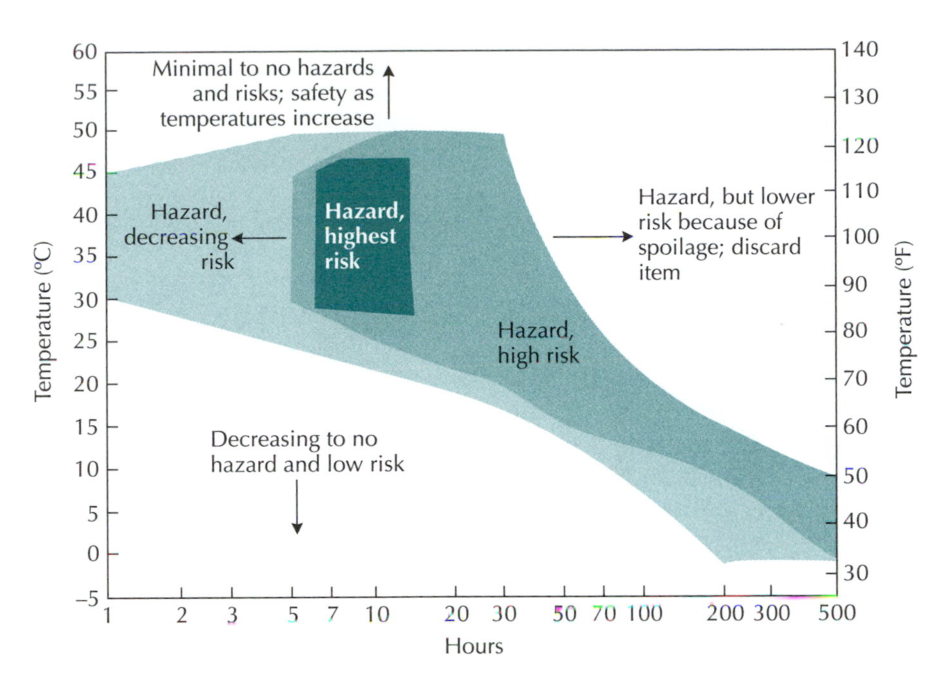

Figure 2.8 A more sophisticated examination of time and temperature allows one to evaluate risk and hazard. Redrawn from an illustration by Frank Bryan, with permission of *Food Safety Magazine*. Note that the x axis is on a log scale. doi: 10.1128/9781555817206.ch02.f02.08

Hurdle technology works best when it uses inhibitors that work by different mechanisms. For example, bacteria must maintain intracellular pH within narrow limits. This is done by using energy to pump out protons, as described below. In low-a_w environments, cells must use energy to accumulate compatible solutes. Membrane fluidity must be maintained as temperatures change through a different mechanism, homeoviscous adaptation. The expenditure of energy to maintain homeostasis is fundamental

Figure 2.9 Relative growth rates of bacteria at different temperatures. Redrawn from M. P. Doyle, L. R. Beuchat, and T. J. Montville (ed.), *Food Microbiology: Fundamentals and Frontiers*, 2nd ed. (ASM Press, Washington, DC, 2001). doi:10.1128/9781555817206.ch02.f02.09

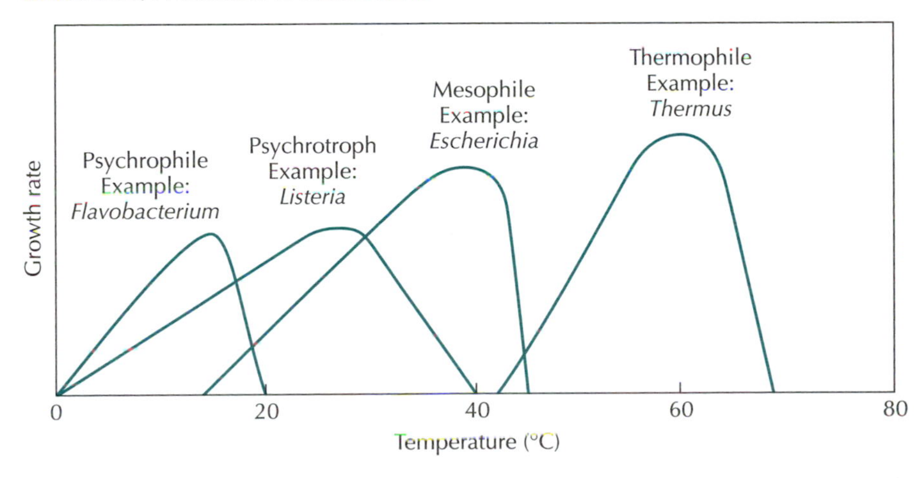

Table 2.8 Interaction of pH and a_w on food stability

pH-and-a_w combination	Recommended storage temp (°C)	Food classification
pH of >5.0 and a_w of >0.95	<4	Easily perishable
pH of 5.0–5.2 and a_w of 0.90–0.95	<10	Perishable
pH of <5.2 and a_w of <0.95 or pH of <5.0 or a_w of <0.90	May not require refrigeration	Shelf stable

to life. When cells channel the energy needed for growth into maintenance of homeostasis, their growth is inhibited. When the energy demands of homeostasis exceed the cell's energy-producing capacity, the cell dies. Hurdle technology can encompass the use of antimicrobial agents and technology, including the use of ozone and the application of irradiation in conjunction with shifts in pH and a_w, to inhibit microbial growth.

Hurdle effects can be additive or synergistic. Claims of synergy should be supported by a quantitative analysis, such as the use of an isobologram (Fig. 2.10). To construct an isobologram, the minimum inhibitory concentration (MIC) of compound A is plotted on the x axis. The MIC of compound B is plotted on the y axis. A line is drawn to connect the points. Experiments determine the MICs of compounds A and B in various concentrations, and these are plotted on the isobologram. If the points fall on the line, the effect is additive. If they fall below the line, the effect is synergistic. If they fall above the line, there is an antagonistic effect.

Novel approaches to hurdle technology include the use of bacteriocins with high pressure, pulsed electric fields, and other antimicrobials such as lysozyme and lactoferrin. Control of *S. aureus* in pasteurized milk can be achieved using phage-encoding endolysin and nisin. Combining the antimicrobials causes 64- and 16-fold reductions in nisin and endolysin MICs, respectively. The mechanism(s) by which the synergistic effect is achieved is not yet known.

Figure 2.10 An isobologram is used to determine the interaction of two antimicrobial inhibitors (see text). doi:10.1128/9781555817206.ch02.f02.10

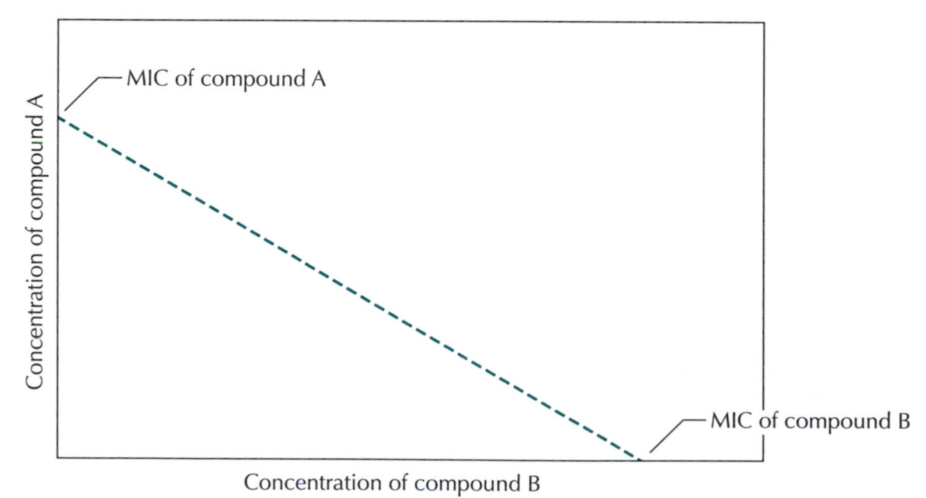

GROWTH KINETICS

The change in bacterial numbers over time is represented by growth curves showing the lag, exponential, stationary, and death phases of the population (Fig. 2.11). These phases and the growth kinetics described below are based on the binary replication of DNA (i.e., one copy makes two copies makes four copies makes eight copies, etc.). In bacteria, which replicate by binary fission, increases in cell number closely correspond to the replication of DNA. Thus, the growth curves can be plotted as the number of cells (CFU per milliliter) on a logarithmic scale or $\log_{10}$ CFU per milliliter versus time. This method may not be appropriate for yeasts which replicate by budding because each cell may have many buds at a given period. In this case, the log of the optical density will give accurate results. Fungi are even more problematic since they grow both by branching, which is exponential, and by hyphal elongation, which is linear. In this case, it is the log of the cell dry weight that must be plotted against time. Growth curve graphs represent the state of microbial populations rather than individual microbes. Thus, both the lag phase and stationary phase of growth represent periods when the growth rate equals the death rate to produce no net change in cell numbers.

Food microbiology is concerned with all four phases of the microbial growth. Microbial inhibitors can extend the lag phase, decrease the growth rate, decrease the size of the stationary-phase population, and increase the death rate. During the lag phase, cells adjust to their new environment by turning genes on or off, replicating their genes, and, in the case of spores, differentiating into vegetative cells (see chapter 3). The lag phase duration depends on the temperature, the inoculum size (larger inocula usually have shorter lag phases), and the physiological history of the organism. If actively growing cells are transferred into an identical fresh medium at the same temperature, the lag phase may vanish. These factors can also be manipulated to extend the lag phase enough so that some other quality attribute of the food (such as proteolysis or browning) becomes unacceptable to humans. Foods are considered microbially safe if the food spoils before pathogens grow

Figure 2.11 All sections of the microbial growth curve are affected by time and temperature. doi:10.1128/9781555817206.ch02.f02.11

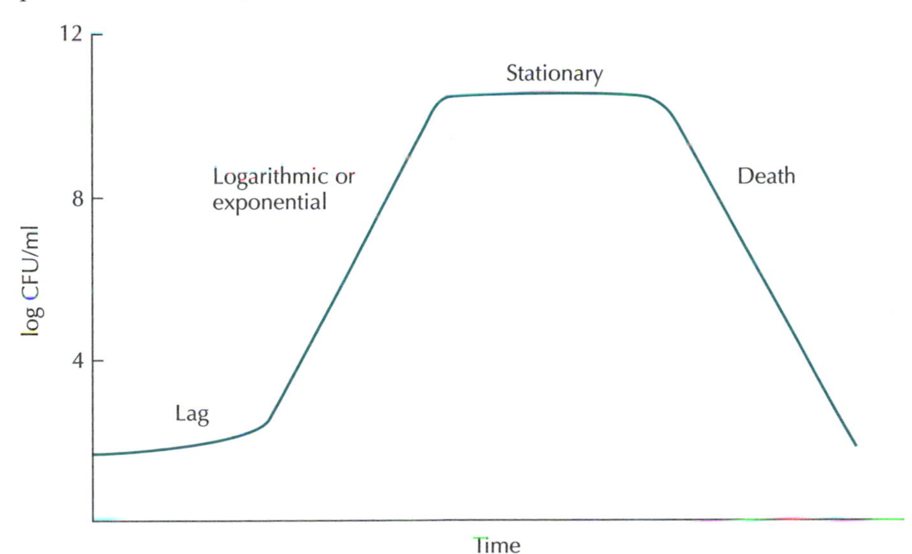

Table 2.9 Representative specific growth rates and doubling times of microorganisms

Organism and conditions	μ (h^{-1})	t_d (h)
Bacteria		
Optimal conditions	2.3	0.3
Limited nutrients	0.20	3.46
Psychrotroph, 5°C	0.023	30
Molds		
Optimal conditions	0.1–0.3	6.9–20

(because people will not eat spoiled food). However, "spoiled" is a subjective and culturally biased concept. It is safer to prevent cell growth, regardless of time (such as reduction of pH to <4.6 to inhibit pathogen growth).

Bacteria reproduce by binary fission during the log, or exponential, phase of growth. One cell divides into two cells that divide into four cells that divide into eight cells, etc. Food microbiologists often use doubling times as the constant to describe the rate of logarithmic growth. Doubling times (t_d), which are also referred to as generation times (t_{gen}), are inversely related to the specific growth rate (μ) as shown in Table 2.9.

Equations can be used to calculate the influence of different variables on a food's final microbial load. The number of organisms (N) at any time (t) is proportional to the initial number of organisms (N_0).

$$N = N_0 e^{\mu t} \quad \textbf{(1)}$$

Thus, decreasing the initial microbial load 10-fold will reduce the cell number at any time 10-fold. Because the specific growth rate (μ) and time are in the power (i.e., exponent) function of the equation, they have a greater effect on the cell number than does the initial cell number. Consider a food where the N_0 is 1×10^4 CFU/g and the μ is 0.2 h^{-1} at 37°C. After 24 h, the cell number would be 1.2×10^6 CFU/g. Reducing the initial number 10-fold reduces the number after 24 h 10-fold to 1.2×10^5 CFU/g. However, reducing the temperature from 37 to 7°C has a more profound effect. If the growth rate decreases twofold with every 10°C decrease in temperature, then the μ will be decreased eightfold to 0.025 h^{-1} at 7°C. When equation 1 is solved using these values (i.e., $N = 10^4 e^{0.025 \times 24}$), the cell number ($N$) at 24 h is 1.8×10^4 CFU/g. Both time and temperature have a much greater influence over the final cell number than does the initial microbial load.

How long it will take a microbial population to reach a certain level can be determined from the following equation:

$$2.3 \log(N/N_0) = \mu \Delta t \quad \textbf{(2)}$$

Consider the case of ground meat manufactured with an N_0 of 1×10^4 CFU/g. How long can it be held at 7°C before reaching a level of 10^8 CFU/g? According to equation 2, $t = [2.3(\log 10^8/10^4)]/0.025$ or 368 h.

The relationship between doubling times (t_d) and μ is more obvious if equation 2 is written using natural logs [i.e., $\ln(N/N_0) = \mu \Delta t$] and solved for the condition where t is t_d and N is $2N_0$. Since the natural log of 2 is 0.693, the solution is

$$0.693/\mu = t_d \quad \textbf{(3)}$$

How many doublings will it take for 4 bacteria to become 20 million?

Reprinted with permission from CartoonStock.

"I wish you'd learn to put the lid on your Petri dish, Harry! We came here with four kids, and now it looks like we've got twenty million. . . . !!"

Some typical specific growth rates and doubling times are given in Table 2.9. Examples of how to calculate doubling times and growth rates are given in Box 2.4.

MICROBIAL PHYSIOLOGY AND METABOLISM

Think about your room. Its natural state is disordered. It takes energy to combat the disorder. This is as it should be. It is a law of the universe. (Tell that to your mother, roommate, or partner the next time they complain about your messy room.) The second law of thermodynamics dictates that *all* things go to a state of maximum disorder in the absence of energy input. Since life is a fundamentally ordered process, all living things must generate energy to maintain their ordered state. Foodborne bacteria do this by oxidizing reduced compounds. Oxidation occurs only when the oxidation of one compound is coupled to the reduction of another. In the case of aerobic bacteria (bacteria that require oxygen to live), the initial carbon source,

BOX 2.4

Calculating specific growth rates and doubling times from plate count data

You have obtained plate count data such as those shown in the table. How do you use these to calculate specific growth rates and doubling times? There are many different ways. If your instructor has a preference, or even a method not shown here, use it! If not, find the way that works for you.

Graphical solution using semilog paper

This method is good for visual learners and the mathematically challenged.

1. Plot the data on semilog paper with CFU/milliliter on the *y* axis and time on the *x* axis.
2. Draw a straight line through the linear portion of the curve.
3. Pick a convenient value on the *y* axis (y_1), and note the corresponding time (t_1).
4. On the *y* axis, find the value that is $2y_1$. Find the time (t_2) that corresponds to $2y_1$.
5. The doubling time is $t_2 - t_1$.
6. Note that you cannot do linear regression on these data since the *y* axis is on a log, not linear, scale.

Calculations using an Excel spreadsheet

This method is more exact, quantitative, and, once you know how to use it, easier than the graphical method shown above. This method can be used with other plotting programs, although the specific commands will be different. You can also do the log conversions, plot them on regular graph paper, and do linear regression with any calculator.

Using an Excel spreadsheet:

1. Convert the data into log CFU per milliliter.
2. Arrange the data in columns of time (first column) and log CFU/milliliter (second column) as shown in the table.
3. Highlight all the rows of data.
4. In the program use the commands "insert" → chart → xy (scatter). This will plot the data, and the linear section will become apparent.
5. Press "cancel" (yes, cancel!), highlight the rows identified in step 4 above as being in the linear section, and repeat the process in step 4.
6. Now you have the chart. With the chart selected, click on the Chart Menu. Choose "add trendline" → linear and from the options, "display equation on chart." The slope of the line is the specific growth rate. Since the data were plotted in $\log_{10}$ and natural log, the 2.3 conversion factor (as explained in equations 2 and 3) must be used. To convert the specific growth rate to doubling times, use the equation $0.693/(2.3)(\mu) = t_d$. Or, in this case, $0.693/(2.3)(0.0105) = 28.7$ min.

Time (min)	CFU/ml (text example)	CFU/ml (problem 16)
0	1.5×10^5	3.0×10^5
30	1.8×10^5	
60	1.4×10^5	3.0×10^5
90	1.9×10^5	
120	2.0×10^5	4×10^5
150	4.0×10^5	
180	8.0×10^5	6.5×10^5
210	2.2×10^6	
240	4.0×10^6	2.0×10^6
270	6.5×10^6	
300	1.8×10^7	1.5×10^6
330	3.3×10^7	
360	4.0×10^7	2.0×10^6
390	4.4×10^7	
420	5.0×10^7	2.0×10^6

for example, glucose, is oxidized to carbon dioxide, oxygen is reduced to water, and 34 ATP molecules are generated. Aerobes generate most of their ATP through oxidative phosphorylation in the electron transport chain. In *oxidative phosphorylation*, energy is generated when oxygen is used as the terminal electron acceptor. This drives the formation of a high-energy bond between free phosphate and adenosine 5′-diphosphate (ADP) to form ATP, which can be thought of as the cell's energy currency that can be "saved" for future use. Anaerobic bacteria, which do not use oxygen, must use an internal organic compound as an electron acceptor in a process of fermentation. They generate only 1 or 2 mol of ATP per mol of glucose used. The ATP is formed by *substrate-level phosphorylation*, and the phosphate group is transferred from a phosphorylated organic compound to ADP to make ATP.

Authors' note

We have written this book assuming no coursework in biochemistry. But microbes cannot be understood without understanding any biochemistry. Consider this section your minimal primer in biochemistry.

Carbon Flow and Substrate-Level Phosphorylation

The EMP Pathway Forms Two ATPs from Six-Carbon Sugars

The Embden-Meyerhof-Parnas (EMP) pathway (Fig. 2.12) is the most common pathway for glucose catabolism (glycolysis). The overall rate of glycolysis is regulated by the activity of phosphofructokinase. This enzyme

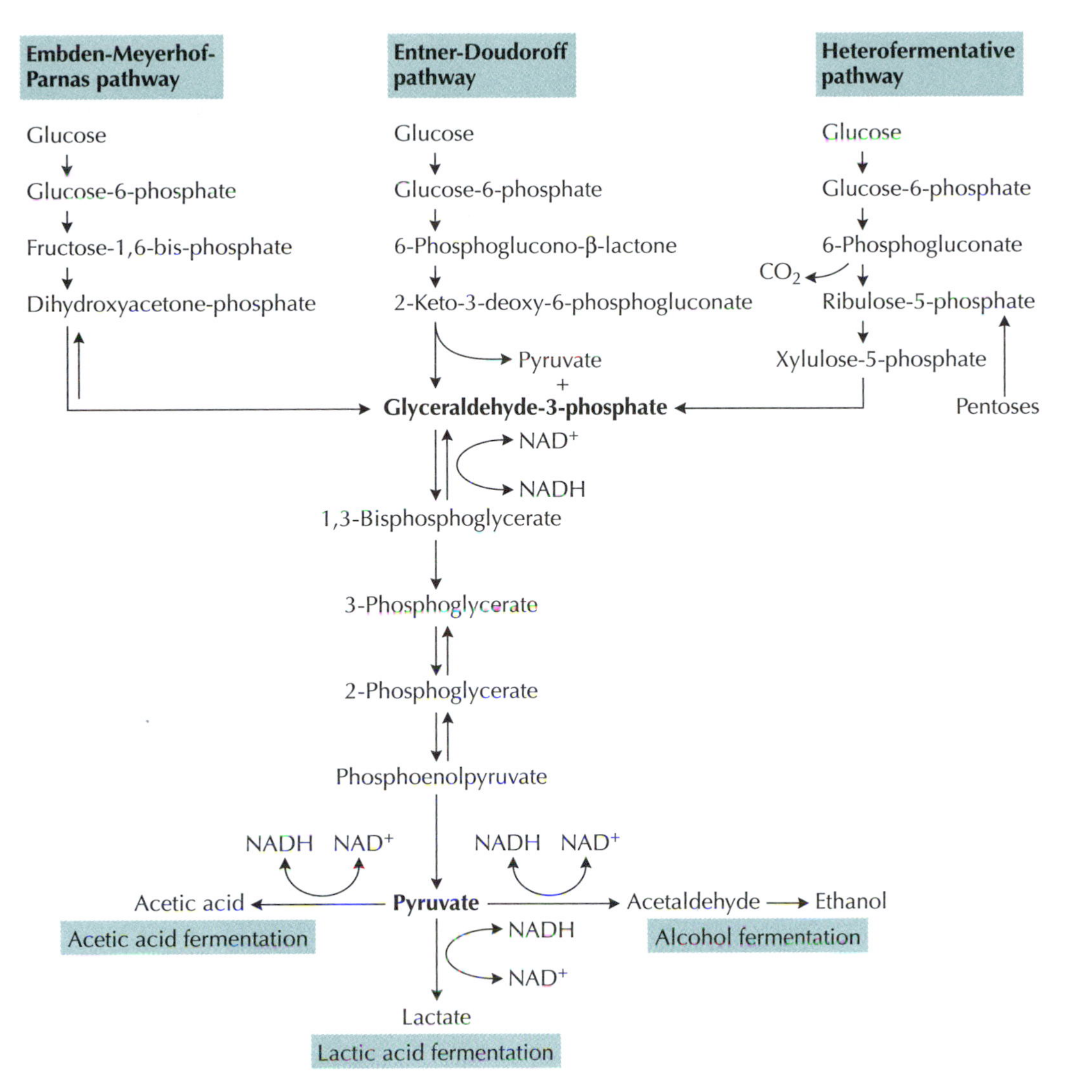

Figure 2.12 Major catabolic pathways used by foodborne bacteria. Redrawn from M. P. Doyle, L. R. Beuchat, and T. J. Montville (ed.), *Food Microbiology: Fundamentals and Frontiers*, 2nd ed. (ASM Press, Washington, DC, 2001). doi:10.1128/9781555817206.ch02.f02.12

converts fructose-6-phosphate to fructose-1,6-bisphosphate. Phosphofructokinase activity is regulated by the binding of adenosine 5′-monophosphate (AMP) or ATP to inhibit or stimulate (respectively) the phosphorylation of fructose-6-phosphate. Fructose-1,6-bisphosphate activates lactate dehydrogenase (see below) so that the flow of carbon to pyruvate is tightly linked to nicotinamide adenine dinucleotide (NAD) regeneration. This occurs when pyruvate is reduced to lactic acid, the only product of this "homolactic" fermentation. Two ATP molecules are formed in this pathway.

The Entner-Doudoroff Pathway Allows Use of Five-Carbon Sugars and Makes One ATP Molecule

The Entner-Doudoroff pathway is an alternate glycolytic pathway that yields one ATP molecule per molecule of glucose and diverts one three-carbon unit to biosynthesis. In aerobes that use this pathway, such as *Pseudomonas* species, the difference between forming one ATP by this pathway and forming two ATPs by the EMP pathway is inconsequential compared to the 34 ATPs formed from oxidative phosphorylation. In the Entner-Doudoroff pathway, glucose is converted to 2-keto-3-deoxy-6-phosphogluconate. The enzyme keto-deoxy phosphogluconate (KDGP) aldolase cleaves this to one molecule of pyruvate (directly, without the generation of an ATP) and one molecule of 3-phosphoglyceraldehyde. The 3-phosphoglyceraldehyde then follows the same reactions as in the EMP pathway. One ATP is made by substrate-level phosphorylation using phosphoenolpyruvate to donate the phosphoryl group.

Homolactic Catabolism Makes Only Lactic Acid

Homolactic bacteria in the genera *Lactococcus* and *Pediococcus* and some *Lactobacillus* species produce lactic acid as the sole fermentation product. The EMP pathway is used to produce pyruvate, which is then reduced by lactate dehydrogenase, forming lactic acid, regenerating NAD, and generating two ATPs.

Heterofermentative Catabolism Gives Several End Products

Heterofermentative bacteria, such as leuconostocs and some lactobacilli, have neither aldolases nor KDPG aldolase. The heterofermentative pathway is based on the catabolism of five-carbon sugars (pentoses). The pentose can be transported into the cell or made by decarboxylating hexoses. In either case, the pentose is converted to xylulose-5-phosphate, with ribulose-5-phosphate as an intermediate. The xylulose-5-phosphate is split to glyceraldehyde-3-phosphate and a two-carbon unit. The two-carbon unit can be converted to acetaldehyde, acetate, or ethanol. Although this pathway yields only one ATP, it offers cells a competitive advantage by allowing them to utilize pentoses that homolactic organisms cannot use.

The TCA Cycle Links Glycolysis to Aerobic Respiration

The tricarboxylic acid (TCA) cycle links glycolytic pathways to respiration. It generates reduced nicotinamide adenine dinucleotide ($NADH_2$) and reduced flavin adenine dinucleotide ($FADH_2$) as substrates for oxidative phosphorylation and makes some ATP through substrate-level phosphorylation. With each turn of the TCA cycle, 2 pyruvate + 2 ADP + 2 FAD + 8 NAD $\rightarrow$ 6 CO_2 + 2 ATP + 2 $FADH_2$ + 8 NADH, and oxygen is used as the terminal electron acceptor to form water. The TCA cycle is used by all aerobes. Anaerobes may have some of the enzymes in the TCA cycle but not enough to complete the full cycle.

The TCA cycle is also the basis for two industrial fermentations important to the food industry. The industrial fermentations for the acidulant citric acid and the flavor enhancer glutamic acid have similar biochemical bases. Both fermentations take advantage of impaired TCA cycles.

CONCLUSION

Microbial growth in foods is complex. It is governed by genetic, biochemical, and environmental factors. Developments in molecular biology and microbial ecology will change or deepen our perspective about the growth of microbes in foods. Some of these developments are detailed in this book. Other developments will unfold over the coming decades, perhaps led by readers whose journey begins now.

Aerobes have 17 times more energy than anaerobes.

Reprinted with permission from CartoonStock.

"The anaerobic ones are just sitting there, but the aerobic bacteria are doing jumping jacks, sit-ups, leg lifts . . ."

Summary

- Food microbiology is a highly specialized subfield of microbiology.
- The amount of energy a microbe makes depends on the metabolic pathways it can use.
- The plate count determines how many organisms can grow on a given medium under the incubation conditions used.
- Selective media allow low numbers of specific pathogens to be enumerated when they are in a larger population of other bacteria.
- Enrichment allows very low numbers of specific pathogens to be detected in the presence of large numbers of other bacteria.
- The most-probable-number (MPN) method allows very low numbers of bacteria to be estimated from statistical tables.
- Injured cells may escape detection by cultural methods but can still cause illness when consumed.
- Viable but nonculturable (VNC) cells are exactly that.
- Intrinsic factors, such as pH and a_w, and extrinsic factors, such as time and temperature, can be manipulated to control microbial growth.
- The amount of available water (a_w) determines which organisms can grow in a food and influences lag time, growth rate, and final cell density.
- An a_w of 0.85 is the lowest a_w that allows growth of pathogens. This makes it a cardinal value in food microbiology.
- Hurdle technology challenges bacteria by several mechanisms to inhibit microbial growth.
- The microbial growth curve consists of lag, log, stationary, and death phases of growth. Each phase is important to microbial food safety and can be manipulated.

- The equation $N = N_0e^{\mu t}$ describes the exponential phase of microbial growth.
- Different bacteria use different metabolic pathways to generate the energy they need to maintain an ordered state.
- Aerobic bacteria make ATP by oxidative phosphorylation.
- Anaerobic bacteria make ATP by substrate-level phosphorylation.

Suggested reading

Barer, M., and C. R. Harwood. 1999. Bacterial viability and culturability. *Adv. Microb. Physiol.* **41:**93–137.

Bryan, F. L. 2004. The "danger zone" reevaluated. *Food Safety Magazine* February/March 2004.

International Commission on Microbiological Specifications for Foods. 1980. *Microbial Ecology of Foods*, vol. 1. *Factors Affecting Life and Death of Microorganisms.* Academic Press, New York, NY.

International Commission on Microbiological Specifications for Foods. 1980. *Microbial Ecology of Foods*, vol. 2. *Food Commodities.* Academic Press, New York, NY.

Kell, D. B., A. S. Kaprelyants, D. H. Weichart, C. R. Harwood, and M. R. Barer. 1998. Viability and activity in readily culturable bacteria: a review and discussion of the practical issues. *Antonie van Leeuwenhoek* **73:**169–187.

Montville, T. J., and K. R. Matthews. 2007. Growth, survival, and death of microbes in foods, p. 3–22. *In* M. P. Doyle and L. R. Beuchat (ed.), *Food Microbiology: Fundamentals and Frontiers*, 2nd ed. ASM Press, Washington, DC.

Questions for critical thought

1. The plate count is based on the assumptions that one cell forms one colony and that every colony is derived from only one cell. For each assumption, list two situations where this would not be true.
2. Given the data in Fig. 2.1, calculate the number of *S. aureus* organisms in the sample using the proper units.
3. Why *can't* you use the standard plate count to enumerate 3.2×10^4 CFU of *S. aureus*/ml in a population of 5.3×10^7 other bacteria?
4. If the dark circles in the diagram below represent test tubes positive for growth, what would the MPN of the sample be?

0.1 ml	0.01 ml	0.001 ml	0.0001 ml	0.00001 ml
●●●	●●●	○●●	○○●	○○○

5. What is the difference between intrinsic and extrinsic factors in food? Name one intrinsic factor and one extrinsic factor that you think are very important for controlling microbes in food. Why do you think that they are important? How do they work?
6. What three mechanisms do cells use to maintain pH homeostasis? Why do they need three? Loss of which mechanism would be worst for the cell? Why?
7. Why do the data for sucrose (but not salt) end at 0.92? (Hint: what is the highest concentration of sucrose you can find in food?)
8. What would happen to the a_w of a 44% sucrose solution if it were completely converted to glucose and fructose?

9. Table 2.5 was constructed using *calculations* to determine the amount of salt needed to achieve a certain a_w. However, when the amount of salt is determined by actual measurement of a_w during salt addition, the amount of salt required is lower than the theoretical value. Speculate on reasons for this. How would you test your hypothesis?
10. If the salt concentration of a soup is reduced from 5 to 3%, what will the effect on a_w be? (Calculate the a_w at both concentrations.)
11. Consider equation 1. Your food product has an initial aerobic plate count of 3.8×10^4 CFU/g. After 1 week of refrigerated storage, the count reaches an unacceptable 3.8×10^6 CFU/g. You decide that an acceptable level of 3.8×10^5 CFU/g could be achieved if the initial number were lower. What starting level of bacteria would you need to achieve this goal? (Hint: put your calculator away and think about it.) In real life, how might a food manufacturer achieve this lower initial count?
12. Your company manufactures ready-to-eat refrigerated meals. Your supervisor asks you to start tracking microbial quality by doing the total plate count with incubation at 37°C. Why is this *not* a good idea? How would you convince your boss that she was wrong?
13. Make a chart listing the three main catabolic pathways and their key regulatory enzyme, carbohydrate-splitting enzyme, amount of ATP generated, and final metabolic product.
14. Both anaerobic lactic acid bacteria and aerobic pseudomonads are naturally present in ground beef. In beef packed in oxygen-permeable cellophane, pseudomonads are the main spoilage organisms. In beef packed in oxygen-impermeable plastic film, lactic acid bacteria predominate. Why?
15. Write a narrative explaining the relationship between time and temperature as illustrated in Fig. 2.7 and 2.8.
16. What is the doubling time for the growth curve in the problem column of the table in Box 2.4?
17. Why is there no simple answer to the question "How many bacteria are in this sample of food?"
18. Although we speak of a "zero tolerance" for a pathogen in food, the actual limit is 0/25 g. Why?
19. What are examples of structures where biofilms could grow? Why would these be food safety problems?
20. If a food safety test is negative for bacterial pathogens, why doesn't that guarantee a safe product?

3 Spores and Their Significance

LEARNING OBJECTIVES

The information in this chapter will help the student to:

- discuss how the canning industry overcomes the challenge of spore heat resistance
- define the traits of low-acid canned foods and how they must be processed
- identify, characterize, and differentiate among spore-forming bacteria that cause illness and spoilage
- understand the fundamental difference between a bacterial spore and a vegetative cell
- correlate the unique properties of spores with the challenge they present for food preservation
- draw a diagram of a spore and compare it to a diagram of a bacterial cell
- explain the physical and chemical bases for spore heat resistance
- compare the cycles of sporulation and germination with the gain and loss of spore resistance characteristics

INTRODUCTION

This chapter is about spores. They are unique life forms of tremendous significance in food processing. The importance of spores and the steps used to control them are discussed first. Then the structure and unique properties of the spore are described. Finally, the complex life cycle of a spore, which consists of sporulation (the conversion of a vegetative cell to a spore) and germination (the conversion of a spore to a vegetative cell), is presented.

SPORES IN THE FOOD INDUSTRY

Spore-forming bacteria and heat-resistant fungi cause big problems for the food industry. Three species of sporeformers, *Clostridium botulinum*, *Clostridium perfringens*, and *Bacillus cereus*, are infamous for producing toxins. Other sporeformer species, such as *Alicyclobacillus*, *Geobacillus*, and *Sporolactobacillus*, cause spoilage. Sporeformers that cause foodborne illness and spoilage are particularly important in low-acid foods (pH ≥ 4.6) packaged in cans, bottles, pouches, or other hermetically (vacuum) sealed containers (i.e., "canned" foods) that are processed by heat. Diseases and spoilage caused by sporeformers are usually associated with thermally processed foods, since heat kills vegetative cells but allows survival and growth of spore-forming organisms. Other sporeformers cause spoilage of high-acid

doi:10.1128/9781555817206.ch03

foods (pH < 4.6), but not illness. Psychrotrophic sporeformers cause spoilage of refrigerated foods. Fungi that produce heat-resistant ascospores cause spoilage of acidic foods and beverages.

Louis Pasteur discovered that spore-forming bacteria cause food spoilage during his studies of butyric acid fermentations in wines. Pasteur isolated an organism he termed *Vibrion butyrique*, which is probably what we now call *Clostridium butyricum*. Spores were also discovered independently by Ferdinand Cohn and Robert Koch in 1876. Pasteur and Koch linked microbial activity with food quality and safety. The famous Koch's postulates (Box 3.1) were born during his investigation of *Bacillus anthracis*. These postulates are still used to prove that a disease is caused by a specific bacterium.

In the late 1700s, Nicolas Appert invented the process of appertization, a predecessor to canning. This process places food in a hermetically (airtight) sealed container and preserves it by heating. Appert (incorrectly) believed that the elimination of air stabilized canned foods. The empirical use of thermal processing gradually developed into modern-day canning industries. In the late 1800s and early 1900s, scientists in the United States developed scientific principles to ensure the safety of thermally processed foods. Because of these studies, thermal processing of foods in hermetically sealed containers became an important industry. Samuel Prescott and William Underwood of the Massachusetts Institute of Technology and Harry Russell of the University of Wisconsin found that spore-forming bacilli caused the spoilage of thermally processed clams, lobsters, and corn. The classic study of J. R. Esty and K. F. Meyer in California provided definitive values for the heat resistance of *C. botulinum* type A and B spores. C. Olin Ball of Rutgers University developed the mathematical foundation for commercial canning. These efforts helped rescue the U.S. canning industry from its near death in the 1940s due to botulism outbreaks caused by commercially canned olives and other foods. Quantitative thermal processes, understanding of spore heat resistance, aseptic processing, and implementation of the Hazard Analysis and Critical Control Point concept all grew out of the canned-food industry.

Low-Acid Canned Foods

The U.S. Food and Drug Administration (FDA) and the U.S. Department of Agriculture (USDA) Food Safety Inspection Service define a low-acid canned food as one with a final pH of >4.6 and a water activity (a_w) of >0.85. The USDA has regulatory control of foods that contain at least 3% raw red meat or 2% cooked poultry. The FDA regulates everything else. The regulations for the thermal processing of canned foods are described in the U.S. *Code of Federal Regulations* (21 CFR, parts 108 to 114). For every food they make, food processors must file a description of the thermal process, facility, equipment, and formulations with the FDA or USDA before they start making the product.

Low-acid canned foods are packaged in hermetically sealed containers. These are usually cans or glass jars but can also be pouches and other types of containers. These "cans" (as collectively defined in this chapter) must be processed by heat to achieve "commercial sterility." Commercial sterility uses heat to inactivate foodborne pathogens and spoilage microorganisms that can grow in the food. Commercial sterility indicates a shelf-stable product with a negligible level of microbial survival. Acidification or a_w reduction can also produce commercial sterility. These techniques are often combined to reduce heat treatments and improve product quality.

BOX 3.1

Koch's postulates for proof for the bacterial origin of a disease

1. The bacteria must be isolated from a sick animal.
2. The bacteria must be identified in pure culture.
3. When the pure culture of bacteria is reintroduced into a healthy animal, the animal becomes sick.

Inactivation of *C. botulinum* spores is the primary processing goal for low-acid canned foods. *C. botulinum* is the most heat-resistant microbial pathogen. (There are, however, spores from other bacteria, such as *Geobacillus stearothermophilus*, that are more heat resistant.) The severity of the heat treatment required depends on the class of food, its spore content, pH, storage conditions, and other factors. For example, canned low-acid vegetables and uncured meats usually receive a 12*D* process (see below) or "botulinum cook." This results in a 12-$\log_{10}$, or 99.9999999999%, reduction of viable *C. botulinum* spores. Milder heat treatments are applied to shelf-stable canned cured meats (where the curing agents inhibit growth) and to foods with reduced a_w or other antimicrobial factors. Certain foods and ingredients such as mushrooms, potatoes, spices, sugars, and starches may contain high spore levels and require more than a 12*D* process to prevent spoilage. When spore loads influence the process, they must be monitored.

Researchers in the early and mid-20th century quantified the thermal processes required to prevent botulism and spoilage in canned foods. The lethality of a thermal process is calculated from the semilogarithmic microbial death model. In this model, the number of surviving cells is plotted on a logarithmic *y* axis against the time of heating on the linear *x* axis. This results in a linear relationship. One over the negative slope of this line is the *D* value. The *D* value is defined as the time it takes to reduce viability by a factor of 10. This reasoning provides the foundation for the 12*D* botulinum cook. The botulinum cook is the thermal process time required for a 12-$\log_{10}$ reduction of viable *C. botulinum* spores. Of course, there would never be 10^{12} spores in a single can. But statistically, the behavior of 10^{12} spores in a single can is the same as that of one spore in each of 10^{12} cans. So, if there were one spore per can and the 12*D* process were used, only one of the 10^{12} cans would contain a viable spore. Why 12*D* (and not 10 or 14)? No one remembers. Two explanations are circulating. One is that 12*D* equals a million million, a ridiculously low probability. The other explanation is that 10^{12} spores is the maximum number that could fit in a 1-cm^3 volume.

Two terms, or values, are used to describe thermal inactivation of bacteria. (i) The *D* value is the time required for a 1-log reduction in viability at a given temperature. (ii) The *z* value is the temperature change required to alter the *D* value by a factor of 10. The *D* value represents the heat sensitivity of an organism at a specific temperature. The *z* value represents how much that heat sensitivity changes as temperatures change. The *z* value is usually expressed as the number of degrees that it takes to change the *D* value by a factor of 10.

The canning industry prioritizes process design to protect against (i) public health hazard from *C. botulinum* spores, (ii) spoilage from mesophilic spore-forming organisms, and (iii) spoilage from thermophilic organisms in containers stored in warm environments. Generally, low-acid canned foods require 3 to 6 min at 121°C (250°F) to achieve a 12*D* treatment. This ensures the inactivation of *C. botulinum* spores having a $D_{121°C}$ of 0.21 min and a *z* value of 10°C (18°F). Economic spoilage is avoided by achieving ~5*D* killing of mesophilic spores that typically have a $D_{121°C}$ of ~1 min. Foods distributed in warm climates require a severe thermal treatment of ~20 min at 121°C to achieve a 5*D* killing of *Clostridium thermosaccharolyticum*, *Geobacillus* (formerly *Bacillus*) *stearothermophilus*, and

Desulfotomaculum nigrificans. These organisms have a $D_{121°C}$ of ~3 to 4 min. Such severe treatments decrease nutrient content and sensory qualities but ensure a shelf-stable food.

Aseptic processing can improve the quality of low-acid canned foods. There are three steps in aseptic processing (Fig. 3.1). (i) The product is commercially sterilized outside of the container. Since the product is in contact with the heat source, it rapidly achieves a uniform process temperature. (ii) The processed product is cooled and transferred into presterilized containers. (iii) The container receives an aseptic hermetic sealing in a sterile environment. This technology was initially used for commercial sterilization of milk and creams in the 1950s. It is now used for other foods such as soups, eggnog, cheese spreads, sour cream dips, puddings, and high-acid products such as boxed fruit drinks. Aseptic processing and packaging systems reduce energy, packaging material, and distribution costs.

Bacteriology of Sporeformers of Public Health Significance

Three species of sporeformers, *C. botulinum*, *C. perfringens*, and *B. cereus*, cause foodborne illness. They are covered extensively in later chapters. There have also been occasional incidences of intestinal anthrax caused by eating contaminated raw or poorly cooked meat. However, *B. anthracis* is not considered a foodborne pathogen. Other sporeformers are associated with food and nonfood environments (Table 3.1). In some cases, advances in molecular biology have led to the division of a genus (e.g., *Geobacillus thermophilus* was once *Bacillus thermophilus*; nothing has changed, except the name). In other cases, such as with *Filobacillus*, the organism does not fit in any known genus, so a new one is initiated. The thermal resistance parameters for many foodborne sporeformers are shown in Table 3.2.

Figure 3.1 Comparison of conventional **(top)** and aseptic **(bottom)** processing of food. In conventional processing, the food is placed in the container, sealed, and sterilized. In aseptic processing, the food and the container are sterilized separately, filled, and sealed under aseptic conditions. doi:10.1128/9781555817206.ch03.f03.01

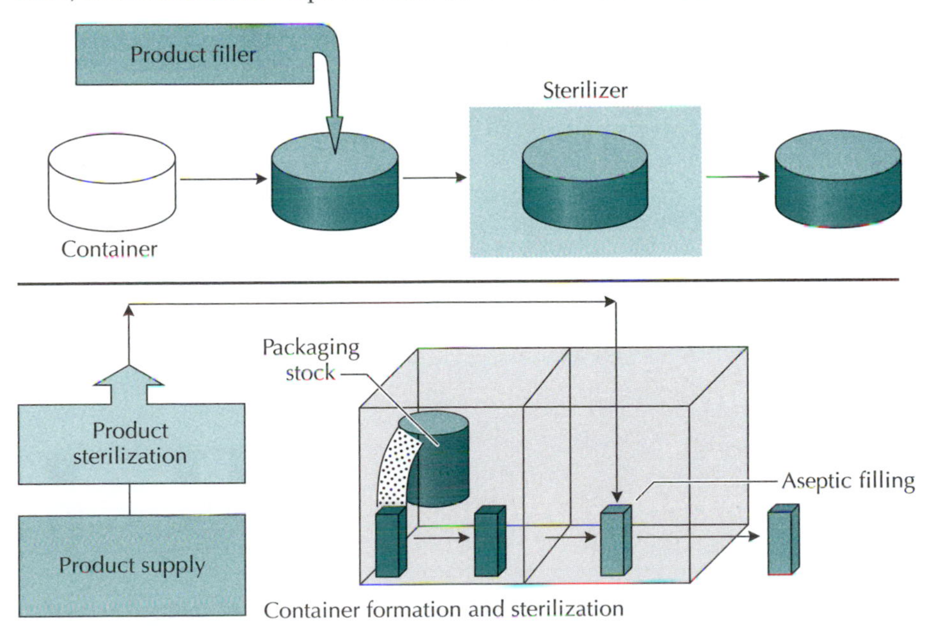

Table 3.1 Genera of gram-positive spore-forming bacteria

Genus	Characteristics
Alicyclobacillus	Thermophilic aciduric spore-forming bacteria that survive hot fill processes and cause spoilage in juices
Amphibacillus	Facultatively anaerobic xylan-degrading spore-forming bacteria
Bacillus	Aerobic spore-forming rod-shaped bacteria
Clostridium	Anaerobic spore-forming rod-shaped bacteria
Desulfotomaculum	Sulfate-reducing spore-forming bacteria
Filobacillus	Halophilic aerobic spore-forming bacteria
Geobacillus	Thermophilic spore-forming rod-shaped bacteria
Sporolactobacillus	Spore-forming lactobacilli
Sulfolobus	Sulfur-oxidizing thermophilic aciduric spore-forming bacteria

C. botulinum is the principal microbial hazard in heat-processed vacuum-packed foods and in minimally processed refrigerated foods. The genus *Clostridium* consists of gram-positive, anaerobic, spore-forming bacilli that obtain energy by fermentation. *C. botulinum* spores swell the mother sporangium, giving a "tennis racket" or club-shaped appearance. The strains of *C. botulinum* are heterogeneous and produce antigenically distinct toxins (A through G) that are strain specific. Organisms that produce these toxin types fall into two larger groups, type I and type II. Type I *C. botulinum* strains are mesophiles, produce heat-resistant spores, and make proteases that cause obvious spoilage. Type II *C. botulinum* strains are

Table 3.2 Heat resistance of sporeformers of importance in foods[a]

Type of spore	Approx $D_{100°C}$ (min)
Spores of public health significance	
Group I *Clostridium botulinum* types A and B	7–30
C. botulinum type E	0.01
Bacillus cereus	3–200
Clostridium perfringens	0.3–18
Mesophilic aerobes	
Bacillus subtilis	7–70
Bacillus licheniformis	13.5
Bacillus megaterium	1
Bacillus polymyxa	0.1–0.5
Bacillus thermoacidurans	2–3
Thermophilic aerobes	
Geobacillus stearothermophilus	100–1,600
Bacillus coagulans	20–300
Mesophilic anaerobes	
Clostridium sporogenes	80–100
Thermophilic anaerobes	
Desulfotomaculum nigrificans	<480
Clostridium thermosaccharolyticum	400

[a]Source: P. Setlow and A. E. Johnson, p. 33–70, *in* M. P. Doyle, L. R. Beuchat, and T. J. Montville (ed.), *Food Microbiology: Fundamentals and Frontiers*, 2nd ed. (ASM Press, Washington, DC, 2001).

nonproteolytic and make spores that are less heat sensitive. But they can grow at low temperatures. Thus, regulators are concerned that minimally processed refrigerated food might be a botulism hazard.

C. perfringens is widespread in soils and in the intestinal tracts of humans and certain animals. *C. perfringens* grows extremely rapidly in high-protein foods. This is especially true for meats that are cooked, killing vegetative cells, and then temperature abused, allowing *C. perfringens* to grow. The time in the 40 to 140°F temperature abuse zone allows growth and production of a diarrheal enterotoxin. *C. perfringens* produces other extracellular toxins and enzymes, but these are mainly important for gas gangrene and animal diseases. Its ubiquitous distribution in foods and food environments, formation of spores that survive cooking, and extremely high growth rate in warm foods (doubling times of 6 to 9 min at 43 to 45°C) are characteristics that lead *C. perfringens* to cause foodborne illness.

The heat resistance of *C. perfringens* spores varies among strains. In general, two classes of heat sensitivity are common. Heat-resistant spores have $D_{90°C}$ ($D_{194°F}$) values of 15 to 145 min and z values of 9 to 16°C (16 to 29°F), while heat-sensitive spores have $D_{90°C}$ values of 3 to 5 min and z values of 6 to 8°C (11 to 14°F). The spores of the heat-resistant class generally require a heat shock of 75 to 100°C (167 to 212°F) for 5 to 20 min in order to germinate. The spores of both classes may survive cooking of foods and may be stimulated by the heat shock that occurs during thermal processing. Both classes cause diarrheal foodborne illness.

The genus *Bacillus* contains only two pathogenic species, *B. anthracis* and *B. cereus*. *B. cereus* can produce a heat-labile enterotoxin causing diarrhea and a heat-stable toxin causing emesis (vomiting) in humans. Generally, the organism must grow to very high numbers ($>10^6$/g of food) to cause human illness. *B. cereus* is closely related to *B. megaterium*, *B. thuringiensis*, and *B. anthracis*. *B. cereus* can be distinguished from these species by biochemical tests and the absence of toxin crystals. Other bacilli, including *B. licheniformis*, *B. subtilis*, and *B. pumilus*, have caused foodborne outbreaks, primarily in the United Kingdom.

Spores of *B. cereus* are located in the central to subterminal part of the vegetative cell. Spore germination occurs over the range of 8 to 30°C (46.4 to 86°F). Spores from strains associated with food poisoning have a heat resistance ($D_{95°C}$ [$D_{203°F}$]) of ~24 min. Other strains have a wider range of heat resistances. The strains involved in food poisoning may have higher heat resistances and therefore be more apt to survive cooking.

Heat Resistance of *C. botulinum* Spores

Heat resistance of spores varies greatly among species, and even among strains (Table 3.2). Group I *C. botulinum* type A and proteolytic B strains produce spores that are extremely heat resistant. They are the most important sporeformers for the public health safety of canned foods. Commercial outbreaks of botulism associated with olives and other canned vegetables led Esty and Meyer to conduct their classic investigations on the heat resistance of botulinal spores. They examined 109 type A and B strains at five heating temperatures from 100 to 120°C (212 to 248°F). They found that the inactivation rate is logarithmic between 100 and 120°C, and the inactivation rate depends on the spore concentration, the pH, and the food they are heated in. Esty and Meyer found that 0.15 M phosphate buffer (Sorensen's buffer), pH 7.0, gives the most consistent heat inactivation results; their standardized system for comparing heat resistances is still used by researchers. The data of Esty and Meyer can be extrapolated to give a maximum value for $D_{121.1°C}$ of 0.21 min for *C. botulinum* type A and B spores in phosphate buffer. The canning industry uses $D_{121°C}$ as a standard in calculating process requirements. Proteolytic type F *C. botulinum* spores have the following heat resistances: a $D_{98.9°C}$ of 12.2 to 23.2 min and a $D_{110°C}$ of 1.45 to 1.82 min. These values are much lower than for type A spores. Spores of nonproteolytic type B and E *C. botulinum* have much lower heat resistances than proteolytic A and B strains. Type E spores have $D_{70°C}$ ($D_{158°F}$) values varying from 29 to 33 min and $D_{80°C}$ ($D_{176°F}$) values from 0.3 to 2 min depending on the strains. The *z* values range from 13 to 15°F. The heat resistance of *C. botulinum* spores depends on environmental and recovery conditions. Heat resistance is markedly affected by acidity. Esty and Meyer found that spores have maximum resistance at pH 6.3 and 6.9 and that resistance decreases markedly at pH values below 5 or above 9. Increased sodium chloride or sucrose concentrations as well as decreased a_w increase the heat resistance of *C. botulinum* spores. *C. botulinum* spores coated in oil are more resistant to heat. Sporulation of *C. botulinum* at higher temperatures results in spores with greater heat resistance, possibly through the formation of heat shock proteins.

C. botulinum spores of groups I and II are highly resistant to irradiation compared with most vegetative cells. It is probably not practical to inactivate them in foods by irradiation. *C. botulinum* spores have a *D* of 2.0 to 4.5 kGy. *C. botulinum* spores are also highly resistant to ethylene oxide but are inactivated by halogen sanitizers and by hydrogen peroxide. Hydrogen peroxide is commonly used for sanitizing surfaces in aseptic packaging. Halogen sanitizers are used in cannery cooling waters.

Spoilage of Acid and Low-Acid Canned and Vacuum-Packaged Foods by Sporeformers

Thermally processed low-acid foods receive a heat treatment adequate to kill spores of *C. botulinum* but not sufficient to kill more heat-resistant spores. Acid and acidified foods with an equilibrium pH of $\leq$4.6 are not heated enough to inactivate all spores. Most species of sporeformers do not grow under acid conditions. Inactivation of these spores is unnecessary (since they cannot grow). A process that inactivated them would decrease food quality and nutrition. Other foods, such as cured meats and hams that are not heated enough to kill spores, must be kept refrigerated to prevent microbial growth. Nonpathogenic sporeformers cause spoilage in these foods.

The spores naturally present in foods and the cannery environment contribute to spoilage problems. Dry ingredients such as sugar, starches, flours, and spices often contain high spore levels. Spore populations can also accumulate in a food-processing plant, such as thermophilic spores on heated equipment and saccharolytic clostridia in plants processing sugar-rich foods such as fruits.

The principal spoilage organisms and spoilage manifestations are presented in Table 3.3. The principal classes of sporeformers causing

Table 3.3 Spoilage of canned foods by sporeformers[a]

Type of spoilage	pH	Major sporeformer(s) responsible	Spoilage defects
Flat-sour	≥5.3	*B. coagulans, B. stearothermophilus*	No gas, pH lowered. May have abnormal odor and cloudy liquor.
Thermophilic anaerobe	≥4.8	*C. thermosaccharolyticum*	Can swells, may burst. Anaerobic end products give sour, fermented, or butyric odor. Typical foods are spinach and corn.
Sulfide spoilage	≥5.3	*D. nigrificans, Clostridium bifermentans*	Hydrogen sulfide produced, giving rotten-egg odor. Iron sulfide precipitate gives blackened appearance. Typical foods are corn and peas.
Putrefactive anaerobe	≥4.8	*C. sporogenes*	Plentiful gas. Disgusting putrid odor. pH often increased. Typical foods are corn and asparagus.
Psychrotrophic clostridia	>4.6		Spoilage of vacuum-packaged chilled meats; production of gas, off-flavors and -odors, and discoloration
Aerobic sporeformers	≥4.8	*Bacillus* spp.	Gas usually absent except for cured meats; milk is coagulated. Typical foods are milk, meats, and beets.
Butyric spoilage	≥4.0	*C. butyricum* and *Clostridium tertium*	Gas, acetic and butyric odors. Typical foods are tomatoes, peas, olives, and cucumbers.
Acid spoilage	≥4.2	*B. thermoacidurans*	Flat (*Bacillus*) or gas (butyric anaerobes). Off-odors depend on organism. Common foods are tomatoes, tomato products, and other fruits.
	<4	*Alicyclobacillus acidoterrestris*	Flat spoilage with off-flavors. Most common in fruit juices and acid vegetables; also reported to spoil iced tea.

[a]Source: P. Setlow and E. A. Johnson, p. 33–68, *in* M. P. Doyle and L. R. Beuchat (ed.), *Food Microbiology: Fundamentals and Frontiers*, 3rd ed. (ASM Press, Washington, DC, 2007).

spoilage are thermophilic flat-sour organisms, thermophilic anaerobes not producing hydrogen sulfide, thermophilic anaerobes forming hydrogen sulfide, putrefactive anaerobes, facultative *Bacillus* mesophiles, butyric clostridia, lactobacilli, and heat-resistant molds and yeasts. *Alicyclobacillus acidoterrestris* and psychrotrophic clostridia can spoil fruit products and meat, respectively. *Alicyclobacillus* is particularly noticeable because it causes a disinfectant smell in juices before they appear visibly spoiled. These sporeformers are controlled by monitoring raw foods to limit the initial spore load, thermal processing appropriate for storage and distribution conditions, cooling products rapidly, chlorination of cooling water, and good manufacturing practices.

Heat-resistant fungi cause spoilage of acidic foods, particularly fruit products. Heating for a few minutes at 60 to 75°C kills most filamentous fungi and yeasts. However, heat-resistant fungi produce thick-walled ascospores that survive heating at ≥85°C for 5 min. The most common heat-resistant spoilage fungi are *Byssochlamys, Neosartorya, Talaromyces*, and *Eupenicillium*. Some heat-resistant fungi also produce toxic secondary metabolites called mycotoxins. To prevent spoilage of heat-treated foods, raw materials should be screened for heat-resistant fungi and strict good manufacturing practices and sanitation programs should be followed. Manipulation of a_w and oxygen and the application of antimycotic agents can prevent fungal growth.

SPORE BIOLOGY

Structure

The spore is released from the mother cell at the end of sporulation. Spores are biochemically, structurally, and physiologically different from vegetative cells. The spore has seven layers: the exosporium, coats, outer membrane, cortex, germ cell wall, inner membrane, and core (Fig. 3.2). Many spore structures have no counterparts in the vegetative cell.

The outermost spore layer, the exosporium, varies in size among species. Underlying the exosporium are the spore coats. The spore coats protect the spore cortex from attack by lytic enzymes. Underlying the spore coats is the outer spore membrane. It helps keep the spore impermeable to small molecules.

The cortex is under the outer membrane. The peptidoglycan that makes up the cortex is structurally similar to cell wall peptidoglycan, but with several differences. The spore peptidoglycan always contains diaminopimelic acid, which is not found in the vegetative cell wall (Fig. 3.3). The spore cortex mechanically dehydrates the spore core and is responsible for much of spore resistance (see below).

Between the cortex and the inner membrane is the germ cell wall. Its structure may be identical to that in vegetative cells. The next structure, the

Figure 3.2 Structure of a dormant spore. The various structures are not drawn precisely to scale, especially the exosporium, the size of which varies tremendously among spores of different species. The relative size of the germ cell wall is also generally smaller than shown. The positions of the inner and outer forespore membranes, between the core and the germ cell wall and between the cortex and coats, are also noted. Redrawn from P. Setlow and E. A. Johnson, p. 33–70, *in* M. P. Doyle, L. R. Beuchat, and T. J. Montville (ed.), *Food Microbiology: Fundamentals and Frontiers*, 2nd ed. (ASM Press, Washington, DC, 2001). doi:10.1128/9781555817206.ch03.f03.02

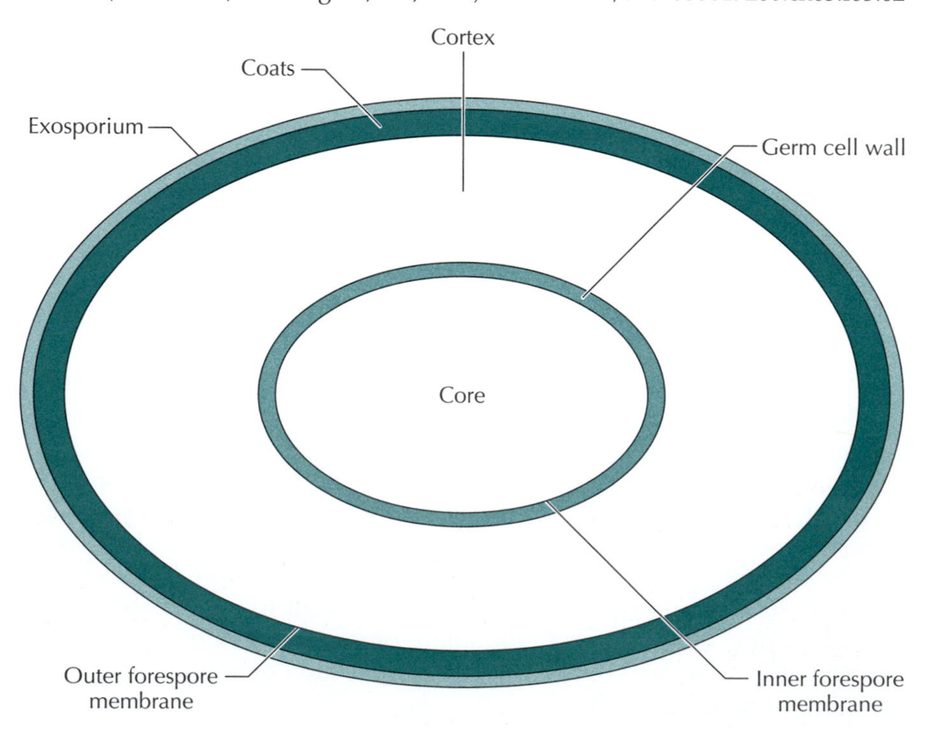

inner spore membrane, is a complete membrane. It is a very strong permeability barrier. This membrane's phospholipid content is similar to that of vegetative cells.

Finally, the spore core contains the spore's DNA, ribosomes, and most enzymes, as well as the deposits of pyridine-2,6-carboxylic acid (dipicolinic acid [DPA]) and divalent cations. There are many unique compounds in the dormant spore, including the large pool of small acid-soluble proteins (SASP), which comprises 10 to 20% of spore protein. Much of the SASP is bound to spore DNA. Spore cores have a very low water content. While vegetative cells have ~4 g of water per g of dry weight, the spore core has only 0.4 to 1 g of water per g of dry weight. The core's low water content is responsible for spore dormancy and spore resistance. The spore cortex generates and maintains spore core dehydration, but the precise mechanism(s) involved is unknown.

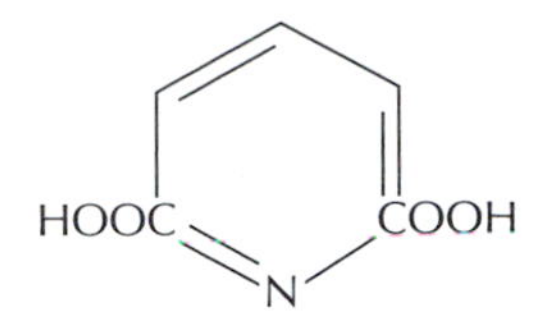

Figure 3.3 Structure of DPA. Note that at physiological pH both carboxyl groups are ionized.
doi:10.1128/9781555817206.ch03.f03.03

Macromolecules

Spores are biochemically different from vegetative cells. Some spore proteins are absent in the vegetative cell. The SASP play a major role in spore resistance. The binding of these proteins to DNA provides resistance to chemical and enzymatic cleavage of the DNA backbone and alters the DNA's UV photochemistry. Conversely, many proteins present in vegetative cells are absent in spores. These include amino acid and nucleotide biosynthetic enzymes. The catabolic enzymes for using amino acid and carbohydrate are present in both spores and cells. Spore DNA appears identical to cell DNA.

Small Molecules

The spore's small molecules, located in the core, are different from those in cells. The small amount of spore core water and the huge deposit of DPA and divalent cations have already been noted (Table 3.4). The ions in the spore core are immobile, since there is no free water. The pH in the spore core is 1 to 1.5 units lower than that in a growing cell. In contrast to growing cells, spores have little, if any, "high-energy" compounds such as deoxynucleoside triphosphates, ribonucleoside triphosphates, reduced pyridine nucleotides, and acyl coenzyme A (acyl-CoA) (Table 3.4).

Dormancy

Spores are metabolically dormant. They have no detectable metabolism. The major cause of the spore's metabolic dormancy is undoubtedly its low water content. This precludes enzyme action. Dormancy is further demonstrated by enzyme-substrate pairs in the core which are stable for months to years but which are degraded in the first 15 to 30 min of germination.

Resistance

The spore's dormancy helps it survive for extremely long periods in the absence of nutrients (Box 3.2). A second factor in long-term spore survival is the spore's extreme resistance to heat, radiation, chemicals, and desiccation. Spores are much more resistant than vegetative cells. Spore resistance is due to many factors, such as spore core dehydration, SASP, and impermeability. Since different factors contribute to different types of resistance, it is not surprising that different types of resistance are gained at different

Table 3.4 Small molecules in cells and spores of *Bacillus* species

	Content (mmol/g [dry wt]) in:	
Molecule[a]	Cells[b]	Spores[c]
ATP	3.6	≤0.005
ADP	1	0.2
AMP	1	1.2–1.3
Deoxynucleotides	0.59[d]	<0.025[e]
NADH	0.35	<0.002[f]
NAD	1.95	0.11[f]
NADPH	0.52	<0.001[f]
NADP	0.44	0.018[f]
Acyl-CoA	0.6	<0.01[f]
CoASH	0.7	0.26[f]
CoASSX	<0.1	0.54[f]
3PGA	<0.2	5–18
Glutamic acid	38	24–30
DPA	<0.1	410–470
Ca^{2+}		380–916
Mg^{2+}		86–120
Mn^{2+}		27–56
H^+	7.6–8.1[g]	6.3–6.9[g]

[a]Abbreviations: ATP, adenosine 5′-triphosphate; ADP, adenosine 5′-diphosphate; AMP, adenosine 5′-monophosphate; NADH, reduced nicotinamide adenine dinucleotide; NAD, nicotinamide adenine dinucleotide; NADPH, reduced nicotinamide adenine dinucleotide phosphate; NADP, nicotinamide adenine dinucleotide phosphate; CoASH, free CoA; CoASSX, CoA in disulfide linkage to CoA or a protein; 3PGA, 3-phosphoglycerate.

[b]Values for *B. megaterium* in mid-log phase.

[c]Values are ranges from spores of *B. cereus*, *B. subtilis*, and *B. megaterium*.

[d]Value is the total of all four deoxynucleoside triphosphates.

[e]Value is the sum of all four deoxynucleotides.

[f]Values are for *B. megaterium* only.

[g]Values are expressed as pHs and are ranges for *B. cereus*, *B. megaterium*, and *B. subtilis*.

Authors' note

Dormant spores that have been trapped in prehistoric amber for 20 million years have germinated and resumed growth as vegetative cells when introduced to microbiological media!

times in sporulation. The mechanisms of spore heat, ultraviolet (UV), and H_2O_2 resistance have been determined. The following discussion of spore resistance concentrates on *B. subtilis* because of the detailed mechanistic data available for this organism. Factors involved in resistance of *B. subtilis* spores are undoubtedly involved in resistance of other species.

BOX 3.2

Jurassic spores

Just how long can spores be dormant and still be viable? Think along geological time lines. The earth was formed about 6 billion years ago. Bacteria came on the scene about 3 billion years ago. Dinosaurs went extinct 65 million years ago. "Humanity" is about 6 million years old. Bacteria as a class clearly have staying power. Individual organisms also have staying power. Raúl Cano, a microbiologist at California Polytechnic University, resurrected a spore that had been dormant for 25 million to 40 million years. *Bacillus* species have a symbiotic relationship with bees, which are frequently embedded in amber. Spores had been seen in these bees using electron microscopes, and spore DNA had been isolated, but no one had ever determined if the spores were still alive. Cano rigorously sterilized the amber samples and all of the experimental material before pulverizing it into broth where spores could normally grow. Grow they did! The spores produced bacilli. Analysis of their DNA showed that they were ancient and unrelated to any modern bacilli that could have been potential contaminants.

Cano, R. J., and M. K. Borucki. 1995. Revival and identification of bacterial spores in 24–40 million year old Dominican amber. *Science* **228:**1060–1064.

Across

2. The structure that is largely responsible for spores' resistance properties
5. One contributor to spore resistance
7. The canning of this type of food is heavily regulated by the U.S. Food and Drug Administration
8. Low-acid canned foods are subject to a _______ cook
9. Although absolute sterility is theoretically impossible, this type of sterility is sufficient in the food industry

Down

1. A notable property of spores
3. When spores undergo this process, they get ready to turn into vegetative cells and lose their resistance properties
4. The process by which vegetative cells form spores
5. This type of acid is found in spores but not in vegetative cells
6. The structure at the center of a spore

Freezing and Desiccation Resistance

Some growing bacteria are killed during freezing (although not enough for freezing to be considered a lethal process). Killing also occurs during drying. The precise mechanism(s) of killing is not clear, but one cause may be DNA damage; freeze-drying cells can cause significant mutagenesis. Spores, however, are resistant to multiple cycles of freeze-drying.

A complete explanation for spore drying resistance is not yet available. However, the SASP contribute to this by preventing DNA damage.

Pressure Resistance

Spores are much more resistant to high pressures ($\geq$12,000 atm) than cells are. Spores are more resistant than cells to lower pressures, but paradoxically, spores are killed more rapidly at lower pressures than they are at higher pressures. This apparent anomaly is because lower pressures promote spore germination; the germinated spores are then rapidly killed at high pressure. High pressures do not promote spore germination.

γ-Radiation Resistance

Spores are generally more resistant to γ-radiation than are vegetative cells. In the few species studied, γ-radiation resistance is gained during sporulation 1 to 2 h before heat resistance is acquired. The precise factors involved in spore γ-radiation resistance are not known, although SASP are not involved. Radiation resistance does not correlate with heat resistance. The low water content in the spore core may provide protection against γ-radiation. Radiation presumably damages spore DNA through an unknown mechanism.

UV Radiation Resistance

Spores are 7 to 50 times more resistant than are vegetative cells to UV light. During sporulation, UV resistance is acquired 2 h before heat resistance, in parallel with synthesis of SASP. The latter proteins are essential for spore UV resistance. Coats, cortex, and core dehydration are not necessary for spore UV resistance.

Chemical Resistance

Spores are much more resistant than cells to many chemicals, pH extremes, and lytic enzymes such as lysozyme. Resistance of spores to chemicals is acquired at different times in sporulation. For some compounds, spore coats play a role in chemical resistance, possibly by providing an initial barrier against attack. This is clearly true for lytic enzymes, as spores with coats removed are sensitive to lysozyme. The inability of most molecules to penetrate into the spore core plays an important role in chemical resistance.

Heat Resistance

The heat resistance of spores is a huge problem for the food industry. It is probably the best-studied form of resistance. Spore heat resistance is truly remarkable; many spores withstand 100°C for several minutes. Heat resistance is often quantified as a D_t value. This is the time, in minutes, at the temperature (t) needed to kill 90% of a population. Generally, D values for spores at a temperature of t + 40°C are about equal to those for their vegetative cell counterparts at temperature t. An often-overlooked feature of spore heat resistance is that the extended survival of spores at elevated temperatures is paralleled by even longer survival times at lower temperatures. Spore D values increase 4- to 10-fold for each 10°C decrease in temperature. Consequently, a spore with a $D_{90°C}$ of 30 min may have a $D_{20°C}$ of many years.

We do not know what target(s) is damaged by heat to kill the spores. The killing of spores by heat is not caused by DNA damage or mutagenesis.

Protein(s) may be the target of spore heat killing. Sublethal heat treatment can also injure spores. This damage can be repaired during spore germination and outgrowth. Several factors that modulate spore heat resistance are discussed below.

Sporulation Temperature

Elevated sporulation temperatures increase spore heat resistance. Indeed, spores of thermophiles are generally more heat resistant than spores of mesophiles. Since spore macromolecules are identical to cell macromolecules, spore macromolecules are not intrinsically heat resistant. Presumably, the total macromolecular content of spores from thermophiles is more heat stable than that of mesophiles. This would account for their higher heat resistance (Fig. 3.4). However, spores of the same strain prepared at various temperatures are most resistant when prepared at the highest temperature.

α/β-Type SASP

Surprisingly, neither general mutagenesis nor DNA damage occurs during the killing of spores by wet heat. Therefore, spore DNA must be well protected against heat damage, and the thermal inactivation of spores must be due to other mechanisms. The major cause of spore DNA protection against heat damage appears to be the saturation of spore DNA by α/β-type SASP. Spores from *B. subtilis* $\alpha^-\beta^-$ mutants are more heat sensitive than those of wild-type strains.

Figure 3.4 Influence of sporulation temperature on *D* values of *Bacillus subtilis* sporulated at 32°C (pale green dots) or 52°C (dark green dots). Redrawn from F. Sala et al., *J. Food Prot.* **58:**239–243, 1995. Copyright International Association for Food Protection, Des Moines, IA. doi:10.1128/9781555817206.ch03.f03.04

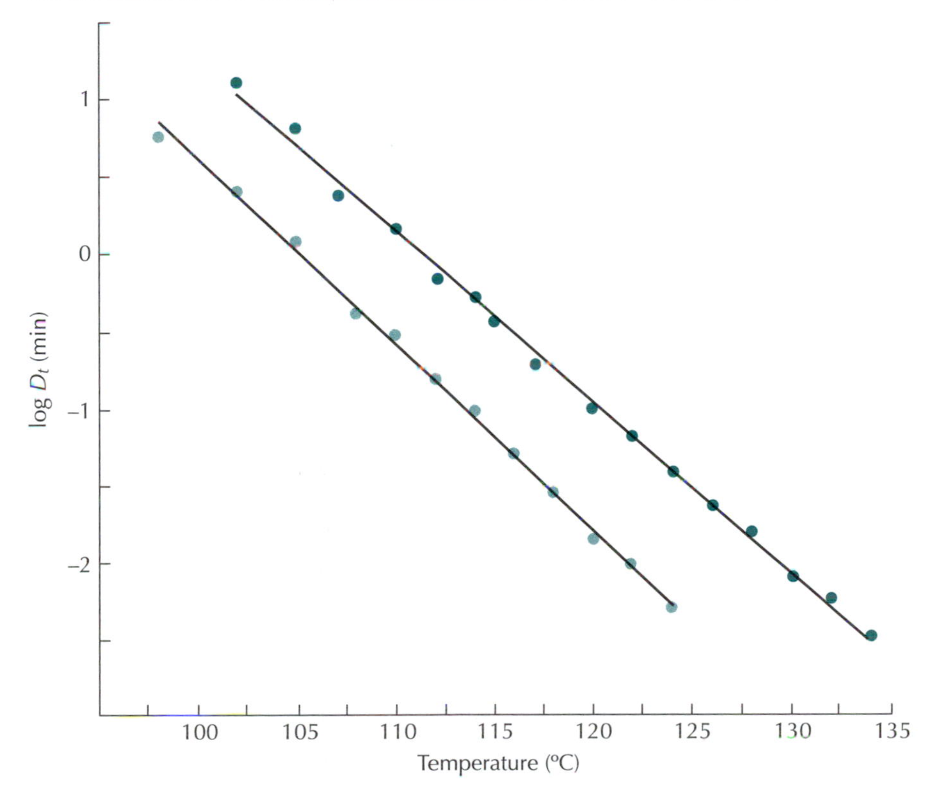

Dry heat affects spores differently than wet heat. First, spores are much more resistant to dry heat than to wet heat. *D* values are 100- to 1,000-fold higher in dry versus hydrated spores. Second, spores have a rather high level of mutagenesis (~12% of survivors) when killed by dry heat. This mutagenesis damages spore DNA.

Core Water Content and Heat Resistance

Low core water content is a major factor in heat resistance. The spore becomes more heat resistant as it is dehydrated. The spore cortex is essential for creating and maintaining the dehydrated state of the spore core. This is undoubtedly due to the cortex's ability to change its volume upon changes in ionic strength and/or pH. If the expansion in cortex volume were restricted to one direction, i.e., towards the spore core, mechanical action in the opposite direction would express water from the core.

There is a good correlation between spore core water content and heat resistance over a 20-fold range of *D* values. However, at the extremes of core water contents, *D* values vary widely, presumably reflecting the importance of other factors such as sporulation temperature, cortex structure, etc. Presumably, water-driven chemical reactions are inhibited by the spore core's low water content. This also contributes to heat resistance. Low water content stabilizes macromolecules such as proteins by restricting their molecular motion.

THE CYCLE OF SPORULATION AND GERMINATION

Sporulation

Spores are made in response to environmental stress or nutrient depletion. The molecular biologies of sporulation and spore resistance have been extensively studied in the genus *Bacillus*. While spores formed by *Alicyclobacillus*, *Clostridium*, *Desulfotomaculum*, and *Sporolactobacillus* are also problematic in foods, much less is known about them.

The first event of sporulation is an unequal cell division. This creates the smaller spore compartment and the larger mother cell compartment. As sporulation proceeds (Fig. 3.5), the mother cell engulfs the forespore,

Figure 3.5 Cycle of sporulation, dormancy, activation, and outgrowth. doi:10.1128/9781555817206.ch03.f03.05

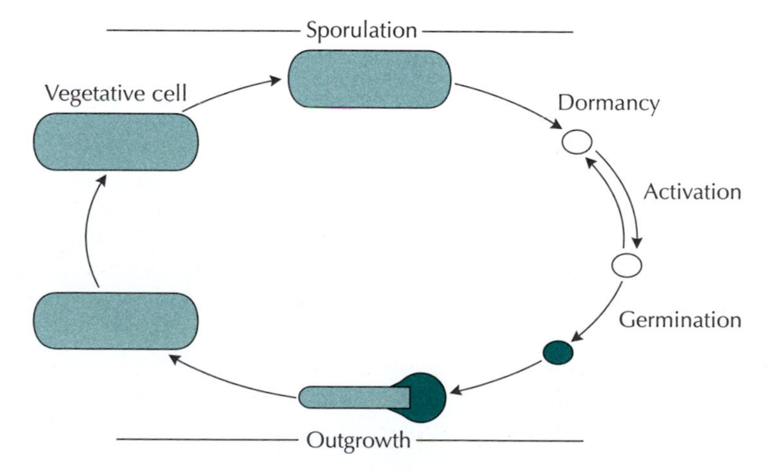

resulting in a cell (the forespore) within a cell (the mother cell). Each has a complete genome. Since the spore is formed within the mother cell, it is called an endospore.

Genes are expressed at specific times and places during sporulation. Some genes are expressed only in the mother cell. Other genes are expressed only in the spore. Gene expression is controlled by the ordered synthesis and activation of new sigma (specificity) factors for RNA polymerase. Many DNA-binding proteins, both repressors and activators, also regulate gene expression during sporulation.

As sporulation proceeds, there are striking morphological and biochemical changes in the developing spore. It becomes encased in two layers, the spore cortex and the spore coat. These are not found in vegetative cells. The spore also accumulates a huge store (≥10% of dry weight) of DPA (Fig. 3.3), found only in spores. It also has a large amount of divalent cations, primarily calcium. The spore is then metabolically dormant and extremely resistant to heat, radiation, and chemicals. Dormant spores are refractile (bright) when viewed under a phase-contrast microscope.

Activation

Dormant spores resume metabolism when they germinate (Fig. 3.5). Spores germinate more rapidly and completely if "activated" prior to exposure to a nutrient which induces germination (i.e., a germinant). There are many ways to activate spores. The most widely used is a sublethal heat shock (e.g., 10 min at 80°C). The precise changes induced by spore activation are not clear. In some species, the activation is reversible. In most species, heat activation releases a small amount of the spore's DPA.

Authors' note

If one enumerates a population of 850,000 spores per ml on agar medium, the colony count may be as low as 75,000. If the spores are heated for 10 min at 80°C (heat shocked), this might increase to 775,000.

Germination

Spores lose most of their unique attributes, including heat resistance, within minutes of encountering a germinant. (A germinant is a compound, such as an amino acid, that induces germination.) The spores lose DPA and SASP. During this time, active metabolism begins and synthesis of large molecules starts. Eventually, the germinated spore differentiates into a growing vegetative cell.

Spore germination occurs during the first 20 to 30 min after mixing of spores and germinant. During this period, a resistant dormant spore with a cortex and a large pool of DPA, minerals, and SASP is transformed to a sensitive, actively metabolizing germinated spore in which the cortex and SASP have been degraded. These degradative changes can occur in the absence of nutrients. However, further conversion of the germinated spore into a growing cell requires biosynthetic reactions and added nutrients.

Spore germination can be triggered by many compounds, including amino acids, sugars, salts, and DPA. How these compounds trigger spore germination is not clear. Metabolism of the germinant is not required. Indeed, even inorganic salts cause spores to germinate. The stereospecificity of germinants (e.g., L-alanine is a germinant, whereas D-alanine often inhibits germination) suggests that some germinants interact directly with specific proteins.

In the first minutes of spore germination, protons and some divalent cations are released. Release of DPA, loss of spore refractility, and cortex degradation follow. During germination, the spore excretes up to 30% of its dry weight and the core water content increases to that of the vegetative cell. These events require large changes in the permeability of the inner spore membrane. However, neither the nature nor the mechanism of these changes is understood. The changes accompanying the initiation of spore germination may be extremely fast. An individual spore can lose refractility in as little as 30 s to 2 min. However, for a spore population, the time can be much longer. Individual spores initiate germination after widely different lag times.

RNA synthesis starts during the first few minutes of germination. It uses nucleotides stored in the spore or generated by breakdown of preexisting spore RNA. Protein synthesis begins shortly after RNA synthesis. Early in germination amino acids produced from SASP breakdown are used for protein synthesis.

Outgrowth

The transition from spore germination to spore outgrowth is not distinct. Outgrowth is the time from ~25 min after the start of germination until the first cell division. Existing compounds are used for most processes during germination. Outgrowth requires the synthesis of new amino acids, nucleotides, and other small molecules.

Germination proceeds for at least 60 min before DNA replication is initiated. But DNA repair can occur well before DNA replicative synthesis. Even in the first minute of germination, spores contain deoxynucleoside triphosphates. During spore outgrowth, the volume of the outgrowing spore continues to increase, requiring the synthesis of membrane and cell wall components.

Summary

- Spores are unique life forms that are resistant to many stresses.
- The heat resistance of spores in food is measured as their *D* value, i.e., the number of minutes at a given temperature required to kill 90% of the spores. The *z* value is the number of degrees that it takes to change the *D* value by a factor of 10.
- Low-acid canned foods are those with a pH of >4.8 and an a_w of >0.85.
- Commercial sterility is achieved through the application of a 12*D* botulinum cook.
- Many sporeformers cause economic spoilage.
- Sporulation is the process by which a vegetative cell produces a spore.
- Germination and outgrowth enable the spore to resume life as a vegetative cell.
- The spore cortex is a unique form of peptidoglycan that contributes to spore resistance.
- Dipicolinic acid (DPA) and small acid-soluble proteins (SASP) also contribute to spore resistance.
- The thermal resistance of a spore crop is influenced by many factors, including its sporulation temperature.
- The relatively low water content of the spore core helps make it resistant.
- Sporulation, germination, and outgrowth are parts of the spore's life cycle.

Suggested reading

Driks, A. 2002. Overview: development in bacteria: spore formation in *Bacillus subtilis*. *Cell. Mol. Life Sci.* **59:**389–391.

Nicholson, W. L., P. Fajardo-Cavazo, R. Rebeil, T. A. Slieman, P. J. Riesenman, J. F. Law, and Y. Xue. 2002. Bacterial endospores and their significance in stress resistance. *Antonie van Leeuwenhoek* **81:**27–32.

Setlow, P., and E. A. Johnson. 2007. Spores and their significance, p. 35–67. *In* M. P. Doyle and L. R. Beuchat (ed.), *Food Microbiology: Fundamentals and Frontiers*, 3rd ed. ASM Press, Washington, DC.

Questions for critical thought

1. You may remember that low-acid canned foods must receive a thermal process equivalent to 2.4 min at 250°F. This is based on a 12-log reduction of *C. botulinum* spores, which have a $D_{250°F}$ of 0.2 min. Consider two process deviations. In deviation A, the temperature is correct, but the time is ~10% too short (i.e., the product receives 2.16 min at 250°F). In deviation B, the time is correct, but the temperature is ~10% too low (i.e., the product receives 2.40 min at 232°F). Which process deviation represents a greater threat to public safety? Be quantitative in your answer (i.e., calculate the log reduction in *C. botulinum* spores as the result of each deviation, assuming a *z* value of 18°F). Show all of your work.
2. Define the *z* value. If an organism has a $D_{80°C}$ of 20 min, a $D_{95°C}$ of 2 min, and a $D_{110°C}$ of 0.2 min, what is its *z* value? (Hint: if you understand the concept, you will not need a calculator.)
3. If you had to choose, would you rather be a spore or a vegetative cell? Why?
4. What are two components found in spores but not vegetative cells? What is their function?
5. Is it okay for cans of tomatoes to contain viable *C. botulinum* spores? Why or why not?
6. Explain what "gene expression is regulated in time and space" means in your own words.
7. Discuss how spore water content influences heat resistance. How does this integrate with what else you know about thermal lethality (i.e., wet versus dry heat, heat resistance at low water activity, etc.)?
8. Figure 3.2 shows an idealized diagram of a spore. How many of the structures can you identify in the real transmission electron micrograph shown here?
9. Many mechanisms in this chapter are described as "not understood," "unknown," or "yet to be determined." Identify one such mechanism and use an online search engine, such as Google Scholar, to find at least one article dealing with that mechanism. Read the introduction and discussion of the paper and explain what is known and what is unknown.
10. Speculate—if spore formation is such a good mechanism for surviving adverse conditions, why don't all bacterial species form spores?
11. Figure 3.4 illustrates the influence of sporulation temperature on *D* value. What is the effect of sporulation temperature on the *z* value in this case?

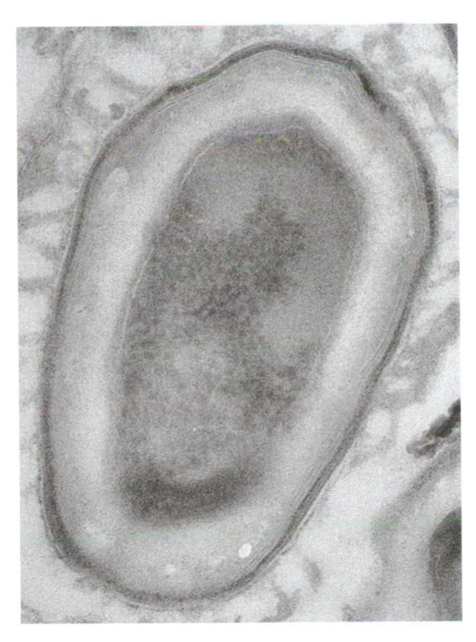

Shown is a micrograph of a section through a spore and sporangium of *Bacillus subtilis*. Courtesy of Lee Simon and the Montville laboratory, Rutgers University. doi:10.1128/9781555817206.ch03.fQ8

4

Detection and Enumeration of Microbes in Food

LEARNING OBJECTIVES

The information in this chapter will enable the student to:

- discuss methods available for microbiological analysis of food
- compare methods of analysis, indicating advantages and disadvantages of each method
- detail procedures for collection and processing of food samples
- calculate the microbial load of a sample
- recognize the difference between conventional and rapid microbiological methods

INTRODUCTION

Microorganisms inhabit our bodies and our environment. Therefore, it is not surprising to find many types of microorganisms (yeasts, molds, viruses, bacteria, and protozoa) in food. Often, food microbiologists refer to microorganisms in food as "the good, the bad, and the ugly." Good microorganisms are those that are used in the production of food (e.g., yogurt, wine, and beer), bad microorganisms cause spoilage, and ugly microorganisms cause human illness. Determining the types and numbers of microorganisms in a food is an important aspect of food microbiology.

The *aerobic plate count* (APC) provides an estimation of the number of microorganisms in a food. As the name implies, strict anaerobes do not grow on the plate and therefore are not included in the APC. The APC changes—increases or decreases—during the processing, handling, and storage of foods. In foods that are held raw, such as refrigerated meats, the APC increases during storage, whereas in dried or frozen foods, the APC remains unchanged or decreases. Depending on the food, the APC can be as low as 10 or as high as 1,000,000 microorganisms per g.

The levels of microorganisms in products from the same origin can differ greatly. For instance, the microbial load of ground meat is much greater than that of whole cuts of meat. During the handling, grinding, and packaging of meat pieces, bacteria may multiply or bacteria on equipment can be transferred to the product. The starting number of microorganisms associated with a food will have a significant impact on the number of microorganisms present in the finished product, even for foods that receive heat treatment. Poor-quality ingredients, poor sanitation, recontamination, and improper handling can result in high levels of microorganisms even in foods that were properly heat treated.

doi:10.1128/9781555817206.ch04

Microbiological Testing of Fresh Produce

David Gombas, United Fresh Produce Association, Washington, D.C.

Fresh produce is probably the most difficult food commodity to establish microbiological criteria and controls for. Let's look at why.

Microbial ecology of fresh produce

Produce that is grown to be eaten fresh is still alive. It has the same full complement of microbes that colonized the fruit or vegetable in the field. The natural microbiota is complex; only a few species would be recognizable to a food microbiologist, or culturable using typical microbiological procedures. Food microbiologists have not paid much attention to these organisms because they are not thought to affect produce safety or quality. We are beginning to understand that this may not be true. Research that is just underway demonstrates that the microflora varies from plant to plant, may be different at different locations on the plant, and may change during the plant's growth cycle. Interestingly, some of these organisms seem to either support survival or compete with organisms more important for spoilage (e.g., *Pseudomonas*) or safety (e.g., *Salmonella*). This may explain the wide variability in microbiological counts and pathogen detection, even from adjacent plants. Hopefully, future research will reveal whether these microbial interactions are predictable and can be harnessed to improve the quality and safety of fresh produce.

Microbiological criteria for fresh produce

This box refers to "fresh produce" as if it were a single commodity. In reality, it includes hundreds of commodities as diverse as apples and onions. Something they all have in common is their variability in microbiological condition. It is not unusual for total plate counts to vary by as much as 5 log units (log 1 to log 6 per gram or cm^2) from item to item, without these counts having any consistent correlation to the initial quality or shelf life potential of the produce. Coliform or fecal coliform counts also vary by orders of magnitude. Coliforms and fecal ("thermotolerant") coliforms are not even useful indicators of the sanitary condition of produce because organisms detected by these tests (e.g., *Klebsiella*) grow naturally on produce. As a consequence, microbiological tests for coliforms and fecal coliforms, which are commonly used to assess the quality or safety of raw and processed products, are not useful for fresh produce. While *Escherichia coli* is not part of the normal produce microflora, its presence is only a weak indicator of fecal contamination and does not correlate well with the presence of pathogens, including *E. coli* O157:H7.

Fresh produce as a human pathogen risk

Prior to the 1990s, fresh produce was not commonly recognized as a vehicle of foodborne infections. Since then, surveillance testing by the FDA, the USDA, and the fresh-produce industry has shown that human pathogens are detectable at a low frequency. FDA data cite over 70 foodborne outbreaks linked to fresh and fresh-cut produce in the United States between 1999 and 2009. Only a few types of human pathogens are involved: the bacteria *Salmonella*, enterohemorrhagic *E. coli* (including O157:H7), and *Shigella*; the parasites *Giardia*, *Cryptosporidium*, and *Cyclospora*; and the viruses hepatitis A virus and norovirus. Human parasites and pathogenic viruses cannot grow on the produce, of course. While the bacterial pathogens can grow on fresh produce under certain conditions, under normal produce handling conditions they only survive. Unlike other enteric pathogens, *Listeria monocytogenes* can grow on several fresh produce commodities under normal conditions, even under refrigeration conditions that prevent the growth of *Salmonella* and pathogenic *E. coli*. *L. monocytogenes* detection has led to several fresh produce recalls. But, at this time, commercially prepared fresh and fresh-cut produce have only been linked to listeriosis cases twice in the United States and Canada: a 1981 Canadian outbreak linked to coleslaw prepared with contaminated cabbage and a 2010 U.S. outbreak linked to fresh-cut celery processed under poor sanitary conditions in a processing plant. This scarcity of epidemiological evidence has resulted in some controversy over whether *L. monocytogenes* should be considered a "significant" food safety hazard in fresh produce. Changes in growing and handling practices during growing ("production") of fresh produce, commonly referred to as good agricultural practices (GAPs), have reduced the likelihood of contamination with certain pathogens. Outbreaks of shigellosis and hepatitis A, which are typically human-vectored pathogens, were more commonly linked to fresh produce in the 1990s. Since then, GAP programs have encouraged worker health and hygiene practices that minimize the opportunity for workers to transmit these pathogens to the produce. Consequently, outbreaks of these pathogens linked to fresh produce have virtually stopped. According to FDA data, there has not been a *Shigella* or hepatitis A outbreak linked to fresh produce since 2003, emphasizing the importance and effectiveness of the GAP programs. Meanwhile, illnesses caused by *Salmonella*, *E. coli* O157:H7, and *Cyclospora*, pathogens more commonly associated with animal vectors, have continued.

Microbiological testing of fresh produce

Because fresh produce has only recently been identified with foodborne disease outbreaks, the science and technology of microbiological testing of fresh produce is still young. Cultural methods that result in an isolated, viable culture are still the most definitive, but they are slow. Faster enzyme-linked

(continued)

Microbiological Testing of Fresh Produce *(continued)*

immunosorbent assays (ELISAs) and polymerase chain reaction (PCR)-based tests can be used to test for pathogens in fresh produce. But these methods need to be validated for each pathogen on each produce commodity. Few of these tests have been validated, resulting in a lack of confidence in both positive and negative test results. Arguably the greatest weakness in testing fresh produce in-field is a lack of statistically validated sampling protocols. While the International Commission on Microbiological Specifications for Foods (ICMSF) established sampling protocols for processed foods many years ago, those protocols were based on the assumption that the contaminant is uniformly distributed across the tested lot. Experience has demonstrated that, with the exception of gross contamination of a field or orchard (e.g., by spraying the crop with contaminated water), contamination of fruits and vegetables is sporadic (isolated contamination by an animal) or directional (along one side of an orchard) and almost never uniform. Consequently, while the ICMSF sampling protocols are still used, they are likely based on false assumptions. However, there are no science-based alternatives at this time.

Conclusions

Today, the spotlight is very much on the microbiological safety of fresh produce. The FDA is now in the process of writing a federal food safety regulation for fresh produce. Regardless of the eventual content of that regulation, attention (and research funding) is now being drawn to the information we are missing, and need, to assess the microbiological condition of fresh fruits and vegetables. With the enabling of such research it is hoped that in years to come, the microbiological criteria and controls for fresh produce will be as routine as they are for many other food commodities today.

In order to determine if a product meets the microbial levels prescribed in specifications, guidelines, or standards, an estimation of the number of microorganisms in or on a food is needed. A rule of thumb is that as the microbial count increases, the quality of the food decreases. Of course there are exceptions to this rule; for instance, in fermented foods the level of microorganisms would be expected to increase. This chapter focuses on some of the methods for determining the microbial levels of foods. Note that for the most part microbiological analysis of food samples refers to determining the levels of bacteria.

SAMPLE COLLECTION AND PROCESSING

The methods used for sample collection and processing vary from food to food and for specific microorganisms. The Food and Drug Administration (FDA) outlines the methods that it uses in the *Bacteriological Analytical Manual*, and the Food Safety and Inspection Service of the U.S. Department of Agriculture (USDA) publishes the *Microbiology Laboratory Guidebook*. Regardless of the protocol used, a sample will only yield significant and meaningful information if it represents the mass of material being examined, if the method of collection used protects against microbial contamination, and if the sample is handled in a manner that prevents changes in microbial numbers between collection and analysis.

Care must be taken during sample collection to prevent the introduction of microbes into the food sample. If possible, individual containers of food should be submitted for analysis. However, this is not always practical, and therefore a representative sample must be collected using an aseptic technique. Instruments used for sample collection should be sterilized in the laboratory rather than at the place of sampling. Sterile containers (plastic bags or widemouthed jars) should be labeled with the name of the food, the date collected, and other information that may be useful in the analysis of results and tracking of the sample.

Authors' note

Bacteria readily stick to fat molecules in food. Therefore, foods high in fat (for example, ground beef [80% lean and 20% fat]) must be properly processed, or else the number of bacteria will be underestimated.

Once the sample is collected, it should be analyzed as quickly as possible to prevent a change in the microbial population. However, this is not always possible, and in that case, the sample should be refrigerated and frozen samples should be kept frozen. Products that are normally refrigerated should not be frozen, since freezing may cause death or damage to some bacterial cells, producing incorrect results.

Before analysis, some preparation of the sample is generally required (Fig. 4.1). For meaningful results, the sample should be processed to produce a homogeneous suspension of bacteria so that they can be pipetted. If the

Figure 4.1 Preparation of sample for determination of microbial load. Milk, which initially contains millions of microbes, is diluted, and samples are spread plated so that individual colonies can be counted. Appropriate calculations are then made to determine the actual number of microbes in the milk. Step 1: Since a liquid sample is being tested, a 1-ml aliquot can be removed and added to 9 ml of buffer for direct preparation of serial 10-fold dilutions. Step 2: 100 μl of each dilution is transferred to replicate plates. Step 3: A glass "hockey stick" is flame sterilized, allowed to cool, and used to spread the sample on the surface of the agar; the plate should be rotated to ensure distribution across the entire surface. Step 4: The plates are incubated under specified conditions. Step 5: Dilutions yielding 30 to 300 colonies per plate are counted and expressed as numbers of CFU per milliliter of milk. doi:10.1128/9781555817206.ch04.f04.01

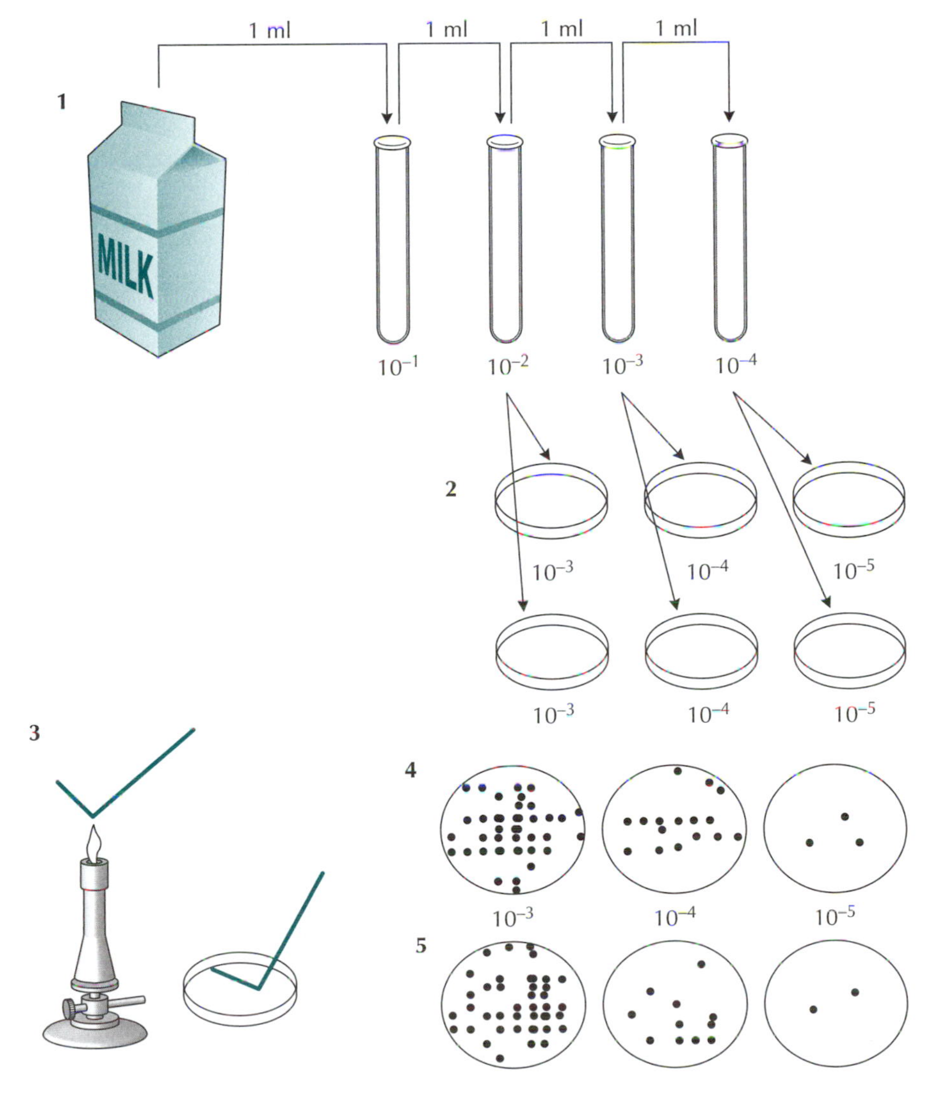

sample is solid, the food is generally mixed with a sterile diluent, such as Butterfield's buffered phosphate or 0.1% peptone water. If an appropriate buffer is not used, the bacteria may multiply in the dilution medium prior to being plated. To make a homogeneous suspension, the sample is added to a sterile plastic stomacher bag, diluent is added, and the sample is processed in a *stomacher* (a device that has two paddles that move rapidly back and forth against the sample). A typical-size food sample of 25 g is added to the sterile bag containing 225 ml of sterile diluent and mixed (placed in the stomacher), and a 1:10 dilution of the food and associated bacteria is obtained.

ANALYSIS

The APC, discussed above, provides a good estimate of the number of microorganisms associated with a sample. Drawbacks of surface plating are the *coalescence* of colonies (two or more colonies growing together) and the growth of spreaders. In the APC, the hardened agar surfaces must be dried thoroughly to prevent the spread of bacteria by moisture droplets. A 0.1-ml aliquot of the selected dilution of the sample is dispensed onto the surface of the agar and spread with the aid of "hockey sticks," or bent glass rods. For the standard plate count (SPC; also called a pour plate), a measured aliquot (1 ml) of the appropriate dilution is dispensed onto the surface of a sterile petri plate, and sterile, melted, and tempered (45°C) agar is added to the plate. The agar and sample are mixed, and the agar is allowed to solidify. The agar plate is then incubated, and isolated colonies known as colony-forming units (CFU) are counted. This method has many disadvantages, including the time required for setup, expense, and lack of accuracy.

A CFU in theory arises from a single bacterial cell; however, this may not always hold true. In all likelihood, colonies arise from clusters or chains of bacteria, resulting in an underestimation of the number of bacteria in a sample. Therefore, samples may need to be processed to separate the cells into individual reproductive units. This is achieved through homogenization of a sample and by making a series of 1:10 dilutions (Fig. 4.1). A 1:10 dilution of a 1:10 dilution is a 1:100 dilution. For samples with unknown microbial populations, an extensive (10^{-1} to 10^{-7}) series of dilutions may need to be made and plated. In general, standard 100-mm-diameter plates should contain between 25 and 250 colonies. Fewer or more colonies per plate can affect accuracy. The total number of bacteria is determined by multiplying the colony count by the dilution factor to yield the number of bacteria per gram of food.

Figure 4.2 A typical roll tube. Notice that the colonies appear in a spiral pattern, a function of the tube containing inoculated medium being rolled while the medium cools.
doi:10.1128/9781555817206.ch04.f04.02

The *roll tube* method is basically the same as the pour plate method, but screw-cap tubes are used in place of petri plates (Fig. 4.2). Test tubes containing 2 to 4 ml of plate count agar or other appropriate medium are sterilized, and the melted agar is allowed to cool to 45°C (tempered). A 0.1-ml aliquot of the appropriate dilution of the sample is added, and the tube is immersed in cold water in a horizontal position and rolled. The agar then solidifies, forming a thin layer on the inner wall of the tube. The tubes are then incubated upside down to prevent the possibility that any condensation which may form will cause smearing of the colonies. A decided advantage of this method is the savings associated with the use of less plate count agar.

Mechanical devices are available to facilitate the determination of bacterial numbers in food samples. The *spiral plater* uses a robotic arm to dispense a sample in an Archimedes spiral onto the surface of a rotating agar plate (Fig. 4.3). A continuously decreasing volume of the inoculum is deposited from near the center of the plate to the outside of the plate so that a concentration range of up to 10,000:1 is achieved on a single plate. The spiral plater compares favorably with the SPC. A specialized counting grid is used to count colonies on a plate, although the colonies are tightly packed together and may be difficult to count. The method is best suited for liquid samples, such as milk, since food particles may cause blocking of the dispenser. As with any method, there are advantages and disadvantages. The advantages include the use of fewer plates, fewer dilution bottles, and less agar and the ability to process a greater number of samples. The disadvantages include limitation of the types of samples that can be processed and difficulty in counting colonies.

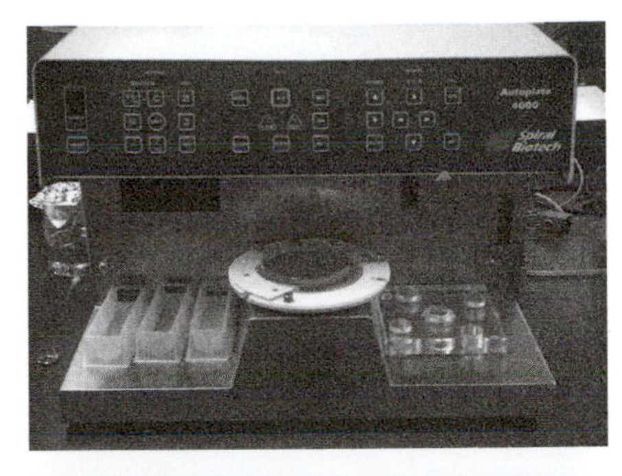

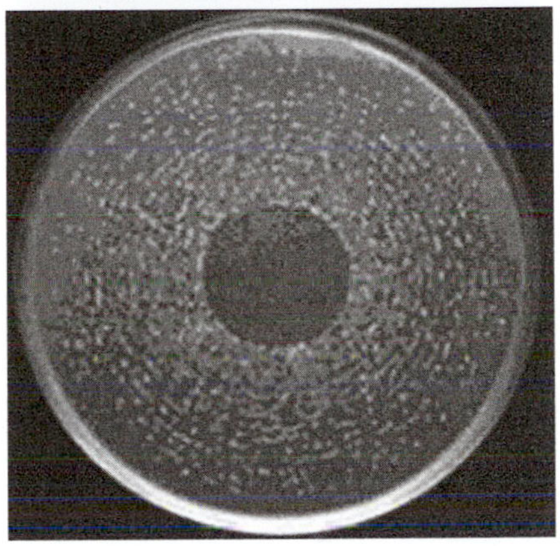

Figure 4.3 **(Top)** A typical spiral plater. These units are fully automated and have a small footprint. **(Bottom)** Colonies formed on a plate in which the test sample was plated with a spiral plater.
doi:10.1128/9781555817206.ch04.f04.03

Fluid samples (water or diluent) can be passed through membrane filters to collect bacteria. Membranes with a pore size of 0.45 μm are typically used. The filter can either be placed onto a plate containing a culture medium of choice or be used in the direct microscopic count (DMC) assay. This *membrane filter technique* is useful for examination of water or other dilute liquid samples that can easily pass through the filter. Therefore, even low numbers of microbes can be detected, since large volumes (1 liter or more) can be passed through a single filter. The method can also be used to determine microbial numbers in air. Small amounts of viscous liquid samples, such as milk or juice, can be processed without clogging the membrane.

A number of commercial methods based on the membrane filter technique are available. Some of these methods are discussed in chapter 5. The *direct epifluorescent filter technique* uses fluorescent dyes to stain bacteria that are then counted using a fluorescence microscope. The number of cells per gram or milliliter is calculated by multiplying the average number per field by the microscope factor. This method has been used for estimating numbers of bacteria in milk, meat, and poultry products and on food contact surfaces. The *hydrophobic grid membrane filter* method is also used to estimate microbial numbers associated with a variety of foods. For this method, specially designed filters containing 1,600 grids that restrict microbial growth and colony size are used. Generally, 1 ml of a 1:10 sample homogenate is passed through the filter. Grids that contain colonies are counted, and a most probable number is determined.

The DMC is faster than most methods, since no incubation period is required for cells to metabolize and grow (multiply). With this method, a smear of a food sample or a liquid sample (0.01 ml) is uniformly spread over a 1-cm^2 area on a microscope slide. Generally, liquid foods can be analyzed directly, but solid foods must be put into a suspension (1:10 dilution) before analysis. Once the sample is fixed, defatted, and stained, the cells are counted. A calibrated microscope must be used, since the diameter of the field to be examined must be known in order to calculate the number of organisms per gram. A stage micrometer can be used to measure the diameter of the circle (on the slide described above) to the nearest 0.001 mm. The average number of cells or clumps is calculated on a per-field basis and divided by the area of the field to determine the number per square millimeter.

The DMC is perhaps the simplest and most rapid method for estimating bacterial numbers in a sample. Therefore, adjustments or changes in processing can be implemented immediately to correct any problems. The test has other advantages, including little processing of samples, limited equipment required, a slide(s) that can be maintained as a record, and the need for only a small amount of sample. The disadvantages include fatigue (eyes becoming tired from peering through a microscope) and detection of live and dead cells without the ability to distinguish live cells from dead ones, and the method has little or no value for foods with low microbial numbers.

Fluorescent dyes are now available that can be used to differentially stain live and dead bacteria. Using such stains, the total bacterial count and a live-dead count can be obtained. Since the stains are intensely fluorescent, they can be used in samples that may contain low numbers of bacteria. The disadvantages are that the fluorescent signal fades over time and food constituents may also stain or fluoresce, interfering with the ability to count the bacterial cells.

Metabolism-Based Methods

In general microbiology, the metabolism of microorganisms is used to determine starch hydrolysis, sugar fermentation, production of hydrogen sulfide and indole, or nitrate reduction. Measurements of microbial metabolism or production of metabolic products in foods can be used to estimate the bacterial population. Today many of these tests have been miniaturized and are the bases of a number of rapid test methods.

Microorganisms obtain energy through chemical reactions, i.e., oxidation-reduction reactions, where the energy source becomes oxidized while another compound is reduced. Oxidation-reduction reactions consist of electron transfers and can therefore be measured electrically with a potentiometer. The measured oxidation-reduction potential is known as the *redox potential*. The redox potential can also be determined with indicators and dyes. The addition of such compounds to metabolizing bacteria results in the transfer of electrons to the indicator and subsequent change in color. Several oxidation-reduction indicators or dyes, including methylene blue, resazurin, and tetrazoliums, can be used. The rate at which the indicator shows a change in color is related to the metabolic rate of the microbial culture (sample). Basically, the larger the number of bacteria, the faster the color change occurs. Tests using these compounds are often referred to as *dye reduction tests*, but this is not strictly correct, since resazurin and tetrazoliums are not dyes but indicators.

Reductase tests can be used for a variety of products, and the results compare favorably with those of the SPC. However, for some food samples, including raw meat, the test may not be appropriate, since meat contains inherently high levels of reductive substances. The test is commonly used in the dairy industry for determining the microbial quality of raw milk. The advantage of reductase tests over the SPC is that they can be completed in a shorter time.

Surface Testing

This chapter has concentrated on methods designed to detect and enumerate microorganisms associated with food samples. However, to ensure that food contact surfaces are being maintained in a hygienic state, they should be tested regularly for microorganisms. The inherent problem with surface testing is the consistent removal of microorganisms from the test surface.

A number of methods are available; however, each may be appropriate only in specified areas of a food-processing plant.

The *swab test* method is perhaps the most widely used method for microbiological examination of surfaces in food-processing facilities. The basics of the method include swabbing a given area with a moistened cotton or calcium alginate swab. The area to be examined can be defined through the use of templates that have predefined openings (usually 1 cm^2). The template should be sterilized prior to use. After the defined area is swabbed, the swab is returned to a test tube containing a suitable diluent and agitated to dislodge the bacteria. To facilitate the process, calcium alginate swabs can be used and dissolved with the addition of sodium hexametaphosphate to the diluent. The inoculated diluent can be used in both conventional and rapid microbiological assays. The SPC is generally used to determine the number of bacteria in the inoculated diluent. The swab test method is ideal for testing rough, uneven surfaces.

Other methods for examining food contact surfaces include direct contact of agar with a test surface and the use of sticky film or tape. In the *replicate organism direct area contact* method, specially designed petri plates are used so that the medium, when poured, produces a raised agar surface. The agar can then make direct contact with the test surface. Once the plate has been used to test the desired surface area, the lid is replaced and the plate is incubated. Selective media can be used to reduce the growth of bacteria that may spread across the surface of the plate. This method is not suitable for heavily contaminated or rough surfaces.

Sterile sticky tape is commercially available for sampling of surfaces from equipment to beef carcasses. The tape and dispenser are similar in appearance to a standard roll of adhesive tape used for sealing packages. The tape is withdrawn from the roll and held over a premeasured (in square centimeters) area located on the tape dispenser. The sticky surface of the tape is then pressed against the area to be tested, and then the tape is pressed onto the surface of an agar plate. The method is comparable to the swab test except on wooden surfaces.

A sponge system has been employed for examination of animal carcasses and food contact surfaces. As with other contact systems, a moistened sponge is wiped across the test area and then placed into a container (test tube, sterile plastic bag, etc.) containing an appropriate diluent. The container is agitated to dislodge the bacteria from the sponge, and an aliquot of the inoculated diluent is used in the SPC or other microbiological assay.

The methods outlined in this chapter are all commonly referred to as conventional methods. To facilitate the detection and enumeration of microorganisms from food or food-processing facilities, selective medium is commonly used and is commercially available. Selective medium may contain antibiotics or other inhibitors to permit the growth of selected microorganisms. Some selective media contain specific substrates that are used by only certain microorganisms. A few of these media are discussed in chapter 5. Often, conventional methods are combined with rapid methods to decrease the time required for identification of microorganisms found in food (Fig. 4.4). Combining conventional and molecular methods facilitates also the identification of specific strains of a bacterium.

Sample preparation

Leafy produce: Weigh 125 g of produce rinsate into 125 ml of double-strength (2×) modified buffered peptone water with pyruvate (mBPWp)

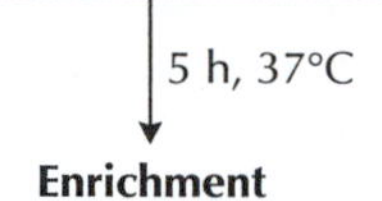

Enrichment

Add 1 ml each of acriflavine-cefsulodin-vancomycin supplements and incubate at 42°C static overnight (18–24 h).

Real-time PCR screening

1 ml of overnight enrichment for preparation of PCR cocktail. Method based on use of either SmartCycler II or Light Cycler 2.0.
Targets: *stx1*, *stx2*, *uidA*.

Negative samples: No further analysis
Probable positive STEC or O157: Continue to culture isolation and identification

Cultural isolation and presumptive isolate screening

1. Serially dilute the overnight sample enrichment in Butterfield's phosphate buffer and spread-plate appropriate dilutions in duplicate onto TC-SMAC and one chromogenic agar (Rainbow Agar O157 or R&F *E. coli* O157:H7 agar).
2. Incubate plates at 37°C for 18–24 h.
3. Screen typical colonies, testing for O157 antigen by latex agglutination.
4. Streak onto TSAYE plates.
5. Place a ColiComplete disk in the heaviest streak area. Incubate the plates for 18–24 h at 37°C.
6. Perform a spot indole test.
7. Typical colonies are shown to be X-Gal positive, MUG negative, and indole positive. Confirm the presence of the O157 and H7 antigens with commercial antisera.
8. Test O157- and H7-positive strains with API20E or Vitek to identify as *E. coli*. Isolates that have been confirmed to be O157:H7 as well as isolates that are O157 positive but H7 negative: Do a 5P multiplex for PCR for confirmation of O157:H7 isolates.

Figure 4.4 The FDA *Bacteriological Analytical Manual* method for detection and enumeration of *Escherichia coli* O157:H7 associated with leafy greens is outlined. The method incorporates many steps using conventional and molecular methods. See http://www.fda.gov/food/scienceresearch/laboratorymethods/bacteriologicalanalyticalmanualbam/ucm070080.htm. STEC, Shiga toxin-producing *Escherichia coli*; TC-SMAC, tellurite cefixime-sorbitol-MacConkey agar; TSAYE, tryptic soy agar-yeast extract; X-Gal, 5-bromo-4-chloro-3-indolyl-β-D-galactopyranoside; MUG, methylumbelliferyl-β-glucuronide. doi:10.1128/9781555817206.ch04.f04.04

Summary

- Samples must be collected using aseptic techniques and held in such a way as to prevent the growth of microorganisms.
- Food samples should be blended or homogenized prior to examination.

- The SPC is typically used to provide an estimate of the number of microorganisms in a sample.
- Surface test methods are essential in determining whether food contact surfaces are being maintained in a hygienic state.
- Rapid methods (miniaturized tests) based on conventional biochemical methods are now available, reducing the time required to identify a microorganism.

Suggested reading

Corry, J. E. L., B. Jarvis, S. Passmore, and A. Hedges. 2007. A critical review of measurement uncertainty in the enumeration of food micro-organisms. *Food Microbiol.* **24:**230–253.

Food and Drug Administration. 2001. *Bacteriological Analytical Manual Online.* http://www.fda.gov/Food/ScienceResearch/LaboratoryMethods/BacteriologicalAnalyticalManualBAM/default.htm.

Questions for critical thought

1. Differential media often contain enzymes linked to chromogenic and fluorogenic compounds or a combination of both. Identify a commercially available medium that contains one or more of those compounds and outline how you would test a food sample using the selected medium. What advantages does the differential medium have over a multipurpose growth medium such as Trypticase soy agar?
2. The DMC method has little or no value for foods with low microbial numbers. Why?
3. What is the difference between the surface and pour plate SPCs?
4. Indicate why the likelihood of underestimating the microbial load of hot dogs would be greater than for water.
5. How can the buffer used for dilution of a sample influence the estimation of bacterial numbers associated with a sample?
6. The swab test and sticky-tape test are both methods for direct sampling of contact surfaces. List advantages and disadvantages for each.
7. Outline a protocol for the collection of a sample of ground beef from a 25-kg block. Ultimately, you want to perform an SPC.
8. The spiral plater is a convenient method for microbial analysis of a sample. Why?
9. Select a food and, using the *Bacteriological Analytical Manual*, available through the FDA website, outline the methodology for determining total bacterial numbers and the tests required for detection of a food pathogen of concern in the selected food.
10. Many infections occur in the home. Do you believe that your home is "clean" and free of pathogens or large numbers of microbes? Evaluate the microbial load in your home, using methods discussed in the chapter. Select several rooms or sites within your home. For example, swab the surface of the kitchen sink and each of the bathroom sinks and either the tub or the shower. Determine which site has the greatest microbial load.
11. On your own: How would you need to add to an agar-based medium to determine whether a microbe produced a hemolytic compound(s)?

5

Rapid and Automated Microbial Methods

LEARNING OBJECTIVES

The information in this chapter will enable the student to:

- identify the various categories of rapid identification methods
- explain the basis of immunological, nucleic acid, and biochemical methods
- recognize when rapid methods are suitable to use
- understand the advantages and disadvantages associated with the use of rapid methods

INTRODUCTION

Microbiological analysis of food is not a "piece of cake." Foods are complex matrices of fats, carbohydrates, proteins, preservatives, and other chemicals. Foods also vary with respect to their physical natures: solid, dry, liquid, or semisolid. Collectively, these attributes can make it difficult to process a sample for microbiological analysis. Even if hurdles associated with processing of a sample are overcome, often foodborne pathogens are present at extremely low levels, further complicating the detection process. To decrease the time required to detect and identify target microbes in a food, an array of rapid microbiological methods have been developed.

Significant advances in the development of rapid methods have been made during the last decade, although rapid methods in some form have been available for more than 25 years. Initially, the medical community led these advances, but as a result of increased consumer awareness of the microbial safety of foods, food microbiologists have closed the gap. Significant advances have been made in the development of real-time sensors for the detection of specific toxins, including Shiga toxin and botulinum toxin, in food, based on concerns about deliberate contamination by bioterrorists. The task was made more difficult by the increased range of microorganisms now associated with food compared to 20 or 30 years ago. In the "old" days, food microbiologists tested for *Salmonella*, *Clostridium botulinum*, and *Staphylococcus aureus*; now *Listeria monocytogenes*, *Escherichia coli* O157:H7, other Shiga toxin-producing *Escherichia coli* organisms, *Campylobacter jejuni*, and *Vibrio parahaemolyticus* are also of concern. The change in the complexity of microorganisms associated with food is linked to many factors, including changes in processing and consumer preference and a global marketplace.

The basis for ensuring that safe and wholesome foods are available for the consumer is the testing of foods for pathogens and spoilage microorganisms. Conventional test methods (see chapter 4), although not lacking in sensitivity and cost-effectiveness, can be laborious and require

doi:10.1128/9781555817206.ch05

several days before results are known. Products that are minimally processed have an inherently short shelf life, which prevents the use of many conventional methods that may require several days to complete, ultimately limiting product time on the shelf. Rapid assays are based on immunological, biochemical, microbiological, molecular, and serological methods for isolation, detection, enumeration, characterization, and identification.

The times required to complete different assays vary greatly. Indeed, "rapid" may imply seconds, minutes, hours, or even days. When considering whether an assay is rapid, all steps in the assay must be included, not just the time required to complete the rapid test itself. For instance, a rapid assay may require only 15 min to complete, but the food sample may need to be mixed with a general growth medium (*preenrichment*) and incubated for several hours, with an aliquot then transferred to a selective enrichment medium and incubated before an aliquot is finally used in the rapid 15-min assay. Most rapid methods are adaptable to a wide range of foods, since they rely on culture methods to recover injured cells and increase the number of target cells.

The fast pace at which rapid methods are being developed precludes discussion of all available methods. In this chapter, the breadth of rapid methods available and the scientific principles of the methods used for detection of specific microbes in foods are examined. Existing rapid methods and those that show promise for the future but that are not yet commercially available are discussed.

SAMPLE PROCESSING

As mentioned above, food samples generally require some type of processing before a rapid method can be employed. The methods mentioned in this chapter are generally accepted for a broad range of foods, including milk, yogurt, apple juice, ground beef, and tomatoes. The composition of the food—the presence of fats, carbohydrates, and protein—can have a dramatic impact on successfully determining whether a target organism is present. The way that the food is processed, such as cooling, drying, heating, and addition of chemical additives, can also influence the ability to detect the presence of a target microorganism. In general, foodborne pathogens are present in low numbers in foods. Therefore, most rapid methods include enrichment, preenrichment, and/or selective enrichment to facilitate detection of the target pathogen.

Rapid tests have detection limits ranging from 10^2 to 10^5 colony-forming units (CFU)/g or 10^2 to 10^5 CFU/ml; therefore, enrichment increases the probability of detecting low levels of a target organism. The duration of the enrichment incubation varies considerably, based on the manufacturer's directions. Often, the period is relatively short, <12 h, but it can be as long as 24 h. The rapid-test manufacturer may specify the medium to be used for enrichment, or a standard growth medium may be appropriate. Prior to selecting a rapid method, the assay protocol should be read carefully to determine whether an enrichment step is required, since this may dramatically increase the time needed to complete the assay.

REQUIREMENTS AND VALIDATION OF RAPID METHODS

Considerable effort goes into the development of rapid methods. Before rapid methods can be discussed, an understanding of how the accuracy and validity of a test is determined is necessary. To develop a reliable method, basic information about the target microbe is required. This facilitates the identification of unique features that can then be used for rapid detection. The cornerstone of any method is its *accuracy*. This consists of the *sensitivity*, or the ability of the assay to detect low numbers of the target microorganism, and the *specificity*, or its capability to differentiate the microbe of interest from other microorganisms. A *false-negative* result occurs when an assay fails to detect a target pathogen that is present upon culture. Similarly, a *false-positive* result occurs when the test system gives a positive result for a culture-negative sample. R. R. Beumer et al. gave standard equations to calculate sensitivity and specificity rates: sensitivity = $(p \times 100)/(p +$ number of false negatives) and specificity = $(n \times 100)/(n +$ number of false positives), where p is the number of true positives and n is the number of true negatives. Without question, the assay must be as sensitive as possible and the detection limit must be as low as possible. For microorganisms that cause disease, the criterion is <1 cell per 25 g of food.

The intent in developing a rapid assay is to reduce the time required to obtain an accurate result. Conventional testing may require several days, whereas an ideal rapid test should provide results within an 8-h time frame, or a typical workday. In an ideal world, a rapid test would provide results nearly instantaneously, but this has yet to be realized. At the end of this chapter, new technologies that may be capable of delivering nearly instantaneous results are discussed, but for now, we must operate within the constraints of the currently available systems. These have limits of detection of 10^2 to 10^5 CFU/g or ml and require enrichment for 6 to 24 h. Another factor in the selection of a rapid method is the number of samples that can be processed and analyzed at one time. Many tests have adopted the use of *microtiter plates* (small rectangular plates having 96 wells) or similar multiwell formats that can perform many tests for a single sample or that can be used to screen multiple samples simultaneously.

Although the method selected may be rapid and able to be completed within a time frame acceptable for a given operation, other factors, including the speed of sample processing, accuracy, and cost, must be considered. With respect to speed, single diagnostic tests may be acceptable if few tests are to be performed. If large numbers of samples must be processed, a high-throughput system, even though more costly, would be the best choice. The cost of the system selected must also be appropriate for the company using it. Many factors can influence cost, including the training of personnel, the purchase of specialized equipment to conduct the assay, service and maintenance contracts, disposable supplies, and reagents.

A significant factor to consider is whether specialized skills are required for laboratory personnel to perform the assay. Ideally, the assay should be technically easy to perform, the equipment should be easy to operate, and the results should be easy to interpret. If you have not guessed by now, all aspects of the assay should be easy. Of course, the assay should be suitable for the food matrix that is being tested. In many cases, food constituents interfere with performance of the assay. The natural microflora of a sample and other debris may also interfere with the accuracy of the test. Finally, the method should be deemed acceptable by industry or government agencies.

There are several organizations, including the International Standards Organization, the International Dairy Federation, and AOAC International (formerly the Association of Official Analytical Chemists), that validate the effectiveness of testing methods for foods. The Food and Drug

Administration (FDA) and U.S. Department of Agriculture, which do not validate methods, have manuals that outline standard methods used by each organization. In some instances, rapid methods are incorporated into the protocols; however, this does not imply an official endorsement or approval of those tests. AOAC International publishes the *FDA Bacteriological Analytical Manual*, and the Food Safety Inspection Service of the U.S. Department of Agriculture publishes the *Microbiology Laboratory Guidebook*.

AOAC International is the most widely recognized and used service that provides third-party performance testing for manufacturers of test kits. There are two programs, the collaborative study program and the peer-verified program, used by AOAC International to validate microbiological and chemical assays designed for the testing of foods. The collaborative study program is more rigorous. Methods that pass examination are reviewed by the AOAC Official Methods Board for approval as a first-action method. After 3 years of use by the scientific community, the methods are eligible for voting for final-action status. Rapid-test kit manufacturers generally list on the product or in product information whether the test has undergone AOAC validation, i.e., AOAC first action or AOAC final action.

RAPID METHODS BASED ON TRADITIONAL METHODS

Traditionally, bacteria were differentiated and identified based on their biochemical profiles. To accomplish this, bacteria needed to be separated from a food matrix. This requires a number of labor-intensive steps to obtain a culture that can then be plated onto or inoculated into various differential media to determine a specific bacterial response. To facilitate the isolation of bacteria from food samples, the samples are mixed with diluent in sterile plastic bags and massaged using a stomacher. Previously, individual sterile blender cups were used, which meant that each cup required cleaning and sterilization after use. Additional dilution of samples can now be done using automated diluters. To further facilitate the process, automated plating systems that eliminate the need to make dilutions and that use a robotic arm to distribute liquid sample onto a rotating plate have been developed. This allows a >1,000-fold dilution range to be counted on a single plate. The process of counting colonies on a plate is time-consuming and open to human error, particularly if a large number of plates must be counted. Colony-counting systems use scanners and specialized software. An added advantage of these systems is that images of plates can be stored for future viewing and analysis.

A variety of products that replace the standard agar plate are available (Table 5.1). One of the most widely used is the Petrifilm system (3M). This system consists of rehydratable nutrients and a gelling agent embedded in disposable cardboard. Petrifilms have been developed for enumerating yeasts, molds, total bacteria, and specific bacteria. To rehydrate the film to support microbial growth, a 1-ml aliquot of sample is dispensed onto the center of the film. The film is incubated under the appropriate conditions, and colonies are counted directly. The Petrifilm system for determining coliforms in foods is comparable to the widely used violet red bile agar and has greater sensitivity. The API *Listeria* system consists of a strip of wells that provides a biochemical profile of the isolate (Fig. 5.1). Wells are inoculated with an aliquot of the target organism to rehydrate the compound in each well; strips are then incubated under appropriate conditions. Change in color of the liquid in the well, for example, from yellow to red, would suggest a positive reaction.

Table 5.1 Representative manual and automated miniaturized biochemical tests

Organism(s)	System	Manufacturer
Enterobacteriaceae	API	BioMérieux
	MICRO-ID	Remel
	BBL Crystal	Becton Dickinson
	Vitek	BioMérieux
Coliforms	Bactometer	BioMérieux
Gram-negative bacteria	Vitek	BioMérieux
	Microlog	Biolog
	Omnilog ID	Biolog
Gram-positive bacteria	Vitek	BioMérieux
	Microlog	Biolog
	Omnilog ID	Biolog
Anaerobes	IDS RapID	Remel
	BBL Crystal	Becton Dickinson
Listeria	MICRO-ID	Remel
	API	BioMérieux
Streptococci	RapID STR	Remel

The *hydrophobic grid membrane filter* (HGMF) (ISO-GRID system; QA Life Science) is more complex than the Petrifilm system and requires specialized equipment to complete. The HGMF requires filtration of the sample through a filter that contains a set of 1,600 grid cells. The HGMF is then placed onto an appropriate medium and incubated, and the colonies are counted. Grids have been designed for *Salmonella, E. coli* O157:H7, coliform *E. coli*, yeast, and mold counts and aerobic plate counts.

Similar to the Petrifilm system and the HGMF system, the SimPlate system is based on the use of specialized dehydrated medium that is reconstituted with sterile water prior to use. The system is a modification of the most-probable-number method. The petri dish device contains numerous wells, and the positive wells are counted and compared to a most-probable-number chart. The system is well suited for testing a variety of products, from fresh vegetables to ice cream.

The methods discussed thus far are basically kits and must be used as such. However, in recent years, a staggering number of media have been developed for the detection, enumeration, and identification of specific bacteria (Table 5.2), and media are now available for the detection and enumeration of yeasts and molds. For instance, media marketed by CHROMagar (Paris, France) contain specific chemicals that permit the differentiation and counting of specific bacteria. On CHROMagar ECC, coliform colonies appear pink and *E. coli* colonies appear blue. For the detection of *E. coli* O157:H7, the list of media just keeps growing. These media include MacConkey sorbitol agar, phenol red sorbitol agar containing 4-methylumbelliferyl-β-D-glucuronide, Levine eosin-methylene blue agar, Fluorcult *E. coli* O157:H7 agar (EM Sciences), BCM O157:H7(+) (Biosynth Biochemica and Synthetica), and MacConkey sorbitol agar containing either 5-bromo-4-chloro-3-indoxyl-β-D-glucuronic acid or 4-methylumbelliferyl-β-D-glucuronide for the detection of β-D-glucuronidase activity.

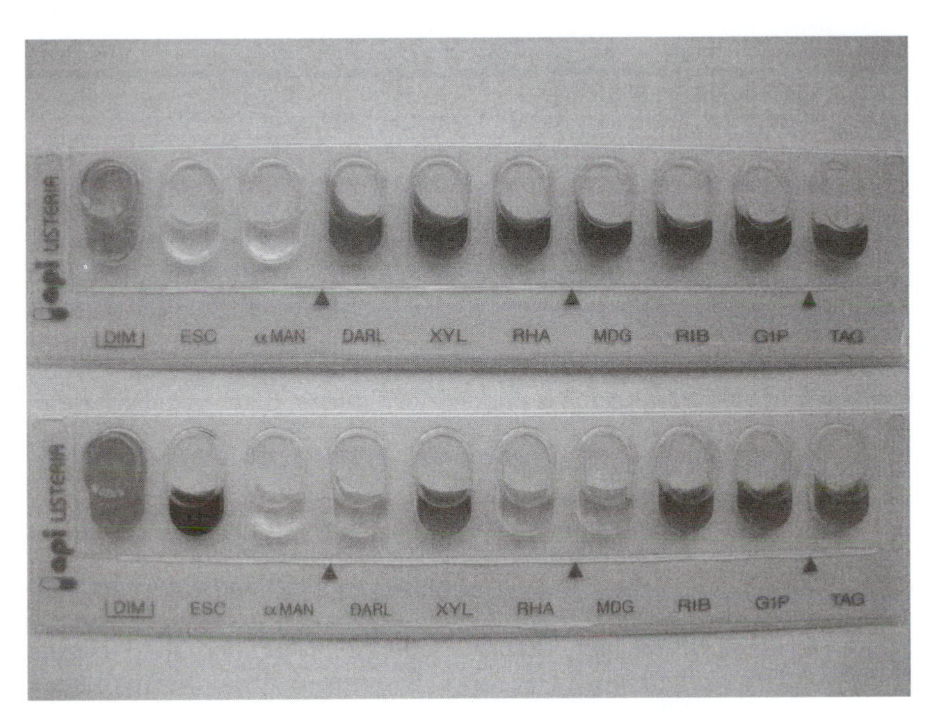

Figure 5.1 The API *Listeria* strip is an example of a rapid method based on traditional methods. The wells serve as test tubes in which biochemical reactions occur. Based on the reaction, a code number is generated that is then used to determine the *Listeria* species. doi:10.1128/9781555817206.ch05.f05.01

Table 5.2 Representative specialty medium rapid kits for detection of bacteria associated with food

Organism(s)	Trade name	Assay format[a]	Manufacturer
E. coli	Petrifilm	Medium film	3M
	Redigel (ColiChrome)	Medium	3M
	Coligel	MUG–X-Gal	Charm Science
	E-Colite	MUG–X-Gal	Charm Science
	Pathogel	MUG–X-Gal	Charm Science
Coliforms	Petrifilm	Medium film	3M
	Redigel (ColiChrome)	Medium	3M
	Redigel (Violet Red Bile)	Medium	3M
	Coligel	MUG–X-Gal	Charm Science
	E-Colite	MUG–X-Gal	Charm Science
	Pathogel	MUG–X-Gal	Charm Science
Salmonella	XLT-4	Medium	Difco
	Rambach	Medium	CHROMagar
	SM ID	Medium	BioMérieux
Enterobacteriaceae	Petrifilm	Medium film	3M
	Pathogel	MUG–X-Gal	Charm Science
Staphylococci	Microdase Disk	Medium	Remel
	Bactistaph kit	Medium	Remel
	Novobiocin Disk	Medium	Remel
S. aureus	Petrifilm	Medium film	3M

[a]MUG, 4-methylumbelliferyl-β-D-glucuronide; X-Gal, 5-bromo-4-chloro-3-indolyl-β-D-galactopyranoside.

Table 5.3 Representative list of antibody-based assays

Organism(s)	Trade name	Format[a]	Manufacturer
EHEC O157[b]	Reveal	Ab-ppt	Neogen
	E. coli O157	LA	Unipath
	Alert	ELISA	Neogen
	VIDAS	ELFA	BioMérieux
	Assurance EHEC EIA	ELISA	BioControl
	Premier O157	ELISA	Meridian
	VIP	Ab-ppt	BioControl
	TECRA	ELISA	TECRA
E. coli O157:H7	DETEX	ELISA	Molecular Circuitry Inc.
		ELISA	Binax
Salmonella spp.	Reveal	Ab-ppt	Neogen
	Alert	ELISA	Neogen
	VIDAS	ELFA	BioMérieux
	1-2 Test	Diffusion	BioControl
	Assurance *Salmonella* EIA	ELISA	BioControl
	VIP	Ab-ppt	BioControl
	Assurance Gold *Salmonella* EIA	ELISA	BioControl
	TECRA	ELISA	TECRA
	UNIQUE	Capture EIA	TECRA
	DETEX	ELISA	Molecular Circuitry Inc.
Listeria	Reveal	ELISA	Neogen
	VIP	Ab-ppt	BioControl
	Assurance	ELISA	BioControl
	TECRA	ELISA	TECRA
	UNIQUE	Capture EIA	TECRA
	DETEX	ELISA	Molecular Circuitry Inc.
L. monocytogenes	DETEX	ELISA	Molecular Circuitry Inc.
Campylobacter	Alert	ELISA	Neogen
	Uni-Lite XCEL	ELISA	Neogen
	VIDAS	ELFA	BioMérieux
	VIP	Ab-ppt	BioControl
	Campylobacter	ELISA	BioControl
	DETEX	ELISA	Molecular Circuitry Inc.
	TECRA	ELISA	TECRA
Bacillus thuringiensis	Agri-Screen	S-ELISA	Neogen
Pseudomonas	TECRA	ELISA	TECRA
S. aureus	Staphyloslide	LA	Becton Dickinson
	Staphlatex	LA	Difco
	TECRA	ELISA	TECRA

[a]EIA, enzyme immunoassay; S-ELISA, sandwich ELISA; Ab-ppt, antibody precipitation; ELFA, enzyme-linked fluorescent assay; LA, latex agglutination.

[b]EHEC, enterohemorrhagic *E. coli*.

IMMUNOLOGICALLY BASED METHODS

The most widely used rapid methods are based on immunoassay technology. The systems are acceptable for the screening and identification of specific bacteria associated with a variety of foods. The antibodies used in these systems may detect either many cellular targets (polyclonal antibodies) or specific targets (monoclonal antibodies). The use of monoclonal antibodies should eliminate some problems associated with cross-reaction with bacteria other than the target organism. The systems range from simply mixing antibody-coated beads with a test sample to a multistep procedure that includes addition of an enzyme-labeled antibody to a well, followed by incubation, washing, and addition of a chromogen that is acted on and that can be detected either visually or using a spectrophotometer. The typical limits of detection for these systems range from 10^3 to 10^5 CFU/ml. Table 5.3 contains a list of some of the commercially available methods.

Coating of latex particles (small beads) or magnetic beads permits screening of samples for a specific bacterium or separation of a target bacterium from a sample, respectively. For the *latex agglutination test*, latex particles are coated with specific antibodies that upon contact with specific antigens of the target organism produce a visible clumping, or agglutination, reaction (Fig. 5.2). The method is not that sensitive, since $>10^6$ CFU are required for a visible reaction. However, the method is rapid and provides a convenient means to screen for a target microbe. *Immunomagnetic separation* differs from latex agglutination in that the method is designed to separate the target organism from a sample. Basically, an aliquot of magnetic beads coated with a specific antibody is added to a sample (usually broth culture), the sample is incubated to facilitate binding of the target cells to the antibody, and then the complex is isolated from the sample by using a magnet. The beads can then be used to inoculate broth, be plated onto selective agar medium, or be used directly in polymerase chain reaction (PCR) assays or in enzyme-linked immunosorbent assays (ELISAs). Immunomagnetic separation systems are available for separation of *E. coli* O157:H7, *Listeria*, and *Salmonella* from a range of samples, including feces, raw milk, and ice cream.

The *Salmonella* 1-2 Test (BioControl) is a popular and widely accepted method based on *immunodiffusion* (the movement of proteins through agar) and the formation of an antibody-antigen complex that forms a visible line of precipitation in a chamber. The antibody is specific for *Salmonella* flagellar antigen; therefore, nonmotile *Salmonella* organisms are not detected. Many immunoassays are self-contained units that are based on the migration of a sample along a chromatographic strip. Systems using this or similar technology include the VIP tests (BioControl), Reveal tests (Neogen Corp.), SafePath (SafePath Laboratories LLC), and PATH-STIK (Lumac). Tests are available for *Salmonella*, *E. coli* O157:H7, *Listeria*, *Campylobacter*, and a variety of toxins. The systems have been used to screen fecal, beef, milk, and environmental samples. The specificity and sensitivity vary with the sample type and assay used.

There are several ELISA-based methods, such as TECRA for detection of *Salmonella*, *Campylobacter*, and *E. coli* O157:H7; Listeria-TEK (Organon Teknika); and the Assurance EIA system (BioControl). ELISA-based systems are also available for the detection and identification of fungi to the genus or species level. Mycotoxins can also be detected by commercially available ELISA systems. The basis for detection of a target microbe or antigen is depicted in Fig. 5.3.

Figure 5.2 Agglutination tests are a convenient means to screen for a target microbe. Note the dark aggregates (arrows) in the positions marked O157 and H7. The control position is void of aggregates. In this instance the isolate was positive for both antigens. doi:10.1128/9781555817206.ch05.f05.02

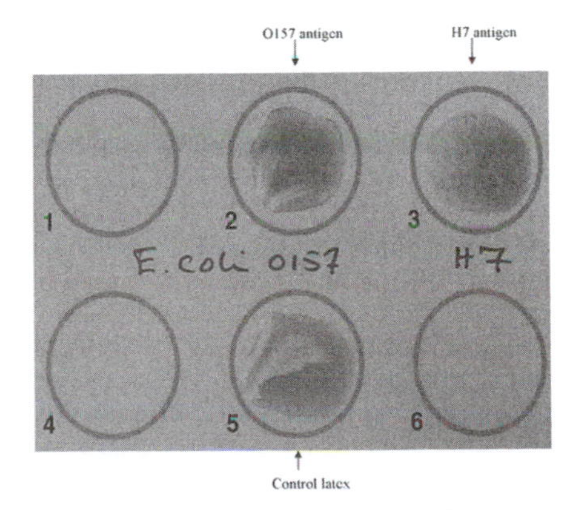

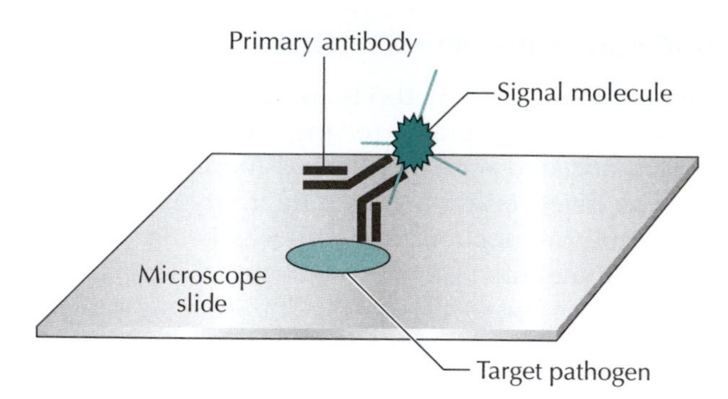

Figure 5.3 Detection of a target foodborne microbe by direct immunoassay. doi:10.1128/9781555817206.ch05.f05.03

MOLECULAR METHODS

There has been an explosion in the past 10 years in the introduction of nucleic acid (RNA and DNA)-based assays for the differentiation and identification of foodborne pathogens (Table 5.4). DNA methods include, but are not limited to, PCR, pulsed-field gel electrophoresis, ribotyping, plasmid typing, randomly amplified polymorphic DNA, and restriction fragment length polymorphism. Some of these methods have been automated, and kits are available to facilitate the recovery of pure DNA. Perhaps the most widely recognized and used method is PCR.

PCR is a basic three-step process that is based on the amplification of a specific segment of cellular DNA (Fig. 5.4). In the first step, double-stranded DNA, also known as *template DNA*, is denatured into single strands, followed by annealing of *primers* (short segments of DNA complementary to a specific region on the template DNA strand) to the template DNA. The annealing temperature is critical, since it determines the *stringency* of the reaction, or how specific the attachment will be. The final step is extension of the primers, which is accomplished using a thermostable DNA polymerase. Within 2 h, PCR can amplify a single copy of DNA a millionfold. The PCR products can be visualized as a band on an agarose gel stained with ethidium bromide.

Currently, there are only a few commercially available PCR kits for the identification of foodborne pathogens. Although the test is extremely

Table 5.4 Representative nucleic acid-based rapid methods

Organism(s)	System	Format	Manufacturer
Salmonella	PCR-Light Cycler	PCR-ELISA	Biotecon Diagnostics
	Gene-Trak	Probe	Gene-Trak
	BAX	PCR	Qualicon
E. coli O157:H7	BAX	PCR	Qualicon
E. coli	Gene-Trak	Probe	Gene-Trak
L. monocytogenes	PCR-Light Cycler	PCR-ELISA	Biotecon Diagnostics
	Gene-Trak	Probe	Gene-Trak
Listeria spp.	BAX	PCR	Qualicon
Campylobacter	Gene-Trak	Probe	Gene-Trak
S. aureus	Gene-Trak	Probe	Gene-Trak

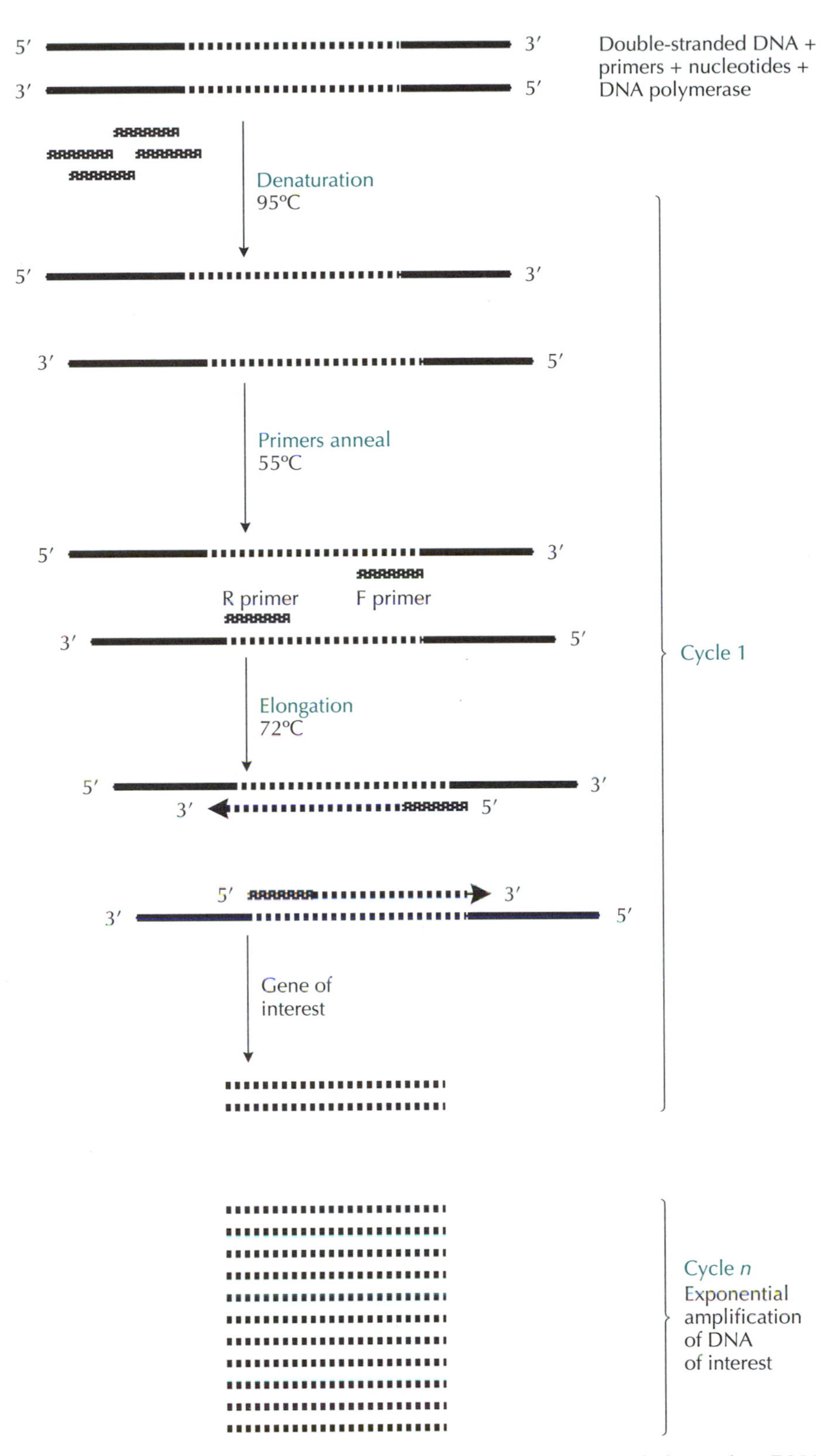

Figure 5.4 Representation of PCR cycle. Initially, double-stranded template DNA (from a microbe) is denatured to form two single-stranded pieces of DNA. Primers designed to amplify a specific region on the template DNA are allowed to anneal to the single-stranded DNA. Elongation and extension of the primer make a complementary copy of the DNA template. These three steps make up a single cycle. The cycle is repeated a specified number of times. doi:10.1128/9781555817206.ch05.f05.04

sensitive, equipment required to run the test is expensive, technical expertise is required, and PCR assays are affected by complex components of food. High levels of fats, protein, humic substances, and iron can interfere with the reaction. To overcome these problems, an enrichment step is often included in the assay, effectively diluting inhibitors and increasing cell numbers. The BAX system (Qualicon) is a commercially available PCR assay for the detection of *E. coli* O157:H7, *Salmonella*, and *Listeria*. The kit is unique in that all reagents required for the PCR (primers, enzyme, and deoxyribonucleosides) are *lyophilized* (freeze-dried) and included in a reaction tube. This eliminates potential problems with the introduction of contaminants or the use of incorrect amounts of a reagent.

There are a number of alternatives to conventional PCR, including reverse transcription-PCR (RT-PCR), real-time PCR, and the nucleic acid sequence-based amplification system. Commercial kits are available for nucleic acid sequence-based amplification (NucliSens Basic Kit; Organon Teknika) and for real-time PCR (Biotecon Diagnostics, Hamilton Square, New Jersey). For assays for which kits are not available, the methods are well defined. Reagent packages that have been optimized for the isolation of RNA and DNA are available. RT-PCR uses the enzyme reverse transcriptase to convert RNA into DNA, which then serves as the template in PCR. RT-PCR is a useful method, since it is based on RNA, which has a very short half-life. When performed properly, RT-PCR detects only live cells, whereas in PCR, DNA from live and dead cells can potentially be detected.

Authors' note

DNA-based methods are extremely sensitive. Caution must be exercised with use of some of these tests, since the DNA detected may be associated with dead bacteria that would present no health hazard.

The commercially available Riboprinter from Qualicon makes it easy to conduct ribotyping for the characterization and fingerprinting of strains and for epidemiological investigation. Results using the automated method can be obtained within 16 h, and riboprint patterns can be stored to create a unique database. Conducting ribotyping without the aid of an automated system is time-consuming and requires a high level of skill. Ribotyping has excellent reproducibility and discriminatory power.

A POTPOURRI OF RAPID METHODS

Only a few of the many rapid methods that exist have been discussed. There are a few others that deserve mention, including luminescence-based assays, flow cytometry, and impedance. Of particular importance to the food industry are the bioluminescence-based methods, which can provide a rapid estimate of total microbial numbers. The presence of specific types of bacteria cannot be determined, but the production of light can be correlated to the number of microbes. The method makes use of the ubiquitous presence of adenosine triphosphate (ATP) in all living cells and the ability of it to react with the luciferase enzyme complex found in fireflies. The Uni-lite system (Biotrace, Inc., Plainsboro, New Jersey), the Lightning bioluminescence system (Idexx Laboratories, Inc., Westbrook, Maine), and the Lumac Hygiene Monitoring kit (Integrated BioSolutions, Inc., Monmouth Junction, New Jersey) all make use of this technology. For example, the Uni-lite system consists of a transportable luminometer reading unit and swabs prepackaged in tubes containing reagents. The surface to be tested is swabbed, the reaction is activated by placing the swab into an enzyme solution, and the swab and tube are inserted into a chamber in the luminometer to obtain a reading. Results are obtained in <1 min.

The *direct epifluorescence filter technique* is another technique that is designed to illuminate bacteria. The method is simple and involves membrane filtration of a food sample to collect cells associated with the sample. The membrane is then processed by using fluorescent antibodies to stain target bacteria. Bacteria are visualized using an epifluorescence microscope. Problems can occur if food material trapped on the filter also fluoresces (i.e., it is difficult to differentiate bacteria from food).

Other methods that are available or in the developmental stages are based on nanotechnology, impedance, flow cytometry, and bacterial ice nucleation. The Bactometer (BioMérieux) and the Malthus System V (Malthus Diagnostics) are commercially available automated systems based on impedance. As microbes grow, they metabolize substrates with low conductivity into products with high conductivity and therefore decrease impedance. Impedance systems are good for processing a large number of samples and can detect low levels of a given microbe.

The future development of rapid methods will likely exploit the use of nanotechnology. Nanotechnology will permit the development of real-time sensors capable of detecting several different target microbes of concern. Although many different formats are available, silicon chips can be imprinted with specific DNA sequences for the detection and identification of target organisms. Chips could be imprinted with antibodies, RNA, or a host of other agents to target microorganisms. Microbial sensors are particularly applicable in fluid systems with limited organic substances. Presently, problems exist with sensor efficacy in food systems containing fats and proteins that coat the sensor and render it inoperable. These problems will be overcome as technology advances.

Concerns related to bioterrorism and the safety of the food supply have already inspired a massive effort for the development of rapid methods capable of detecting not only typical foodborne pathogens but also microbes of concern for human health that may be intentionally used to contaminate food. Handheld PCR units have been developed for the detection of *Bacillus anthracis* and *Yersinia pestis*. Research is ongoing to modify the units for detection of *Salmonella*, *E. coli* O157:H7, and *L. monocytogenes*. These units have potential applications in the screening of food at ports of entry and border crossings and on farms.

Summary

- Specialized media that target specific biochemical reactions of the target microbe are available.
- The most widely used format for rapid methods is immunologically based assays.
- An enrichment step is required for most assays to increase the number of bacteria to a detectable level.
- Most assays require 10^2 to 10^5 CFU per g or ml.
- PCR-based assays are extremely sensitive but are subject to interference from the presence of inhibitors in food.
- Future rapid assays will likely be based on nanotechnology.

Suggested reading

Beumer, R. R., E. Brinkman, and F. M. Rombouts. 1991. Enzyme-linked immunoassays for the detection of *Salmonella* spp.: a comparison with other methods. *Int. J. Food Microbiol.* **12:**363–374.

Feng, P. 1995. Rapid methods for detecting foodborne pathogens, p. App.1.01–App.1.16. *In* R. I. Merker (ed.), *FDA Bacteriological Analytical Manual*, ed. 8A. AOAC International, Gaithersburg, MD.

Food and Drug Administration. 2001. *Bacteriological Analytical Manual Online.* http://www.fda.gov/Food/ScienceResearch/LaboratoryMethods/BacteriologicalAnalyticalManualBAM/default.htm.

Matthews, K. R. 2003. Rapid methods for microbial detection in minimally processed foods, p. 151–163. *In* J. S. Novak, G. M. Saper, and V. K. Juneja (ed.), *Microbial Safety of Minimally Processed Foods*. CRC Press, Boca Raton, FL.

Questions for critical thought

1. Why is it important to process a food sample properly before analysis?
2. The FDA now incorporates the use of rapid methods in the screening of food samples for the presence of diarrheagenic *E. coli*. They state that ". . . use of other platforms and protocols must first be validated." Indicate, in brief, what you believe would be required to validate that a PCR protocol you wish to use is equivalent to the platforms/protocols recommended by the FDA.
3. The FDA has requested proposals for the development of PCR-based methods for the purpose of providing rapid confirmatory identification of microbial pathogens that are not usually associated with food and foodborne illness. Develop an outline for a proposal and include the pathogens that you would focus on and the type of PCR method you would use. Justify each of your responses.
4. Select a rapid assay listed in one of the tables. Gather literature from the company and find peer-reviewed journal articles related to the assay's efficacy. Based on your review of the literature, write a one-page abstract of the assay. Pretend that the abstract is for a rapid-methods newsletter.
5. Using a table format, compare PCR, RT-PCR, and real-time PCR. Provide a brief narrative concerning when each method would be most appropriate to use.
6. Food constituents often interfere with the function of a rapid assay. For example, fat in a sample may coat a biosensor, inhibiting the binding of bacteria present in a sample to the sensor. Using immunologically based assays as an example, indicate from a scientific viewpoint why a given food constituent would interfere with the assay. Now, provide a protocol that would eliminate or prevent such interference.
7. You are working for a small development group that is interested in marketing a new rapid method for detection of foodborne pathogens. The only problem is that they do not have a new rapid method and have asked you to develop one. What type of method would you develop? It must be novel.
8. Many of the rapid bioluminescence assays are based on the detection of ATP. Since ATP is found in all living cells, what is a potential problem(s) associated with this type of assay in determining whether a surface is contaminated with bacteria?

6

Indicator Microorganisms and Microbiological Criteria

LEARNING OBJECTIVES

The information in this chapter will enable the student to:

- differentiate among the various microbiological criteria
- identify national and international agencies involved in establishing microbiological criteria
- recognize how indicator organisms are used in microbiological criteria
- understand why some sampling plans are more stringent than others
- identify organisms and types of foods for which zero tolerance has been established
- identify and list steps required to manage microbiological hazards in foods

INTRODUCTION

The Purpose of Microbiological Criteria

Microorganisms cause foodborne illness and spoil food. Trained food microbiologists could probably isolate microbes from most raw and finished food products. Does that mean that the product is not safe to eat or will spoil rapidly? Not necessarily. Indeed, it is okay to have certain bacteria, even pathogenic bacteria, in food, depending on the product. Microbiological criteria ensure that a product has been produced under sanitary conditions and is microbiologically safe to consume.

Microbiological criteria are used to distinguish between an acceptable and an unacceptable product or between acceptable and unacceptable food-processing practices. The numbers and types of microorganisms associated with a food may be used to judge its microbiological safety and quality. Safety is determined by the absence, presence, or level of pathogenic microorganisms or their toxins and their expected control or destruction. The level of spoilage microorganisms reflects the microbiological quality, or wholesomeness, of a food, as well as the effectiveness of measures used to control or destroy such microorganisms. Indicator organisms may be used to assess either the microbiological quality or safety. Specifically, microbiological criteria are used to assess (i) the safety of food, (ii) adherence to good manufacturing practices (GMPs), (iii) the keeping quality (shelf life) of perishable foods, and (iv) the suitability of a food or ingredient for a particular purpose. Appropriately applied microbiological criteria ensure the safety and quality of foods. This, in turn, increases consumer confidence.

doi:10.1128/9781555817206.ch06

The Need To Establish Microbiological Criteria

A microbiological criterion should be established only in response to a need and when it is both effective and practical. There are many considerations to be taken into account when establishing microbiological criteria. Listed below are a few important factors for assessing the need for microbiological criteria.

- Evidence of a health hazard based on epidemiological data or a hazard analysis
- The nature of the food's normal microbial makeup and the ability of the food to support microbial growth
- The effect of processing on the microflora of the food
- The potential for microbial contamination and/or growth during processing, handling, storage, and distribution
- Spoilage potential, utility, and GMPs

Definitions

The National Research Council (NRC) of the U.S. National Academy of Sciences addressed the issue of microbiological criteria in their 1985 report entitled *An Evaluation of the Role of Microbiological Criteria for Foods and Food Ingredients*. The report indicates that a microbiological criterion should state what microorganism, group of microorganisms, or toxin produced by a microorganism is covered. The criterion should also indicate whether it can be present or present in only a limited number of samples or a given quantity of a food or food ingredient. In addition, a microbiological criterion should include the following information:

- Statement describing the identity of the food
- Statement identifying the contaminant
- Analytical method to be used for the detection, enumeration, or quantification of each contaminant
- Sampling plan (discussed below)
- Microbiological limits considered appropriate to the food and commensurate with the sampling plan

Criteria may be either mandatory or advisory. A *mandatory criterion* is a criterion that may not be exceeded. Food that does not meet the specified limit is required to be rejected, destroyed, reprocessed, or diverted. An *advisory criterion* permits acceptability judgments to be made. It serves as an alert to deficiencies in processing, distribution, storage, or marketing. For application purposes, there are three categories of criteria: standards, guidelines, and specifications. The following definitions were recommended by the NRC Subcommittee on Microbiological Criteria for Foods and Food Ingredients.

Standard: A microbiological criterion that is part of a law, ordinance, or administrative regulation. A standard is a mandatory criterion. Failure to comply constitutes a violation of the law, ordinance, or regulation and will be subject to the enforcement policy of the regulatory agency having jurisdiction.

Guideline: A microbiological criterion often used by the food industry or regulatory agency to monitor a manufacturing process. Guidelines function as alert mechanisms to signal whether microbiological conditions prevailing at critical control points (CCPs) or in the finished product are within the normal range. Hence, they are used to assess processing efficiency at CCPs and conformity with GMPs. A microbiological guideline is advisory.

Specifications: Microbiological criteria that are used as a purchase requirement whereby conformance becomes a condition of purchase between the buyer and the vendor of a food ingredient. A microbiological specification may be advisory or mandatory.

Who Establishes Microbiological Criteria?

Different scientific organizations are involved in developing general principles for the application of microbiological criteria. The scientific organizations which have most influenced the U.S. food industry include the Joint Food and Agricultural Organization and World Health Organization Codex Alimentarius International Food Standards Program, the International Commission on Microbiological Specifications for Foods (ICMSF), the U.S. National Academy of Sciences, and the U.S. National Advisory Committee on Microbiological Criteria for Foods. The Codex Alimentarius Program first formulated *General Principles for the Establishment and Application of Microbiological Criteria* in 1981; it has since been revised. In 1984, the NRC Subcommittee on Microbiological Criteria for Foods and Food Ingredients formulated general principles for the application of microbiological criteria to food and food ingredients as requested by four U.S. regulatory agencies. The *Codex Alimentarius* contains standards for all principal foods in the form (i.e., processed, semiprocessed, or raw) in which they are delivered to the consumer. Fresh perishable commodities not traded internationally are excluded from these standards. The World Trade Organization provides a framework for ensuring fair trade and harmonizing standards and import requirements for food traded through the Agreements on Sanitary and Phytosanitary Measures and Technical Barriers to Trade. Countries are required to base their standards on science, to base programs on risk analysis methodologies, and to develop ways of achieving equivalence among the different methods of inspection, analysis, and certification used by various trading countries. The World Trade Organization recommends the use of standards, guidelines, and recommendations developed by the Codex Alimentarius Program to make all standards similar.

SAMPLING PLANS

A sampling plan includes both the sampling procedure and the decision criteria. To examine a food for the presence of microorganisms, a representative sample is examined by defined methods. A *lot* is a quantity of product produced, handled, and stored within a limited time under uniform conditions. Since it is impractical to examine the entire lot, statistical concepts of population probability and sampling are used to determine the number and size of sample units required from the lot and to provide conclusions drawn from the analytical results. The sampling plan is designed so that inferior lots will be rejected.

Authors' note

The Codex Alimentarius *is a set of food standards and guidelines developed to protect the health of consumers and to ensure fair trade practices in the food trade.*

In a simplified example of a sampling plan, let us assume that 10 samples were taken and analyzed for the presence of a particular microorganism. Based on the decision criterion, only a certain number of the sample units could be positive for the presence of that microorganism for the lot to be considered acceptable. If in the criterion the maximum allowable number of positive units had been set at 2 ($c = 2$), then a positive result for >2 of the 10 sample units ($n = 10$) would result in rejection of the lot. Ideally, the decision criterion is set to accept lots that are of the desired quality and to reject lots that are not. However, since only part of the lot is examined, there is always the risk that an acceptable lot will be rejected or that an unacceptable lot will be accepted. The more samples examined, or the larger the size of n, the lower the risk of making an incorrect decision about the lot quality. However, as n increases, sampling becomes more time-consuming and costly. Generally, a compromise is made between the size of n and the level of risk that is acceptable.

Types of Sampling Plans

Sampling plans are divided into two main categories: variables and attributes. A variables plan depends on the frequency distribution of organisms in the food. For correct application of a variables plan, the organisms must have a lognormal distribution (i.e., counts transformed to logarithms are normally distributed). When the food is from a common source and it is produced and/or processed under uniform conditions, a lognormal distribution of the organisms present is assumed. Attributes sampling is the preferred plan when microorganisms are not homogeneously distributed throughout the food or when the target microorganism is present at low levels. This is often the case with pathogenic microorganisms. Attributes plans are also widely used to determine the acceptance or rejection of products at ports or other points of entry, since there is little or no knowledge of how the food was processed and past performance records are not available. Attributes sampling plans may also be used to monitor performance relative to accepted GMPs. Attributes sampling, however, is not appropriate when there is no defined lot or when random sampling is not possible, as might occur when monitoring cleaning practices. Because attributes sampling plans are used more frequently than variables sampling plans, this text does not cover variables sampling in detail.

Attributes Sampling Plans

Choosing an appropriate sampling plan does not need to be an onerous task. Generally, two-class plans are used almost exclusively for pathogens and three-class plans are used for examining hygiene indicators. Using a decision tree can facilitate the selection process (Fig. 6.1).

Two-class plans. A two-class attributes sampling plan assigns the concentration of microorganisms of the sample units tested to a particular attribute class, depending on whether the microbiological counts are above or below some preset concentration, represented by the letter m. The decision criterion is based on (i) the number of sample units tested, n, and (ii) the maximum allowable number of sample units yielding unsatisfactory test results, c. For example, when n is 5 and c is 2 in a two-class sampling plan designed to make a presence-absence decision about the lot (i.e., $m = 0$), the lot is rejected if more than two of the five sample units tested are positive. As n increases for the set number (c), the stringency of the sampling plan also increases. Conversely, for a set sample size n, as c increases, the stringency of the sampling plan decreases, allowing a higher probability of accepting, P_a, food lots of a given quality. Two-class plans are applied most often in qualitative (semiquantitative) pathogen testing in which the results are expressed as the presence or absence of the specific pathogen per sample weight analyzed.

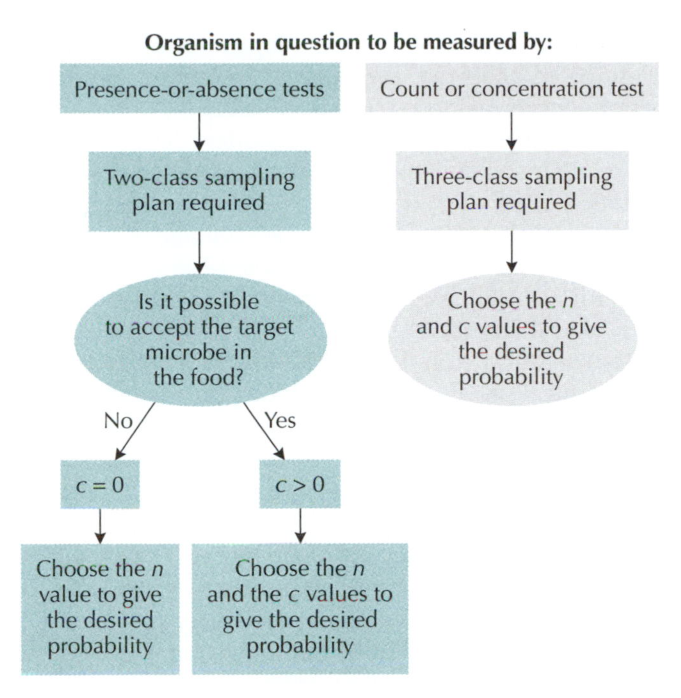

Figure 6.1 Guide for choosing an appropriate sampling plan. doi:10.1128/9781555817206.ch06.f06.01

Three-class plans. Three-class sampling plans use the concentrations of microorganisms in the sample units to determine levels of quality and/or safety. Counts above a preset concentration M for any of the n sample units tested are considered unacceptable, and the lot is rejected. The level of the test organism acceptable in the food is represented by m. This concentration in a three-class attributes plan separates acceptable lots (i.e., those with counts less than m) from marginally acceptable lots (i.e., those with counts greater than m but not exceeding M). Counts above m and up to and including M are not desirable, but the lot can be accepted provided the number (n) of samples that exceed m is no greater than the preset number, c. Thus, in a three-class sampling plan, the food lot will be rejected if any one of the sample units exceeds M or if the number of sample units with contamination levels above m exceeds c. Like with the two-class sampling plan, the stringency of the three-class sampling plan is dependent on the two numbers represented by n and c. The larger the value of n for a given value of c, the better the food quality must be to have the same chance of passing, and vice versa. From n and c, it is then possible to find the probability of acceptance, P_a, for a food lot of a given microbiological quality.

ESTABLISHING LIMITS

Microbiological limits, as defined in a criterion, represent the level above which action is required. Levels should be realistic and should be determined based on knowledge of the raw materials and the effects of processing, product handling, and storage and the end use of the product. Limits should also take into account the likelihood of uneven distribution of microorganisms in the food, the inherent variability of the analytical procedure, the risk associated with the organisms, and the conditions under which the food is handled and consumed. Microbiological limits should include the sample weight to be analyzed, the method reference, and the confidence limits of the referenced method where applicable.

The shelf life of a perishable product is often determined by the number of microorganisms initially present. As a general rule, a food containing a large population of spoilage organisms will have a shorter shelf life than the same food containing fewer of the same spoilage organisms. However, the relationship between total counts and shelf life is not carved in stone. Some types of microorganisms have a greater impact on the sensory characteristics of a food than others.

Foods produced and stored under GMPs may be expected to have a better microbiological profile than those foods produced and stored under poor conditions. The use of poor-quality materials, improper handling, or unsanitary conditions may result in higher bacterial numbers in the finished product. However, low counts in the finished product do not necessarily mean that GMPs were adhered to. Processing steps, such as heat treatment, fermentation, freezing, or frozen storage, can reduce the counts of bacteria that have resulted from noncompliance with GMPs. In addition, some products, such as ground beef, may normally contain high microbial counts even under the best conditions of manufacture due to the growth of psychrotrophic bacteria during refrigeration.

The use of quantitative risk assessment techniques to scientifically determine the probability of occurrence and the severity of known human exposure to foodborne hazards is rapidly gaining acceptance. The process consists of (i) hazard identification, (ii) hazard characterization, (iii) exposure assessment, and (iv) risk characterization. Though quantitative risk assessment techniques are well established for chemical agents, their application to food safety microbiology is relatively new.

INDICATORS OF MICROBIOLOGICAL QUALITY

Examination of a product for indicator organisms can provide simple, reliable, and rapid information about process failure, postprocess contamination, contamination from the environment, and the general level of hygiene under which the food was processed and stored.

Ideal indicators of product quality or shelf life should meet the following criteria.

- They should be present and detectable in all foods whose quality is to be assessed.
- Their growth and numbers should have a direct negative correlation with product quality.
- They should be easily detected and enumerated and be clearly distinguishable from other organisms.
- They should be enumerable in a short period, ideally within a workday.

Indicator Microorganisms

Indicator microorganisms can be used in microbiological criteria. The presence of indicator microbes may suggest a microbial hazard. For example, the presence of generic *Escherichia coli* in a sample indicates possible fecal contamination. These criteria might be used to address existing product quality or to predict the shelf life of the food. Some examples of indicators and the products in which they are used are shown in Table 6.1. Those

Table 6.1 Organisms highly correlated with product quality[a]

Organism(s)	Product(s)
Acetobacter spp.	Fresh cider
Bacillus spp.	Bread dough
Byssochlamys spp.	Canned fruits
Clostridium spp.	Hard cheeses
Flat-sour spores	Canned vegetables
Lactic acid bacteria	Beers, wines
Lactococcus lactis	Raw milk (refrigerated)
Leuconostoc mesenteroides	Sugar (during refining)
Pectinatus cerevisiiphilus	Beers
"*Pseudomonas putrefaciens*"	Butter
Yeasts	Fruit juice concentrates
Zygosaccharomyces bailii	Mayonnaise, salad dressing

[a]Source: J. M. Jay, *Modern Food Microbiology*, 4th ed. (Springer, New York, NY, 1991).

microorganisms listed in the table are the primary spoilage organisms of the specific products listed. Loss of quality in other products may not be limited to one organism but may involve a variety of organisms. In those types of products, it is often more practical to determine the counts of groups of microorganisms most likely to cause spoilage in each particular food.

The aerobic plate count (APC) is commonly used to determine the total number of microorganisms in a food product. By modifying the environment of incubation or the medium used, the APC can be used to preferentially screen for groups of microorganisms, such as those that are thermoduric, mesophilic, psychrophilic, thermophilic, proteolytic, or lipolytic. The APC may be a component of microbiological criteria assessing product quality when those criteria are used to (i) monitor foods for compliance with standards or guidelines set by various regulatory agencies, (ii) monitor foods for compliance with purchase specifications, and (iii) monitor adherence to GMPs.

Microbiological criteria as specifications are used to determine the usefulness of a food or food ingredient for a particular purpose. For example, specifications are set for thermophilic spores in sugar and spices intended for use in the canning industry. Lots of sugar that fail to meet specifications may not be suitable for use in low-acid canning but could be diverted for other uses. The APCs of refrigerated perishable foods, such as milk, meat, poultry, and fish, may be used to indicate the condition of the equipment and utensils used, as well as the time-temperature profile of storage and distribution of the food.

When evaluating the results of an APC for a particular food, it is important to remember that (i) APCs measure only live cells, and therefore, a grossly spoiled product may have a low APC if the spoilage organisms have died; (ii) APCs are of little value in assessing sensory quality, since high microbial counts are generally required for sensory-quality loss; and (iii) since the biochemical activities of different bacteria vary, quality loss may also occur at low total counts. With any food, specific causes of unexpectedly high counts can be identified by examination of samples at control points and by inspection of processing plants. Interpretation of the APC of a food requires knowledge of the expected microbial population at the point where the sample is collected. If counts are higher than expected, this indicates the need to determine why there has been a violation of the criterion.

The direct microscopic count (DMC) is used to give an estimate of the numbers of both viable and nonviable cells in samples containing a large number of microorganisms (i.e., $>10^5$ colony-forming units [CFU]/ml). Considering that the DMC does not differentiate between live and dead cells (unless a fluorescent dye, such as acridine orange, is employed) and that it requires that the total cell count exceed 10^5, the DMC has limited value as a part of the microbiological criteria for quality issues. The use of the DMC as a part of microbiological criteria for foods or ingredients is restricted to a few products, such as raw non-grade A milk, dried milk, liquid and frozen eggs, and dried eggs.

Other methods commonly used to indicate the quality of different food products include the Howard mold count, yeast and mold count, heat-resistant-mold count, and thermophilic-spore count. The Howard mold count (microscopic analysis) is used to detect the inclusion of moldy material in canned fruit and tomato products, as well as to evaluate the sanitary condition of processing machinery in vegetable canneries. Yeasts and molds grow on foods when conditions for bacterial growth are less favorable. Therefore, they can be a problem in fermented dairy products, fruits, fruit beverages, and soft drinks. Yeast and mold counts are used as parts of the microbiological standards for various dairy products, such as cottage cheese and frozen cream and sugar. Heat-resistant molds (e.g., *Byssochlamys fulva* and *Aspergillus fischerianus*) that may survive the thermal processes used for fruit and fruit products may need limits in purchase specifications for ingredients such as fruit concentrates. The canning industry is concerned about thermophilic spores in ingredients because they cause defects in foods held at elevated temperatures (i.e., due to inadequate cooling and/or storage at too-high temperatures). Purchase specifications and verification criteria are often used for thermophilic-spore counts in ingredients intended for use in low-acid, heat-processed canned foods.

Metabolic Products

Bacterial levels in a food product can sometimes be estimated by testing for metabolic products made by the microorganisms in the food. When a correlation is established between the presence of a metabolic product and product quality loss, tests for the metabolite may be part of a microbiological criterion.

An example of the use of a test for metabolic products as part of a microbiological criterion is the sensory evaluation of imported shrimp. Trained personnel are able to classify the degree of decomposition (i.e., quality loss) into one of three classes through sensory examination. The shrimp are placed in one of the following quality classes: class 1, passable; class 2, decomposed (slight but definite); and class 3, decomposed (advanced). The limits of acceptability of a lot are based on the number of shrimp in a sample that are placed in each of the three classes. Other commodities in which organoleptic examination (examination for changes in appearance, color, texture, and odor) is used to determine quality deterioration include raw milk, meat, poultry, and fish and other seafoods. The food industry also uses these examinations to classify certain foods into quality grades. Other examples of metabolic products used to assess product quality are listed in Table 6.2.

Table 6.2 Some microbial metabolic products that correlate with food quality[a]

Metabolite(s)	Applicable food product(s)
Cadaverine and putrescine	Vacuum-packaged beef
Diacetyl	Frozen juice concentrate
Ethanol	Apple juice, fishery products
Histamine	Canned tuna
Lactic acid	Canned vegetables
Trimethylamine	Fish
Total volatile bases, total volatile nitrogen	Seafoods
Volatile fatty acids	Butter, cream

[a]Source: J. M. Jay, *Modern Food Microbiology*, 4th ed. (Springer, New York, NY, 1991).

INDICATORS OF FOODBORNE PATHOGENS AND TOXINS

Microbiological criteria for product safety should be developed only when the application of a criterion can reduce or eliminate a foodborne hazard. Microbiological criteria verify that the process is adequate to eliminate the hazard. Each food type should be carefully evaluated through risk assessment to determine the potential hazards and their significance to consumers. When a food is repeatedly implicated as a vehicle in foodborne disease outbreaks, the application of microbiological criteria may be useful. Public health officials and the dairy industry responded to widespread outbreaks of milk-borne disease that occurred around the turn of the 20th century in the United States. By imposition of controls on milk production, development of safe and effective pasteurization procedures, and setting of microbiological criteria, the safety of milk supplies was greatly improved.

Microbiological criteria may also be applied to food products, such as shellfish, which are frequently subject to contamination by harmful microorganisms. The National Shellfish Sanitation Program utilizes microbiological criteria in this manner to prevent the use of shellfish from polluted waters, which may contain various intestinal pathogens. Depending on the type and level of contamination anticipated, the imposition of microbiological criteria may or may not be justified. Contamination of food with pathogens that cannot grow to harmful levels does not warrant microbiological criteria. However, microbiological criteria would be warranted for a pathogen that has a low infectious dose. Though fresh vegetables are often contaminated with small numbers of *Clostridium botulinum*, *Clostridium perfringens*, and *Bacillus cereus* organisms, epidemiological evidence indicates that this contamination presents no health hazard. Thus, imposing microbiological criteria for these microorganisms would not be beneficial. However, criteria may be appropriate for enteric pathogens on produce, since there have been several outbreaks of foodborne illness resulting from fresh produce contaminated with enteric pathogens.

Often food processors alter the intrinsic or extrinsic parameters of a food (nutrients, pH, water activity, inhibitory chemicals, gaseous atmosphere, temperature of storage, and presence of competing organisms) to prevent the growth of undesirable microorganisms. If control over one or more of these parameters is lost, then there may be risk of a health hazard. For example, in the manufacture of cheese or fermented sausage, a lactic acid starter culture is used to produce acid quickly enough to inhibit the growth of *Staphylococcus aureus* to harmful levels. Process CCPs, such as the rate of acid formation, are implemented to prevent the growth of harmful microorganisms or their contamination of the food and to ensure that process control is maintained.

Depending on the pathogen, low levels of the microorganism in the food product may or may not be of concern. Some microorganisms have such a low infectious dose that their mere presence in a food presents a significant public health risk. For such organisms, the concern is not the ability of the pathogen to grow in the food but its ability to survive in the food. Foods having intrinsic or extrinsic factors that prevent the survival of pathogens or toxigenic microorganisms may not be candidates for microbiological criteria related to safety. For example, the acidity of certain foods, such as fermented meat products, might be assumed to be sufficient for pathogen control. However, while the growth and toxin production of *S. aureus* might be prevented, enteric pathogens, such as *E. coli* O157:H7, could survive and produce a product that is unsafe for consumption.

An important and often overlooked variable in microbiological risk assessment is the consumer. More rigid microbiological requirements may be needed if a food is intended for use by infants, the elderly, or immunocompromised people, since they are more susceptible to infectious agents than are healthy adults. The ICMSF proposed a system for classification of foods according to risk into 15 hazard categories called cases, with suggested appropriate sampling plans, shown in Table 6.3.

The stringency of sampling plans is based on either the hazard to the consumer from pathogenic microorganisms and their toxins or toxic metabolites or the potential for quality deterioration. It should take into account the types of microorganisms present and their numbers. Foodborne pathogens are grouped into one of three categories based on the severity of the

Table 6.3 Plan stringency (case) in relation to degree of health hazard and conditions of use[a]

Degree of concern relative to utility and health hazard	Conditions in which food is expected to be handled and consumed after sampling, in the usual course of events[b]		
No direct health hazard	**Increased shelf life**	**No change**	**Reduced shelf life**
Utility (e.g., general contamination, reduced shelf life, and incipient spoilage)	Case 1 3-class $n = 5, c = 3$	Case 2 3-class $n = 5, c = 2$	Case 3 3-class $n = 5, c = 1$
Health hazard	**Reduced hazard**	**No change**	**Increased hazard**
Indicator: low indirect hazard	Case 4 3-class $n = 5, c = 3$	Case 5 3-class $n = 5, c = 2$	Case 6 3-class $n = 5, c = 1$
Moderate hazard: direct limited spread	Case 7 3-class $n = 5, c = 2$	Case 8 3-class $n = 5, c = 1$	Case 9 3-class $n = 5, c = 0$
Serious hazard: incapacitating, usually not life threatening	Case 10 2-class $n = 5, c = 0$	Case 11 2-class $n = 10, c = 0$	Case 12 2-class $n = 20, c = 0$
Severe hazard: general or restricted population	Case 13 2-class $n = 15, c = 0$	Case 14 2-class $n = 30, c = 0$	Case 15 2-class $n = 60, c = 0$

[a]Based on International Commission on Microbiological Specifications for Foods, *Microorganisms in Foods 7: Microbiological Testing in Food Safety Management* (Springer Science, New York, NY, 2002).

[b]*n*, number of sample units drawn from lot; *c*, maximum allowable number of positive results.

Table 6.4 Hazard grouping of foodborne pathogens or toxins[a]

Severe hazards: life threatening for general population
Enterohemorrhagic *E. coli*
Shigella dysenteriae I
Salmonella enterica serovars Typhi and Paratyphi
Vibrio cholerae O1
Brucella abortus, Brucella suis, Brucella melitensis
Botulinal neurotoxin
Severe hazards: for restricted populations
Hepatitis A virus
Vibrio vulnificus
Listeria monocytogenes
Enterotoxigenic *E. coli*
Cryptosporidium parvum
Campylobacter jejuni serovar O:19
Enterobacter sakazakii (*Cronobacter* sp.)
Serious hazards: incapacitating, not life threatening
Yersinia enterocolitica
Cyclospora cayetanensis
Salmonella enterica serovar Enteritidis
Shigella flexneri
Cryptosporidium parvum
Ochratoxin A
Moderate hazards: usually not life threatening
Bacillus cereus
Clostridium perfringens
Staphylococcus enterotoxins (*Staphylococcus aureus*)
Vibrio cholerae, non-O1
Vibrio parahaemolyticus
Biogenic amines (e.g., histamine)

[a]Based on International Commission on Microbiological Specifications for Foods, *Microorganisms in Foods 7: Microbiological Testing in Food Safety Management* (Springer Science, New York, NY, 2002).

potential hazard (i.e., severe hazards, serious hazards, and moderate hazards [Table 6.4]). Pathogens with the potential for extensive spread are often initially associated with specific foods. An example is fresh beef contaminated with *E. coli* O157:H7. One or a few contaminated pieces of meat can lead to widespread contamination of the product during processing, such as grinding to produce ground beef. There can also be cross-contamination if the fresh beef is improperly stored with ready-to-eat (RTE) foods. Microbial pathogens in the lowest-risk group (moderate hazards with limited spread) are found in many foods, usually in small numbers. Generally, illness is caused only when ingested foods contain large numbers of a pathogen, e.g., *C. perfringens*, or have at some point contained large enough numbers of a pathogen, e.g., *S. aureus*, to produce sufficient toxin to cause illness. Outbreaks are usually restricted to consumers of a particular meal or a particular kind of food.

The effects of conditions to which the food will likely be exposed in relation to the growth or death of a relevant organism(s) must be considered before the appropriate case (Table 6.3) can be determined. In general, if a food will be subjected to treatment that would permit growth of the microorganism (e.g., *C. perfringens* in cooked meats held at inappropriate temperature) and thereby increase the hazard, then case 3, 6, 9, 12, or 15 would apply. Case 2, 5, 8, or 11 would apply if no change in the number of organisms present were expected, for example, in freezing of the product. Finally, case 1, 4, 7, or 10 would apply if the product was intended to be fully cooked, reducing the microbial hazard. Consider the case of foods derived from dried whole eggs which are cooked and eaten immediately (scrambled) or made into products (pasta). This would diminish the microbiological hazard. If the dried eggs were consumed dry or reconstituted and consumed immediately, the microbiological hazard would not change (no opportunity was provided for microorganisms present to multiply). A third scenario involves reconstituting the dried eggs and waiting for an extended period before consuming them, in which case any microorganisms present would have an opportunity to multiply and potentially cause illness.

Indicator Organisms

Microbiological criteria for food safety may use tests for indicator organisms that suggest the possibility of a microbial hazard. *E. coli* in drinking water, for example, indicates possible fecal contamination and therefore the potential presence of other enteric pathogens.

Microbial indicators should have the following characteristics.

- They should be easily and rapidly detectable.
- They should be readily distinguishable from microbes commonly associated with the food.
- They should have a history of constant association with the pathogen whose presence they are to indicate.
- They should be present when the pathogen of concern is present.
- They should have numbers that ideally correlate with those of the pathogen of concern.
- Their growth requirements and growth rate should equal those of the pathogen.
- They should have a die-off rate that at least parallels that of the pathogen and, ideally, have a slightly longer persistence than the pathogen of concern.
- They should be absent from foods that are free of the pathogen except perhaps at certain minimum numbers.

Additional criteria for fecal indicators used in food safety have been suggested and include the following. Ideally, the bacteria selected should demonstrate specificity, occurring only in intestinal environments. They should occur in very high numbers in feces so as to be encountered in high dilutions. They should possess a high resistance to the external environment, the contamination of which is to be assessed. They should permit relatively easy and fully reliable detection even when present in low numbers.

With the exception of those for *Listeria monocytogenes*, *E. coli* O157:H7, *Salmonella*, and *S. aureus*, most tests for ensuring safety use indicator organisms rather than direct tests for the specific hazard. An overview of some of the more common indicator microorganisms used for ensuring food safety is given below.

Fecal Coliforms and *E. coli*

Fecal coliforms, including *E. coli*, are easily destroyed by heat and may die during freezing and storage of frozen foods. Microbiological criteria involving *E. coli* are useful in those cases where it is desirable to determine if fecal contamination may have occurred. Contamination of a food with *E. coli* implies a risk that other enteric pathogens may also be present. Fecal coliform bacteria are used as a component of microbiological standards to monitor the wholesomeness of shellfish and the quality of shellfish-growing waters. The purpose is to reduce the risk of harvesting shellfish from waters polluted with fecal material. The fecal coliforms have a higher probability of containing organisms of fecal origin than do coliforms which are comprised of organisms of both fecal and nonfecal origins. Fecal coliforms can become established on equipment and utensils in the food-processing environment and contaminate processed foods. At present, *E. coli* is the most widely used indicator of fecal contamination.

The value of testing for fecal coliforms has recently come under scrutiny. Fecal coliforms include *Klebsiella*, *Enterobacter*, and *Citrobacter* species, but these organisms may be considered false-positive indicators of fecal contamination since they can grow in nonfecal niches, including water, food, and waste. The current interpretation is that *E. coli* is the only valid indicator microbe for monitoring food containing fresh vegetables for fecal contamination. Using only the *E. coli* assay should serve to eliminate misinterpretation of results of the fecal coliform assay by some physicians and public health officials.

The presence of *E. coli* in a heat-processed food means either process failure or, more commonly, postprocess contamination. In the case of refrigerated RTE products, such as shrimp and crabmeat, coliforms are indicators of process integrity with regard to the reintroduction of pathogens from environmental sources and maintenance of adequate refrigeration. The source of coliforms after thermal processing appears to be the processing environment, inadequate sanitation procedures, and/or poor temperature control. Coliforms are recommended over *E. coli* and APCs, because coliforms are often present in higher numbers than *E. coli* and the levels of coliforms do not increase over time when the product is stored properly.

Metabolic Products

Certain microbiological criteria related to safety rely on tests for metabolites that indicate a potential hazard. Examples of metabolites as a component of microbiological criteria include (i) tests for thermonuclease or thermostable deoxyribonuclease in foods containing or suspected of containing $\geq 10^5$ *S. aureus* organisms per ml or g; (ii) illuminating grains under ultraviolet (UV) light to detect the presence of aflatoxin produced by *Aspergillus* spp.; and (iii) assaying for the enzyme alkaline phosphatase, a natural constituent of milk that is inactivated during pasteurization, to determine whether milk has been pasteurized or whether pasteurized milk has been contaminated with raw milk.

APPLICATION AND SPECIFIC PROPOSALS FOR MICROBIOLOGICAL CRITERIA FOR FOOD AND FOOD INGREDIENTS

In this section, examples of the application of microbiological criteria to various foods are presented. There are no general criteria suitable for all foods. The relevant background literature that relates to the quality and

Table 6.5 Microbiological criteria for *Listeria monocytogenes* in RTE foods[a]

	Limit (per g)			
Country	*n*	*c*	*m*	*M*
United States	5	0	0 (negative/25 g)	
United Kingdom		5	100	
Canada				
Category 1, 2	5	0	0 (0/25 g)	
Category 3	5	0	100	
Germany			10,000	

[a]Based on International Commission on Microbiological Specifications for Foods, *Microorganisms in Foods 7: Microbiological Testing in Food Safety Management* (Springer Science, New York, NY, 2002).

safety of a specific food product should be consulted before implementation of microbiological criteria.

The utilization of microbiological criteria, as stated above, may be either mandatory (standards) or advisory (guidelines or specifications). One obvious example of a mandatory criterion (standard) is the zero tolerance set for *Salmonella* in all RTE foods. The U.S. Department of Agriculture (USDA) Food Safety and Inspection Service has mandated zero tolerance for *E. coli* O157:H7 in fresh ground beef. Since the emergence of *L. monocytogenes* as a foodborne pathogen, the Food and Drug Administration and USDA have also applied zero tolerance for the organism in all RTE foods. However, there is considerable debate as to whether zero tolerance is warranted for *L. monocytogenes*. Table 6.5 provides examples of microbiological criteria for *L. monocytogenes* in RTE foods. The microbiological criteria vary considerably for different countries. These recommendations are examples of advisory criteria. Other advisory criteria that have been established to address the safety of foods are shown in Table 6.6.

Mandatory criteria in the form of standards have also been employed for quality issues. Presented in Table 6.7 are examples of foods and food ingredients for which federal, state, and city, as well as international, microbiological standards have been developed.

CURRENT STATUS

The ICMSF recommends that a series of steps be taken to manage microbiological hazards for foods intended for international trade. These steps include conducting a risk assessment and an assessment of risk management options, establishing a food safety objective (FSO), and confirming that the FSO is achievable by application of GMPs and Hazard Analysis and CCPs (HACCPs).

An FSO is a statement of the frequency or maximum concentration of a microbiological hazard in a food that is considered acceptable for consumer protection. Examples of FSOs include staphylococcal enterotoxin levels in cheese that must not exceed 1 μg/100 g or an aflatoxin concentration in peanuts that should not exceed 15 μg/kg. FSOs are broader in scope than microbiological criteria and are intended to communicate the acceptable level of hazard. *Control measures* are actions and activities that can be used to prevent or eliminate a food safety hazard or to reduce it to an acceptable level. The food industry is responsible for applying GMPs and establishing appropriate control measures, including CCPs, in their processes and HACCP plans. For foods in international commerce, quantitative FSOs

Table 6.6 Examples of various food products for which advisory microbiological criteria have been established

Product category	Test parameter(s)	Case	Plan class	Limit (per g)			
				n	*c*	*m*	*M*
Roast beef[a]	*Salmonella*	12	2	20	0	0	
Pâté	*Salmonella*	12	2	20	0	0	
Raw chicken	APC	1	3	5	3	5×10^5	10^7
Cooked poultry, frozen, RTE	*S. aureus*	8	3	5	1	10^3	10^4
Cooked poultry, frozen, to be reheated	*S. aureus*	8	3	5	1	10^3	10^4
	Salmonella	10	2	5	0	0	
Chocolate or confectionery	*Salmonella*	11	2	10[b]	0	0	
Dried milk	APC	2	3	5	2	3×10^4	3×10^5
	Coliforms	5	3	5	1	10	10^2
	Salmonella[c] (normal routine)	10	2	5	0	0	
		11	2	10	0	0	
		12	2	20	0	0	
	Salmonella[c] (high-risk populations)	10	2	15	0	0	
		11	2	30	0	0	
		12	2	60	0	0	
Fresh cheese[d]	*S. aureus*			5	2	10^2	10^3
	Coliforms			5	2	10^2	10^3
Soft cheese[d]	*S. aureus*			5	2	10^2	10^3
	Coliforms			5	2	10^2	10^3
Pasteurized liquid; frozen and dried egg products	APC	2	3	5	2	5×10^4	10^6
	Coliforms	5	3	5	2	10^3	10^3
	Salmonella[b] (normal routine)	10	2	5	0	0	
		11	2	10	0	0	
		12	2	15	0	0	
	Salmonella[b] (high-risk populations)	10	2	15	0	0	
		11	2	30	0	0	
		12	2	60	0	0	
Fresh and frozen fish, to be cooked before being eaten	APC	1	3	5	3	5×10^3	10^7
	E. coli		4		3	11	500
	Salmonella[e]	10	2	5	0	0	
	Vibrio parahaemolyticus[e]	7	3	5	2	10^2	10^3
	S. aureus[e]	7	3	5	2	10^3	10^4
Coconut	*Salmonella*	1	3	5	3	5×10^5	10^7
	Growth not expected	11	2	10	0	0	
	Growth expected	12	2	20	0	0	

[a]Product is cooked, not a roast of beef.
[b]The 25-g analytical unit may be composited.
[c]The case is to be chosen based on whether the hazard is expected to be reduced, unchanged, or increased.
[d]Requirements only for fresh and soft cheese made from pasteurized milk.
[e]For fish known to derive from inshore or inland waters of doubtful bacteriological quality or where fish are to be eaten raw, additional tests may be desirable.

Table 6.7 Examples of various food products for which mandatory microbiological criteria have been established

Product category	Test parameter(s)	Comment
United States		
Dairy products		
Raw milk	Aerobic bacteria	Recommendations of U.S. Public Health Service
Grade A pasteurized milk	Aerobic bacteria Coliforms	Recommendations of U.S. Public Health Service
Grade A pasteurized (cultured) milk	Aerobic bacteria Coliforms	Recommendations of U.S. Public Health Service
Dry milk (whole)	SPC[a] Coliforms	Recommendations of U.S. Public Health Service
Dry milk (nonfat)	SPC Coliforms	Standards of Agricultural Marketing Service (USDA)
Frozen desserts	SPC Coliforms	Recommendations of U.S. Public Health Service
Starch and sugars	Total thermophilic-spore count Flat-sour spores Thermophilic anaerobic spores Sulfide spoilage spores	National Canners Association (National Food Processors Association)
Breaded shrimp	APC *E. coli* *S. aureus*	Food and Drug Administration compliance policy guide
International		
Caseins and caseinates	Total bacterial count Thermophilic organisms Coliforms	Europe
Natural mineral waters	Aerobic mesophilic count Coliforms *E. coli* Fecal streptococci Sporulating sulfite-reducing anaerobes *Pseudomonas aeruginosa*, parasites, and pathogenic organisms	*Codex Alimentarius*
Hot meals served by airlines	*E. coli* *S. aureus* *B. cereus* *C. perfringens* *Salmonella*	Europe
Tomato juice	Mold count	Canada
Fish protein	Total plate count *E. coli*	Canada
Gelatin	Total plate count Coliforms *Salmonella*	Canada

[a]SPC, standard plate count.

provide an objective basis to identify what an importing country is willing to accept in relation to the microbiological safety of its food supply. A more detailed discussion on FSOs can be found in chapter 29.

Summary

- Microbiological criteria are most effectively applied as part of quality assurance programs in which HACCP and other prerequisite programs are in place.
- Microbiological criteria for foods continue to evolve as new information becomes available.
- Quantitative risk assessment facilitates the establishment of microbiological criteria for foods in international trade and of guidelines for national standards and policies.
- Microbiological criteria differ based on the food and microorganism of concern.
- Microbiological criteria may be based on the presence or absence of specific metabolites.

Suggested reading

Doyle, M. P., and M. C. Erickson. 2006. Closing the door on the fecal coliform assay. *Microbe* **1:**162–163.

International Commission on Microbiological Specifications for Foods. 1994. Choice of sampling plan and criteria for *Listeria monocytogenes*. *Int. J. Food Microbiol.* **22:**89–96.

International Commission on Microbiological Specifications for Foods. 1986. *Microorganisms in Foods 2. Sampling for Microbiological Analysis: Principles and Applications,* 2nd ed. University of Toronto Press, Toronto, Canada.

International Commission on Microbiological Specifications for Foods. 2002. *Microorganisms in Foods 7: Microbiological Testing in Food Safety Management.* Springer Science, New York, NY.

Jay, J. M. 2006. *Modern Food Microbiology,* 7th ed. Springer, New York, NY.

Pierson, M. D., D. L. Zink, and L. M. Smoot. 2007. Indicator microorganisms and microbiological criteria, p. 69–86. *In* M. P. Doyle and L. R. Beuchat (ed.), *Food Microbiology: Fundamentals and Frontiers,* 3rd ed. ASM Press, Washington, DC.

Pierson, M. D., and D. A. Corlett, Jr. (ed.). 1992. *HACCP: Principles and Applications.* Van Nostrand Reinhold, New York, NY.

Randell, A. W., and A. J. Whitehead. 1997. Codex Alimentarius: food quality and safety standards for international trade. *Rev. Sci. Tech.* **16:**313–321.

Questions for critical thought

1. Principles of microbiological risk management include FSO, performance objective, and performance criterion. Only FSOs were discussed in this chapter. Define each term (hint: ftp://ftp.fao.org/codex/Publications/ProcManuals/Manual_20e.pdf) and then describe how they are used as tools in the risk management of microbiological hazards of foodstuffs.
2. Violation of which of the following could result in government action: a microbiological standard, guideline, or specification?
3. Fecal coliform counts are more informative under many circumstances than coliform counts. Briefly describe the difference between the two indicator methods and provide a scenario for the use of each.

4. Using the pathogens *Campylobacter jejuni* and *S. aureus* as targets, select a food and, using the decision tree in Fig. 6.1, outline parameters for which the organisms are measured by presence or absence and enumeration.
5. The fecal coliform assay may not accurately reflect whether a sample contains fecal material. Why? Aside from the information provided in the chapter, perform a literature search to provide support for this statement.
6. The microbiological criteria for *L. monocytogenes* in shrimp are as follows: $n = 5$, $c = 0$, and $m = M = 0$. Indicate what each term represents. Based on the information provided, if one sample had 10 CFU, would the product be accepted or rejected?
7. Explain why it is acceptable for some products to be positive for *S. aureus* but not for *Salmonella*.
8. A variety of microbes are associated with the spoilage of foods. Why would one food spoil more rapidly than another food even though the numbers of spoilage organisms associated with the two foods are the same?
9. How can the APC be used to screen specifically for a group of microorganisms? Explain your answer.
10. Explain why foods produced and stored under GMPs may be expected to have microbiological profiles different from those of foods produced and stored without the benefit of GMPs.
11. *C. botulinum* is a dangerous pathogen. Why is it acceptable for fresh vegetables to be contaminated with low numbers of this pathogen?
12. Why are microbiological criteria needed for products that will ultimately be cooked?
13. Failure to comply with which microbiological criterion—standard, guideline, or specification—would result in legal action? Explain your answer.
14. Zero tolerance is set for several pathogens. To what pathogens, and under what conditions, does zero tolerance apply? Speculate why zero tolerance was established for each of the pathogens.
15. How do microbiological criteria affect a food processor? Provide an example of a food and describe the microbiological criteria associated with the product and how those criteria could be met.

SECTION

II

Foodborne Pathogenic Bacteria

7 Regulatory Issues

LEARNING OBJECTIVES

The information in this chapter will enable the student to:

- identify the major players, divisions, and branches of the U.S. food regulatory agencies
- appreciate the role of regulations and their historical significance
- understand the process of an outbreak investigation
- appreciate the role of surveillance in the development of new regulations
- outline the process for recognition of potential acts of bioterrorism affecting the food supply
- appreciate the U.S. role in global regulations

INTRODUCTION

According to the Centers for Disease Control and Prevention (CDC), about one in every six people in the United States gets sick each year from eating contaminated food. This is equivalent to 48 million individuals who are affected as part of the more than 1,000 outbreaks that happen each year. These outbreaks of foodborne illness are caused by unintentional contamination mostly from viruses, various bacteria, and some protozoa. (While biological hazards cause the majority of illnesses, some may be due to chemical contamination as well, but as this is a food microbiology text, the focus here is on biological hazards.) This textbook is filled with information about the leading foodborne pathogens. This chapter addresses the role of regulatory authorities in enhancing the safety of the U.S. food supply. Over the past 15 years, the United States has seen a reduction in some aspects of contamination and a reduction in the numbers of cases of foodborne illness. Collaborations among regulatory bodies, industry, and academic scientists have helped to make this possible. For example, over the past 15 years infections caused by *Escherichia coli* O157:H7 have been reduced by half, in particular those associated with consumption of ground beef. Unfortunately, *Salmonella* infection numbers have not declined. *Salmonella* causes more hospitalizations and deaths than any other type of microorganism found in food, plus $365 million in direct medical costs annually. These data were generated by the CDC in 2011, based on surveillance information and mathematical modeling. Surveillance is extremely important in understanding the burden of foodborne illness. Effective surveillance is essential, yet it is really difficult to do within the United States due to lack of funding and the inherent nature of underreporting of foodborne illness. You can probably relate to the latter aspect. Have you ever felt nauseated, with vomiting

doi:10.1128/9781555817206.ch07

and/or diarrhea? Did you visit your family physician or the school health center? No? You are not alone; many people do not visit a physician with these symptoms. Scientists and mathematicians have developed sophisticated calculations using data obtained from specific testing sites within the United States, population surveys, and estimations of illness.

U.S. AGENCIES INVOLVED IN FOOD REGULATION

Two federal agencies work together in preventing the occurrence of foodborne illness within the United States. These are the U.S. Food and Drug Administration (FDA) and the U.S. Department of Agriculture (USDA). Within each agency there are several important branches that all contribute to this effort. Hazards in the food supply may be effectively controlled in a collaborative effort among regulatory agencies, with effective use of tools by industry (see chapter 29), and with support in following food safety and good health directives by consumers.

The emphasis in this chapter is on prevention of foodborne illness, and the majority of the focus here concerns biological hazards. It is important as well to recognize that chemical and physical hazards exist and that these are also of concern in food production and distribution. The FDA and USDA rank foodborne illness as the primary food safety concern according to risk. Information gathered through science-based risk assessment and research is used to develop guidance and regulations to reduce contamination or cross-contamination of foods. A regulation is defined as "an authoritative rule dealing with details or procedures issued by an executive authority or regulatory agency of a government and having the force of law." Regulations use words like "shall" and "must." A guidance principle is defined as "the action provided by a guide providing leadership and direction." Guidance documents use words like "should." The current food safety regulatory arena is in a state of flux as leaders learn more and risks change. The following provides an overview of the history of the FDA and the USDA and a framework for understanding the current regulations and guidelines at work protecting food safety in the 21st century. Of course, these are not the only two agencies with important roles in protecting the U.S. food supply, but they play major roles which are described here. This is a complex system and often criticized for the many players and found to be frustrating and confusing for consumers to determine the appropriate regulatory agency for specific situations.

The USDA

The USDA is a cabinet level department within the U.S. government (Fig. 7.1). The mission statement of the USDA is to provide leadership on food, agriculture, natural resources, and related issues based on sound public policy, the best available science, and efficient management. The department is composed of 17 agencies and 15 offices which work in various areas related to agriculture. A listing of the agencies and their responsibilities is shown in Table 7.1. There are seven mission areas: farm and foreign agricultural services; food, nutrition, and consumer services; food safety; marketing and regulatory programs; natural resources and environment; research, education, and economics; and rural development. Concerning food, the USDA offers several programs and services. These include meat and poultry safety, food safety information, food preservation and home

Figure 7.1 The USDA logo.

Table 7.1 List of USDA agencies, indicating the depth and breadth of the work of the USDA

Agency name	Agency abbreviation[a]	Brief description
Agricultural Marketing Service	AMS	Facilitates the marketing of agricultural products in domestic and international markets while ensuring fair trading practices and promoting a competitive and efficient marketplace; works to develop new marketing services
Agricultural Research Service	ARS	USDA's principal in-house research agency; conducts agricultural research and provides information on agriculture
Animal and Plant Health Inspection Service	APHIS	Works to ensure the health and care of animals and plants, to improve agricultural productivity and competitiveness, and to contribute to the national economy and public health
Center for Nutrition Policy and Promotion	CNPP	Works to improve the health and well-being of Americans by developing and promoting dietary guidance that links scientific research to the nutritional needs of consumers
Economic Research Service	ERS	USDA's principal social science research agency; communicates research results and socioeconomic indicators via briefings, analyses for policymakers and their staffs, market analysis updates, and major reports
Farm Service Agency	FSA	Implements agricultural policy, administers credit and loan programs, and manages conservation, commodity, disaster, and farm marketing programs through a national network of offices
Food and Nutrition Service	FNS	Works with other organizations to increase food security and reduce hunger by providing children and low-income people access to food, a healthy diet, and nutrition education while supporting American agriculture
Food Safety and Inspection Service	FSIS	Works to protect the public from foodborne illness; inspects meat, poultry, and egg products for safety and proper packaging
Foreign Agricultural Service	FAS	Works to improve foreign market access for U.S. products; operates programs designed to build new markets and improve the competitiveness of U.S. agriculture
Forest Service	FS	Works to sustain the health, diversity, and productivity of the nation's forests and grasslands to meet the needs of present and future generations
Grain Inspection, Packers and Stockyards Administration	GIPSA	Facilitates the marketing of and ensures open and competitive markets for livestock, poultry, meat, cereals, oilseeds, and related agricultural products while promoting fair and competitive trading practices for the overall benefit of consumers and American agriculture
National Agricultural Library	NAL	Ensures and enhances access to agricultural information for a better quality of life
National Agricultural Statistics Service	NASS	Serves the basic agricultural and rural data needs of the country by providing objective, important, and accurate statistical information and services to farmers, ranchers, agribusinesses, and public officials
National Institute of Food and Agriculture	NIFA	In partnership with land-grant universities and other public and private organizations, provides the focus to advance a global system of extramural research, extension, and higher education in the food and agricultural sciences
Natural Resources Conservation Service	NRCS	Provides leadership in a partnership effort to help people conserve, maintain, and improve our natural resources and environment
Risk Management Agency	RMA	Helps to ensure that farmers have the financial tools necessary to manage their agricultural risks; provides coverage through the Federal Crop Insurance Corporation, which promotes national welfare by improving the economic stability of agriculture
Rural Development	RD	Helps rural areas to develop and grow by offering federal assistance that improves quality of life. RD targets communities in need and then provides financial and technical resources.

[a]Abbreviations are used so often with federal agencies and regulations that they are often jokingly referred to as "alphabet soup."

canning assistance, child nutrition programs, supplemental nutrition assistance programs, economic research on food safety through the Economic Research Service, and basic and applied aspects of food safety through the Agricultural Research Service.

The USDA's Food Safety and Inspection Service (FSIS) is responsible for the safety and labeling of traditional meats and poultry. The FDA regulates game meats, such as venison, ostrich, and snake. The FSIS has worked to protect the U.S. food supply under the direction of the Federal Meat Inspection Act (FMIA) for over 100 years. In 1862, President Abraham Lincoln signed legislation creating the USDA to stimulate food production by providing seed and agricultural information to farmers and help them receive a fair price for their crops. As the United States expanded towards the west after the Civil War, the development of the refrigerated rail car helped spur the livestock industry. Meat packing and international trade followed. In 1884, President Chester Arthur created the Bureau of Animal Industry (BAI), the forerunner to the FSIS. This agency was responsible for preventing diseased animals from being used as food. The BAI enforced the Meat Inspection Act, which was created in 1890 and amended 1 year later to cover inspection and certification of all live cattle for export, including those that were to be slaughtered and their meat exported. In 1905, with the publication of Upton Sinclair's *The Jungle*, the BAI confronted its first challenge. This book, which is still widely read today, was meant to raise political issues to a new level and in the end did so, but in a different way than the author had first planned. *The Jungle* exposed the unsanitary conditions in the Chicago meatpacking industry. This ignited public outrage. President Theodore Roosevelt commissioned a governmental inspection which confirmed the horrid tale described by Sinclair. The FMIA was passed by Congress in June 1906 in response. This act established the sanitary requirements for the meatpacking industry and is as important today as it was more than 100 years ago. These include mandatory inspection of livestock before slaughter, mandatory postmortem inspection of every carcass, and explicit sanitary standards for abattoirs. In addition to these, the act granted the agency the right to conduct inspections at slaughter and processing operations and enforce food safety regulatory requirements. The agency grew in size, hiring more than 2,000 inspectors to carry out inspection activities at 700 establishments by 1907.

Prior to 1920, consumers typically bought poultry direct from a farmhouse. This changed in 1920 when an outbreak of avian influenza occurred in New York City, the hub of poultry distribution in the United States. This outbreak led to poultry inspection programs.

Authors' note

New York City is still a large hub for live-bird markets. Since 2004 the USDA has been reworking current poultry biosecurity efforts to reduce the movement of potentially infected birds back and forth between other birds and farms.

Poultry consumption increased especially after World War II. The Poultry Products Inspection Act was passed in 1957 and required the inspection of poultry products before slaughter, after slaughter, before processing, and before interstate commerce. The Poultry Products Inspection Act mirrored the requirements of the FMIA. Facilities were sanitary, and processing operations and product labels were accurate and truthful after these federal regulations were in place.

The scope of the inspections was expanded with the passage of the Agricultural Marketing Act in 1946, which granted inspection of exotic and game animals on a fee-for-service basis. Grading and quality identification on products were added to this under the Agricultural Marketing Service in 1981. In 1958 Congress passed the Food Additive Amendment and the

Humane Methods of Slaughter Act in response to public concerns; however, this act was not amended until 1978. This act amended the FMIA of 1906. Today this remains an important area for public concern as well as ensuring quality meat and safe foods. In 1967 and 1968 Congress passed the Wholesome Meat Act and the Wholesome Poultry Products Act to ensure that state inspection programs would be equal in rigor to federal inspection programs. In 1970 the Egg Products Inspection Act mandated continuous inspection of liquid, dried, and frozen egg products. This was carried out by the Agricultural Marketing Service, which may sound surprising. In 1953 the BAI was abolished and parts of USDA were reorganized. Meat inspection was transferred to the Agricultural Research Service in 1965 and then to the Animal and Plant Health Inspection Service in 1972 after this agency was created in 1971. In 1977 the newly created Food Safety and Quality Service agency began performing meat and poultry grading and some inspection activities that were transferred from the Animal and Plant Health Inspection Service. In 1981 the last change was made with the creation of the FSIS, and this agency now is in charge of all inspections. The FSIS has worked diligently in defining new inspection programs. The framework for the current regulations is the Pathogen Reduction/Hazard Analysis and Critical Control Points system, discussed in detail in chapter 29. The FSIS continues to work with risk-based inspection systems that rely on science-based policies. The FSIS is flexible enough to meet challenges as they develop and will continue to modernize to meet the ever-changing threats and food safety challenges.

The FDA

The FDA (Fig. 7.2) was born during the development of the USDA. The FDA grew from a single chemist in his own agency, the Bureau of Chemistry, within the USDA in 1862 to a staff of more than 9,000 employees in 2011. The modern FDA began with the passing of the Federal Food and Drugs Act in 1906, and in 1927, the Bureau of Chemistry changed its name to the Food, Drug, and Insecticide Administration. The name was shorted to the FDA 3 years later. Throughout the 19th century, states had authority over food distribution and production, but there were problems with misbranding and adulteration. By the late 19th century, science had advanced and frauds of this type were more easily identifiable. In 1883 Harvey Wiley was chief chemist and took charge of determining adulteration and misbranding of food and drugs with his "poison squad," a self-proclaimed group of chemists who conducted experiments to determine the impact of questionable food additives on health (Fig. 7.3). Upton Sinclair's *The Jungle* was the driving force for the 1906 Food and Drugs Act signed into law by President Roosevelt. Adulteration of drugs and false label claims were serious problems in that era. In the 1930s numerous women were blinded by the eyelash dye Lash-Lure, and in 1937 more than 100 individuals died from a new sulfa wonder drug, Elixir Sulfanilamide, that was marketed for pediatric patients in particular.

The public outcry against these and similar issues led to passage of the Food, Drug, and Cosmetic Act on 25 June 1938. This law brought better control for cosmetics and medical devices as well as food additives. The first food standards to be issued under the new act were for canned tomato products. These standards were followed by many others, and by the 1960s nearly half of the U.S. food supply was subject to standards. As food

Figure 7.2 The FDA logo.

Figure 7.3 The Wiley Poison Squad at work. They would be unconventional in today's terms but were very effective in their time. doi:10.1128/9781555817206.ch07.f07.03

technology grew along with the numbers of food ingredients, the need for standard recipes became evident. If a food varied from its recipe, it would be labeled an imitation. Tolerance for food additives changed under the influence of Representative James Delaney, with new laws coming in place for pesticide residues in 1954, food additives in 1958, and color additives in 1960. The Delaney clause in 1958 banned any carcinogenic additive from the food supply. In 1968 the Animal Drug Amendments combined veterinary drugs and additives under one authority, the Bureau of Animal Drugs within the FDA.

The FDA overall has a myriad of responsibilities. The FDA is the federal agency responsible for ensuring that foods are safe, wholesome, and sanitary; human and veterinary drugs, biological products, and medical devices are safe and effective; cosmetics are safe; and electronic products that emit radiation are safe. The FDA also works to ensure that these products are honestly, accurately, and informatively represented to the consuming public. An interesting statistic is that the FDA oversees items that account for 25 cents of every dollar spent in the United States. For all of these responsibilities, the FDA is a relatively small agency. The FDA is under the directive of the Department of Health and Human Services and has several branches. For foods specifically, the FDA is responsible for regulating labeling and the safety of all food products (except meat and poultry) and bottled water. The FDA and Environmental Protection Agency share the regulation of water. The FDA also regulates livestock and pet feeds.

Regulatory reform in the 1990s included the FDA, with numerous Congressional investigations and reports but no changes until 2010. In December 2010, after much Congressional debate, President Barack Obama signed the Food Safety Modernization Act (FSMA) into law. While several other regulatory changes for the FDA affected oversight of drugs and cosmetics, the FSMA was a response to the 48 million Americans who suffer from foodborne illness each year. The FSMA is touted as the great law of the future that puts focus on prevention. The FSMA will give the FDA more effective enforcement tools. Under the new law the FDA will establish science-based standards for the safe production and harvesting of fruits and vegetables and for the safe transportation of foods. At this time the FDA relies on guidance documents for these standards, but under the FSMA these will be new laws. A new aspect of the FSMA is the risk-based inspection of food-processing facilities within 5 years of law enactment and no less than every 3 years thereafter. The FSMA enhances the FDA's ability to oversee and inspect food imports, which is an important point since an estimated 15% of the U.S. food supply is imported, including 50% of all fresh fruits, 20% of fresh vegetables, and 80% of seafood. Both consumer advocacy groups and food company trade associations were in favor of the FSMA. Specific points of interest to you, the consumer, from the FSMA are listed below.

Issuing recalls. The FDA will have the authority to order a recall of food products if necessary. Before the FSMA, the FDA had to rely on food manufacturers and distributors to recall food voluntarily (with the exception of infant formula).

Conducting inspections. The FSMA calls for more frequent inspections using science-based risk to determine which foods and facilities pose a greater risk to food safety.

Importing food. The FSMA enhances the FDA's ability to oversee food produced outside the United States. The FDA also has the authority to prevent food from entering the United States if the facility has refused U.S. inspection.

Preventing problems. Food facilities must have a written plan for how they will first avoid and then confront possible problems that affect the safety of their products.

Focusing on science and risk. The FSMA establishes science-based standards for the safe production and harvesting of fruits and vegetables. (Prior to the FSMA, only guidance documents existed for fresh produce.)

Respecting the role of small business and farms. As is often the case with U.S. federal laws, small farms that sell directly to consumers at roadside stands or farmers' markets or through community-supported agriculture programs have exemptions.

Over the past decade, the number of multistate outbreaks has increased; many of those outbreaks involved multi-ingredient products. Several companies acquire ingredients or have processing plants outside of the United States. In response to this, the FDA established an office in China and has become active outside of the United States in other regions, including Mexico, India, Europe, the Middle East, north Africa, and South and Central America. A chain of incidents increased the need for the FDA to have a greater presence in China. In 2007 many pet foods were recalled in the United States due to ingredients from China that were contaminated with the chemical melamine, which may be added to food in order to increase the protein content of the product and thereby increase the economic value of the product. After this, another incident occurred involving melamine in infant formula and various candies sold in the United States and abroad. These incidents reflect the globalization of the food supply and demand that regulatory organizations like the FDA act on a more global level. As stated above, the FSMA will allow the FDA greater authority regarding imported products. This is quite a daunting task and is sure to change with time.

The CDC

While the CDC is not a regulatory agency, it is an important federal agency in dealing with food safety. Around the world the CDC is synonymous with public health. The CDC was organized in Atlanta, Georgia, in 1946 as part of the Public Health Service, which like the FDA is under the Department of Health and Human Services. The CDC quickly grew into a mecca for distinguished scientists. In 1949 the medical epidemiology branch was started, and shortly thereafter the CDC's Epidemiological Intelligence Service was created. This is the agency that is involved in the investigations of foodborne illness outbreaks. The CDC established several famous public health programs, including the venereal disease program in 1957 and the tuberculosis program in 1960, and it was an integral part of the eradication of smallpox in 1977. The CDC has been successful in enhancing the detection and tracking of unknown diseases. The CDC works closely with the FDA, USDA, and state public health laboratories in conducting foodborne-illness surveillance.

SURVEILLANCE

Authors' note

Foodborne illness can result from infection or intoxication. An infection results from ingesting living pathogenic bacteria. An intoxication results from ingesting a preformed toxin.

The CDC has maintained surveillance programs monitoring foodborne illness within the United States since 1973. A foodborne disease outbreak is defined as two or more cases of a similar illness resulting from the ingestion of a common food. There are two exceptions to this definition: one case of intoxication from a chemical or marine toxin and one case of botulism caused by *Clostridium botulinum*.

Surveillance information for specific organisms is given throughout this book. Surveillance information is essential to determine when an outbreak occurs, as that is when the surveillance detection is over the baseline data. For example, if routine surveillance indicates that within one week in September at a university 50 people visit health services for diarrhea, then this is the baseline. The following year during a week in September 300 people visit health services for diarrhea. This increase in surveillance above the baseline data may indicate an outbreak of gastroenteritis related to foodborne illness. It is important that epidemiologists and public health officials be aware of *real* increases in surveillance and what might be an *artificial* increase. Artificial increases in surveillance could be a result of increased testing, an increase in the student population, or a new testing strategy by health services.

The main surveillance sites within the United States are connected through a network known as FoodNet, short for Foodborne Diseases Active Surveillance Network. FoodNet conducts surveillance for *Campylobacter*, *Cryptosporidium*, *Cyclospora*, *Listeria*, *Salmonella*, Shiga toxin-producing *Escherichia coli* O157 and non-O157, *Shigella*, *Vibrio*, and *Yersinia* infections diagnosed by laboratory testing of samples from patients. The network was established in 1995 as a collaborative program among the CDC, the FDA, the USDA-FSIS, and state health departments from 10 states (California, Colorado, Connecticut, Georgia, Maryland, Minnesota, New Mexico, New York, Oregon, and Tennessee) (Fig. 7.4). FoodNet accomplishes its work through active surveillance (FoodNet investigators collect information on lab-confirmed cases of diarrheal illness); surveys of laboratories, physicians, and the general population; and population-based epidemiological studies. Information gathered from FoodNet is used to assess the

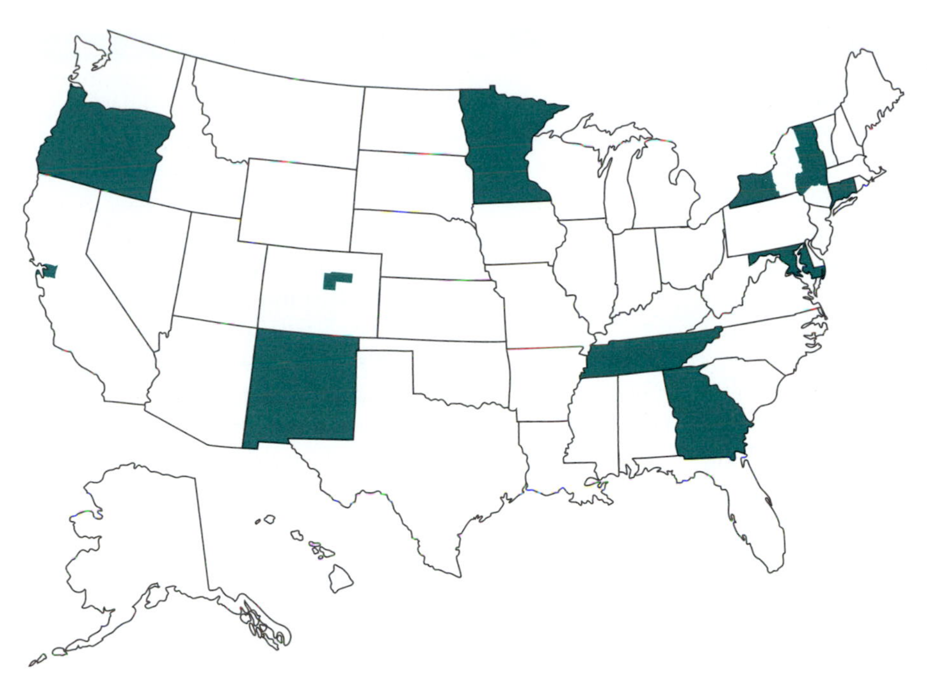

Figure 7.4 The FoodNet sites are highlighted in this map. These include state health departments as listed in the text. doi:10.1128/9781555817206.ch07.f07.04

impact of food safety guidelines, initiatives, and regulatory changes on the burden of foodborne illness. The four main objectives of FoodNet are as follows:

1. determine the burden of foodborne illness in the United States
2. monitor trends in the burden of specific foodborne illnesses over time
3. attribute the burden of foodborne illness to specific food and settings
4. develop and assess interventions to reduce the burden of foodborne illness

Both active and passive surveillance has helped identify changes in outbreak scenarios within the United States. Passive surveillance occurs when health agencies are contacted by physicians or laboratories, which report illnesses or laboratory results to them. In active surveillance, the health agencies regularly contact physicians and laboratories to make sure that reportable diseases have been reported and required clinical specimens or isolates have been forwarded to state laboratories for further analysis.

The classic "church supper" or Sunday picnic outbreaks still occur once in a while, but they have been replaced more often by large multistate outbreaks associated with multi-ingredient food products. There are two general types of foodborne-outbreak scenarios. The first is the traditional outbreak scenario that follows a church supper, family picnic, wedding reception, or other event affecting a discrete population. This is an acute and highly localized outbreak, often with a high inoculum dose (Box 7.1) (that is, the implicated product tends to be highly contaminated) and with a high attack rate (that is, most people who ate the implicated product become ill). More recently, we have faced the "newer" outbreak scenario

BOX 7.1

Infective dose information

Most chapters include a statement on infectious dose. These numbers should be viewed cautiously because (i) they are extrapolated from outbreaks, (ii) they may have been done by human feeding studies on healthy young adult volunteers (such studies are rarely done these days), and (iii) they are worst-case estimates. Because of the following variables, they cannot be directly used to assess risk.

Variables of the parasite or microorganism:

- Variability of gene expression
- Injured, stressed, or viable but nonculturable microorganisms
- Interaction of organism with food
- Immunologic "uniqueness" of the organism
- Interactions with other organisms

Variables of the host:

- Age
- General health
- Pregnancy
- Medications (over the counter or prescription)
- Metabolic disorders
- Alcoholism, cirrhosis, and hemochromatosis
- Malignancy
- Amount of food consumed (dose)
- Gastric acidity variation: antacids, natural variation
- Nutritional status
- Immunocompetence

Because of the complexity of factors involved in making risk decisions, the multidisciplinary Health Hazard Evaluation Board judges each situation on all available facts.

This information is adapted from the FDA's Health Hazard Evaluation Board.

that has been reported with increasing frequency. This occurs as a result of low-level or intermittent contamination of a widely distributed commercial food product. This type of outbreak tends to be much more difficult to detect and investigate.

Foodborne outbreaks may be detected at the local, state, or national level. Most are initially picked up at the local level—a person who became ill after eating at a party or restaurant may call their local health department, for example. In addition, the local clinical laboratories and, at least in theory, the health care workers fill out reports on each reportable illness and submit them to the local health department. The local health department may detect an increase in a particular enteric organism, or an unusual clustering of cases, in that manner. The state health department collects foodborne-outbreak reports. In addition, the local health departments are usually very conscientious about reporting larger or unusual outbreaks to the state. For example, the state public health labs have the capability to serotype *Salmonella* isolates and can thereby tell if there is an increase in the number of a particular *Salmonella* serotype and alert epidemiologists. Using a technique called pulsed-field gel electrophoresis (PFGE), a method of fingerprinting of foodborne disease-causing microorganisms (Fig. 7.5), strains can be distinguished at the DNA level. DNA "fingerprints," or patterns, are submitted electronically to a dynamic database at the CDC. These databases are available on demand to participants, which allows for rapid comparison of the patterns of potential outbreak strains and information shared through a database with laboratories across the United States and even around the world via PulseNet International to encourage rapid detection of outbreaks. Finally, there is PulseNet, which is the CDC-coordinated molecular subtyping network for surveillance of foodborne bacterial diseases. PulseNet is a national network of public health and regulatory agency laboratories that are instrumental in communication of the PFGE types and sharing these resources with other labs.

Figure 7.5 This pulsed-field gel image compares clinical, food, and environmental isolates. Can you find matching samples? Samples should be considered indistinguishable from each other to be considered to be part of the outbreak. If there are three or fewer bands different, then the samples may be related; if more than three bands are different, then the samples are not likely to be related. doi:10.1128/9781555817206.ch07.f07.05

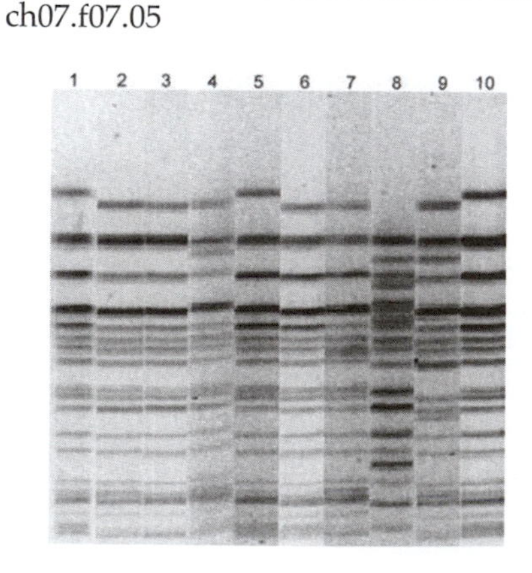

WHEN AN OUTBREAK OCCURS

An outbreak investigation is quite complex and occurs in various stages. In describing the investigation, there are three main aspects. The first is the epidemiological investigation. The second is the laboratory investigation. The third is the environmental investigation, which includes traceback to where the product was grown, harvested, produced, or distributed. It is important to understand that the outbreak investigation may contain these three parts, but the investigation process is often not linear in fashion and may jump back and forth among these three aspects. Unlike a fictional investigation that occurs in a 1-hour television show, a real-life investigation may take months.

An outbreak may be first detected by a noticeable increase in laboratory testing or reporting of an illness or by other means of surveillance. Increasingly, outbreaks are being detected by syndromic surveillance. One example of this is when many individuals search Google for information related to specific symptoms. Another example is the widespread purchasing of antidiarrheal medication. When these actions occur more than usual, they may be signs of an outbreak. The state epidemiologist responds to a noticeable increase in laboratory reporting or one of the items mentioned above. Epidemiology is the study of the incidence, distribution, and control of disease in a population. The first part of an outbreak investigation is the epidemiological investigation. As noted previously, surveillance of foodborne illness has limitations due to the underreporting of illness in the general population. For this reason, estimations are important. It is generally believed that for every one case of illness due to *Salmonella* that is diagnosed, 38 cases are undiagnosed. It is important that during an outbreak investigators think "outside the box," since not all cases of illness are directly related to consumption of contaminated food or water. Some primary or secondary cases may be due to person-to-person transmission or contact with animals. The state epidemiologist and his or her team will actively search for cases. They will collect information on demographics, clinical symptoms, and identifying risk factors. They will organize all of this information in a line listing that can quickly be reviewed and updated as the investigation progresses (Table 7.2).

Table 7.2 Example of a line listing for acute hepatitis A virus[a]

				Signs and symptoms							Demographics	
Case no.	Report date (mo/day/yr)	Onset date (mo/day/yr)	Physician diagnosis	Nausea	Vomiting	Elevated aminotransferase	Fever	Discrete onset	Jaundice	Lab test result (HAV IgM)	Sex	Age (yr)
1	10/12/11	10/05/11	Hepatitis A	Y	Y	Y	Y	Y	Y	Y	M	21
2	10/12/11	10/04/11	Hepatitis A	Y	N	Y	Y	Y	Y	Y	M	63
3	10/13/11	10/04/11	Hepatitis A	Y	N	Y	Y	Y	Y	Y	F	41
4	10/13/11	10/09/11	ND	N	N	Y	N	?	N	ND	F	18
5	10/15/11	Unsure	Hepatitis A	Y	Y	Y	Y	Y	N	Y	F	37
6	10/16/11	10/06/11	Hepatitis A	N	N	Y	Y	Y	Y	Y	M	29

[a]A line listing includes components of the case definition: case name or identifying number, date of symptom onset (or specimen collection date), and demographic information like age, gender, race, occupation, and risk factors. Abbreviations: Y, yes; N, no; ND, not determined; F, female; M, male; HAV IgM, hepatitis A immunoglobulin M antibody test.

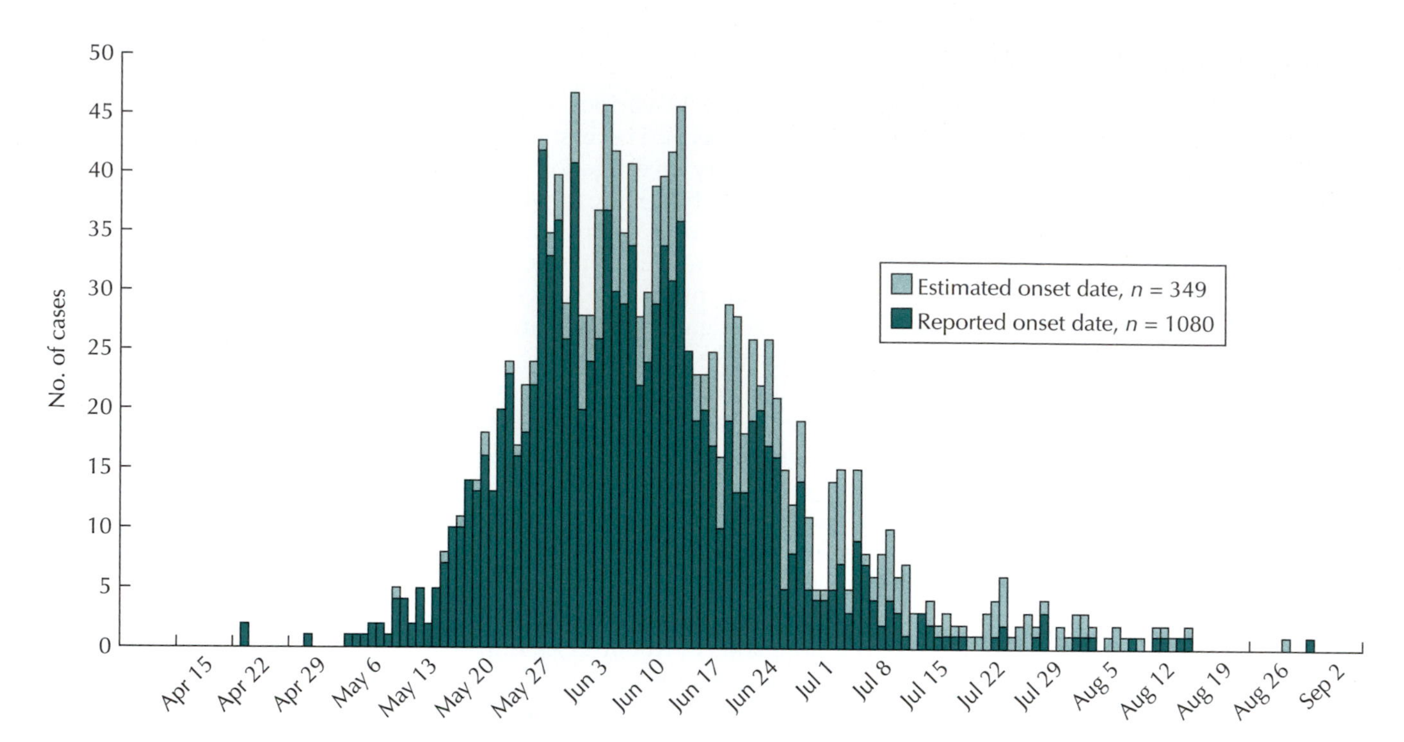

Figure 7.6 An epidemiological curve can provide information on the following characteristics of an outbreak: pattern of spread, magnitude of the outbreak, case outliers, and the time trend. doi:10.1128/9781555817206.ch07.f07.06

The line listing will provide information on the frequency distributions of demographics which may provide information about exposure and/or risk of disease. It will also provide frequency distributions of potential exposures that may yield information about source or route of transmission. A case definition will be developed from this information and used to include or exclude cases as they arise. The case definition may be revised during the outbreak investigation. Most importantly, the line listing will be used to generate an epidemic curve, which is a graphical depiction of the number of cases of illness by the date of illness onset (Fig. 7.6).

The laboratory investigation is ongoing through all parts of the outbreak. This is when microbiologists are working to identify the cause of the illness. Laboratory samples may include clinical samples like diarrhea and vomitus, environmental swabs, and food samples. It is important to identify the causative agent in a clinical sample and match it using PFGE to a sample found in an unopened food product. The unopened food product will ensure that the cause of the outbreak was not intentional or due to cross-contamination.

The environmental investigation is usually initiated by the local board of health. The objectives of this part of the outbreak investigation are twofold: (i) identify the reason for or sources of contamination, and (ii) initiate corrective actions, if necessary, to eliminate contaminated foods or poor food-handling practices which may result in contaminated foods. You may have already guessed that the most important part of any outbreak investigation is the steps involved in control and prevention. These are not easy tasks. A realistic goal may be to document opportunities for contamination and for the growth and survival of pathogens for the purpose of developing interventions and steps to prevent recurrences. Ideally an environmental investigation is initiated within 24 to 48 h of the receipt of a foodborne-illness complaint, but this may not happen. Environmental investigators use lab results and the epidemiological findings as their guides. They investigate the places where foods are produced, distributed, or cooked; the people involved in all processes; and the equipment used to perform these steps. Often when an environmental investigation begins the fields have been plowed under or the food products are no longer available. This is due in part to the incubation periods of microbial pathogens before they cause disease (as you will read about in the following chapters) and compounded by the wide distribution of our global food supply.

Figure 7.7 Food security is a large umbrella covering food safety. doi:10.1128/9781555817206.ch07.f07.07

AGROTERRORISM

Food security includes food safety and ensuring food growth and development for the global population (Fig. 7.7). Concerns about food security have increased over the past decade. In preparing for unintentional contamination, we can also prepare for intentional contamination. The events of 11 September 2001 reinforced the need to enhance the security of the United States. Congress responded by passing the Public Health Security and Bioterrorism Preparedness and Response Act of 2002 (the Bioterrorism Act), which President George W. Bush signed into law 12 June 2002. This regulation required that all food businesses that supply the United States be registered with the FDA. The act focuses on biosecurity to reduce the risk of bioterrorism in the food supply. The act also enhances controls on dangerous biological agents and toxins. There is no substitute for a strong public health infrastructure, which must be able to prevent bioterrorist attacks (Box 7.2), detect an attack as early as possible, respond rapidly, and have strong research and training programs. There are three layers of public health infrastructure within the United States that all work together in promoting food safety and security.

1. The public health service: federal, state, and local in nature, including the FDA, USDA, and CDC

BOX 7.2

Historical bioterrorist acts

In the United States, several incidents of intentional contamination have occurred. These have many similarities to incidents caused by accidental contamination. Thorough investigations identified them as intentional acts. Through monitoring and basic surveillance for accidental outbreaks, intentional contamination can also be identified.

- In 1970 four university students became seriously ill after eating a meal contaminated with the pig roundworm identified as *Ascaris suis*. The meal was contaminated by a colleague.
- In 1984 an outbreak of *Salmonella enterica* serovar Typhimurium was traced to a salad bar in Oregon. There were 751 cases of illness. The group that contaminated the salad bar did so to try to sway an election. This plan did not work.
- In 1996, in a Texas medical center, a worker baked muffins laced with *Shigella dysenteriae* and served them to a number of coworkers. Everyone recovered.

2. A network of health care providers: physicians, dentists, nurses, pharmacists, and others
3. The general public: an informed public is crucial to success of the public health infrastructure

The challenges that these groups face continue to increase. These include a global food supply, increasing numbers of biological and chemical agents that can affect foods in different ways, and an inadequate number of epidemiologists to monitor and respond to food outbreaks. The FSMA, mentioned above, may help address these areas. The Bioterrorism Act also provides the FDA with counterterrorism programs through a four-pronged approach: (i) threat assessment, (ii) surveillance, (iii) deterrence and prevention, and (iv) containment through rapid response. Various agents may be considered bioterrorism agents. Many of them are discussed in this book. In order to combat these threats, it is essential that industry appreciate these risks and the biological and chemical agents. Biosecurity is important in all food production environments. Communication is equally important in addressing strategies to reduce capability of attack and address vulnerabilities. Combating potential agroterrorism threats is certainly a collaborative effort among industries, federal and state authorities, and consumers.

WHAT'S NEXT?

No one knows for sure what the future holds. We can say with some certainty that there will likely continue to be new regulations and modifications to our current ones. There will likely be more commodity-specific guidance documents. Perhaps over the next few years the U.S. regulatory agencies will attempt to tackle some of the issues that cross food safety and social boundaries, including local food production and distribution, as the world population continues to grow and natural resources like water grow in short supply.

GLOBAL PERSPECTIVE

The global commitment to food security is documented in several programs. Food-processing companies and global regulatory authorities are working towards a global regulation system that ensures the safety and quality of food regardless of the country of origin. The *Codex Alimentarius*, which is Latin for "food code," is a collection of internationally recognized standards targeting the whole food chain and includes codes of practice, guidelines, and other recommendations relating to foods, food production, and food safety. The *Codex Alimentarius* is provided by the Food and Agriculture Organization of the United Nations (FAO) and the World Health Organization (WHO).

The Global Harmonization Initiative (GHI) was launched in 2005 by the International Division of the Institute of Food Technologists and the European Federation of Food Science and Technology. The GHI is a network of scientific organizations and individual scientists working together to promote harmonization of global food safety regulations and legislation. The three objectives of the GHI are to (i) provide the foundation for sound, sensible, science-based regulations; (ii) create a forum for scientists and technologists to interact globally with regulatory authorities; and

(iii) provide industry, regulators, and consumers an independent authoritative information resource. Only the future will tell how effective these programs are at enhancing global food security.

"IT TAKES A VILLAGE" AND MAYBE MORE

Reducing foodborne illness takes a great deal of work and collaboration. It is hoped that you see your role in this cause. Everyone plays an important role in reducing the incidence of illness. Be sure to follow guidelines for food preparation and always demonstrate good personal hygiene. Specifically, your role includes following these guidelines (according to http://www.foodsafety.gov).

- **Clean.** Wash hands, cutting boards, utensils, and countertops.
- **Separate.** Keep raw meat, poultry, and seafood separate from ready-to-eat foods.
- **Cook.** Use a food thermometer to ensure that foods are cooked to a safe internal temperature: 145°F for whole meats (allowing the meat to rest for 3 min before carving or consuming), 160°F for ground meats, and 165°F for all poultry.
- **Chill.** Keep your refrigerator below 40°F, and refrigerate food that can spoil.
- **Report** suspected illness from food to your local health department.
- **Do not prepare food for others** if you have diarrhea or vomiting.
- **Be especially careful** when preparing food for children, pregnant women, those in poor health, and older adults.

As you have learned, government and industry will work with you in this cause. The government can implement policies and regulations to ensure that food production and food service facilities have the tools and best practices required for eggs, meat, poultry, fruits and vegetables, and processed and imported foods. Investigations of outbreaks have helped to identify sources and improve control strategies through tracking trends, reporting progress, and making sure policies are reflective of these. Researchers in academia and private industry work together to provide the best information for regulatory authorities to reduce contamination when raising livestock, food animals, and all food crops. Health care providers diagnose and treat infections using best practices to help identify the start of an outbreak. It is important that we make this our crusade to strengthen food safety, reduce illness, and save lives.

Summary

- Each year, 48 million Americans suffer from foodborne illness due to accidental contamination. Outbreak investigations can help identify accidental and intentional contamination. Mechanisms in place to protect the food supply from accidental contamination can help protect against intentional contamination as well.
- The lead regulatory agencies are the FDA and the USDA. These two agencies are quite complex and fulfill many responsibilities.
- Upton Sinclair's *The Jungle* spurred the predecessors to our current regulations, including the FMIA (1906) and the FDA's Federal Food, Drug, and Cosmetic Act (1938).

- The FSMA is the newest regulation and is an amendment to the Federal Food, Drug, and Cosmetic Act. The FSMA increases the FDA's ability to issue recalls, conduct inspections on imported products, and add science-based policies for fresh fruits and vegetables.
- The CDC is a federal nonregulatory agency that actively works in conducting surveillance of foodborne illness. Surveillance is essential to detect when an outbreak occurs, which is when there is an increase in illness over baseline information. FoodNet is an active-surveillance system run by the CDC along with other regulatory agencies and public health laboratories. PFGE is used by the CDC in PulseNet to compare microbes found in foods and those found in related clinical samples.
- Food safety "takes a village." The *Codex Alimentarius* and the GHI are examples of how food safety regulations are being developed on a global scale. You also play an important role in ensuring food safety.

Suggested reading

Centers for Disease Control and Prevention. 2011. Making food safer to eat. *Vital Signs.* http://www.cdc.gov/vitalsigns.

Scallan, E., R. M. Hoekstra, F. J. Angulo, R. V. Tauxe, M.-A. Widdowson, S. L. Roy, J. L. Jones, and P. M. Griffin. 2011. Foodborne illness acquired in the United States—major pathogens. *Emerg. Infect. Dis.* **17:**7–15.

Scallan, E., P. M. Griffin, F. J. Angulo, R. V. Tauxe, and R. M. Hoekstra. 2011. Foodborne illness acquired in the United States—unspecified agents. *Emerg. Infect. Dis.* **17:**16–22.

U.S. Food and Drug Administration. 2010. H.R. 2751. Food Safety Modernization Act. An amendment to the Federal Food, Drug, and Cosmetic Act (21 U.S.C. 301 et seq.).

Questions for critical thought

1. The origin and development of the USDA and FDA were quite monumental, especially for this time in history. Explain why each agency was developed.
2. Why did Upton Sinclair's *The Jungle* play such a pivotal role in the development of new regulations in the USDA? In your opinion, is there a piece of contemporary literature that plays a similar role?
3. There has been criticism that the United States is limited by not having one solid public health and regulatory agency controlling food safety. Read about these criticisms in current literature and explain your opinion in agreement for today's system or not.
4. What is the current state of the FSMA? How is it affecting food production companies, including those that produce fruits and vegetables?
5. Describe the roles of active and passive surveillance in the United States. Who is involved in these activities? Why are both forms of surveillance important?
6. Using the four objectives of FoodNet as your guide, describe real-life examples that demonstrate how these objectives are being met and how the four play off of each other. *Hint:* Think about current outbreaks; what you read in the following chapters may help you.
7. You work for a company that produces chocolate milk. How can you help protect your company and your products from potential agroterrorism?

8. Describe the roles of programs like the *Codex Alimentarius* and the GHI in food security.
9. Have you ever thought about your own behaviors when cooking dinner at home, or when eating out at a restaurant? What behaviors can you change? Can you identify ways in which you can help your friends and family with their behaviors?
10. What role do you play in enhancing food security around the globe?

8 Bacillus cereus

LEARNING OBJECTIVES

The information in this chapter will enable the student to:

- use basic biochemical characteristics to identify *Bacillus cereus*
- begin to appreciate the fluidity of bacterial nomenclature and etiology
- understand what conditions in foods favor the growth of *B. cereus*
- recognize, from symptoms and time of onset, a case of foodborne illness caused by *B. cereus*
- choose appropriate interventions (heat, preservatives, and formulation) to prevent *B. cereus* from causing foodborne illness
- identify environmental sources of *B. cereus*
- understand the role of spores and toxins in causing foodborne illness

Outbreak

Even though fried rice is the classic example used to illustrate *B. cereus* food poisoning, it still causes problems. The *Morbidity and Mortality Weekly Report* (the Centers for Disease Control and Prevention's weekly listing of how many people are getting sick or dying of various diseases) summarized a *B. cereus* food poisoning outbreak at two day care centers in Virginia. Sixty-seven children ate a "special" catered international lunch. Fourteen of the people who ate the lunch became ill with nausea, abdominal cramps, and diarrhea. The median time of onset for the illness was 2 h after eating. The median time for recovery was 4 h after onset. None of the 13 children who did not eat the lunch got sick.

Chicken fried rice was associated with the outbreak. One-third of the people who ate the rice became ill. None of the people who did not eat the rice became ill. *B. cereus* at $>10^5$ organisms per g was isolated from one child's vomit. Levels in the leftover chicken were $>10^6$ organisms per g.

The investigation uncovered a classic pattern of food mishandling. The rice was cooked the night before the lunch, cooled at room temperature, and refrigerated overnight. The next morning, the rice and chicken were fried with oil, delivered to the day care center, held without refrigeration, and served without heating. Neither the restaurant staff nor the day care center staff knew that these were dangerous food-handling practices.

INTRODUCTION

With only ~27,000 cases per year in the United States, *B. cereus* is considered a minor foodborne pathogen, but one that is increasingly important. The seemingly low incidence may reflect that *B. cereus* foodborne disease is not notifiable.

doi:10.1128/9781555817206.ch08

While *B. cereus* causes <2% of foodborne cases of identifiable origin in the United States, the prevalence is as high as 30% in Europe. *B. cereus* produces two toxins. An emetic (vomiting) toxin is produced by cells growing in the food and acts rapidly (0.5 to 6 h). This is due to the release of an extracellular toxin that is made by amino acid synthetases (i.e., by enzymatic linkage of amino acids rather than by protein synthesis). In contrast, the diarrheal illness is caused by a protein enterotoxin(s) produced during vegetative growth of *B. cereus* in the small intestine. The diarrheal toxin that causes diarrhea, cramps, and rectal tenesmus acts longer after ingestion (6 to 14 h). Both types of illness are mild, brief, and self-limiting. This contributes to their being underreported. *B. cereus* does not compete well with other vegetative cells. But when the other cells are killed by cooking, *B. cereus* spores survive, grow, and cause illness. Furthermore, if that food is heated again prior to serving, those cells are killed and the emetic toxin remains active. Outbreaks are commonly associated with meats, gravies, fried rice, pasta, sauces, puddings, and dairy products. Fish is also being recognized as an implicated food. A recent study isolated *B. cereus* from 17% of fish sampled, with half of these producing both the diarrheal and emetic toxins.

Authors' note

Notifiable diseases are those diseases that must be reported by local health officials to the Centers for Disease Control and Prevention. These include botulism, salmonellosis, shigellosis, listeriosis, and Shiga toxin-producing Escherichia coli *disease.*

CHARACTERISTICS OF THE ORGANISM

Genus and species designations were originally based on biochemical tests and what the organism looked like (on agar plates and under the microscope). Thus, *any* spore-forming, aerobic rod-shaped bacterium was classified as a *Bacillus* species. Molecular biology and genetic characterization have now revealed that many of these species identifications by phenotypic characteristics are inaccurate. The genus *Bacillus* as described in *Bergey's Manual of Systematic Bacteriology* (the dictionary of microbial classification) is too diverse for a single genus. According to 16S ribosomal RNA (rRNA) sequences, there are at least five genera or rRNA groups within the genus *Bacillus*. With the subsequent identification of many new species, the number of proposed genera has increased to approximately 16.

Bacillus anthracis, *B. cereus*, *B. mycoides*, *B. thuringiensis*, *B. pseudomycoides*, and *B. weihenstephanensis* belong to the *B. cereus* "group" (an informal but useful designation). These species are differentiated by the toxins they make. *Bacillus anthracis* makes a plasmid-mediated (i.e., the gene is carried on a plasmid rather than the chromosome) toxin that causes pneumonia. *B. cereus* makes three plasmid-mediated toxins that cause foodborne illness. *B. thuringiensis* makes a plasmid-mediated insecticide. That species determination is based on plasmid-mediated factors muddies the water. Plasmids are unstable and can be lost or transferred. A *B. cereus* strain that loses its plasmid has no defining characteristic. Or worse, a *B. cereus* isolate that gains a *B. anthracis* plasmid becomes more dangerous and difficult to treat in a clinical setting. These bacteria have very similar 16S and 23S rRNA sequences, indicating that they have diverged from a common evolutionary line. Extensive studies of DNA from *B. cereus* and *B. thuringiensis* suggest that there is no scientific reason for them to be in separate species. Indeed, *B. thuringiensis* strains used as a natural bioinsecticide have caused cases of "*B. cereus*" food poisoning, and some "*B. cereus*" strains produce anthrax toxins. There have been proposals to group *Bacillus anthracis*, *B. cereus*, and *B. thuringiensis* into a single species. However, psychrotrophic *B. cereus* (which grows at 7°C but not at 43°C) has been given a species name of its own, *B. weihenstephanensis*.

Most *B. cereus* strains are unable to grow below 10°C or in milk stored between 4 and 8°C. Some strains are motile via peritrichous flagella, while others are not motile at all. *B. cereus* strains can produce the emetic and/or diarrheal toxins. Generally, strains that produce the emetic toxin are associated with starchy foods like rice. Strains that produce the diarrheal toxin are associated with proteinaceous foods like fish.

ENVIRONMENTAL SOURCES

Authors' note

Microbiologists study bacteria in the lab as populations of bacteria of a single type and species. However, in nature bacteria usually exist as multispecies populations in biofilms or other organized structures.

B. cereus is widespread in nature, associated with decaying matter and frequently isolated from soil and growing plants. It has a life of its own in soil, where the vegetative cells sporulate. With the onset of favorable conditions, the spores become vegetative cells. In the soil, these planktonic (free-living single cells) bacteria converge to filamentous multicellular forms that swarm in a coordinated direction.

B. cereus is easily spread from the environment to foods. *B. cereus* is especially problematic in dairy foods. The spores can be present in soil at 10^5/g and spread to the cows' udders and then into raw milk. Because *B. cereus* spores are especially hydrophobic, they attach to pipelines and other solid surfaces in the dairy, making the spores difficult to remove. The spores survive milk pasteurization. After germination, free from competition from other vegetative cells, they grow, causing problems in many milk products. Psychrotrophic *B. cereus* strains make the situation worse because they can grow while the milk is refrigerated.

FOODBORNE OUTBREAKS

B. cereus outbreaks of foodborne illness are highly underestimated. This can be due to the relatively short duration of illness (usually <24 h), the similarity of symptoms to those caused by other foodborne pathogens, and the likelihood that illness involving *B. cereus*-contaminated milk is limited to one or two cases (due to a few incidences of temperature abuse) which are not identified as an outbreak. The dose required to cause illness is 10^3 to 10^4 organisms. *B. cereus* spores can survive pasteurization. Near the end of milk's shelf life, there frequently are enough *B. cereus* cells to cause illness. Fortunately, *B. cereus* proteases cause off-flavors. One would hope that this spoilage would keep consumers from drinking highly contaminated milk.

Authors' note

The argument that "the food would spoil before it could make someone sick" is invalidated by actual experience. Investigators are often led to the cause of the outbreak by victims who tell them to test a particular food because "it tasted really bad."

It is not possible to compare the incidences of outbreaks in different countries because their methods of tracking foodborne illnesses differ greatly. The percentages of outbreaks and cases attributed to *B. cereus* in Japan, North America, and Europe vary from 1 to 47% for outbreaks and from ~0.7 to 33% for cases. The largest numbers of reported *B. cereus* outbreaks and cases are from Iceland, The Netherlands, and Norway. In The Netherlands, *B. cereus* caused 27% of outbreaks in which a causative agent was identified. However, the actual incidence of *B. cereus* was only 2.8% of the total number of cases because most cases of foodborne illness were of unknown etiology.

CHARACTERISTICS OF DISEASE

There are two types of *B. cereus* foodborne illness. The first type, caused by an emetic toxin, results in vomiting. The second type, caused by an enterotoxin(s), results in diarrhea. A few *B. cereus* strains make both toxins and cause both symptoms. The incubation time (>6 h; average, 12 h) for

Table 8.1 Characteristics of the two types of illness caused by *B. cereus*[a]

Characteristic	Diarrheal syndrome	Emetic syndrome
Dose causing illness	10^5–10^7 cells (total)	10^5–10^8 cells per g
Toxin produced	In the small intestine of the host	Preformed in foods
Type of toxin	Protein; enterotoxin(s)	Cyclic peptide; emetic toxin
Incubation period	8–16 h (occasionally >24 h)	0.5–5 h
Duration of illness	12–24 h (occasionally several days)	6–24 h
Symptoms	Abdominal pain, watery diarrhea, and occasionally nausea	Nausea, vomiting, and malaise (sometimes followed by diarrhea, due to production of enterotoxin)
Foods most frequently implicated	Meat products, soups, vegetables, puddings, sauces, milk, milk products	Fried and cooked rice, pasta, pastry, and noodles

[a]Reprinted from P. E. Granum, p. 373–381, *in* M. P. Doyle, L. R. Beuchat, and T. J. Montville (ed.), *Food Microbiology: Fundamentals and Frontiers*, 2nd ed. (ASM Press, Washington, DC, 2001).

diarrheal illness is too long for the illness to be caused by a toxin premade in the food. (Preformed toxins act rapidly when ingested.) However, the emetic toxin can be formed in food when the *B. cereus* population is at least 100-fold higher than that necessary for causing the diarrheal illness. Products with such large populations of *B. cereus* are no longer acceptable to the consumer, although food containing $>10^7$ *B. cereus* organisms/ml may not always appear spoiled. The two types of *B. cereus* foodborne illness are characterized in Table 8.1.

DOSE

After the first diarrheal outbreak of *B. cereus* foodborne illness in Oslo, Norway (from vanilla sauce), S. Hauge isolated the causative agent, grew it to 4×10^6 cells/ml, and drank the culture. (Although the use of oneself as a human subject was accepted and even considered courageous in the early days of microbiology, this practice would be unethical by current standards. Do not try this at home.) About 13 h later, he developed abdominal pain and watery diarrhea that lasted for approximately 8 h. Counts of *B. cereus* ranging from 200 to 10^9 organisms per g (or per milliliter) have been reported for foods incriminated in outbreaks, indicating that the total dose (cells/milliliter times milliliters consumed) ranges from approximately 5×10^4 to 1×10^{11} cells. The total number of *B. cereus* cells required to produce enough toxin to produce illness is probably 10^5 to 10^8 cells. Food containing $>10^3$ *B. cereus* organisms/g is not safe to eat.

VIRULENCE FACTORS AND MECHANISMS OF PATHOGENICITY

Very different types of toxins cause the two types of *B. cereus* foodborne illness. The emetic toxin, causing vomiting, has been isolated and characterized, whereas the diarrheal disease is caused by one or more enterotoxins.

Table 8.2 Properties of the emetic toxin cereulide[a]

Trait	Property, activity, or value
Molecular mass	1.2 kilodaltons
Structure	Ring-shaped peptide
Isoelectric point	Uncharged
Antigenic	No
Biological activity in living primates	Vomiting
Cytotoxic	No
Heat stability	90 min at 121°C
pH stability	Stable at pH 2–11
Effect of proteolysis (trypsin, pepsin)	None
Conditions under which toxin is produced	In food (rice and milk) at 25–32°C

[a]Adapted from P. E. Granum, p. 373–381, *in* M. P. Doyle, L. R. Beuchat, and T. J. Montville (ed.), *Food Microbiology: Fundamentals and Frontiers*, 2nd ed. (ASM Press, Washington, DC, 2001).

The Emetic Toxin

The emetic toxin cereulide (Table 8.2) causes emesis (vomiting). Cereulide is not a conventional protein. Cereulide has a ring structure of three repeats of four amino acids and/or oxyacids: $[\text{D-}O\text{-Leu–D-Ala–L-}O\text{-Val–L-Val}]_3$. The structure suggests that cereulide is an enzymatically synthesized peptide and not a gene product. The emetic toxin's biosynthetic pathway is unknown, but the required genes are plasmid mediated. The mechanism of action for cereulide is also unknown. Emetic toxin is resistant to autoclaving, a wide range of pH values (2 to 11), and digestion by proteases. It is not antigenic, so immunology-based methods cannot be used for its detection.

Enterotoxins

Cloning and sequencing studies show that *B. cereus* produces at least five different proteins (or protein complexes) known as enterotoxins. The extent to which the different proteins are involved in foodborne illness might be determined through microslide diffusion tests (Fig. 8.1).

Three enterotoxins are probably involved in *B. cereus* foodborne illness. Two of the enterotoxins are multicomponent and related. A third enterotoxin is a single 34-kDa protein. A three-component hemolysin (Hbl) has enterotoxin, dermonecrotic, and vascular permeability activities. Hbl1 causes fluid accumulation in ligated rabbit ileal loops. It may be the virulence factor in *B. cereus* diarrhea. All three proteins of the Hbl are transcribed from one operon (*hbl*). The second important agent is a nonhemolytic enterotoxin, Nhe. It is also a complex of three proteins. These toxins and cereulide are tightly regulated at the genetic level through an autoinduction mechanism similar to that of nisin (see chapter 26).

Authors' note

In the rabbit ileal loop assay, a rabbit is anesthetized, its surgically exposed intestine is tied off (i.e., ligated) into loops, and the sample is injected. If the sample causes fluid accumulation in the ligated loop, it is positive for toxin.

B. cereus as a Medical Pathogen

As food microbiologists, we view microbes from the perspective of their life in food. However, many "foodborne" pathogens cause a variety of other illnesses. *Bacillus cereus* produces four hemolysins, three phospholipases, three pore-forming enterotoxins, the emetic toxin, hemolysin B, nonhemolytic enterotoxin, cytotoxin B, and, as mentioned

A

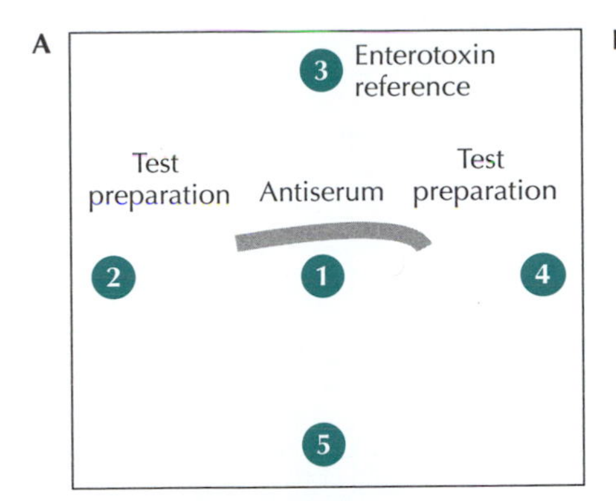

B

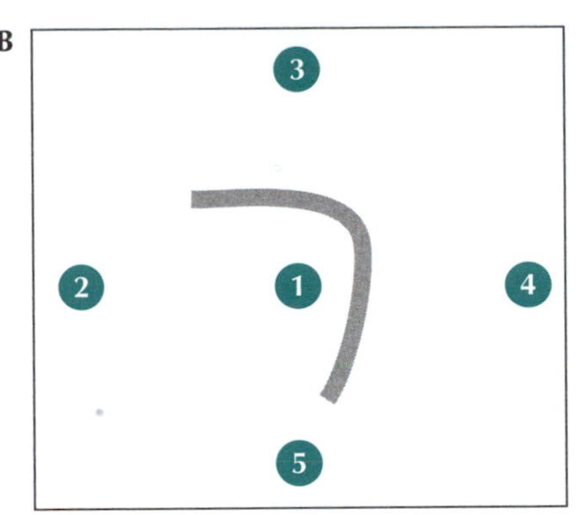

C

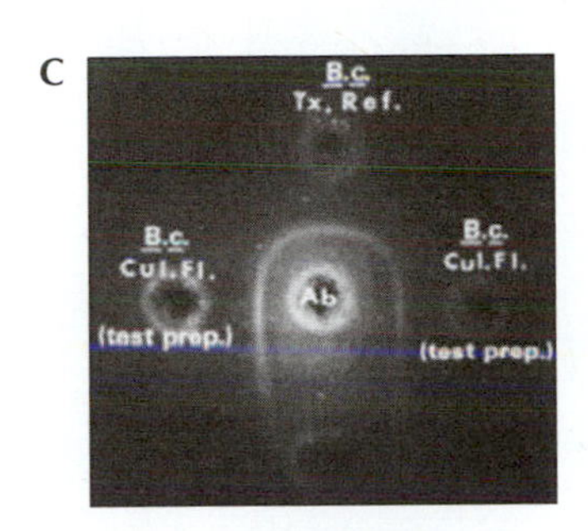

Figure 8.1 Microslide diffusion tests are used to detect *B. cereus* diarrheagenic and staphylococcal enterotoxins. Antiserum to the toxin is put into well 1, and reference toxin (a positive control) is put into well 3. Test samples are put into wells 2, 4, and 5. When the antiserum and toxin diffuse from their respective wells and meet, a precipitation line is formed. **(A)** Test where the food samples are negative but a line is formed with the known enterotoxin; **(B)** test where the sample in well 4 contains enterotoxin; **(C)** photo of actual results, where samples in wells 2 and 4 (but not 5) are positive for toxin. Photo courtesy of R. Bennett, U.S. Food and Drug Administration; diagrams redrawn from those supplied by R. Bennett. doi:10.1128/9781555817206.ch08.f08.01

above, cereulide. Because it is ubiquitous in nature and has a reputation as a *foodborne* pathogen, its isolation from clinical specimens is often dismissed as contamination. Only recently has its kaleidoscopic nature been appreciated. Some *B. cereus*-mediated diseases are listed in Table 8.3.

Table 8.3 Clinical illnesses caused by *Bacillus cereus*

Illness	Characteristic
Respiratory infection	Due to the lethal toxin gene normally carried by *Bacillus anthracis* on the pXO1 plasmid
Nosocomial infections	The organism is ubiquitous in hospitals and is transferred on linens, gloves, ventilators, and catheters and even in alcohol
Endophthalmitis	Vision-threatening eye infections after cataract surgery or traumatic eye injury
Central nervous system infections	Hemorrhagic necrosis, meningoencephalitis, brain abscess
Gas gangrene	Sepsis, exudate, tissue destruction

Across

3. Italian food that can be a source of spores
4. Type of food especially problematic for outbreaks of *Bacillus cereus*
8. Rectal equivalent of "dry heaves"
9. Structure of both *Bacillus cereus* toxins
10. *Bacillus cereus* spores are ________ , which makes them difficult to remove from equipment
11. Genus of most aerobic spore-forming rods

Down

1. Resistant, dormant life form found in the genera *Bacillus* and *Clostridium*
2. A pathogen that makes emetic and diarrheal toxins (2 words)
5. Type of toxin that causes vomiting
6. Pathogenic species in the genus *Bacillus* that is not normally found in food
7. Food microbiologists call pathogens that produce emetic and ________ toxins "double-bucket" bacteria

The Spore

The *B. cereus* spore is important in foodborne illness. It is in most respects similar to the spores of other foodborne pathogens (see chapter 3). However, *B. cereus* spores are more hydrophobic than any other *Bacillus* species' spores. This enables them to adhere to surfaces. They are difficult to remove during cleaning and are a difficult target for disinfectants. *B. cereus* spores also contain appendages and/or pili that are involved in adhesion. These spore properties enable them to withstand sanitation and remain on surfaces to contaminate foods. They also aid in adherence to epithelial cells during clinical illnesses.

Summary

- *B. cereus* is a sporeformer that is especially problematic in dairy foods.
- It is a normal inhabitant of soil and is isolated from a variety of foods.
- Two different toxins cause an emetic or a diarrheal type of foodborne illness.
- Desserts, meat dishes, and dairy products are frequently associated with diarrheal illness.
- Rice and pasta are the most common vehicles of emetic illness.
- Some strains of the *B. cereus* group grow at refrigeration temperatures.
- *B. cereus* foodborne illness is probably highly underreported.

Suggested reading

Bottone, E. J. 2010. *Bacillus cereus*, a volatile human pathogen. *Clin. Microbiol. Rev.* **23:**382–398.

Centers for Disease Control and Prevention. 1994. *Bacillus cereus* food poisoning associated with fried rice in two child day care centers—Virginia, 1993. *MMWR Morb. Mortal. Wkly. Rep.* **43:**177–178.

Granum, P. E. 2007. *Bacillus cereus*, p. 445–455. *In* M. P. Doyle and L. R. Beuchat (ed.), *Food Microbiology: Fundamentals and Frontiers*, 3rd ed. ASM Press, Washington, DC.

Stenfors Arnesen, L. P., A. Fagerlund, and P. E. Granum. 2008. From soil to gut: *Bacillus cereus* and its food poisoning toxins. *FEMS Microbiol. Rev.* **32:**579–605.

Questions for critical thought

1. Speculate as to why *B. cereus* is considered a larger problem in Europe than in the United States.
2. What is "rectal tenesmus"?
3. Create a case study of a *B. cereus* foodborne outbreak.
4. Some people have suggested that *B. cereus* be used as a surrogate to research the deadly biothreat agent *B. anthracis*. What are the pros and cons of this proposal?
5. Why are *B. cereus* foodborne illnesses more likely to be underreported than other foodborne illnesses?
6. Describe the genetics of enterotoxin production by *B. cereus*.

9 *Campylobacter* Species

LEARNING OBJECTIVES

The information in this chapter will enable the student to:

- use basic biochemical characteristics to identify *Campylobacter* spp.
- understand what conditions in foods favor *Campylobacter* sp. growth
- recognize, from symptoms and time of onset, a case of foodborne illness caused by a *Campylobacter* sp.
- choose appropriate interventions (heat, preservatives, and formulation) to prevent the growth of *Campylobacter* spp.
- identify environmental sources of the organism
- understand the roles of *Campylobacter* sp. toxins and virulence factors in causing foodborne illness

Outbreak

Visiting a farm can be a fun learning experience. In 2002, a group of 86 preschoolers and adults visited a farm in Kansas. Aside from walking through the barns and petting the cows, they were given the opportunity to sample raw milk. Sixty-five of the 86 visitors consumed raw milk, and all of those individuals reported becoming ill. They suffered from campylobacteriosis, experiencing diarrhea and dehydration. In a separate outbreak, also in Kansas, 67 of 101 individuals who consumed fresh cheese at a fair and picnic developed campylobacteriosis. The cheese was made from unpasteurized milk. Fresh cheese (e.g., cottage cheese) does not go through an aging process. Indeed, the cheese was made and consumed in a single day. These outbreaks underscore the potential danger of consuming raw milk or products made with raw milk. Finding *Campylobacter jejuni* in a farm environment is not unusual, and there are many routes by which milk can be contaminated. Pasteurization effectively kills *C. jejuni*.

INTRODUCTION

Campylobacter jejuni subsp. *jejuni* (hereafter referred to as *C. jejuni*) is one of many species and subspecies within the genus *Campylobacter*, family *Campylobacteraceae*. Since the 1980s, an explosion of information on these bacteria has been published. The 1984 edition of *Bergey's Manual of Systematic Bacteriology* listed only eight species and subspecies within the genus *Campylobacter*. Since then, many investigators have become interested in studying the taxonomy and clinical importance of campylobacters. This has greatly expanded the number of genera and species associated with this group of bacteria.

doi:10.1128/9781555817206.ch09

CHARACTERISTICS OF THE ORGANISM

The genera *Campylobacter* and *Arcobacter* are included in the family *Campylobacteraceae*. This family continues to expand as researchers identify new species and subspecies in the genera *Campylobacter* (the type genus of the family) and *Arcobacter*. The organisms are curved, S-shaped, or spiral rods that are 0.2 to 0.9 μm wide and 0.5 to 5 μm long. They are gram-negative, non-spore-forming rods that may form spherical or coccoid bodies in old cultures or cultures exposed to air for prolonged periods. The organisms are motile by means of a single polar unsheathed flagellum at one or both ends. The various species are microaerobic (5% oxygen and 10% carbon dioxide), with a respiratory type of metabolism. Some strains grow aerobically or anaerobically. An atmosphere containing increased hydrogen may be required by some species for microaerobic growth. As determined by pulsed-field gel electrophoresis (PFGE), *C. jejuni* and *Campylobacter coli* have genomes ~1.7 megabases in size, which is about one-third the size of the *Escherichia coli* genome.

Authors' note

Campylobacter *is a fragile organism. It cannot tolerate drying and can be killed by oxygen. It grows only if there is less than the atmospheric amount of oxygen present. Freezing reduces the number of* Campylobacter *bacteria present on raw meat.*

ENVIRONMENTAL SUSCEPTIBILITY

C. jejuni is susceptible to a variety of environmental conditions that make it unlikely to survive for long periods outside the host. The organism does not grow at temperatures below 30°C, is microaerobic, and is sensitive to drying, high-oxygen conditions, and low pH. The decimal reduction time for campylobacters varies and depends on the food source and temperature. Thus, the organism should not survive in food products brought to adequate cooking temperatures. The organisms are susceptible to gamma irradiation (1 kilogray), but the rate of killing depends on the type of product being processed. Irradiation is less effective for frozen materials than for refrigerated or room temperature meats. Early-log-phase cells are more susceptible than cells grown to log or stationary phase. *Campylobacter* spp. are more radiation sensitive than other foodborne pathogens, such as salmonellae and *Listeria monocytogenes*. Irradiation treatment that is effective against salmonellae and *L. monocytogenes* should be sufficient to kill *Campylobacter* spp.

Campylobacter spp. are susceptible to low pH and are killed at pH 2.3. Campylobacters remain viable and grow in bile at 37°C and survive better in feces, milk, water, and urine held at 4°C than in material held at 25°C. The maximum periods of viability of *Campylobacter* spp. at 4°C are in the range of 3 weeks in feces, 4 weeks in water, and 5 weeks in urine. Freezing reduces the number of *Campylobacter* organisms in contaminated poultry, but even after being frozen to −20°C, small numbers of *Campylobacter* organisms can be recovered. In spite of its fragility, *Campylobacter* still ranks close to *Salmonella* in number of laboratory-confirmed cases of foodborne illness.

RESERVOIRS AND FOODBORNE OUTBREAKS

Many animals serve as reservoirs of *C. jejuni* for human disease. Reservoirs for infection include rabbits, rodents, wild birds, sheep, horses, cows, pigs, poultry, and domestic pets (Table 9.1). Contaminated vegetables and shellfish may also be vehicles of infection. *Campylobacter* is frequently isolated from water and water supplies, and water has been the source of infection

Table 9.1 Reservoirs and disease-associated species in the family *Campylobacteraceae*[a]

Organism	Reservoir(s)	Disease, sequelae, or comments	
		Humans	Animals
C. jejuni subsp. *jejuni*	Humans, other mammals, birds	Diarrhea, systemic illness, GBS	Diarrhea in primates
C. jejuni subsp. *doylei*	Unknown	Diarrhea	
C. fetus subsp. *fetus*	Cattle, sheep	Systemic illness, diarrhea	Abortion
C. fetus subsp. *venerealis*	Cattle		Infertility
C. coli	Pigs, birds	Diarrhea	
C. lari	Birds, dogs	Diarrhea	
C. upsaliensis	Domestic pets	Diarrhea	Diarrhea
C. hyointestinalis subsp. *hyointestinalis*	Cattle, pigs, hamsters, deer	Rare; proctitis, diarrhea	Proliferative enteritis
C. hyointestinalis subsp. *lawsonii*	Pigs		
C. mucosalis	Pigs	Rare; diarrhea	Proliferative enteritis
C. hyoilei	Pigs		
C. sputorum biovar sputorum	Humans	Oral-cavity abscesses	Genital tract of bulls; abortion in sheep
C. sputorum biovar paraureolyticus	Cattle	Diarrhea	
C. sputorum biovar faecalis	Cattle, sheep		Enteritis
C. concisus	Humans	Periodontal disease	
C. curvus	Humans	Periodontal disease	
C. rectus	Humans	Periodontal disease, pulmonary infections	
C. showae	Humans	Periodontal disease	
C. helveticus	Domestic pets		Diarrhea
C. gracilis	Humans	Infections of head, neck, other sites	
Arcobacter butzleri	Cattle, pigs	Diarrhea	Diarrhea, abortion
Arcobacter cryaerophilus	Cattle, sheep, pigs	Diarrhea, bacteremia	Abortion
Arcobacter skirrowii	Cattle, sheep, pigs		Abortion, diarrhea; isolated from genital tract of bulls
Arcobacter nitrofigilis	Plants		

[a]Some data are from M. B. Skirrow, *J. Comp. Pathol.* **111:**113–149, 1994.

in some outbreaks. *C. jejuni* can remain dormant in water in a state that has been called *viable but nonculturable* (VBNC). Under unfavorable conditions, the organism essentially remains dormant and cannot be easily recovered on growth media. The role of these forms as a source of infection for humans is not clear.

Campylobacteriosis is not confined to the United States; over 30,000 cases were associated with 900 outbreaks worldwide from 1978 to 2003. The vehicles of *Campylobacter* outbreaks have changed over the past few decades. Water and unpasteurized milk were responsible for over one-half of the outbreaks between 1978 and 1987, whereas other foods accounted for >80% of outbreaks between 1988 and 1996. In a 10-year review of outbreaks from 1981 to 1990, 20 outbreaks occurred, affecting 1,013 individuals who drank raw milk (Fig. 9.1). The attack rate was 45%. At least one outbreak occurred each year, and most of the outbreaks involved children who had gone on field trips to dairy farms.

Figure 9.1 Drinking a large glass of cold milk can be refreshing. However, if the milk has not been pasteurized, it may contain a number of microbes, including *Campylobacter jejuni*, that when ingested result in illness (diarrhea, abdominal cramps, and fever). The risks of drinking unpasteurized milk far outweigh any benefits that may be attributed to consumption of such milk. doi:10.1128/9781555817206.ch09.f09.01

The seasonal distribution of outbreaks is somewhat different from that of sporadic cases. Milk- and waterborne outbreaks tend to occur in the spring and fall but do not occur frequently in the summer. Sporadic cases are usually most frequent during the summer. In contrast to the relatively low occurrence of outbreaks caused by campylobacters, the bacteria have been identified in many studies as being among the most common causes of foodborne illness in the United States. In 1995, the Centers for Disease Control and Prevention, the U.S. Department of Agriculture, and the Food and Drug Administration developed an active surveillance system for foodborne diseases, including *Campylobacter* infections, called the Foodborne Diseases Active Surveillance Network, also known as FoodNet. On the basis of a variety of data, including FoodNet incidence data, U.S. Census Bureau data, and data from other sources, it is estimated that there are 800,000 *Campylobacter* infections in the United States each year. Since campylobacteriosis is self-limiting, the true number of cases that occur each year can only be estimated. This is similar to the incidence of infection in the United Kingdom and other developed nations. Cases occur more often during the summer months and usually follow the ingestion of improperly handled or cooked food, primarily poultry products. Other exposures include drinking raw milk, contaminated surface water, overseas travel, and contact with domestic pets (Table 9.2). FoodNet data show that the incidence of *Campylobacter* infections decreased by 30% from 1996 to 2008 in the United States. The decrease is likely related to several factors, including implementation of government programs (Hazard Analysis and Critical Control Points in the meat and poultry industry), enhanced monitoring of finished product by industry, and consumer knowledge.

Table 9.2 Isolation of *Campylobacter* from different food sources[a]

Product	% Positive samples
Chicken	14–98
Turkey	3–25
Duck	48
Goose	38
Cow's milk	0–12.3
Goat's milk	0
Ewe's milk	0
Beef	0–23.6
Pork	1–23.5
Lamb	0–15.5
Sheep	3
Offal	47
Mussels	47–69
Oysters	6–27
Vegetables	<5

[a]Source: W. Jacobs-Reitsma, p. 467–481, *in* I. Nachamkin and M. J. Blaser (ed.), *Campylobacter*, 2nd ed. (ASM Press, Washington, DC, 2000).

CHARACTERISTICS OF DISEASE

C. jejuni and *C. coli*

Most *Campylobacter* species are associated with intestinal tract infection. Infections in other areas of the body are common with some species, such as *Campylobacter fetus* subsp. *fetus* (hereafter referred to as *C. fetus*). *C. jejuni* and *Campylobacter coli* have been recognized since the 1970s as agents of gastrointestinal tract infection.

C. jejuni and *C. coli* are the most common *Campylobacter* species associated with diarrheal illness, and they are indistinguishable. The ratio of *C. jejuni* to *C. coli* infections is not known, because most laboratories do not routinely distinguish between the organisms. In the United States, an estimated 5 to 10% of cases attributed to *C. jejuni* are actually due to *C. coli*; this figure may be higher in other parts of the world.

A spectrum of illness can occur during *C. jejuni* or *C. coli* infection, and patients may be *asymptomatic* (show no sign of illness) to severely ill. Symptoms and signs usually include fever, abdominal cramping, and diarrhea (with or without blood) that lasts several days to >1 week. Symptomatic infections are usually self-limiting, but relapses may occur in 5 to 10% of untreated patients. Recurrence of abdominal pain is common. Other infections and diseases include bacteremia, bursitis, urinary tract infection, meningitis, endocarditis, peritonitis, erythema nodosum, pancreatitis, abortion and neonatal sepsis, reactive arthritis, and Guillain-Barré syndrome (GBS). Deaths directly attributable to *C. jejuni* infection are rarely reported.

C. jejuni and *C. coli* are susceptible to many antibiotics, including macrolides, fluoroquinolones, aminoglycosides, chloramphenicol, and tetracycline. Erythromycin is the drug of choice for treating *C. jejuni* gastrointestinal tract infections, but ciprofloxacin is a good alternative drug. Early therapy of *Campylobacter* infection with erythromycin or ciprofloxacin is effective in eliminating the campylobacters from stool and may also reduce the duration of symptoms associated with infection.

C. jejuni is generally susceptible to erythromycin, with resistance rates of <5%. Rates of erythromycin resistance in *C. coli* vary, with up to 80% of strains having resistance in some studies. Although ciprofloxacin is effective in treating *Campylobacter* infections, resistant strains have emerged. The resistance may be linked to the use of similar antibiotics (fluoroquinolones) in poultry. In 2005, the Food and Drug Administration Center for Veterinary Medicine banned the use of fluoroquinolones, such as enrofloxacin, in poultry.

Other *Campylobacter* Species

In contrast to *C. jejuni*, *C. fetus* is primarily associated with bacteremia and non-intestinal-tract infections. *C. fetus* is also associated with abortions, septic arthritis, abscesses, meningitis, endocarditis, mycotic aneurysm, thrombophlebitis, peritonitis, and salpingitis. Although intestinal tract infections occur with this species, the number of cases is low, because the organism does not grow well at 42°C. *C. fetus* is susceptible to the antibiotic cephalothin. *Campylobacter fetus* subsp. *venerealis* is not associated with human infection.

EPIDEMIOLOGICAL SUBTYPING SYSTEMS USEFUL FOR INVESTIGATING FOODBORNE ILLNESSES

Many typing systems have been devised to study the epidemiology of *Campylobacter* infections. These methods include biotyping, serotyping, bacteriocin sensitivity, detection of preformed enzymes, lectin binding, phage typing, multilocus enzyme electrophoresis, and molecularly based methods, such as PFGE, ribotyping, and restriction fragment length polymorphism combined with polymerase chain reaction (PCR) methods (see chapter 5).

The most frequently used phenotypic systems are biotyping and serotyping. The two major serotyping schemes used worldwide detect heat-labile and O antigens. Serotyping systems are simple to perform and have good ability to discriminate among strains.

During the past 5 years, molecularly based methods have come to the forefront for studying the epidemiology of *Campylobacter* infections. PFGE is one of the most powerful methods for molecular analysis of *Campylobacter* species. It has excellent discriminatory power. A variety of PCR-based methods, including ribotyping, random amplified polymorphic DNA, amplified fragment length polymorphism, multiplex PCR-restriction fragment length polymorphism, and real-time PCR, have all been used in various applications to study *Campylobacter* spp. Since the sequencing of the genome, the use of microarray technology has been developed as a typing system (see chapter 5 for explanations of these methods).

INFECTIVE DOSE AND SUSCEPTIBLE POPULATIONS

C. jejuni is susceptible to low pH, and hence, stomach acidity kills most campylobacters. The infective dose of *C. jejuni* is not high, with <1,000 organisms causing illness. The ability to cause infection may vary greatly among strains. Human volunteers (18%) became ill when infected with 10^8 colony-forming units (CFU) of a strain designated A3249, but 46% of the volunteers became ill when infected with another strain, 81-176. In an outbreak of *Campylobacter* infection after the ingestion of raw milk, the number of individuals who became ill and the severity of illness seemed to be linked to the amount of milk ingested (the more milk consumed, the greater the number of *Campylobacter* organisms consumed) (Fig. 9.1).

The greatest numbers of infections in the United States occur in young children and in young adults 20 to 40 years of age. The incidence of *Campylobacter* infection in developing countries, such as Mexico and Thailand, may be 10 to 100 times higher than in the United States. In developing countries, in contrast to developed countries, campylobacters are frequently isolated from individuals who may or may not have diarrheal disease. Most infections occur in infancy and early childhood, and the incidence decreases with age. As a child matures, his or her immune system can fight off *Campylobacter* infections, thereby decreasing the number of illnesses. Travelers to developing countries may become infected with *Campylobacter*.

The elderly are more likely to get *bacteremia* (presence of bacteria in the blood) than any other group (infants, young adults, etc.). Continual diarrheal illness and bacteremia may occur in immunocompromised individuals, such as patients with human immunodeficiency virus infection.

VIRULENCE FACTORS AND MECHANISMS OF PATHOGENICITY

Little is known about the mechanism by which *C. jejuni* causes human disease. *C. jejuni* can cause an enterotoxigenic-like illness with loose or watery diarrhea or an inflammatory intestinal disease with fever. The mechanism of disease may be *invasive* (entering into intestinal cells), since some individuals have blood in their feces and may develop bacteremia. A major problem in understanding the pathogenesis of *Campylobacter* infection is the lack of suitable animal models.

Cell Association and Invasion

The uptake of *C. jejuni* by host cells occurs by intimate binding of the bacterium with the host cell surface by using many types of adhesins. Following contact, *C. jejuni* produces at least 14 new proteins. Production corresponds to a rapid increase in uptake and changes in the host cell membrane. At a later point, additional proteins are made and secreted into the cytoplasm of host target cells. Following attachment and internalization, *Campylobacter*-infected cells release molecules that promote the recruitment of white blood cells to the site of the infection. Induction of host cell death by *C. jejuni* may enhance the survival and spread of the pathogen.

Flagella and Motility

Campylobacter species are motile and have a single polar, unsheathed flagellum at one or both ends. Motility and flagella are important determinants for the entry process. *Campylobacter* colonization and/or infection in a variety of animal models is dependent on intact motility and full-length flagella; however, other factors may be involved. Two genes are involved in the expression of the flagellar filament, and motility is essential for colonization. Components of intestinal mucin, particularly L-fucose, attract

C. jejuni. Movement of *C. jejuni* toward such compounds may be important in the pathogenesis of infection.

Toxins

C. jejuni may produce an enterotoxin similar to cholera toxin. No genetic evidence is available to support the presence of this toxin. The role of cytolethal distending toxin in pathogenesis is unknown. Most strains of *C. jejuni* have *cdt* genes, but the amounts of toxin produced by the strains vary. The toxin may have several roles in pathogenesis. More information on toxins of the pathogen should be forthcoming, since the genome has been sequenced.

Other Factors

The regulation of genes in response to environmental changes is becoming an important area of *Campylobacter* research. Campylobacters are microaerobic, and *C. jejuni* has a higher temperature for optimal growth, 42°C, than other intestinal bacterial pathogens. Understanding the effects of environmental signals on the growth, metabolism, and pathogenicity of *Campylobacter* strains will have a major impact on the ability to control campylobacters in the environment and food chain. Such pathways include response to iron; oxidative stress; temperature regulation, including cold and heat shock responses; and starvation.

Autoimmune Diseases

Campylobacter infection is a major trigger of GBS, an acute immune-mediated paralytic disorder affecting the peripheral nervous system. The pathogenesis of GBS induced by *C. jejuni* is not clear. The bacterial lipopolysaccharide causes an immune response. The antibodies produced recognize not only the lipopolysaccharide but also peripheral nerve tissue. This is likely a major mechanism of *Campylobacter*-induced GBS.

IMMUNITY

Infection with *Campylobacter* species results in protective immunity and is likely antibody mediated. Rechallenge with the same strains 28 days after the initial challenge resulted in protection against illness but not necessarily against colonization by *C. jejuni*. People who are regularly exposed to the pathogen through drinking raw milk are less likely to become ill. In developing countries, where *Campylobacter* infections are endemic, immunity to *Campylobacter* infection appears to be age dependent. As a child matures, he or she is less likely to become ill.

Flagellin is an important immunogen during *Campylobacter* infection. Antibodies against this protein correlate to some degree with protective immunity. Breast-feeding provides protection against *Campylobacter* infection in developing countries. Antibodies against flagellin present in breast milk appear to be associated with protection of infants against infection.

Other strategies to prevent the spread of infection to humans include improved hygiene practices during broiler chicken production, such as decontamination of water supplies; use of competitive exclusion floras, which may prevent *C. jejuni* colonization of chicks; and immunological approaches through the use of animal vaccines.

Summary

- Campylobacteriosis linked to consumption of contaminated food is the most prevalent form of foodborne illness in the United States.
- *C. jejuni* is commonly isolated from poultry.
- Thermoprocessing easily inactivates campylobacters.
- *Campylobacter* infection is a major trigger of GBS.
- Motility is essential for colonization.
- Five to 10% of cases attributed to *C. jejuni* are actually due to *C. coli*.
- The infective dose of *C. jejuni* may be <1,000 organisms.

Suggested reading

Blaser, M. J. 2000. *Campylobacter jejuni* and related species, p. 2276–2285. *In* G. L. Mandell, J. E. Bennett, and R. Dolin (ed.), *Principles and Practice of Infectious Diseases*, 5th ed. Churchill Livingstone, Philadelphia, PA.

Friedman, C. R., J. Neimann, H. C. Wegener, and R. V. Tauxe. 2000. Epidemiology of *Campylobacter jejuni* infections in the United States and other industrialized nations, p. 121–138. *In* I. Nachamkin and M. J. Blaser (ed.), *Campylobacter*, 2nd ed. ASM Press, Washington, DC.

Jacobs-Reitsma, W. 2000. *Campylobacter* in the food supply, p. 467–481. *In* I. Nachamkin and M. J. Blaser (ed.), *Campylobacter*, 2nd ed. ASM Press, Washington, DC.

Levin, R. E. 2007. *Campylobacter jejuni*: a review of its characteristics, pathogenicity, ecology, distribution, subspecies characterization and molecular methods of detection. *Food Biotechnol.* **21:**271–347.

Moore, J. E., D. Corcoran, J. S. G. Dooley, S. Fanning, B. Lucey, M. Matsuda, D. A. McDowell, F. Megraud, B. C. Millar, R. O'Mahony, L. O'Riordan, M. O'Rourke, J. R. Rao, P. J. Rooney, A. Sails, and P. Whyte. 2005. Campylobacter. *Vet. Res.* **36:**351–382.

Murphy, C., C. Carrol, and K. N. Jordan. 2006. Environmental survival mechanisms of the foodborne pathogen *Campylobacter jejuni*. *J. Appl. Microbiol.* **100:**623–632.

Nachamkin, I. 2007. *Campylobacter jejuni*, p. 237–248. *In* M. P. Doyle and L. R. Beuchat (ed.), *Food Microbiology: Fundamentals and Frontiers*, 3rd ed. ASM Press, Washington, DC.

Ransom, G. M., B. Kaplan, A. M. McNamara, and I. K. Wachsmuth. 2000. *Campylobacter* prevention and control: the USDA-Food Safety and Inspection Service role and new food safety approaches, p. 511–528. *In* I. Nachamkin and M. J. Blaser (ed.), *Campylobacter*, 2nd ed. ASM Press, Washington, DC.

Questions for critical thought

1. Campylobacters are microaerobic, and *C. jejuni* has a higher temperature for optimal growth, 42°C, than other intestinal bacterial pathogens. Given this information, how do they survive/grow on poultry carcasses or other foods?
2. Both *Campylobacter* and *Salmonella* are associated with poultry. Compare and contrast *C. jejuni* to *Salmonella enterica* serovar Enteritidis. How does oxygen concentration influence the growth of *C. jejuni*?
3. How is *C. jejuni* infection linked to GBS?
4. What is the impact of antibiotic-resistant *C. jejuni* on human health? How might the use of antibiotics in poultry production practices influence resistance?

5. Campylobacters can exist in a VBNC state. In what environments do they enter this state? Can the organism recover from this state? Why is entry into the VBNC state important for survival of the organism?
6. *Campylobacter* infections are endemic in some developing countries. As a child matures, he or she is less likely to become ill. In the United States, why is this not the outcome? What bacterial moiety is likely associated with development of antibodies?
7. What practices can be implemented to control campylobacters in food?
8. Describe how you would differentiate *C. jejuni* from *E. coli* O157:H7 microbiologically and based on signs and symptoms of illness.
9. With respect to food safety and food-handling practices, why would a person that is asymptomatic present a risk?
10. The incidence of *Campylobacter* infection in developing countries, such as Mexico and Thailand, is severalfold higher than in the United States. What factors may account for such a great difference in the infection rates?
11. *C. jejuni* is zoonotic. What does this statement mean?
12. You are a new employee at a start-up biotechnology firm and have been asked to develop a vaccine against *C. jejuni*. Outline your plan of attack. Consider what you know about the pathogen and the illness it causes and what others have done in the past with respect to vaccine development.
13. Based on your knowledge of microbes, why do you believe *C. jejuni* survives better in feces, milk, water, and urine held at 4°C than in material held at 25°C?
14. Explain how the banning of fluoroquinolone antibiotics in poultry may affect resistance of *Campylobacter* species. What, if any, impact might this have on poultry health?

10 Clostridium botulinum

LEARNING OBJECTIVES

The information in this chapter enables the student to:

- use basic biochemical characteristics to identify *Clostridium botulinum*
- understand what conditions in foods favor the growth of *C. botulinum*
- recognize, from symptoms and time of onset, a case of botulism
- choose appropriate interventions (heat, preservatives, and formulation) to prevent the growth of *C. botulinum*
- identify environmental sources of *C. botulinum*
- understand the roles of spores, anaerobic conditions, and heat sensitivity of the toxin in causing or preventing botulism

INTRODUCTION

Botulism is a rare but sometimes deadly disease. There are an average of 24 cases of foodborne botulism, 3 cases of wound botulism, and 71 cases of intestinal (or "infant") botulism reported annually to the Centers for Disease Control and Prevention (CDC). Although improper home canning of vegetables causes ~40% of botulism cases in the United States (Box 10.1), many countries report relatively frequent outbreaks from other sources (Table 10.1).

Four Faces of Botulism

Botulism is traditionally associated with canned foods, usually home canned. Commercial cases are very rare (9% of outbreaks) and usually involve small companies that just don't know any better. A famous outbreak involving canned food occurred in New York State in the early 1970s. An elderly couple ate some vichyssoise soup on a hot summer night. By the next morning, they had started to see double, had difficulty swallowing, and experienced paralysis in their arms and legs. If they had not been treated, toxin would have paralyzed their diaphragms and suffocated them. Fortunately, they were quickly diagnosed with botulism, were put on ventilators, and lived. They never fully recovered. The investigation revealed that the workers responsible for operating the retorts (canning machines) had misinterpreted a new process schedule. They underprocessed the soup, allowing *Clostridium botulinum* to grow and make toxin. In response to this incident, the Food and Drug Administration (FDA) instituted good manufacturing practices (GMPs) for low-acid foods (i.e., those with a pH of >4.6 and a water activity [a_w] of >0.86). The GMPs require all retort operators to attend a better processing control school and be certified. Processing schedules and changes to them can be made only by recognized processing authorities.

doi:10.1128/9781555817206.ch10

BOX 10.1

Self-inflicted botulism

People in urban areas may consider home canning an antiquated form of food preservation. But one in five U.S. households uses it to stretch its food budget, maintain connection with "the land," or preserve the large harvests from gardens. In New Jersey, for example, the plethora of home gardeners makes it impossible to give away tomatoes in August.

Health officials recently described three botulism case studies between 2008 and 2009 and compared them with the 116 outbreaks of botulism that occurred between 1999 and 2008. The outbreaks below are from that study, albeit with some literary license.

In the first outbreak, a family ate a blend of home-canned carrots and green beans. That very day, two of the members developed symptoms of botulism, were hospitalized, and were put on ventilators (i.e., intubated with breathing machines). Four more family members went to the hospital the next day. One was intubated, another had milder symptoms, and two were okay. Toxin was isolated from the patients and the food. The blend was obviously spoiled. One victim ate it even though it tasted bad, the second ate just a forkful (still enough to cause botulism), and another ate it because he did not want to waste food. The second outbreak was from a meal with home-canned green beans, tomatoes, and pears. One victim noted that the green beans "smelled like cat litter," but the second person thought that they were okay, ate a lot of them, and became the sickest. The third case involved two men who ate home-canned asparagus which had no signs of spoilage but had lids that had popped.

The victims in all three cases had a range of symptoms typical of botulism. These included nausea, vomiting, dry mouth, and neurological involvement such as blurred or double vision, bilateral paralysis, slurred speech, fatigue, difficulty swallowing, difficulty breathing, dizziness, and change in the sound of the voice.

The root cause of these botulism outbreaks was the failure to use proper pressure cooker processes. The home canners followed procedures that were 50 years old, relied on the advice of neighbors, ignored correct instructions, did not have pressure cookers, cut processes short because the kitchen was so hot, and were ignorant of the botulism risk associated with home canning. It did not help that the victims ignored signs of spoilage (would you eat something that smelled like cat litter?) and failed to fully heat the food, which might have inactivated some of the heat-sensitive botulinal toxin.

Date, K., R. Fagan, S. Crossland, D. MacEachern, B. Pyper, R. Bokanyi, Y. Houze, E. Andress, and R. Tauxe. 2011. Three outbreaks of foodborne botulism caused by unsafe home-canning of vegetables—Ohio and Washington, 2008–2009. *J. Food Prot.* **74:**2090–2096.

In 2006, improperly processed carrot juice caused four people to develop the classic symptoms of botulism. The administration of antitoxin stopped the progression of respiratory failure caused by descending flaccid paralysis. Botulinum toxin type A was found in the patients' stools and in leftover juice. The juice had a pH of 6.0 and was not processed to a 12*D*

Table 10.1 Reported foodborne botulism cases

Country	Period	No. of cases	Usual type	Usual food
Argentina	1979–1997	277	A	Preserved vegetables
Belgium	1988–1998	10	B	Meats
Canada	1985–1999	183	E	Traditional Inuit fermented marine mammal meat
China	1958–1989	2,861	A, B	Fermented bean products
France	1988–1998	72	B	Home-cured ham
Germany	1988–1998	177	B	Meats
Iran	1972–1974	314	E	Fish
Italy	1988–1998	412	B	Vegetables preserved in oil or water
Japan	1951–1987	479	E	Fish or fish products
Norway	1975–1997	26	E	*Rakfisk* (traditional fermented fish)
Poland	1988–1998	1,995	B	Home-preserved meats
Russia	1988–1992	2,300	B	Home-preserved mushrooms, fish
Spain	1988–1998	92	B	Vegetables
United States	1950–1996	1,087	A	Vegetables

botulinum cook. (Low-acid canned foods must be processed to achieve a 12-log reduction in the viability of botulinal spores.) Although the product relied on refrigeration for its safety, none of the victims had refrigerated their juice. Surprisingly, the company still manufactures carrot juice. Commercially bottled garlic in oil has also caused many outbreaks. As a result, garlic in oil being sold in North America must have a second barrier, such as acidification, and must be refrigerated.

Foodborne botulism is often associated with temperature abuse. Figure 10.1 shows spikes in the incidence of botulism which were caused by temperature abuse. An unusually large outbreak caused by temperature abuse occurred in Clovis, New Mexico. Over 40 people who ate at a local salad bar started seeing double, became short of breath and weak, had slurred speech, and showed various signs of paralysis. The local doctor, to his credit, quickly diagnosed botulism. Remembering its association with canned foods, he (prematurely) proclaimed the three-bean salad to be the culprit. Public health authorities from the CDC and the FDA did an exhaustive investigation that cleared the beans and implicated the potato salad. They discovered that it is a common practice in restaurants to put leftover potatoes in a box and store them at room temperature. When enough potatoes accumulate, they are used for potato salad with no further heating. This turns out to be a perfect scenario for the development of botulism poisoning. Botulinal spores are normally found on potatoes. Baking kills any competing bacteria and drives off oxygen, creating the anaerobic conditions under which *C. botulinum* thrives. The surviving spores turn into cells that grow in the potatoes and make botulinal toxin. Since the potato salad is not cooked, the toxin is not destroyed. It can be lethal. Victims of this outbreak required years to recover.

In northern Canada, Alaska, Scandinavia, and northern Japan, most botulism outbreaks involve fish. *C. botulinum* type E is implicated in most outbreaks involving northern native foods. Many of these foods are

Figure 10.1 Reported cases of foodborne botulism, United States, 1982 to 2002. The peaks in the graph demonstrate the role of temperature abuse in outbreaks. Courtesy of the CDC. doi:10.1128/9781555817206.ch10.f10.01

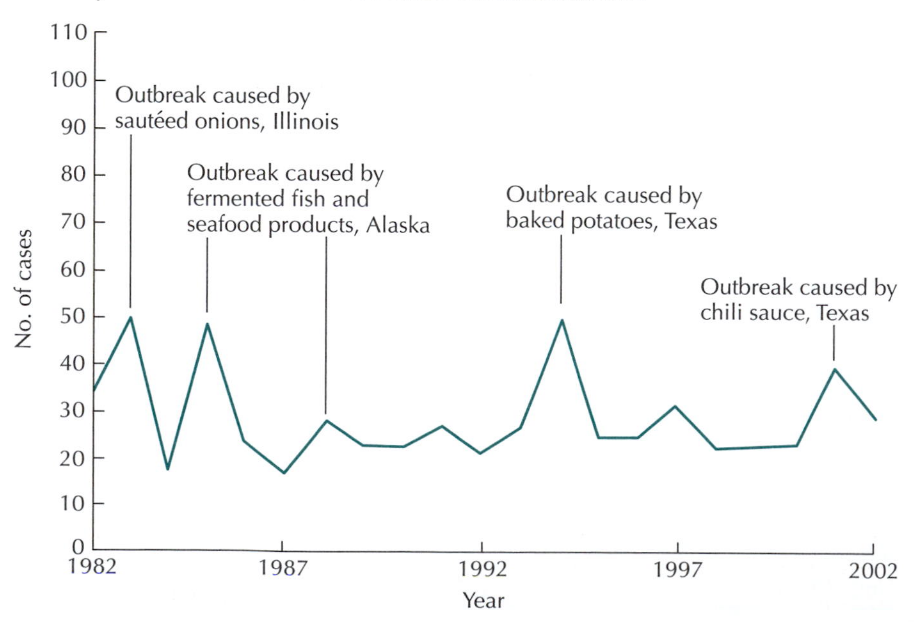

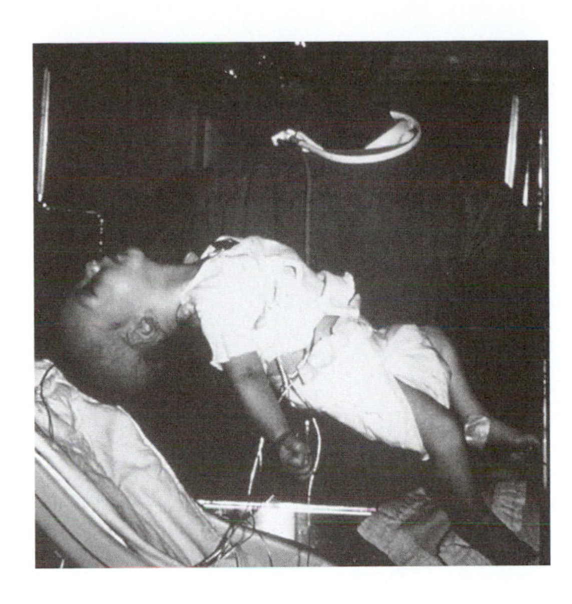

Figure 10.2 An infant with flaccid paralysis that is characteristic of human botulism. Courtesy of the CDC. doi:10.1128/9781555817206.ch10.f10.02

fermented products. However, the level of fermentable carbohydrates is too low to ensure the rapid acidification required to inhibit botulinal growth.

Scientists in the California Department of Public Health discovered the second form of botulism, infant botulism. They noticed infants who were not thriving, could not lift themselves up, and had poor muscle tone (Fig. 10.2). Their deaths resembled sudden infant death syndrome. Early investigation of *hospitalized* infants revealed an association with breast-feeding. It was soon realized, however, that breast-feeding was protective and kept the infants alive long enough to be diagnosed and hospitalized. Infants who were not breast-fed were much sicker and usually died before being diagnosed. Eventually, an association with eating honey and other raw agricultural products led to a new scenario for botulism. We all consume botulinal spores, but we are protected by the bacteria colonizing our gut. When infants, who have no protective intestinal microbiota, consume spores, these spores can germinate, become cells, colonize the gut, and produce toxin in place. Adults who are immunocompromised or who have undergone antibiotic therapy that kills their intestinal bacteria can also get infant botulism, so it has been renamed "intestinal botulism." To prevent infant botulism, the American Academy of Pediatrics recommends that children under the age of 2 years not be fed any raw agricultural products. As shown in Fig. 10.3, cases of infant botulism occur more frequently in the first 6 months of life than in the second 6 months. However, botulism in infants has virtually disappeared.

The third form of botulism, wound botulism, is not foodborne but can still be fatal. It is caused when spores are introduced to body tissue below the skin (Fig. 10.4). In one recent year, the number of botulism cases caused by drug users injecting contaminated heroin was larger than the number caused by food. A fourth face of botulism is the intentional use of *C. botulinum* as an agent of bioterrorism (Box 10.2).

CHARACTERISTICS OF THE DISEASE

Foodborne botulism can range from a mild illness, which may be disregarded or misdiagnosed, to a serious disease that can kill within a day. Symptoms typically appear 12 to 36 h after ingestion of the neurotoxin but

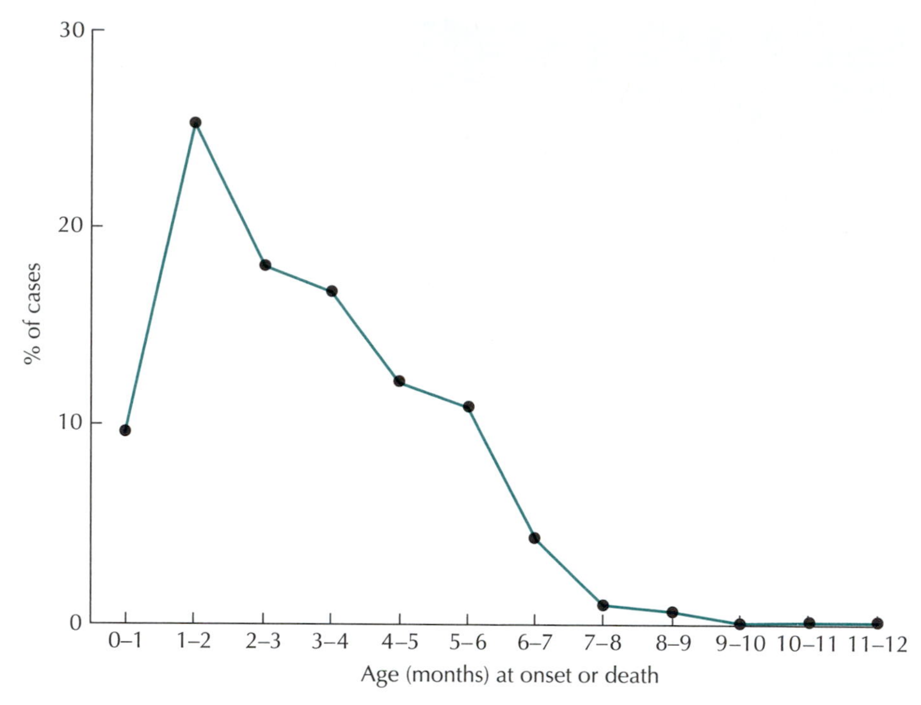

Figure 10.3 Age distribution of infant botulism. The data are for 1,428 cases in the United States from 1976 to 1995. Courtesy of the CDC. doi:10.1128/9781555817206.ch10.f10.03

may appear within a few hours or not until up to 14 days later. The earlier the symptoms appear, the more serious the illness. The first symptoms are generally nausea and vomiting. These are followed by neurological signs and symptoms, including visual impairments (blurred or double vision, drooping eyelids, and fixed and dilated pupils), loss of normal mouth and throat functions (difficulty in speaking and swallowing; dry mouth, throat, and tongue; and sore throat), general fatigue and lack of muscle coordination, and respiratory impairment. Other symptoms may include abdominal pain, diarrhea, or constipation, which is usually the initial symptom of infant botulism. Respiratory failure and airway obstruction are the main causes of death. Fatality rates in the early 1900s were ≥50%. The availability of antisera and modern respiratory support systems has decreased the fatality rate to about 10%. However, it may take 3 or 4 months on a ventilator and then years of therapy to fully recover. If botulinal toxin were

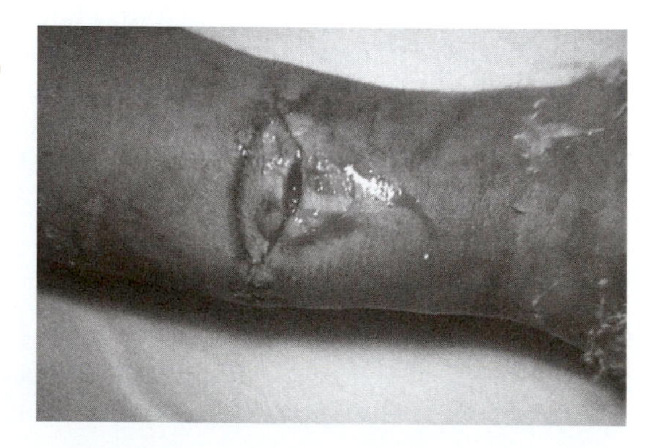

Figure 10.4 Wound botulism. Courtesy of the CDC. doi:10.1128/9781555817206.ch10.f10.04

BOX 10.2

Biosecurity and biosafety in the age of terrorism

Clostridium botulinum produces the most potent toxin known. Yet there was a time when food microbiologists traded *C. botulinum* cultures like baseball cards. If they had nothing to trade, they could buy cultures at minimal cost with minimal red tape. That time ended in October 2003, when a terrorist sent spore-laden letters to politicians and celebrities. *Bacillus anthracis* became a household word: a word associated with terror. Microbiology was suddenly not such a noble field. Microbiologists who worked with spores were "visited" by Federal Bureau of Investigation agents.

By the end of January 2002, Congress passed, and President George W. Bush signed, the Providing Appropriate Tools Required To Intercept and Obstruct Terrorism Act (i.e., the "Patriot Act") and the Public Health Security and Bioterrorism Preparedness and Response Act. Together, these laws give the federal government strong oversight of microbiologists who work with microbes that could be used for bioterrorism, now known as select agents. Select agents that food microbiologists might normally work with include *C. botulinum*, botulinum toxin, *Clostridium perfringens* epsilon toxin, staphylococcal enterotoxin, and T-2 toxin. It is now a federal offense to possess any of these unless the laboratory has been registered, inspected, and approved by the CDC. The Office of the Attorney General of the United States must run background checks and approve scientists who work with select agents. "Restricted persons," i.e., anyone convicted of a felony or another crime that could carry a penalty of more than 1 year in jail, citizens of countries hostile to the United States, and people who are "mentally defective" or were committed (voluntarily or involuntarily) to a mental institution, are prohibited from having access to select agents.

The select agent act requires that the agents be inventoried, kept locked in a room accessible only to people cleared by the Department of Justice, and handled under a biosafety cabinet using biosafety level 3 (BSL-3) work practices. In addition to adhering to the normal BSL guidelines, the laboratory must conduct a risk assessment, have a security plan in place, and maintain detailed records of laboratory entry, culture maintenance, and experimental procedures.

BSLs are designated by how dangerous an organism is and how easy it is to treat the illness it may cause. Microbes that do not consistently cause sickness in healthy adults can be used in a BSL-1 laboratory. BSL-1 labs must use standard microbiological practices such as refraining from smoking, eating, or drinking in the lab and refraining from mouth pipetting. No other special equipment is required. Most foodborne pathogens need a BSL-2 laboratory. These organisms can make people sick if ingested or introduced through cuts or through membranes. These sicknesses are relatively easy to cure. In addition to the BSL-1 requirements, access to BSL-2 labs is restricted to lab personnel, there is a biohazard warning sign, procedures that produce aerosols are done under a biosafety cabinet, and lab personnel must wear lab coats and gloves. In some cases, a personal breathing apparatus may provide an additional level of protection (see figure). BSL-3 labs are much more specialized and are required for microbes that can be transmitted through aerosols. Organisms causing serious lethal illnesses require BSL-3. In addition to BSL-2 requirements, BSL-3 requires controlled access, decontamination of lab waste and lab clothing, and engineering safeguards. The lab cannot be located off a common hallway and must have negative airflow that is exhausted without recirculation to other parts of the building. Biosafety cabinets are used for all culture manipulations. There are also BSL-4 laboratories for organisms, like Ebola virus, that are highly infectious and have no cure. There are only a few BSL-4 laboratories in the United States. Fortunately, no foodborne organisms require BSL-4 precautions.

These security and safety precautions will undoubtedly tighten over the next decade. They will make it more difficult, costly, and restrictive to conduct research on select agents that are also contaminants of food. The long time required for Department of Justice clearance and immunizations will probably prevent graduate students and postdoctoral associates from working with these organisms. This will reduce the number of scientists trained to respond to bioterroristic attacks or to the much more probable outbreaks of foodborne botulism. Hopefully, future legislation will make it easier to conduct legitimate research while still protecting us from the misuse of microbes.

A microbiologist wearing a personal air-purifying apparatus and working in a class IIa biosafety cabinet. doi:10.1128/9781555817206.ch10.fBox10.02

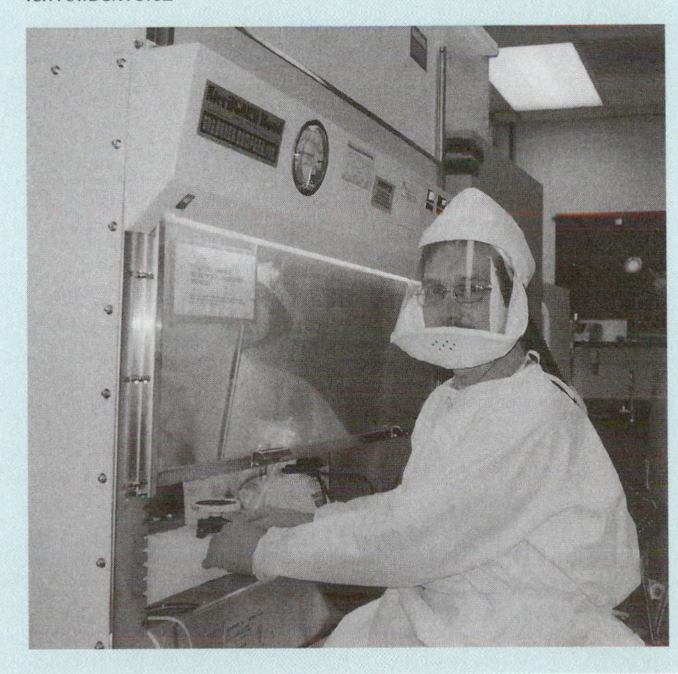

used as an agent of biowarfare, there would not be enough ventilators or antitoxin available to save many people. While an attack might involve 40,000 people, there are perhaps ~4,000 unused ventilators available nationwide at any given time.

Botulism is often incorrectly diagnosed as other illnesses, including other forms of food poisoning, stroke, poliomyelitis, organophosphate poisoning, tick paralysis, myasthenia gravis, carbon monoxide poisoning, and, most commonly, Guillain-Barré syndrome.

Initial treatment of botulism involves removing or inactivating the neurotoxin by (i) neutralizing the circulating neurotoxin with antiserum, (ii) using enemas to remove residual neurotoxin from the bowel, and (iii) gastric lavage (stomach pumping) or treating with emetics if the food might still be in the stomach. Antiserum is most effective in the early stages of illness. Subsequent treatment is mainly mechanical ventilation to counteract paralysis of the respiratory muscles. Optimal treatment for infant botulism consists primarily of high-quality supportive care.

TOXIC DOSES

Since toxicity experiments using humans are frowned on, little is known about the minimum toxic dose of botulinal neurotoxins or the efficacy of immunizations. From a food safety perspective, the presence of any neurotoxin or conditions permitting *C. botulinum* growth cannot be tolerated. Botulinal toxin is the most toxic natural substance known. The LD_{50} (amount required to kill 50% of the subjects) for botulinal toxin injected under the skin in monkeys is ~0.4 ng/kg of body weight. This suggests that the LD_{50} for a 150-pound human would be 0.000000001 (1×10^{-9}) ounce! This toxicity, and the relative ease with which *C. botulinum* can be grown, makes it a potential agent for biological warfare. Indeed, in 2003, the U.N. Special Commission reported that Iraq had produced about 50 gallons of concentrated botulinal toxin. This was allegedly weaponized in 16 missiles and 100 bombs, but no such weapons were ever confirmed by forces on the ground. Immunization of high-risk populations has been considered, but it is not cost-effective. Currently, only botulinum researchers and certain members of the U.S. military are immunized.

CHARACTERISTICS OF *C. BOTULINUM*

Classification

C. botulinum is a gram-positive, rod-shaped, obligately anaerobic (oxygen kills it) bacterium. Anaerobic chambers (Fig. 10.5) or other methods of oxygen exclusion must be used when *C. botulinum* is studied in the laboratory. The organism forms oval endospores in stationary-phase cultures (Fig. 10.6). There are seven types (A through G) of antigenically distinct *C. botulinum* toxins. Human botulism is caused by toxin types A, B, E, and, very rarely, F. Types C and D cause botulism in animals. To date, there is no direct evidence linking type G to disease.

The species is also divided into four groups based on physiological differences (Table 10.2). Group I contains all type A strains and proteolytic strains (i.e., those that produce enzymes that break down proteins) of types B and F. Group II encompasses all type E strains and nonproteolytic strains of types B and F. Group III contains type C and D strains. Group IV comprises

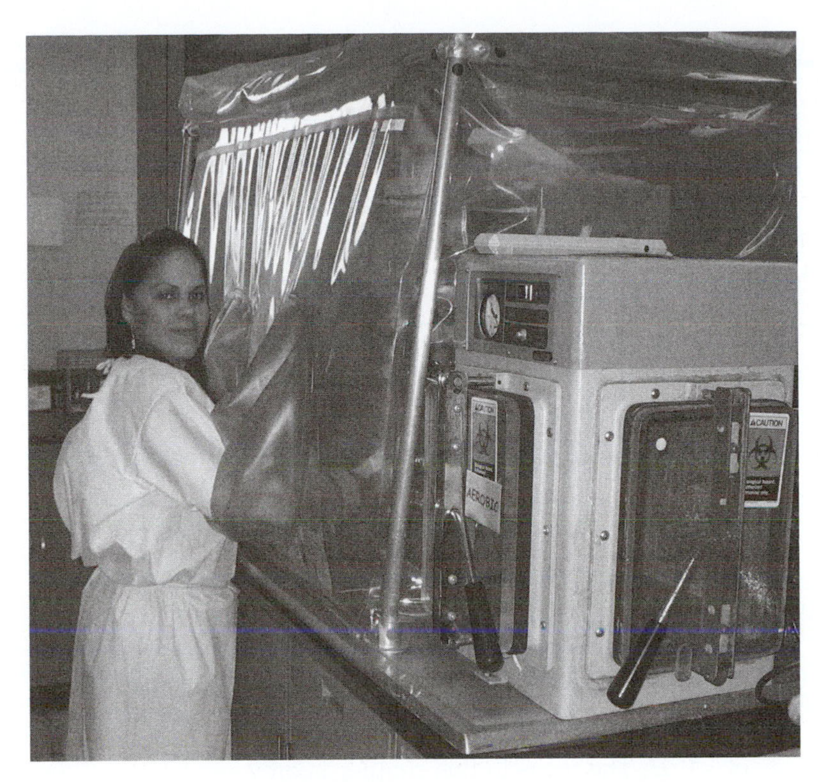

Figure 10.5 Anaerobic chambers contain atmospheres devoid of oxygen, which is lethal to *C. botulinum*. doi:10.1128/9781555817206.ch10.f10.05

C. botulinum type G. These groupings agree with results of DNA homology studies. Sequence studies using 16S and 23S ribosomal RNA (rRNA) show a high degree of relatedness among strains within each group but little relatedness among groups.

Group I strains are proteolytic and are typified by strains that produce neurotoxin type A. Their optimal growth temperature is 37°C, with growth occurring between 10 and 48°C. These cultures typically produce high levels of neurotoxin (10^6 mouse LD_{50}s/ml [1 mouse LD_{50} is the amount of neurotoxin required to kill 50% of injected mice within 4 days]). Spores in this group have a high heat resistance, with $D_{100°C}$ values of approximately 25 min (the *D* value is the time required to inactivate 90% of the population

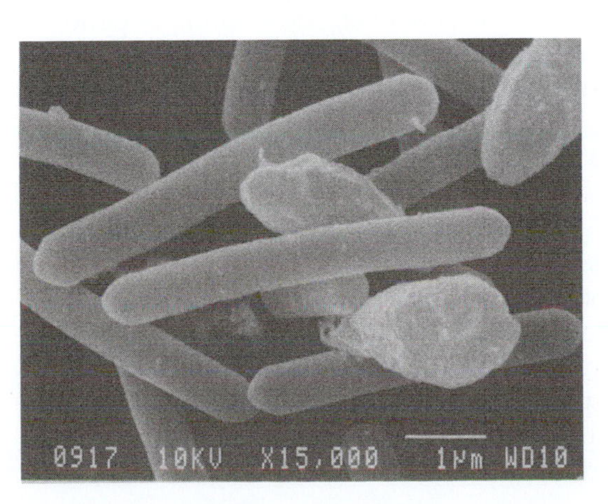

Figure 10.6 Scanning electron micrograph of *C. botulinum* vegetative cells (rods) and spores (irregular ovals). Originally taken by B. Maleeff and kindly provided by P. Cooke, both of the U.S. Department of Agriculture (USDA) Eastern Regional Research Center. doi:10.1128/9781555817206.ch10.f10.06

Table 10.2 Grouping and characteristics of strains of *C. botulinum*[a]

Characteristic	Toxin group			
	I	II	III	IV
Neurotoxin type(s)	A, B, F	B, E, F	C, D	G
Minimum temp (°C) for growth	10	3	15	ND[b]
Optimum temp (°C) for growth	35–40	18–25	40	37
Minimum pH for growth	4.6	ca. 5	ND	ND
Inhibitory NaCl concn (%)	10	5	ND	ND
Minimum a_w for growth	0.94	0.97	ND	ND
$D_{100°C}$ of spores (min)	25	<0.1	0.1–0.9	0.8–1.12
$D_{121°C}$ of spores (min)	0.1–0.2	<0.001	ND	ND

[a]Reprinted from J. W. Austin, p. 329–349, *in* M. P. Doyle, L. R. Beuchat, and T. J. Montville (ed.), *Food Microbiology: Fundamentals and Frontiers*, 2nd ed. (ASM Press, Washington, DC, 2001).

[b]ND, not determined.

at a given temperature). To inhibit growth, the pH must be below 4.6, the salt concentration above 10%, or the a_w below 0.94.

Group II strains are nonproteolytic, have a lower optimum growth temperature (30°C), and grow at temperatures as low as 3°C. The spores have a much lower heat resistance, with $D_{100°C}$ values of less than 0.1 min. Group II strains are inhibited at a pH of <5.0, a salt concentration of >5%, or an a_w of <0.97. The focus in this chapter is on groups I and II, because these strains cause human illness.

Tolerance to Preservation Methods

Temperature, pH, a_w, oxidation-reduction (redox) potential (E_h), added preservatives, and the presence of competing microorganisms are the main factors controlling *C. botulinum* growth (Table 10.2). These factors, used alone, are rarely effective. Usually these factors are used in combination with other inhibitors. These often have synergistic rather than additive effects. (In synergistic combinations, if two treatments which each give a 1-log reduction are used together, a 4-log, rather than a 2-log, reduction may result.) This synergistic effect is the conceptual basis of the multiple-hurdle theory for food preservation (see chapter 2).

Low Temperature

Refrigerated storage prevents *C. botulinum* growth. The lower limits are 10°C for group I and 3°C for group II. However, these limits depend on otherwise optimal growth conditions. Irrespective of the actual minimum growth temperature, production of neurotoxin generally requires weeks at the lower temperature limits. The optimum growth temperatures are from 35 to 40°C for group I and <25°C for group II.

Thermal Inactivation

Thermal processing inactivates *C. botulinum* spores and is the most common way to produce shelf-stable foods. *C. botulinum* spores of group I are very heat resistant, having $D_{121°C}$ values between 0.1 and 0.2 min. A reference temperature of 121°C may seem arbitrary; however, it is the temperature of industrial steam at 15 pounds per square inch. Thus, it has great relevance in canning factories. Group I spores are the main target for the commercial sterilization of canned low-acid foods. The canning industry has adopted a

D value of 0.2 min at 121°C as a standard for calculating thermal processes. The *z* value (the temperature change necessary to cause a 10-fold change in the *D* value) for the most resistant strains is approximately 10°C. Despite variations in *D* and *z* values, the adoption of a 12*D* process as the minimum thermal process for low-acid foods provides a remarkable record of safety.

Strains of group II are considerably less heat resistant ($D_{100°C}$ < 0.1 min). Spores of *C. botulinum* group II can be inactivated at moderate temperatures (40 to 50°C) when heating is combined with high pressures, up to 827 megapascals. The FDA has recently approved the use of high pressure augmented with heat for commercial sterilization of low-acid foods. The survival of spores of *C. botulinum* group II in pasteurized, refrigerated products could be dangerous because they can grow at refrigeration temperatures. These products require special attention. $D_{82°C}$ values of *C. botulinum* type E in neutral phosphate buffer are in the range of 0.2 to 1.0 min. Values ranging from 0.15 to >4.90 min have been reported for type E strains, depending on the heating menstruum, strain, plating agar, and presence of lysozyme (an enzyme that aids the growth of spores that are injured but not killed).

pH

The minimum pH for growth of *C. botulinum* group I is 4.6. For group II, it is about pH 5. Many fruits and vegetables are sufficiently acidic to inhibit *C. botulinum*, whereas acidulants are used to preserve other products. The substrate, temperature, nature of the acidulant, presence of preservatives, a_w, and E_h all influence the acid tolerance of *C. botulinum*. Acid-tolerant microorganisms such as yeasts and molds may grow in acidic products and raise the pH in their immediate vicinity to a level that allows *C. botulinum* growth.

Salt and a_w

Sodium chloride is one of the most important factors used to control *C. botulinum* in foods. It acts primarily by decreasing the a_w. Consequently, the concentration of salt in the water phase, called the brine concentration, is critical.

$$\% \text{ brine} = \frac{\% \text{ NaCl}}{\% \text{ NaCl} + \% \text{ H}_2\text{O}} \times 100$$

The growth-limiting brine concentrations are about 10% for group I and 5% for group II. These concentrations correspond to the limiting a_ws of 0.94 for group I and 0.97 for group II in foods in which NaCl is the main a_w depressant. The solute used to control a_w may influence these limits. Generally, NaCl, KCl, glucose, and sucrose show similar effects, whereas growth occurs in glycerol at lower a_ws. The limiting a_w may be increased substantially by other factors, such as increased acidity or the use of preservatives.

Oxygen and Redox Potential

While one might think that *C. botulinum* cannot grow in foods exposed to O_2, the E_h of most foods is low enough to permit growth. *C. botulinum* grows optimally at an E_h of −350 mV, but growth initiation may occur in the E_h range of 30 to 250 mV. Once growth is initiated, the E_h declines rapidly. Unfortunately, E_h is extremely difficult to measure, so there are few data on this factor. There have been many outbreaks of botulism caused by foods exposed to oxygen when the foods are heated enough to drive out the dissolved oxygen. Oxygen diffusion back into these foods is so slow that they can remain anaerobic for hours.

Across

3. There would probably be more cases of botulism if the toxin were not ________ by heat
5. "Good" botulinal toxin
8. Canned high-acid foods are the perfect place for *C. botulinum* to grow because they are ________
9. Group II strains are characteristically ________
10. Place where spores are found in the environment
11. Type of botulism caused by in situ toxin production

Down

1. Botulinal toxin is the only ________ produced by a foodborne bacterium
2. The most prevalent type of botulism in the year that heroin was contaminated by spores
4. Symptoms caused by botulinal toxin
6. Compound used to inhibit *C. botulinum* in cured meats
7. Paralysis of this part of the body by toxin causes death

CO_2 is used in modified-atmosphere packaging (MAP) to inhibit spoilage and pathogenic microorganisms, but it can stimulate *C. botulinum* growth. Pressurized CO_2 is lethal to *C. botulinum,* with lethality increasing with the pressure of CO_2. The safety of specific atmospheres with respect to *C. botulinum* in foods should be carefully investigated before use.

Preservatives

Nitrite in cured meats such as bacon, ham, and hot dogs inhibits *C. botulinum*. It also contributes to color and flavor. Nitrite's effectiveness depends upon complex interactions among pH, sodium chloride, heat treatment, time and temperature of storage, and the composition of food. Nitrite reacts with many cellular constituents and appears to inhibit *C.*

botulinum by more than one mechanism, including reaction with essential iron-sulfur proteins to inhibit energy-yielding systems in the cell. The reaction of nitrite, or nitric oxide, with secondary amines in meats produces nitrosamines, some of which are carcinogenic. This has led to federal regulations that limit the amount of nitrite used in the manufacture of cured meats. In reality, the levels of nitrites consumed from eating green vegetables are much greater than from eating cured meats.

Sorbates, parabens, nisin, phenolic antioxidants, polyphosphates, ascorbates, ethylenediaminetetraacetate (EDTA), metabisulfite, *n*-monoalkyl maleates and fumarates, and lactate salts are also active against *C. botulinum*.

Competitive and Growth-Enhancing Microorganisms

The growth of competitive and growth-promoting microorganisms in foods influences the fate of *C. botulinum*. Acid-tolerant molds such as *Cladosporium* spp. or *Penicillium* spp. can elevate the pH of acidic foods, allowing *C. botulinum* growth. Other microorganisms may inhibit *C. botulinum* by acidifying the environment, by producing specific inhibitory substances, or both. Lactic acid bacteria, including *Lactobacillus*, *Pediococcus*, and *Streptococcus*, can inhibit *C. botulinum* growth in foods, largely by reducing the pH but also by the production of bacteriocins. The use of lactic acid bacteria and a fermentable carbohydrate, the "Wisconsin process," is permitted in the United States for producing bacon with less nitrite. If the bacon is temperature abused, the lactic acid bacteria ferment the sugar and drop the pH to below 4.5 before toxin can be produced.

Inactivation by Irradiation

C. botulinum spores are probably the most radiation-resistant spores of public health concern. *D* values (irradiation dose required to inactivate 90% of the population) of group I strains at −50 to −10°C are between 2.0 and 4.5 kilograys in neutral buffers and in foods. Spores of type E are more sensitive, having *D* values between 1 and 2 kilograys. Radappertization ("canning" using radiation as the energy source) is designed to reduce the number of viable *C. botulinum* spores by 12 log cycles. Different environmental conditions such as the presence of O_2, change in irradiation temperature, and irradiation and recovery environments can affect the *D* values of spores. Spores in the presence of O_2 or preservatives and at temperatures above 20°C generally have greater irradiation sensitivity.

SOURCES OF *C. BOTULINUM*

Occurrence of *C. botulinum* in the Environment

The contamination of food depends on the incidence of *C. botulinum* in the environment. *C. botulinum* spores are common in soils and sediments, but their numbers and types vary depending on the location (Table 10.3). Type A spores predominate in soils in the western United States, China, Brazil, and Argentina. Type B spores predominate in soils in the eastern United States, the United Kingdom, and much of continental Europe. Most American type B strains are proteolytic, whereas most European type B strains are nonproteolytic. Type E predominates in northern regions and in most temperate aquatic regions. Types C and D are present more frequently in warmer environments.

Table 10.3 Incidence of *C. botulinum* in soils and sediments[a]

Location and sample type	% Positive samples	MPN/kg	% Type[b] A	B	C/D	E	G
Eastern United States, soil	19	21	12	64	12	12	0
Western United States, soil	29	33	62	16	14	8	0
Green Bay, Wisconsin, sediment	77	1,280	0	0	0	100	0
Alaska, soil	41	660	0	0	0	100	0
Britain, soil	6	2	0	100	0	0	0
Scandinavian coast, sediment	100	>780	0	0	0	100	0
Baltic Sea, Finland, offshore	88	1,020	0	0	0	100	0
The Netherlands, soil	94	2,500	0	22	46	32	0
Switzerland, soil[c]	44	48	28	83	6	0	27
Rome, Italy, soil	1	2	86	14	0	0	0
Caspian Sea, Iran, sediment	17	93	0	8	0	92	0
Sinkiang, China, soil	70	25,000	47	32	19	2	0
Ishikawa, Japan, soil	56	16	0	0	100	0	0
Brazil, soil	35	86	57	7	29	0	7
South Africa, soil	3	1	0	100	0	0	0

[a]Adapted from J. W. Austin, p. 329–349, *in* M. P. Doyle, L. R. Beuchat, and T. J. Montville (ed.), *Food Microbiology: Fundamentals and Frontiers*, 2nd ed. (ASM Press, Washington, DC, 2001).

[b]Percentage for each type represents percentage out of all types identified.

[c]The total exceeds 100% because some samples contained multiple serotypes.

Occurrence of *C. botulinum* in Foods

Many surveys have determined the incidence of *C. botulinum* spores in foods (Table 10.4). Food surveys have focused largely on fish, meats, and infant foods, especially honey (due to its association with infant botulism). *C. botulinum* type E spores are common in fish and aquatic animals. Meat products generally have low levels of contamination. They are less likely

Table 10.4 Prevalence of *C. botulinum* spores in food[a]

Product	Origin	% Positive samples	MPN/kg	Type(s) identified
Eviscerated whitefish chubs	Great Lakes	12	14	E, C
Vacuum-packed frozen flounder	Atlantic Ocean	10	70	E
Dressed rockfish	California	100	2,400	A, E
Salmon	Alaska	100	190	A
Vacuum-packed fish	Viking Bank, North Sea	42	63	E
Smoked salmon	Denmark	2	<1	B
Salted carp	Caspian Sea	63	490	E
Fish and seafood	Osaka, Japan	8	3	C, D
Raw meat	North America	<1	0.1	C
Cured meat	Canada	2	0.2	A
Raw pork	United Kingdom	0–14	<0.1–5	A, B, C
Random honey samples	United States	1	0.4	A, B
Honey samples associated with infant botulism	United States	100	8×10^4	A, B

[a]Reprinted from J. W. Austin, p. 329–349, *in* M. P. Doyle, L. R. Beuchat, and T. J. Montville (ed.), *Food Microbiology: Fundamentals and Frontiers*, 2nd ed. (ASM Press, Washington, DC, 2001).

than fish to be contaminated with spores because there is considerably less contamination of the farm environment than the aquatic environment. In North America, the average most probable number (MPN) is ~0.1 spore per kg of meat products, whereas in Europe, the average MPN is ~2.5 spores per kg. The spore types most often associated with meats are A and B.

C. botulinum spores, usually types A and B, can contaminate fruits and vegetables, particularly those in close contact with the soil. Products often contaminated include asparagus, beans, cabbage, carrots, celery, corn, onions, potatoes, turnips, olives, apricots, cherries, peaches, and tomatoes. The overall incidence of *C. botulinum* spores in commercially available precut MAP vegetables is low, approximately 0.36%. Cultivated mushrooms are of special concern, because they are grown in compost and are especially favorable to botulinal growth. Cultivated mushrooms contain up to 2.1×10^3 type B spores per kg. Have you ever wondered why there are holes in the packages of some fresh mushrooms? Anaerobic conditions are created if oxygen diffusion through the package cannot keep up with the oxygen consumed by the respiring mushrooms. These anaerobic conditions, combined with the high incidence of botulinal spores in mushrooms, create a genuine safety hazard. (There are some packing films with high oxygen diffusion rates. Don't panic if your mushroom packages don't have holes in them.)

VIRULENCE FACTORS AND MECHANISMS OF PATHOGENICITY

Botulinal toxin has been called the most poisonous poison known. *C. botulinum* produces eight antigenically distinct toxins, designated types A, B, C_1, C_2, D, E, F, and G. All of the toxins except C_2 are neurotoxins. C_2 and exoenzyme C_3 are adenosine 5′-diphosphate (ADP)-ribosylating enzymes.

All seven of the neurotoxins are similar in structure and mode of action. *C. botulinum* neurotoxins are large (150-kilodalton) two-chain proteins. The neurotoxins block neurotransmission at nerve endings by preventing acetylcholine release. They do this through their action as zinc metalloproteases. Fortunately, botulinal toxins are very heat sensitive. Heating a sample to 80°C for 10 min or bringing it to a boil completely inactivates the toxin.

Structure of the Neurotoxins

Botulinal toxins (Fig. 10.7) are water-soluble proteins produced as a single polypeptide with an approximate molecular weight (M_r) of 150,000. They are cleaved by a protease approximately one-third of the distance from the N terminus. This produces an active neurotoxin composed of one heavy-chain (M_r = 100,000) and one light-chain (M_r = 50,000) protein linked by a single disulfide bond. The heavy chain contains the zinc binding site.

Bacterial proteases or proteases such as trypsin can cause the proteolytic cleavage. When cleaved by a protease, the light chain remains bound to the N-terminal half of the heavy chain by noncovalent bonds and a disulfide bond between Cys-429 and Cys-453. The resultant molecule has increased toxicity relative to the untrypsinized molecule. When the two chains are separated (by breaking of the disulfide bonds), they become nontoxic.

Genetic Regulation of the Neurotoxins

Complete gene sequences have been determined for the neurotoxins produced by *C. botulinum* types A, B (proteolytic and nonproteolytic), C, D, E, F, and G; *Clostridium baratii* type F; and *Clostridium butyricum* type E. (Note that

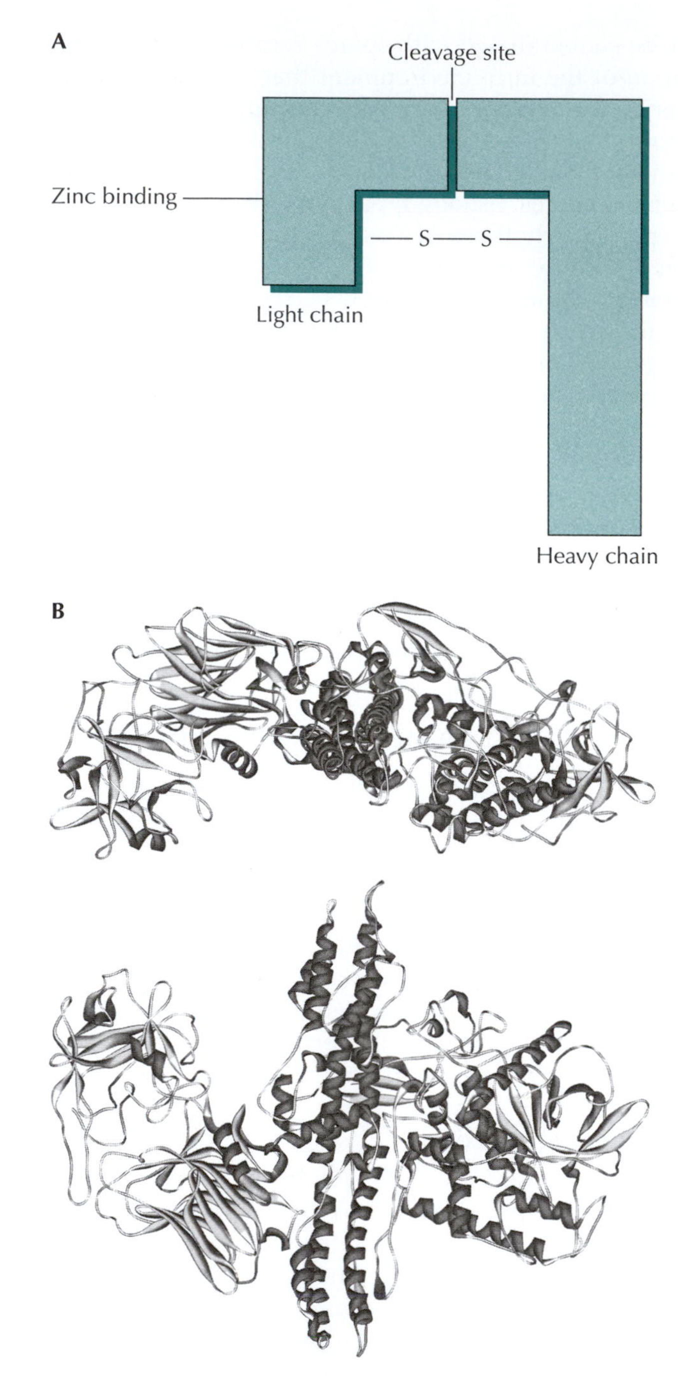

Figure 10.7 Botulinum toxin consists of a single protein that can be activated by proteolyic cleavage into its heavy (carboxy terminus) and light (amino terminus) chains. The disulfide (S-S) bonds that hold it together are not very strong and account for its heat sensitivity. **(A)** Simplified depiction; **(B)** crystalline structure (reprinted from D. B. Lacy, W. Tepp, A. C. Cohen, B. R. DasGupta, and R. C. Stevens, *Nat. Struct. Biol.* **5**:898–902, 1998, with permission). doi:10.1128/9781555817206.ch10.f10.07

"botulinal" toxins can be produced by other clostridial species, although this is rare.) The degree of relatedness of the various neurotoxins is determined by the similarity of the amino acid sequence (i.e., their homology). The locations of the genes coding for botulinal toxin and the associated nontoxic proteins vary depending upon the serotype. The genes coding for botulinal toxins A, B, E, and F and the associated nontoxic proteins are located on the chromosome.

Mode of Action of the Neurotoxins

Botulinal toxin's proteolytic action blocks acetylcholine release from nerve terminals. It has recently been discovered that the toxin is a zinc metalloprotease. It enters the neuron and cleaves synaptobrevin (a protein embedded in the membrane), thereby preventing the release of neurotransmitters. This causes the flaccid paralysis of botulism poisoning. The paralysis can be used to treat several neurological disorders. Under the trade name Botox, botulinal toxin is also used as an alternative to cosmetic surgery to "erase" facial wrinkle lines. When Botox is injected under the skin near the wrinkles, flaccid paralysis of the underlying muscles causes the wrinkles to disappear. What is the difference between the deadly agent of bioterrorism and the benign cosmetic of the rich and famous? As toxicologists like to say, "The dose makes the poison."

Summary

- *Clostridium botulinum* is an anaerobic sporeformer.
- Botulism is a rare but very serious disease. It can be caused by improper canning, temperature abuse, deep puncture wounds, or, in the case of infants, colonization of the digestive tract.
- Symptoms of botulism poisoning are neurological and muscular, such as double vision, slurred speech, and paralysis.
- *C. botulinum* spores are the target of regulations for the canning of low-acid foods.
- Canned foods having a pH of >4.6 and an a_w of >0.86 must receive a "12*D* botulinum cook."
- *C. botulinum* strains produce seven antigenically different neurotoxins.
- Botulinal toxin is a protein that is the most toxic substance known but is easily inactivated by heat.

Suggested reading

Hauschild, A. H. W., and K. L. Dodds. 1993. Clostridium botulinum—*Ecology and Control in Foods*. Marcel Dekker, Inc., New York, NY.

Johnson, E. A. 2007. *Clostridium botulinum*, p. 401–421. *In* M. P. Doyle and L. R. Beuchat (ed.), *Food Microbiology: Fundamentals and Frontiers*, 3rd ed. ASM Press, Washington, DC.

Richmond, J. Y., and R. B. McKinney (ed.). 1999. *Biosafety in Microbiological and Biomedical Laboratories*, 4th ed. U.S. Government Printing Office, Washington DC.

Questions for critical thought

1. Why did the soup in the outbreak have to be vichyssoise soup? (If you know the ingredients of vichyssoise soup and how it is served, the answer will be obvious. If you do not know what vichyssoise is, look it

up.) If creamy Cheddar cheese soup were similarly underprocessed, would you expect people to get sick from it?

2. Create a scenario for a botulism outbreak caused by temperature abuse.
3. Why do most recipes for home-canned foods end with instructions to boil for 10 min before serving?
4. Does it matter that high-acid foods, such as tomato sauce, may contain viable botulinal spores after processing?
5. Over the last 100 years, about 10% of the botulism cases have been attributed to tomatoes or other high-acid foods. How can this be if *C. botulinum* does not grow at pH values of <4.6? Feel free to speculate, but don't use an acid-loving superstrain of *C. botulinum* as your answer.
6. Imagine that it is 2020 and you are a college professor researching botulism. Hundreds of wild geese have died on the college pond. The stench of their floating carcasses is hindering student recruitment efforts. The state Wildlife and Game Commission has bagged some of the dead geese and taken them to their lab for testing. They have just called the dean with the results: type C botulism. The dean (your classmate who got a C in food microbiology) remembers that botulinal toxin is the most poisonous substance known. She calls you in a panic: should she evacuate the campus? What do you tell her?
7. Estimate how many people might be killed if 1 ounce of botulinal toxin were placed in a warhead that dispersed it over a large area. State all your assumptions. Show all your work. Be sure to include an appropriate unit next to each numerical value.
8. You are the secretary of defense. How might you neutralize the threat described in question 7?
9. Write a detailed hypothetical scenario about what factors may have caused the carrot juice outbreak described in the text. Consider manufacturing practices, extrinsic and intrinsic factors, and consumer behavior. Given the large quantity of carrot juice produced, why did the outbreak affect less than a dozen people?
10. After reading the chapter, what question do you have? Exchange your question with a classmate, and research your classmate's answer.
11. Why should BSL-3 labs have negative pressure? What (if any) is the consequence of venting these laboratories directly to the outside?
12. You are an extension specialist at Little State University. Your job is to help small food processors. A few years back, Chef Charles called you about processing his tomato sauce, and you told him about high-acid foods and how to process them in a boiling water bath. Now he has been so successful that he wants to start processing beef gravy in a similar fashion. Charles calls you to see if there is anything else he should know or do. What do you tell him?
13. Read the following and respond.

    WHO: Two couples, Harry and Hilda; Thelma and Louise

    WHAT: Shortness of breath, double vision, slurred speech requiring hospitalization on a respirator and treatment with botulinal antitoxin

    WHEN: About 8 h after a party at Harry and Hilda's house

    WHERE: At home

FACTS ON INVESTIGATION: Both couples had consumed a homemade bean dip which contained garlic. The garlic was Ollie's Olive Garden garlic in oil, packaged in 100% virgin olive oil, with instructions to "keep refrigerated." Botulinal toxin was isolated from the remaining garlic in the jar.

Harry did not refrigerate the product but claimed that it was not refrigerated at Superlow Supermarket when he purchased it. A visit to Superlow revealed that the product was being sold in the condiment aisle, without refrigeration. The manager says that it was delivered on an unrefrigerated truck.

Ollie's Olive Garden, Inc., claims that they are not responsible since the product was actually packed by George's Great Gourmet Gifts. Ollie says that all he did was supply the olive oil. When asked if their distribution system was refrigerated, Ollie replies that the garlic is refrigerated at the place of manufacture, placed on refrigerated trucks, and delivered cold to food wholesalers (middleman), but they have no control of what the wholesaler does with it.

George's Great Gourmet Gifts is actually run by two guys named Joe. They say that they are not responsible because the product was not refrigerated in the correct manner. They also point out that they were packing the garlic according to the specifications of Ollie's Olive Garden, Inc. A discussion with Joe and Joe reveals that they were investors who had always wanted to run their own business. They did not know the reason for refrigerating the garlic, but they did so and kept it on the label because that is what George did when he sold them the company.

At the beginning of the trial, the judge calls all parties into her chambers and tells them, "Someone is responsible for the suffering of these good people. We are going to find out who and make them pay."

You are called on to be the expert witness. What do you tell the judge?

11

Clostridium perfringens

LEARNING OBJECTIVES

The information in this chapter will enable the student to:

- use basic biochemical characteristics to identify *C. perfringens*
- understand which conditions and foods favor *C. perfringens* growth
- recognize, from symptoms, time of onset, and environmental conditions, a case of foodborne illness caused by *C. perfringens*
- choose appropriate interventions (heat, preservatives, and formulation) to prevent *C. perfringens* growth
- identify environmental sources of the organism
- understand the role of spores and the sporulation cycle in *C. perfringens* foodborne illness

THE FOODBORNE ILLNESS

A Spore's-Eye View of *Clostridium perfringens* Toxicoinfections

We spores are part of "the great circle of life." Most of the time, we just wait. Dormant. Patient. Then there is a change. Heat kills our competitors, the vegetative cells. The temperature changes to that of a warm incubator, and we morph into vegetative cells and multiply at an incredibly rapid rate. There are millions of us. A poor human eats us, and we are in perfringens heaven. We have warm temperatures, lots of nutrients, and anaerobic conditions. Our growth produces so much carbon dioxide that the poor human thinks she's having a "gas attack." Oh, the cramps. Then, as nutrients run out, we sporulate back into our dormant form. During sporulation, as we cells break apart, we release a protein toxin that causes diarrhea in our victims. Expelled from our host, we return to the great circle of life. We wait. Dormant. Patient.

Cruel and Unusual Punishment

On the day after Thanksgiving, almost half of the staff and residents of a juvenile detention center became ill with diarrhea (100%), abdominal cramps (90%), nausea (63%), and vomiting (30%). *C. perfringens* at 10^6 colony-forming units (CFU)/g was isolated from five of the nine stool specimens tested. Fortunately, the victims recovered in 24 to 48 h. Most of the people who ate the Thanksgiving feast were stricken. None of those who ate at a separate off-campus event became ill. The consumption of turkey, bread, potatoes, and gravy was implicated statistically. However, because the lunch ladies had served all of the implicated foods to everybody, the outbreak could not be attributed to a single food.

doi:10.1128/9781555817206.ch11

The investigation found that inmates had thawed frozen turkey breasts the day before the meal and then cooked them until they "looked" done. The cooked turkeys were stacked in large pots and placed in a walk-in cooler. Large pots of the gravy were also placed in the cooler. The next day the turkey and gravy were heated and served. Some of the kitchen help remembered that the turkey was still warm when it was removed from the cooler.

This outbreak illustrates the classic attributes of a *C. perfringens* outbreak. Large pieces of meat were inadequately cooked. Their cooling was slow, and reheating was probably inadequate. This outbreak could have been prevented if the turkey had been adequately cooked and cooled to below 40°C in less than 2 h. Slicing it into smaller portions facilitates rapid heating and cooling.

Incidence

C. perfringens type A toxicoinfection is the third most common foodborne disease in the United States. From 1993 to 1997, there were 40 outbreaks (representing 4.1% of total foodborne disease outbreaks) of *C. perfringens* type A toxicoinfection in the United States. (A toxicoinfection results when bacteria simultaneously produce toxins and cause infections.) These accounted for 2,772 cases (6.3% of total cases of bacterial foodborne diseases). For the period from 1998 to 2002, there were 6,742 cases, with four deaths, in the United States. Other countries report a higher proportion; in Norway 30% of foodborne illness cases are attributed to *C. perfringens*. However, like all foodborne illnesses, most cases of *C. perfringens* type A toxicoinfection are not recognized or reported. By conservative estimates, there are 250,000 cases of *C. perfringens* type A foodborne illness per year in the United States. These result in an average of seven deaths. The economic costs associated with *C. perfringens* type A toxicoinfections probably exceed $240 million. There are also huge economic losses from *C. perfringens* fatal infections of cattle, sheep, and pigs.

Outbreaks of *C. perfringens* type A toxicoinfections are usually large (~50 to 100 cases) and often occur in institutional settings. The large size of outbreaks is due to two factors. First, large institutions often prepare food in advance and hold it for later serving. This lets *C. perfringens* grow if the food is temperature abused. Second, given the relatively mild and routine symptoms of most *C. perfringens* cases, public health officials usually become involved only when many people get sick. *C. perfringens* foodborne illness occurs throughout the year but, like most foodborne illness, is more common during summer months. The warmer weather contributes to temperature abuse during cooling and holding. Summer also provides more occasions (like picnics, fairs, and carnivals) where food is prepared by people who are having too much fun to worry about safe food handling.

Vehicles for *C. perfringens* Foodborne Illness

C. perfringens cannot make 13 of the 20 amino acids required for its growth. Thus, high-protein foods rich in amino acids are the most common vehicles for *C. perfringens*. *C. perfringens* foodborne illness in the United States is most commonly associated with meats and poultry. There are recent reports of *C. perfringens* being isolated from fresh and processed seafood. Other products that are high in protein and subject to temperature abuse (e.g., gravies and stews) are also significant sources of the organism.

Factors Contributing to *C. perfringens* Type A Foodborne Illness

C. perfringens foodborne illness usually results from temperature abuse during the cooking, cooling, or holding of foods. Its optimum growth temperature of 43°C gives it an edge over other mesophilic pathogens in temperature-abused foods. The Centers for Disease Control and Prevention (CDC) reports that improper storage or holding temperatures contributed to 100% of recent *C. perfringens* outbreaks. Improper cooking was a factor in ~30% of these outbreaks. Contaminated equipment contributed to ~15% of the outbreaks.

The importance of temperature abuse in *C. perfringens* foodborne illness is not surprising. *C. perfringens* vegetative cells are relatively heat tolerant. However, *C. perfringens* spores are much more heat resistant than vegetative cells. Incomplete cooking can induce the germination of the *C. perfringens* spores which survive the heating. If the food is improperly cooled or stored, the germinated spores can produce the vegetative cells that can multiply rapidly. Under optimum conditions *C. perfringens* vegetative cells can double in number every 10 min.

Preventing *C. perfringens* Type A Foodborne Illness

Thorough cooking is the best intervention against *C. perfringens* foodborne illness. This is particularly important for large roasts and turkeys. Because of their size, it is hard to reach the high internal temperatures needed to kill *C. perfringens* spores. The difficulty of cooking large pieces of meat to such high internal temperatures helps explain why those foods are such common vehicles for *C. perfringens* outbreaks. A second, and perhaps even more important, intervention is to rapidly cool and store cooked foods to temperatures at which *C. perfringens* vegetative cells cannot grow (e.g., below 40°F or above 140°F).

Authors' note

The U.S. Food Code suggests that food be held at below 40°F or above 140°F to prevent microbial growth. U.S. Department of Agriculture regulations dictate that meats should pass through this range within 2 h.

Identification of *C. perfringens* Type A Foodborne Illness Outbreaks

Public health agencies use criteria such as incubation time and illness symptoms or type and history of food vehicles (e.g., is temperature-abused meat or poultry involved?) for identifying *C. perfringens* outbreaks. However, the similarities between the onset times and symptoms of *C. perfringens* foodborne illness and other illnesses such as *Bacillus cereus* toxicoinfection make it unwise to completely rely on clinical and epidemiological features for identifying outbreaks. The presence of large numbers of spores in the feces or the detection of specific toxins is more reliable in identifying outbreaks. The bacteriological criteria used by the CDC to identify an outbreak are the presence of either 10^5 *C. perfringens* organisms/g of stool from two or more ill persons, or 10^5 *C. perfringens* organisms/g of the suspect food. Because *C. perfringens* is widely distributed in the environment, and <5% of environmental samples make toxin, simply finding the organism in the food or feces is not proof of an outbreak. Most strains of *C. perfringens* in food or feces do not carry the gene for toxin production and are unable to cause illness. The CDC and Food and Drug Administration (FDA) now use the detection of *C. perfringens* enterotoxin in feces of multiple people who are sick to identify *C. perfringens* outbreaks.

Since *C. perfringens* enterotoxin can be present in feces of people suffering from nonfoodborne gastrointestinal (GI) diseases (such as antibiotic-associated diarrhea), the presence of *C. perfringens* enterotoxin in feces from a single individual is not sufficient to identify an outbreak. However, detection of *C. perfringens* enterotoxin in feces from several people provides strong evidence for an outbreak. This is particularly true when they consumed a common food, developed illness within typical incubation times, and had the symptoms characteristic of *C. perfringens* toxicoinfection. The use of fecal *C. perfringens* toxin detection for identifying *C. perfringens* outbreaks is limited because fecal samples must be collected soon after the onset of symptoms. It is often difficult to recruit donors under these conditions. However, several commercially available serologic kits are available to detect *C. perfringens* enterotoxin in feces.

CHARACTERISTICS OF *C. PERFRINGENS* TYPE A FOODBORNE ILLNESS

Symptoms of *C. perfringens* type A toxicoinfection develop 8 to 16 h after consumption of contaminated food. They last about 12 to 24 h. Victims of *C. perfringens* toxicoinfection usually suffer only diarrhea and severe abdominal cramps. There is usually no fever. While death rates from *C. perfringens* type A toxicoinfection are low, there are fatalities in debilitated or elderly populations.

A typical *C. perfringens* type A toxicoinfection is illustrated in Fig. 11.1. Initially, temperature abuse stimulates *C. perfringens* cells to rapidly multiply in food. Large numbers of the bacteria are consumed. Some vegetative cells survive passage through the stomach and remain viable when they enter the small intestine. In the small intestine they multiply, sporulate, and release enterotoxin. *C. perfringens* enterotoxin is made during the sporulation of *C. perfringens* cells in the small intestines. After being released in the intestine, *C. perfringens* enterotoxin quickly binds to and damages intestinal epithelial cells. This intestinal damage initiates fluid loss, i.e., diarrhea.

Two factors help explain why *C. perfringens* toxicoinfection is usually relatively mild and self-limited. First, diarrhea probably flushes (this word is not chosen lightly) unbound *C. perfringens* toxin cells from the small intestine. Second, *C. perfringens* toxin preferentially affects villus tip cells. These are the oldest intestinal cells and are rapidly replaced in young, healthy individuals by normal cell turnover.

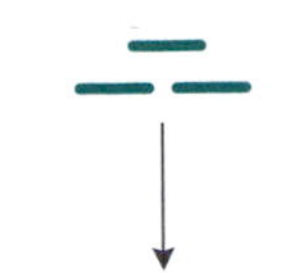

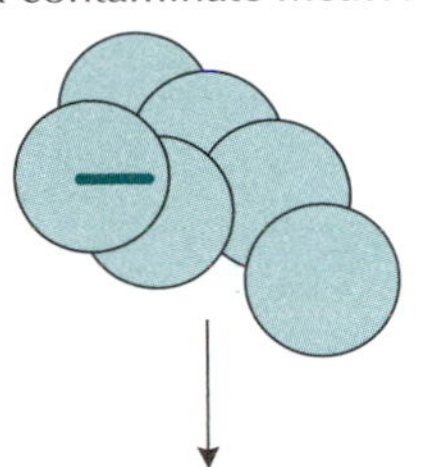

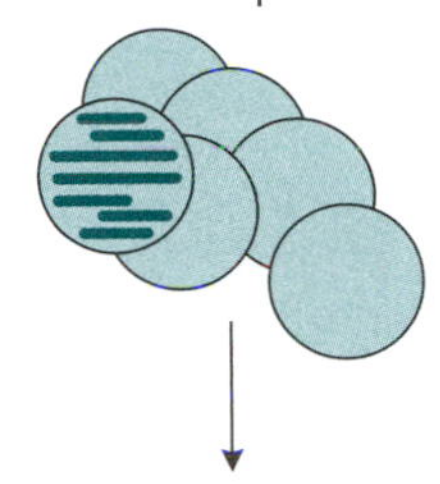

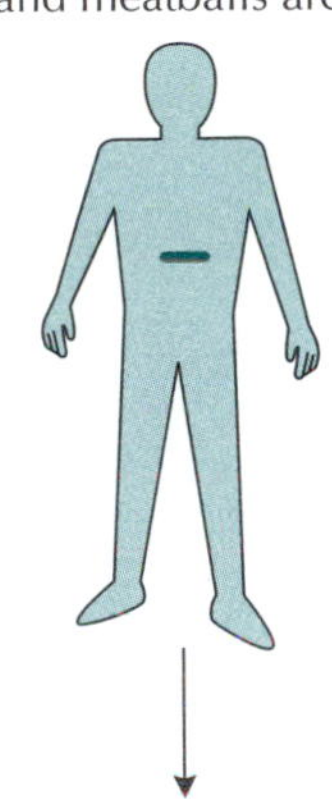

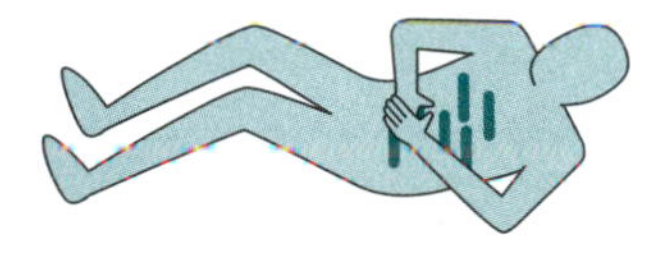

Figure 11.1 Schematic representation of *C. perfringens* food poisoning. Vegetative cells or spores contaminate a meat product and multiply rapidly when food is "incubated." A person consumes the vegetative cells, which then multiply rapidly in the small intestine (producing lots of gas) and sporulate, releasing the toxin at the same time as the spores. The victim is very sick with gas, cramps, and diarrhea but recovers in 24 to 48 h. doi:10.1128/9781555817206.ch11.f11.01

INFECTIOUS DOSE FOR *C. PERFRINGENS* TYPE A FOODBORNE ILLNESS

Most *C. perfringens* cells are killed by stomach acidity. Hence, cases of *C. perfringens* toxicoinfection usually develop only when a heavily contaminated food (i.e., a food containing $>10^6$ to 10^7 *C. perfringens* vegetative cells/g) is eaten. The *C. perfringens* toxin is produced in the victim, and then *C. perfringens* organisms sporulate in their intestines. Therefore, this illness is considered a toxicoinfection.

Everyone is susceptible to *C. perfringens* toxicoinfection, but the illness is most serious in elderly or debilitated individuals. Many people develop a transient serum antibody response to *C. perfringens* toxin following illness. However, previous exposure provides no future protection.

THE ORGANISM

Overview

Clostridium perfringens, formerly called *C. welchii*, became recognized as a foodborne pathogen in the 1940s and 1950s, following the pioneering work of K. Knox, K. Mat-Donald, L. S. McClung, and E. Hobbs. *C. perfringens* is a gram-positive, rod-shaped, encapsulated, nonmotile bacterium that causes a

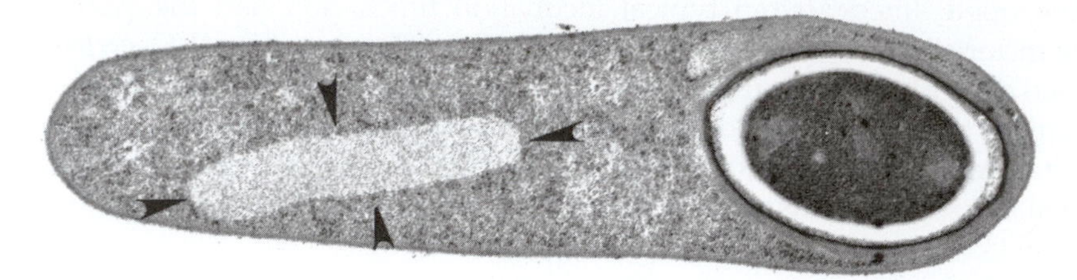

Figure 11.2 Transmission electron micrograph of a sporulating *C. perfringens* cell. Note the terminal fully developed spore and the paracrystalline inclusion body (arrowheads). Courtesy of Ron Labbe, University of Massachusetts. doi:10.1128/9781555817206.ch11.f11.02

broad spectrum of human and veterinary diseases. *C. perfringens* is considered a "facultative anaerobe," since it can be plated on an open bench if it is then incubated anaerobically. *C. perfringens* virulence results from its prolific toxin-producing ability, including production of at least two toxins, i.e., *C. perfringens* enterotoxin and β-toxin, that are active on the human GI tract. These cause *Clostridium perfringens* toxicoinfection and necrotic enteritis. Foodborne necrotic enteritis has a 15 to 25% fatality rate but is extremely rare in industrialized societies. We do not discuss it further here.

In addition to producing GI tract-active toxins, *C. perfringens* has other traits that help it cause foodborne disease. *C. perfringens* grows rapidly, doubling in number in <10 min. This allows *C. perfringens* to quickly multiply in food. *C. perfringens* forms spores (Fig. 11.2) that are resistant to stresses such as radiation, desiccation, and heat. More specifically, they survive in incompletely cooked or inadequately reheated food.

Classification: Toxin Typing of *C. perfringens*

There are at least 14 different *C. perfringens* toxins. Fortunately, no individual *C. perfringens* cell carries the gene(s) encoding all of these toxins. This limitation provides the basis for a toxin typing system. This system classifies *C. perfringens* isolates into five types (A through E), depending upon an isolate's ability to make 4 (alpha, beta, epsilon, and iota) of the 14 toxins (Table 11.1). Table 11.1 identifies the toxins' biological activities and their

Table 11.1 Toxin typing and characteristics of *C. perfringens* toxins[a]

	Biological activity of toxin produced by each type[b]				
***C. perfringens* type classification**	**Alpha (phospholipase C)**	**Beta (necrotizing)**	**Epsilon (permease)**	**Iota (ADP ribosylating)**	**Enterotoxin**
A	+	–	–	–	+
B	+	+	+	–	+
C	+	+	–	–	+
D	+	–	+	–	+
E	+	–	–	+	+
Gene	*plc*	*cpb1, cpb2*	*ext*	*iap, iba*	*cpe*
Gene location	Chromosome	Plasmid	Plasmid	Plasmid	Chromosome, plasmid

[a]Compiled from S. Brynestad and P. E. Granum, *Int. J. Food Microbiol.* **74:**195–200, 2002; C. L. Hatheway, *Clin. Microbiol. Rev.* **3:**66–98, 1990; and B. A. McClane, p. 423–444, *in* M. P. Doyle and L. R. Beuchat (ed.), *Food Microbiology: Fundamentals and Frontiers*, 3rd ed. (ASM Press, Washington, DC, 2007).

[b]+, positive for this phenotype; –, negative for this phenotype.

genes. The epsilon toxin is on the U.S. Department of Homeland Security's list of biothreat agents. It can be aerosolized, damages cell walls, and causes potassium leakage. There is no vaccine to protect against it.

The two foodborne diseases caused by *C. perfringens* are associated with different *C. perfringens* types. Necrotic enteritis is caused by type C isolates which produce β-toxin. As implied by its name, *C. perfringens* type A toxicoinfection is associated with type A isolates of *C. perfringens*, even though other *C. perfringens* types sometimes make a toxin with similar properties. *C. perfringens* enterotoxin-producing type A isolates are more widely distributed than other types of *C. perfringens* and are thus responsible for most of the illness.

Control of *C. perfringens*

The growth of pathogens in food is affected by factors such as temperature, oxidation-reduction potential (E_h), pH, and water activity (a_w) levels. The effects of these factors on *C. perfringens* growth are briefly discussed below.

Temperature

Due to their heat resistance, *C. perfringens* spores survive in undercooked foods. Spore heat resistance is influenced by environmental and genetic factors. The environment in which a *C. perfringens* spore is heated affects its heat resistance. Spores of some *C. perfringens* strains survive boiling for an hour or longer. There are also genetic variations in heat resistance. Perhaps due to selective pressure, spores of toxicoinfectious isolates generally have much greater heat resistance than spores of other *C. perfringens* isolates. Finally, incomplete cooking of foods not only may fail to kill *C. perfringens* spores but also can actually favor *C. perfringens* type A toxicoinfection by inducing spore germination.

Vegetative cells of *C. perfringens* are somewhat heat tolerant. Although not truly thermophilic, they have a relatively high optimal growth temperature (43 to 45°C) and can often grow at 50°C. Vegetative cells of toxicoinfectious isolates are about twofold more heat resistant than vegetative cells of other *C. perfringens* isolates.

Growth rates of *C. perfringens* cells decrease rapidly at temperatures below ~15°C. No growth occurs at 6°C. In contrast, spores of *C. perfringens* are cold resistant. Toxicoinfection can result if viable spores present in refrigerated or frozen foods germinate when that food is warmed for serving.

Other Factors

Growth of *C. perfringens* in food is also influenced by a_w, E_h, pH, and (probably) the presence of curing agents, such as nitrites. *C. perfringens* is less tolerant of low-a_w environments than is *Staphylococcus aureus*. The lowest a_w supporting vegetative growth of *C. perfringens* is 0.93 to 0.97, depending upon the solute used to control the a_w.

C. perfringens does not require an extremely reduced environment for growth. If the environmental E_h is suitably low for initiating growth, *C. perfringens* can then decrease the E_h of its environment (by producing reducing molecules such as ferredoxin) to produce more optimal growth conditions. The E_h of many common foods (e.g., raw meats and gravies) is low enough to support *C. perfringens* growth.

C. perfringens growth is also pH sensitive. Optimal growth occurs at pH 6 to 7, whereas *C. perfringens* grows poorly, if at all, at pH values of ≤ 5 and ≥ 8.3.

Preservation factors such as pH, a_w, and, perhaps, curing agents control *C. perfringens* by inhibiting the outgrowth of *C. perfringens* spores. However, ungerminated spores may remain viable in foods even when cell growth is prevented. Those spores may germinate later if the growth-limiting factor(s) is removed during food preparation.

RESERVOIRS FOR *C. PERFRINGENS* TYPE A

C. perfringens is present in natural environments, including soil (at levels of 10^3 to 10^4 CFU/g), foods (e.g., approximately 50% of raw or frozen meat contains *C. perfringens*), dust, and the intestinal tracts of humans and domestic animals (e.g., human feces usually contain 10^4 to 10^6 *C. perfringens* organisms/g). The widespread distribution of *C. perfringens* is linked to its frequent occurrence in foodborne illness. However, recent studies revealed that the ubiquitous distribution of *C. perfringens* in nature is not especially relevant for understanding the reservoir(s) for this pathogen. This is because <5% of all *C. perfringens* isolates harbor the *cpe* (*C. perfringens* enterotoxin) gene, required for toxin production. Only *C. perfringens* toxin-positive isolates (i.e., with the *cpe* gene) cause foodborne illness.

The *cpe* gene is located on the chromosome or on a plasmid. However, the *cpe* gene of most toxicoinfectious isolates is located on the chromosome. This strong association between foodborne illness and the chromosomal *cpe* gene provides another criterion for identifying toxicoinfectious isolates in nature. By learning the location of the *cpe* gene in environmental isolates, microbiologists will be able to answer critical questions about the ecology of *C. perfringens* toxicoinfectious isolates. Questions to be answered include the following. Are those isolates present, in low numbers, in some healthy human carriers? Are they present in some food animals? Do they enter foods during processing? Do they enter foods during final handling, cooking, or holding?

VIRULENCE FACTORS CONTRIBUTING TO *C. PERFRINGENS* TYPE A FOODBORNE ILLNESS

Heat Resistance

As mentioned above, most cases of *C. perfringens* toxicoinfections are caused by isolates carrying a chromosomal *cpe* gene. This association between isolates with chromosomal *cpe* and toxicoinfections is odd; *C. perfringens* isolates carrying chromosomal or plasmid *cpe* genes make similar levels of the same toxin. However, cells with chromosomal *cpe* have greater heat resistance than cells with plasmid *cpe*. Since cooked meats cause most outbreaks, the greater heat resistance of isolates with chromosomal *cpe* favors their survival during incomplete cooking.

C. perfringens Enterotoxin

Evidence that *C. perfringens* Enterotoxin Is Involved in Foodborne Illness

C. perfringens toxin is classified as an enterotoxin because it induces the loss of fluid from the intestinal tract. There is strong epidemiological evidence that *C. perfringens* enterotoxin causes *C. perfringens* type A foodborne illness. This evidence includes the following.

1. The presence of *C. perfringens* toxin in a victim's feces is strongly correlated to illness.

2. The toxin causes serious intestinal effects in experimental animals.
3. Human volunteers fed purified *C. perfringens* toxin (yum. . .) get sick with the symptoms of *C. perfringens* foodborne illness.
4. The intestinal inflammation caused by *C. perfringens* toxin-positive isolates in experimental animals can be neutralized with *C. perfringens* toxin-specific antisera.

More recently, experiments that fulfilled Koch's postulates on a molecular basis confirmed the importance of enterotoxin for *C. perfringens* pathogenesis. Sporulating (but not vegetative) culture lysates of a *C. perfringens cpe*-positive strain induced fluid accumulation in animals. This is consistent with *C. perfringens* toxin (whose expression is sporulation associated) being necessary for toxicity. However, *cpe*-negative mutants did not make animals sick. Finally, when the *cpe* gene was reintroduced into the mutant, virulence was restored.

C. perfringens type A toxicoinfection is not the only disease involving *C. perfringens* toxin. *C. perfringens* toxin-producing isolates also cause nonfoodborne human GI illnesses. These include antibiotic-associated diarrhea, sporadic diarrhea, and some veterinary diarrheas. The *cpe*-positive *C. perfringens* isolates causing nonfoodborne GI diseases are genetically distinct from those causing toxicoinfections. That is, *cpe* is located on the chromosome of toxicoinfectious isolates but on a plasmid in nonfoodborne disease isolates.

Expression and Release of *C. perfringens* Enterotoxin

C. perfringens enterotoxin expression and release have three interesting features. (i) The toxin is made only by sporulating vegetative cells. (ii) The toxin is not *secreted* by the sporulating cells but is released when the mother cell lyses. (iii) Many *C. perfringens* toxin-producing isolates make extremely large amounts of toxin.

Synthesis of *C. perfringens* enterotoxin. Synthesis of *C. perfringens* toxin begins when cells sporulate and increases for the next 6 to 8 h. After 6 to 8 h of sporulation, *C. perfringens* toxin can represent up to 30% of the total cell protein. Why do some *C. perfringens* strains produce so much toxin during sporulation? The amount of *C. perfringens* toxin made by an isolate is not influenced by the location of the *cpe* gene. Nor is the amount of toxin made related to a gene dosage effect; all *cpe*-positive isolates carry only a single copy of the *cpe* gene. There is a general relationship between an isolate's sporulation ability and its toxin production; i.e., the better a *C. perfringens* isolate sporulates, the more *C. perfringens* toxin is produced. However, this correlation is not absolute.

Authors' note

Cells can have multiple copies of a given gene. One would expect that a higher gene copy number would result in a larger amount of the gene product (i.e., toxin). This is not always true.

Release of *C. perfringens* enterotoxin from *C. perfringens*. *C. perfringens* toxin accumulates in cells during its synthesis, often reaching concentrations high enough to form paracrystalline inclusion bodies (see Fig. 11.2). Intracellular *C. perfringens* toxin is released into the intestines at the end of sporulation, when the mother cell lyses to free its mature spore. The need for the mother cell to lyse and release the *C. perfringens* toxin explains, at least in part, why (despite *C. perfringens* toxin's quick intestinal action) symptoms develop only 8 to 24 h after ingestion. *C. perfringens* cells must grow in the intestine and complete sporulation before toxin is released. This takes 8 to 12 h.

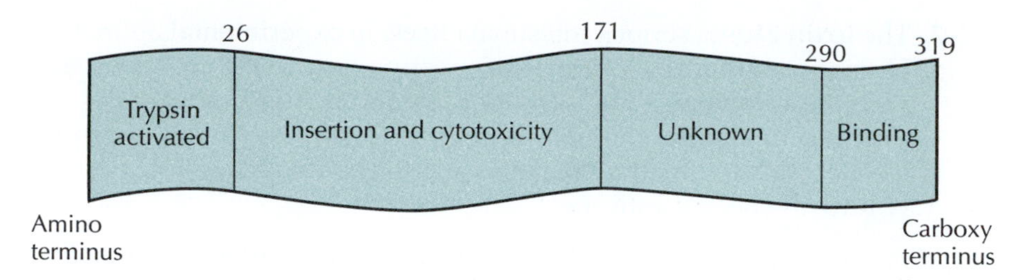

Figure 11.3 Stylized representation of *C. perfringens* enterotoxin showing the region that is cleaved by trypsin, the region responsible for insertion and cytotoxicity, and the binding site. The amino acid sequence numbers are given at the top. Modified from S. Brynestad and P. E. Granum, *Int. J. Food Microbiol.* **74:**195–200, 2002. doi:10.1128/9781555817206.ch11.f11.03

Biochemistry of *C. perfringens* Enterotoxin

C. perfringens toxin is heat sensitive. Its biological activity can be inactivated by heating for 5 min at 60°C. The toxin is also quite sensitive to pH extremes but is resistant to some proteases. In fact, limited treatment with either of the proteolytic enzymes trypsin and chymotrypsin increases *C. perfringens* toxin activity two- to threefold. This suggests that human intestinal proteases may activate *C. perfringens* toxin.

The enterotoxin is a single 3.5-kilodalton protein with specific regions conferring trypsin activation, binding, insertion, and cytotoxicity (Fig. 11.3). The enterotoxin acts through a multistep process. First it binds to a specific receptor on the intestinal epithelial cells. Then the entire enterotoxin molecule is inserted into the epithelial cell's membrane. This releases ions and large amounts of water.

C. perfringens Enterotoxin Action on the GI Tract

The principal target for *C. perfringens* enterotoxin is the small intestine. Several features differentiate the biological activity of *C. perfringens* toxin from those of cholera and *Escherichia coli* heat-labile toxins. *C. perfringens* toxin induces direct cellular damage to the small intestine. The villus tips are particularly sensitive. While other bacterial toxins (e.g., Shiga toxin and *Clostridium difficile* toxins) are also cytotoxic and cause intestinal tissue damage, *C. perfringens* toxin is unique with respect to how quickly it damages intestinal tissue. *C. perfringens* toxin-induced intestinal damage can develop in 15 to 30 min.

Summary

- *C. perfringens* is an anaerobic spore-forming bacterium commonly found on proteinaceous foods.
- The toxin is released as a paracrystalline protein when vegetative cells release their spores in the gut.
- Improper heating or cooling is the major cause of *C. perfringens* type A toxicoinfection.
- The *cpe* gene encodes the enterotoxin; it is located chromosomally in toxicoinfectious isolates but on the plasmids of isolates not associated with foods.
- Victims of *C. perfringens* type A toxicoinfections usually suffer only diarrhea and severe abdominal cramps.

Suggested reading

Food and Drug Administration. 2011. *The Bad Bug Book: Foodborne Pathogenic Microorganisms and Natural Toxins.* http://www.fda.gov/Food/FoodSafety/FoodborneIllness/FoodborneIllnessFoodbornePathogensNaturalToxins/BadBugBook/default.htm (accessed 17 December 2011).

Labbe, R. G. 1989. *Clostridium perfringens,* p. 192–234. *In* M. P. Doyle (ed.), *Microbial Foodborne Pathogens.* Marcel Dekker, Inc., New York, NY.

McClane, B. A. 2001. *Clostridium perfringens,* p. 423–444. *In* M. P. Doyle and L. R. Beuchat (ed.), *Food Microbiology: Fundamentals and Frontiers,* 3rd ed. ASM Press, Washington, DC.

Questions for critical thought

1. Create an imaginary *C. perfringens* outbreak. Describe how the food was prepared and handled and the steps that led to the outbreak. Also include the victims' symptoms.
2. Would you rather be a spore or a vegetative cell? Why?
3. You are an eminent professor of food microbiology, awaiting delivery of food for your son's graduation party. The caterer arrives with 10 trays of Swedish meatballs, lasagna, and chicken parmesan. She assures you that "they are good and cold, just heat them up for a couple of hours on the steam tables." What do you do?
4. What is the difference between a "case" and an "outbreak"? Calculate the number of cases per outbreak for *C. perfringens* and some other foodborne pathogens. What can you infer from these numbers?
5. What is the biggest single causative factor in the occurrence of *Clostridium perfringens* type A toxicoinfections?
6. What characteristics of *C. perfringens* contribute to its ability as a foodborne pathogen?
7. Both *C. perfringens* and *C. botulinum* foodborne diseases are linked to spores. How are these diseases different? How are they similar? Why are their frequencies so different?
8. When foods are heated from 40 to 140°F or cooled from 140 to 40°F, it is important to get through this interval as rapidly as possible to prevent microbial growth. Rapid heating is easily achieved by obvious means. However, improper cooling causes as many outbreaks as improper heating and is much harder to control. Think of three ways to cool foods through the danger zone rapidly.
9. What is a paracrystalline inclusion body?
10. Here are several facts. Multiple copies of the *cpe* gene do not cause increased pathogenicity. Isolates from foodborne disease usually have *cpe* on the chromosome rather than the plasmid. Spores from isolates with a chromosomal *cpe* gene are more heat resistant than those from isolates that have *cpe* on the plasmid. How do you relate these to the role of *cpe* in *C. perfringens* foodborne illness? What experiments could you do to refine this understanding?

12 Enterohemorrhagic *Escherichia coli*

LEARNING OBJECTIVES

The information in this chapter will enable the student to:

- use basic biochemical characteristics to identify *Escherichia coli* O157:H7
- understand what conditions in foods favor *E. coli* O157:H7 growth
- recognize, from symptoms and time of onset, a case of foodborne illness caused by *E. coli* O157:H7
- choose appropriate interventions (heat, preservatives, and formulation) to prevent the growth of *E. coli* O157:H7
- identify environmental sources of *E. coli* O157:H7
- understand the role of *E. coli* O157:H7 toxins and virulence factors in causing foodborne illness
- become familiar with newly emerging Shiga toxin-producing *E. coli* strains

Outbreak

Many children and even adults like to consume raw cookie dough while waiting for cookies to bake. During the summer of 2009, 65 persons became infected with *Escherichia coli* O157:H7 after eating refrigerated prepackaged cookie dough. Twenty-five persons were hospitalized, and 7 developed hemolytic-uremic syndrome (HUS). More than 70% of the persons who became ill were <19 years of age. This has been the only outbreak of *E. coli* O157:H7 illness linked to prepackaged cookie dough. This problem is easily solved by waiting 30 min and then consuming the cookies when they are freshly baked.

Cheese, anyone? Not so fast. Thirty-eight people from five states became ill following the consumption of Dutch-style Gouda cheese. The outbreak strain was isolated from unopened packages of the Dutch-style Gouda cheese purchased at a retail store in New Mexico. No deaths were associated with the outbreak, although one person developed HUS. Twenty years ago, outbreaks of *E. coli* O157:H7 illnesses were predominantly associated with consumption of contaminated undercooked ground beef. However, food microbiologists must now be aware that any type of food can carry this diarrheal pathogen.

Consuming a fresh, cool, crisp spinach salad on a hot summer day pleases the palate and cools the body. However, during the summer of 2006, consuming fresh spinach resulted in 206 cases of illness and three deaths in 26 states. Investigators determined that the spinach was contaminated with *E. coli* O157:H7. How, when, and where the spinach became contaminated could not be conclusively determined. Wild pigs captured in the area where the spinach was grown tested positive for the strain of *E. coli* O157:H7 that was associated with the bagged

doi:10.1128/9781555817206.ch12

spinach and isolated from patients. The U.S. Food and Drug Administration emphasized the need to implement good agricultural practices.

The newbies! Based on reported illnesses and concern for human health, the list of Shiga toxin (Stx)-producing *E. coli* (STEC) serotypes that are of concern in the United States now includes O26, O103, O111, O121, O45, and O145. Not on this list is O104, which was associated with a large outbreak in Germany in 2011. Approximately 4,000 cases, including 900 HUS cases and 50 deaths, were reported. The outbreak was linked to the consumption of bean sprouts. The sprouts were produced from fenugreek seeds imported from Egypt; the seeds became contaminated at some point before leaving the exporter. The outbreak generated international interest and political storms associated not with the magnitude of persons becoming ill but, rather, with the assertion by the Germans that cucumbers imported from Spain were the cause of the outbreak. Consumer desire to have a variety of fresh produce year-round may usher a greater number of unfortunate international events similar to the one described.

INTRODUCTION

Large outbreaks of *E. coli* O157:H7 infection involving hundreds of cases have occurred in the United States, Canada, Japan, and the United Kingdom. The largest outbreak worldwide occurred from May to December 1996 in Japan, involving >11,000 reported cases. In the same year, 21 elderly people died in a large outbreak involving 501 cases in central Scotland. Although *E. coli* O157:H7 is still the predominant serotype of enterohemorrhagic *E. coli* (EHEC) in the United States, Canada, the United Kingdom, and Japan, an increasing number of outbreaks and sporadic cases related to other EHEC serotypes have occurred. A large epidemic, involving several thousand cases, of *E. coli* O157:NM infection occurred in Swaziland and South Africa following consumption of contaminated surface water. In continental Europe, Australia, and Latin America, non-O157 EHEC infections are more common than *E. coli* O157:H7 infections. Details of many reported foodborne and waterborne outbreaks of EHEC infection are provided in Table 12.1. There are no distinguishing biochemical phenotypes for non-O157 EHEC serotypes, making screening for these bacteria difficult and labor-intensive. The prevalence of non-O157 EHEC infections may be greatly underestimated.

Categories of *E. coli*

Diarrheagenic *E. coli* is categorized into specific groups based on virulence properties, mechanisms of pathogenicity, clinical syndromes, and distinct O:H serotypes (Table 12.2). *E. coli* isolates are serologically differentiated based on three major surface antigens that allow serotyping: the O (somatic), H (flagellar), and K (capsule) antigens. The categories of diarrheagenic *E. coli* are enterotoxigenic *E. coli* (ETEC), enteroinvasive *E. coli* (EIEC), enteropathogenic *E. coli* (EPEC), enteroaggregative *E. coli* (EAEC), diffusely adhering *E. coli* (DAEC), and EHEC. All categories of *E. coli* described may be shed in the feces of infected humans, creating the potential for spread to other humans, animals, and the environment (Box 12.1). This chapter focuses primarily on EHEC, which causes the most severe illness.

ETEC

ETEC is a major cause of infantile diarrhea in developing countries. It is also the bacterium most frequently responsible for traveler's diarrhea. ETEC colonizes the small intestine via fimbrial colonization factors and produces

Table 12.1 Representative foodborne and waterborne outbreaks of *E. coli* O157:H7 and other EHEC infections[a]

Yr	Mo	Location(s)	No. of cases (deaths)	Setting	Vehicle(s)
1982	February	Oregon	26	Community	Ground beef
1982	May	Michigan	21	Community	Ground beef
1985		Canada	73 (17)	Nursing home	Sandwiches
1987	June	Utah	51	Custodial institution	Ground beef; person to person
1988	October	Minnesota	54	School	Precooked ground beef
1989	December	Missouri	243	Community	Water
1990	July	North Dakota	65	Community	Roast beef
1991	November	Massachusetts	23	Community	Apple cider
1991	July	Oregon	21	Community	Swimming water
1992	December	Oregon	9	Community	Raw milk
1993	January	California, Idaho, Nevada, Washington	732 (4)	Restaurant	Ground beef
1993	March	Oregon	47	Restaurant	Mayonnaise?
1993	August	Oregon	27	Restaurant	Cantaloupe
1994	November	Washington, California	19	Home	Salami
1995	October	Kansas	21	Wedding	Punch, fruit salad
1995	July	Montana	74	Community	Leaf lettuce
1995	September	Maine	37	Camp	Lettuce
1996	May, June	Connecticut, Illinois	47	Community	Mesclun lettuce
1996	July	Osaka, Japan	7,966 (3)	Community	White radishes, sprouts
1996	October	California, Washington, Colorado	71 (1)	Community	Apple juice
1996	November	Central Scotland	501 (21)	Community	Cooked meat
1997	July	Michigan	60	Community	Alfalfa sprouts
1997	November	Wisconsin	13	Church banquet	Meatballs, coleslaw
1998	June	Wyoming	114	Community	Water
1998	July	North Carolina	142	Restaurant	Coleslaw
1998	September	California	20	Church	Cake
1999	August	New York	900 (2)	Fair	Well water
2000	November	Iowa, Minnesota, Wisconsin	52	Community	Lettuce
2002	July	10 states	38	Community	Ground beef
2002	August	Washington	32	Camp	Ground beef
2005	September, October	Minnesota	23	Community	Prepackaged lettuce
2006	August–October	26 states, Canada	206 (3)	Community	Prepackaged fresh spinach
2007	April	8 states	40	Community	Ground beef patties
2008	March	7 states	49	Community	Ground beef
2009	March–June	30 states	72	Community	Prepackaged cookie dough
2010	October	5 states	38	Community	Gouda cheese
2011[b]	May, June	Germany, France	>4,236 (50)	Community	Prepackaged sprouts

[a]*E. coli* O157:H7 unless otherwise noted.

[b]*E. coli* O104:H4.

Table 12.2 Some properties and symptoms associated with pathogenic *E. coli* subgroups[a]

Property or symptom	ETEC	EPEC	EHEC	EIEC
Toxin	LT/ST[b]	–	Stx or Vero toxin	–
Invasive	–	–	–	+
Intimin	–	+	+	–
Enterohemolysin	–	–	+	–
Stool	Watery	Watery, bloody	Watery, very bloody	Mucoid, bloody
Fever	Low	+	–	+
Fecal leukocytes	–	–	–	+
Intestine involved	Small	Small	Colon	Colon, lower small
Serology	Various	O26, O111, others	O157:H7, O26, O111, others	Various
I_D[c]	High	High	Low	High

[a]Derived from http://www.fda.gov/food/scienceresearch/laboratorymethods/bacteriologicalanalyticalmanualbam/ucm070080.htm.
[b]LT, labile toxin; ST, stable toxin.
[c]I_D, infective dose.

BOX 12.1

Conditions that can lead to the spread of pathogenic *E. coli*

This historic illustration of an old farm outhouse (toilet) underscores what science has demonstrated: pathogens present in human feces such as *E. coli* (EPEC, ETEC, and EHEC) can readily be spread to farm animals, humans, and water through improper controls. In this case, the placement of a dilapidated outhouse near the main residence allows easy access to livestock that may eventually become dinner.

doi:10.1128/9781555817206.ch12.fBox12.01

a heat-labile or heat-stable enterotoxin that causes fluid accumulation and diarrhea. The most frequent ETEC serogroups include O6, O8, O15, O20, O25, O27, O63, O78, O85, O115, O128ac, O148, O159, and O167. Humans are the principal reservoir of ETEC strains that cause human illness.

EIEC

EIEC causes nonbloody diarrhea and dysentery similar to those caused by *Shigella* spp. by invading and multiplying within intestinal epithelial cells. Like that of *Shigella*, the invasive capacity of EIEC is associated with the presence of a large plasmid (ca. 140 megadaltons [MDa]) that encodes several outer membrane proteins involved in invasiveness. The antigenicities of these outer membrane proteins and the O antigens of EIEC are closely related. The principal site of bacterial localization is the colon, where EIEC organisms invade and grow within epithelial cells, causing cell death. Humans are a major reservoir, and the serogroups most frequently associated with illness include O28ac, O29, O112, O124, O136, O143, O144, O152, O164, and O167. Among these serogroups, O124 is the most commonly encountered.

EPEC

EPEC can cause severe diarrhea in infants, especially in developing countries. EPEC was previously associated with outbreaks of diarrhea in nurseries in developed countries. The major O serogroups associated with illness include O55, O86, O111ab, O119, O125ac, O126, O127, O128ab, and O142. Humans are an important reservoir. EPEC organisms have been shown to induce lesions in cells to which they adhere, and they can invade epithelial cells.

EAEC

EAEC is associated with persistent diarrhea in infants and children in several countries worldwide. These *E. coli* strains are different from the other types of pathogenic *E. coli* because of their ability to produce a characteristic pattern of aggregative adherence on HEp-2 cells. EAEC adheres with the appearance of stacked bricks to the surfaces of HEp-2 cells. Serogroups associated with EAEC include O3, O15, O44, O77, O86, O92, O111, and O127. A gene probe derived from a plasmid associated with EAEC strains has been developed to identify *E. coli* strains of this type. More epidemiological information is needed to determine the significance of EAEC as a cause of diarrheal disease.

DAEC

DAEC has been associated with diarrhea primarily in young children who are older than infants. The relative risk of DAEC-associated diarrhea increases with age from 1 to 5 years. The reason for this age-related infection is unknown. DAEC strains are most commonly of serogroup O1, O2, O21, or O75. Typical symptoms of DAEC infection are mild diarrhea without blood. DAEC strains adhere in a random fashion to HEp-2 or HeLa cell lines. DAEC does not usually produce heat-labile or heat-stable enterotoxin or elevated levels of Stx.

EHEC

EHEC was first recognized as a human pathogen in 1982, when *E. coli* O157:H7 was identified as the cause of two outbreaks of bloody diarrhea. Since then, other serotypes of *E. coli*, such as O26, O111, and sorbitol-fermenting

O157:NM, also have been associated with cases of bloody diarrhea and are classified as EHEC. Serotype O157:H7 is the main cause of EHEC-associated disease in the United States and many other countries. All EHEC strains produce factors *cytotoxic* (deadly) to African green monkey kidney (Vero) cells. Thus, they are named verotoxins or Stxs because of their similarity to the Stx produced by *Shigella dysenteriae*. STEC infections are linked to a severe and sometimes fatal condition, HUS. *E. coli* organisms of many different serotypes produce Stxs; hence, they have been named STEC (or EHEC). More than 200 serotypes of EHEC have been isolated from humans, but only those strains that cause bloody diarrhea are considered to be EHEC. Major non-O157:H7 EHEC serotypes include O26:H11, O111:H8, and O157:NM. Several outbreaks of infection with the EHEC serogroup O111 have occurred worldwide. In June 1999, an outbreak of *E. coli* O111:H8 infection involving 58 cases occurred at a teenage cheerleading camp in Texas. Contaminated ice was the implicated vehicle. In Australia, from January to February 1995, a cluster of 23 cases of HUS were associated with consumption of *E. coli* O111:NM-contaminated semidry fermented sausage. Since O157:H7 is the most common serotype of EHEC and because more is known about this serotype than about other serotypes of EHEC, this chapter focuses on *E. coli* O157:H7 (Box 12.2).

CHARACTERISTICS OF *E. COLI* O157:H7 AND NON-O157 EHEC

E. coli is a common part of the normal microbial population in the intestinal tracts of humans and other warm-blooded animals. Most *E. coli* strains are harmless; however, some strains are pathogenic and cause diarrheal disease. At present, a total of 167 O antigens, 53 H antigens, and 74 K antigens have been identified. It is considered necessary to determine only the O and H antigens in order to serotype strains of *E. coli* associated with diarrheal disease. The O antigen identifies the serogroup of a strain, and the H antigen identifies its serotype. The application of serotyping to isolates associated with diarrheal disease has shown that particular serogroups often fall into one category of diarrheagenic *E. coli*. However, some serogroups, such as O55, O111, O126, and O128, appear in more than one category.

E. coli O157:H7 was first identified as a foodborne pathogen in 1982, although there had been prior isolation of the organism, which was identified later among isolates at the Centers for Disease Control and Prevention. The earlier isolate was from a California woman who had bloody diarrhea in 1975. In addition to production of Stx(s), most strains of *E. coli* O157:H7 also have several characteristics uncommon in most other *E. coli* strains, i.e., the inability to grow well, if at all, at temperatures of ≥44.5°C in *E. coli* broth; inability to ferment sorbitol within 24 h; inability to produce β-glucuronidase (i.e., inability to hydrolyze 4-methylumbelliferyl-D-glucuronide); possession of an attaching-and-effacing (AE) gene (*eae*); and carriage of a 60-MDa plasmid. Non-O157 EHEC strains do not share these growth and metabolic characteristics, although they all produce Stx(s) and most carry the *eae* gene and large plasmid.

BOX 12.2

E. coli O157:H7

E. coli O157:H7 emerged as a pathogen in 1982. The initial cases of illness were predominantly linked to the consumption of beef products, but recently, illnesses have been linked to contaminated fruits and vegetables, recreational water (lakes, swimming pools, and water parks), and person-to-person spread. The organism has acquired many tools (e.g., production of Stx and the ability to produce AE lesions) that facilitate its survival in a host but that can cause debilitating illness. As with *Shigella*, illness may be caused by <100 cells. In the United States, *E. coli* O157:H7 is the predominant EHEC strain. In other parts of the world, this is not necessarily the case.

Acid Tolerance

Unlike most foodborne pathogens, many strains of *E. coli* O157:H7 are unusually tolerant of acidic environments. The minimum pH for *E. coli* O157:H7 growth is 4.0 to 4.5, but this is dependent upon the interaction of the pH with other factors. For instance, organic acid sprays containing acetic, citric, or lactic acid do not affect the level of *E. coli* O157:H7 on beef. *E. coli* O157:H7, when inoculated at high levels, has survived fermentation, drying, and storage of fermented sausage (pH 4.5) for up to 2 months at 4°C and has survived in mayonnaise (pH 3.6 to 3.9) for 5 to 7 weeks at 5°C and for 1 to 3 weeks at 20°C and in apple cider (pH 3.6 to 4.0) for 10 to 31 days or for 2 to 3 days at 8 or 25°C, respectively.

Three systems in *E. coli* O157:H7 are involved in acid tolerance: an acid-induced oxidative system, an acid-induced arginine-dependent system, and a glutamate-dependent system. The oxidative system is less effective in protecting the organism from acid stress than the arginine-dependent and glutamate-dependent systems. The alternate sigma factor RpoS is required for oxidative acid tolerance but is only partially involved with the other two systems. Once induced, the acid-tolerant state can persist for a prolonged period (≥28 days) at refrigeration temperatures. More importantly, induction of acid tolerance in *E. coli* O157:H7 can also increase tolerance of other environmental stresses, such as heating, radiation, and antimicrobials.

Antibiotic Resistance

When *E. coli* O157:H7 was first associated with human illness, the pathogen was susceptible to most antibiotics affecting gram-negative bacteria. However, it is becoming increasingly resistant to antibiotics. *E. coli* O157:H7 strains isolated from humans, animals, and food have developed resistance to multiple antibiotics, with streptomycin-sulfisoxazole-tetracycline being the most common resistance profile. Non-O157 EHEC strains isolated from humans and animals also have acquired antibiotic resistance, and some are resistant to multiple antimicrobials commonly used in human and animal medicine. Antibiotic-resistant EHEC strains possess a selective advantage over other bacteria colonizing the intestines of animals that are treated with antibiotics (therapeutically or subtherapeutically). Therefore, antibiotic-resistant EHEC strains may become the primary *E. coli* strains present under antibiotic selective pressure and thus more prevalent in feces. Perhaps fortunately, antibiotic therapy is not recommended in cases of human illness, since the disease may become worse through the release of endotoxin and Stxs from the dead bacteria.

Inactivation by Heat and Irradiation

E. coli O157:H7 is not more heat resistant than other pathogens. The presence of certain compounds in food can protect the organism. The presence of fat protects *E. coli* O157:H7 in ground beef, with *D* values (see chapter 2) for lean (2.0% fat) and fatty (30.5% fat) ground beef of 4.1 and 5.3 min at 57.2°C and 0.3 and 0.5 min at 62.8°C, respectively. Pasteurization of milk (72°C for 16.2 s) is an effective treatment that kills $>10^4$ *E. coli* O157:H7 organisms per ml. Proper heating of foods of animal origin, e.g., heating foods to an internal temperature of at least 68.3°C for several seconds, is an important critical control point to ensure inactivation of *E. coli* O157:H7.

Many countries have approved the use of irradiation to eliminate foodborne pathogens in food. Unlike other processing technologies, irradiation eliminates foodborne pathogens yet maintains the raw character of foods. In the United States, 4.5 kilograys (kGy) is approved for refrigerated raw ground beef and 7.5 kGy is approved for frozen raw ground beef. Foodborne bacterial pathogens are relatively sensitive to irradiation. Doses required to reduce the *E. coli* O157:H7 population by 90% in raw ground beef patties range from 0.241 to 0.307 kGy, depending on the temperature, with values significantly higher for patties irradiated at −16 than at 4°C. Hence, an irradiation dose of 1.5 kGy should be sufficient to eliminate *E. coli* O157:H7 at the levels that may occur in ground beef. Following the large produce-related outbreaks in the United States in 2006, the U.S. Food and Drug Administration in 2008 approved irradiation of iceberg lettuce and spinach to help protect consumers from disease-causing bacteria.

RESERVOIRS OF *E. COLI* O157:H7

Detection of *E. coli* O157:H7 and Other EHEC Strains on Farms

Undercooked ground beef and, less frequently, unpasteurized milk have caused many outbreaks of *E. coli* O157:H7 infection, and hence, cattle have been the focus of many studies of their role as a reservoir of *E. coli* O157:H7. The first link between *E. coli* O157:H7 and cattle was observed in Argentina in 1977 and was associated with a <3-week-old calf. Rates of EHEC carriage as high as 60% have been found in bovine herds in many countries, but in most cases, the rates range from 10 to 25%. The rates of isolation of *E. coli* O157:H7 are much lower than those of non-O157 EHEC. Rates of EHEC carriage by dairy cattle on farms in Canada range from 36% of cows and 57% of calves in 80 herds tested. In the United States, 31 (3.2%) of 965 dairy calves and 191 (1.6%) of 11,881 feedlot cattle tested were positive for *E. coli* O157:H7. Young, weaned animals more frequently carry *E. coli* O157:H7 than do adult cattle. Cattle are more likely to test positive for *E. coli* O157:H7 during the warmer months of the year. This agrees with the seasonal variation in human disease.

Factors Associated with Bovine Carriage of *E. coli* O157:H7

There may be an association between fecal shedding of *E. coli* O157:H7 and feed or environmental factors. For example, some calf starter feed regimens or environmental factors and feed components, such as whole cottonseed, are linked with a reduction of *E. coli* O157:H7. In contrast, keeping calves in groups before weaning, sharing of calf-feeding utensils without sanitation, and early feeding of grain are associated with increased carriage of *E. coli* O157:H7. Grain and hay feeding programs may influence *E. coli* O157:H7 carriage by cattle.

Environmental factors, such as water and feed sources, or farm management practices, such as manure handling, may play important roles in influencing the prevalence of *E. coli* O157:H7 on dairy farms. The pathogen is frequently found in water troughs on farms. *E. coli* O157:H7 can survive for weeks or months in bovine feces and water.

The susceptibility of cattle to intestinal colonization with *E. coli* O157:H7 is largely a function of age. Young animals are more likely to be positive than older animals in the same herd. For many cattle, *E. coli* O157:H7 is transiently carried in the gastrointestinal tract and is intermittently excreted for a few weeks to months by young calves and heifers. Cows can carry more than one strain of *E. coli* O157:H7.

Cattle Model for Infection by *E. coli* O157:H7

E. coli O157:H7 is not a pathogen of weaned calves and adult cattle; hence, animals that carry the pathogen are not ill. However, there is evidence that *E. coli* O157:H7 can cause diarrhea and lesions in newborn calves. The initial sites of localization of *E. coli* O157:H7 in cattle are the forestomachs (rumen, omasum, and reticulum).

Domestic Animals and Wildlife

Although cattle are thought to be the main source of EHEC in the food chain, EHEC strains have also been isolated from other domestic animals and wildlife, such as sheep, goat, deer, dogs, horses, swine, and cats. *E. coli* O157:H7 was also isolated from seagulls and rats. The prevalence of *E. coli* O157:H7 and other EHEC strains in sheep is generally higher than in other animals. In a survey of seven animal species in Germany, EHEC strains were isolated most frequently from sheep (66.6%), goats (56.1%), and cattle (21.1%), with lower prevalence rates in chickens (0.1%), pigs (7.5%), cats (13.8%), and dogs (4.8%). Carriage of the pathogen by feral animals represents a significant risk for the contamination of food crops (lettuce, spinach, and leafy greens) in the field.

Humans

Fecal shedding of *E. coli* O157:H7 by patients with hemorrhagic colitis or HUS usually lasts 13 to 21 days following the onset of symptoms. However, in some instances, the pathogen can be excreted in feces for >3 weeks. A child infected during a day care center outbreak excreted the pathogen for 62 days. People living on dairy farms have elevated titers of antibody against *E. coli* O157:H7; however, the pathogen was not isolated from feces. An asymptomatic long-term carrier state has not been identified. Fecal carriage of *E. coli* O157:H7 by humans is significant because of the potential for person-to-person spread of the pathogen. A factor contributing to person-to-person spread is the bacterium's extraordinarily low infectious dose. Fewer than 100 cells, and possibly as few as 10 cells, can cause illness. Inadequate attention to personal hygiene, especially after using the bathroom, can transfer the pathogen through contaminated hands, resulting in secondary transmission.

DISEASE OUTBREAKS

Geographic Distribution

E. coli O157:H7 is a cause of many major outbreaks worldwide. At least 30 countries on six continents have reported *E. coli* O157:H7 infection in humans. In the United States, 350 outbreaks of *E. coli* O157:H7 infection were documented in 49 states from 1982 to 2002, accounting for 8,598 cases. Although the number of outbreaks has increased, the median outbreak size declined from 1982 to 2002. The increase may be due in part to improved recognition of *E. coli* O157:H7 infection following the publicity of a large multistate outbreak in the western United States in 1993. Since 1 January 1994, individual cases of *E. coli* O157:H7 infection have been reportable to the National Notifiable Diseases Surveillance System. The exact number of *E. coli* O157:H7 illnesses in the United States is not known because infected persons with mild or no symptoms and persons with nonbloody diarrhea often do not seek medical attention, and these cases would not be reported. The Foodborne Diseases Active Surveillance Network (FoodNet [http://www.cdc.gov/ncidod/dbmd/foodnet/]) reports that the annual rate of *E. coli* O157:H7 infection at several surveillance sites in the United States ranged from 2.1 to 2.8 cases per 100,000 population for 1996 to 1999.

However, the number of cases declined to 0.9 case per 100,000 population from 2000 to 2004. In a 2011 report, the Centers for Disease Control and Prevention estimated that *E. coli* O157:H7 causes 63,153 illnesses and 26 deaths annually in the United States and that non-O157 STEC strains account for an additional 112,757 cases, with 20 deaths. Approximately 75% of these cases are due to foodborne transmission.

Seasonality of *E. coli* O157:H7 Infection

Outbreaks and clusters of *E. coli* O157:H7 infection peak during the warmest months of the year. Approximately 89% of outbreaks and clusters reported in the United States occur from May to November. The reasons for this seasonal pattern are unknown but may include (i) an increased prevalence of the pathogen in cattle during the summer, (ii) greater human exposure to ground beef or other *E. coli* O157:H7-contaminated foods during the cookout months, and/or (iii) more improper handling (temperature abuse and cross-contamination) or incomplete cooking of products such as ground beef during warm months than during other months.

Age of Patients

All age groups can be infected by *E. coli* O157:H7, but the very young and the elderly most frequently experience severe illness with complications. HUS usually occurs in children, whereas thrombotic thrombocytopenic purpura (TTP) occurs in adults. Children 2 to 10 years of age are more likely to become infected with *E. coli* O157:H7. The high rate of infection in this age group is likely the result of increased exposure to contaminated foods, contaminated environments, and infected animals, as well as more opportunities for person-to-person spread between infected children with relatively undeveloped hygiene skills and undeveloped immune systems.

Transmission of *E. coli* O157:H7

Many foods are identified as vehicles of *E. coli* O157:H7 infection. Examples include ground beef, roast beef, cooked meats, venison jerky, salami, raw milk, pasteurized milk, yogurt, cheese, ice cream bars and cake, lettuce, spinach, unpasteurized apple cider and juice, cantaloupe, potatoes, radish sprouts, alfalfa sprouts, fruit or vegetable salad, and coleslaw. Among the 183 outbreaks reported in the United States for which a vehicle has been identified, ground beef was the most frequent food vehicle (41%), while produce accounted for 21% and 23% were of unknown origin (Fig. 12.1). Contact of foods with meat or feces (human or bovine) contaminated with *E. coli* O157:H7 is a likely source of cross-contamination. Outbreaks attributed to person-to-person (including secondary) transmission (25.5%) and waterborne (particularly recreational water) transmission (12.4%) have occurred.

Examples of Foodborne and Waterborne Outbreaks

The Original Outbreaks

The first documented outbreak of *E. coli* O157:H7 infection occurred in Oregon in 1982, with 26 cases and 19 persons hospitalized. All of the patients had bloody diarrhea and severe abdominal pain. This outbreak was associated with eating undercooked hamburgers from fast-food restaurants of a specific chain. *E. coli* O157:H7 was recovered from the stools of patients. A second outbreak occurred 3 months later and was associated

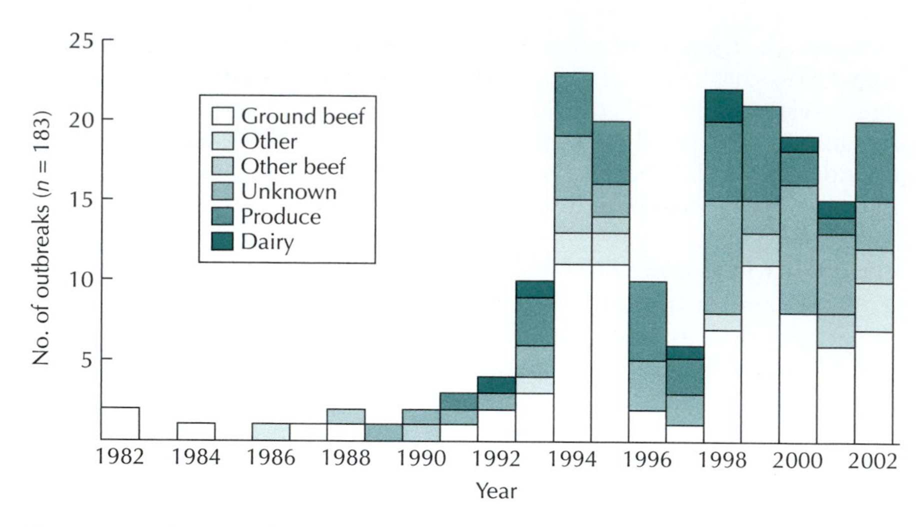

Figure 12.1 Epidemiology of *E. coli* O157:H7 outbreaks, United States, 1982 to 2002. Adapted from J. M. Rangel, P. H. Sparling, C. Crowe, P. M. Griffin, and D. L. Swerdlow, *Emerg. Infect. Dis.* **11:**603–609, 2005. doi:10.1128/9781555817206.ch12.f12.01

Authors' note

E. coli *O157:H7 infections are associated with more than food and water. Each year, children develop* E. coli *O157:H7 infection following a visit to a local farm. Children under the age of 5 years should be extra cautious around cattle (including those in petting zoos). Washing hands thoroughly with running water and soap after contact with cattle is recommended.*

with the same fast-food restaurant chain in Michigan, with 21 cases and 14 persons hospitalized, with an age range of 4 to 58 years. Contaminated hamburgers again were the cause. *E. coli* O157:H7 was isolated both from patients and from a frozen ground beef patty. *E. coli* O157:H7 was identified as the cause by its association with the food and by recovery of bacteria with identical microbiological characteristics from both the patients and the meat from the implicated supplier.

Waterborne Outbreaks

Reported waterborne outbreaks of *E. coli* O157:H7 infection have increased alarmingly in recent years. Swimming water, drinking water, well water, and ice have all been implicated. Investigations of lake-associated outbreaks revealed that in some instances the water was likely contaminated with *E. coli* O157:H7 by toddlers who defecated while swimming. Swallowing lake water was subsequently identified as the cause. A large waterborne outbreak of *E. coli* O157:H7 among attendees of a county fair in New York occurred in August 1999. More than 900 persons were infected, 65 of whom were hospitalized. Two persons, including a 3-year-old girl, died from HUS (kidney failure), and a 79-year-old man died from HUS-TTP. Unchlorinated well water used to make beverages and ice was identified as the vehicle, and *E. coli* O157:H7 was isolated from samples of well water. Waterborne outbreaks of *E. coli* O157 infections have also been reported in Scotland, southern Africa, and Japan.

Outbreaks from Apple Cider and Juice

The first confirmed outbreak of *E. coli* O157:H7 infection associated with apple cider occurred in Massachusetts in 1991, involving 23 cases. In 1996, three outbreaks of *E. coli* O157:H7 infection associated with unpasteurized apple juice or cider were reported in the United States. The largest of the three occurred in three western states (California, Colorado, and Washington) and British Columbia, Canada, with 71 confirmed cases and one death. *E. coli* O157:H7 was isolated from the apple juice involved.

An outbreak also occurred in Connecticut, with 14 cases. Manure contamination of apples was the suspected source of *E. coli* O157:H7 in several of the outbreaks. Using apple drops (i.e., apples picked up from the ground) for making apple cider is a common practice. Apples can become contaminated by resting on soil contaminated with manure. Apples also can become contaminated if transported or stored in areas that contain manure or if they are treated with contaminated water. Investigation of the 1991 outbreak in Massachusetts revealed that the cider press processor also raised cattle that grazed in a field adjacent to the cider mill. These outbreaks led to the U.S. Food and Drug Administration Hazard Analysis and Critical Control Point regulations for juice. These require processors to achieve a 5-log-unit reduction in the numbers of the most resistant pathogen in their finished products.

A Large Multistate Outbreak

A large multistate outbreak of *E. coli* O157:H7 infection in the United States occurred in Washington, Idaho, California, and Nevada in early 1993. Approximately 90% of the primary cases were associated with eating at a single fast-food restaurant chain (chain A), at which *E. coli* O157:H7 was isolated from hamburger patties. The number of people who became ill increased because of secondary spread (48 patients in Washington alone) from person to person. One hundred seventy-eight people were hospitalized, 56 developed HUS, and four children died. The outbreak resulted from insufficient cooking of hamburgers by chain A restaurants. Hamburgers cooked according to chain A's cooking procedures in Washington State had internal temperatures below 60°C, which is substantially less than the minimum internal temperature of 68.3°C required by the state of Washington. Cooking the patties to an internal temperature of 68.3°C would have killed the low numbers of *E. coli* O157:H7 organisms in the ground beef.

Outbreaks Associated with Vegetables

Raw vegetables, particularly lettuce and alfalfa sprouts, have been implicated in outbreaks of *E. coli* O157:H7 infection in North America, Europe, and Japan. In the United States, outbreaks associated with lettuce have continued to increase during the past decade. In May 1996, mesclun lettuce was associated with a multistate outbreak of 47 cases in Illinois and Connecticut. In 2006, a large outbreak in multiple states and Canada was associated with consumption of contaminated prebagged spinach.

Between May and December 1996, multiple outbreaks of *E. coli* O157:H7 infection occurred in Japan, involving 11,826 cases and 12 deaths. The largest outbreak, in Osaka in July 1996, affected 7,892 schoolchildren and 74 teachers and staff, among whom 606 individuals were hospitalized, 106 had HUS, and 3 died. Contaminated white radish sprouts were the source of *E. coli* O157:H7.

CHARACTERISTICS OF DISEASE

The spectrum of human illness caused by *E. coli* O157:H7 infection includes nonbloody diarrhea, bloody diarrhea (hemorrhagic colitis), kidney disease (HUS), and TTP (the adult form of HUS). Some persons may be infected but exhibit no signs or symptoms of illness (this is known as *asymptomatic infection*). Ingestion of the organism is followed by a 3- to 4-day incubation period (range, 2 to 12 days), during which colonization of the large intestine occurs. Illness begins with nonbloody diarrhea and severe abdominal cramps for 1 to 2 days. This progresses in the second or third day of illness to bloody diarrhea that lasts for 4 to 10 days. Many outbreak investigations revealed that >90% of microbiologically documented cases of diarrhea caused by *E. coli* O157:H7 were bloody, but in some outbreaks 30% of cases have involved nonbloody diarrhea. Symptoms usually end after 1 week, but ~6% of patients progress to HUS. The case fatality rate from *E. coli* O157:H7 infection is ~1%. Typically 2 to 3 weeks may elapse before a case of *E. coli* O157:H7 is confirmed (Fig. 12.2).

HUS largely affects children, for whom it is the leading cause of acute kidney failure. The syndrome is characterized by three features: acute renal insufficiency, microangiopathic hemolytic anemia, and thrombocytopenia. TTP affects mainly adults and resembles HUS histologically. It is accompanied by distinct neurological abnormalities resulting from blood clots in the brain.

INFECTIOUS DOSE

The infectious dose of *E. coli* O157:H7 is thought to be extremely low. For example, between 0.3 and 15 colony-forming units (CFU) of *E. coli* O157:H7 per g were enumerated in lots of frozen ground beef patties associated with a 1993 multistate outbreak in the western United States. Similarly, 0.3 to 0.4 CFU of *E. coli* O157:H7 per g was detected in several intact packages of salami that were associated with a foodborne

Figure 12.2 Confirmation of a case of *E. coli* O157:H7 may require 2 to 3 weeks. A typical timeline, with associated processing steps, is shown. Redrawn from http://www.cdc.gov/ecoli/reportingtimeline.htm. doi:10.1128/9781555817206.ch12.f12.02

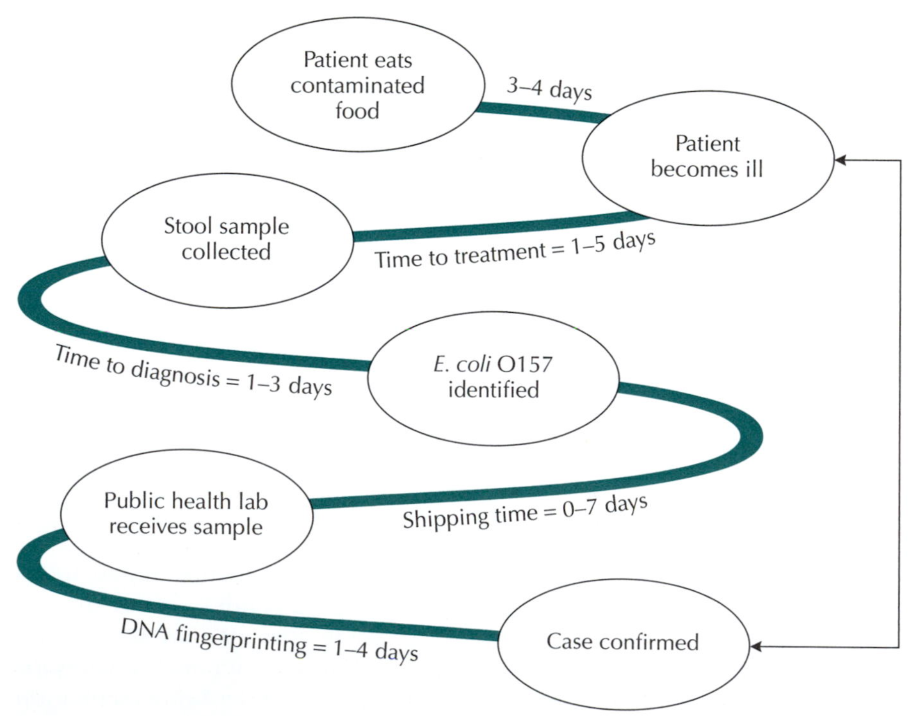

outbreak. These data suggest that the infectious dose of *E. coli* O157:H7 may be <100 cells. Additional evidence for a low infectious dose is the capability for person-to-person and waterborne transmission of EHEC infection. Age, immune status, and preexisting debilitating health conditions can individually or collectively influence whether an individual becomes ill.

MECHANISMS OF PATHOGENICITY

The exact mechanism of pathogenicity of EHEC is not known. General knowledge about the pathogenicity of EHEC indicates that the bacteria cause disease by their ability to adhere to the host cell membrane and colonize the large intestine. They then produce one or more Stxs. However, the mechanisms of intestinal colonization are not well understood. The roles of various other potential virulence factors and host factors remain to be determined. The virulence factors involved in the pathogenesis of EHEC are summarized in Table 12.3.

Attaching and Effacing

Adherence factors and their genes have been characterized, but the mechanisms of adherence and colonization by *E. coli* O157:H7 and other EHEC strains have not been characterized completely. Studies with animal models revealed that the pathogen can colonize the colons of orally infected animals by an AE mechanism (Fig. 12.3). The AE lesion is characterized by intimate attachment of the bacteria to intestinal cells. This results in loss of microvilli on the epithelial cells and accumulation of actin (F actin) in the cytoplasm. It is believed that similar mechanisms regulate the processes of AE formation by *E. coli* O157:H7.

The Locus of Enterocyte Effacement

Genes on the locus of enterocyte effacement of the EPEC chromosome are involved in producing the AE lesion. The locus of enterocyte effacement is not present in *E. coli* strains that are part of the normal intestinal flora, *E. coli* K-12, or ETEC strains. The locus of enterocyte effacement is found in those EPEC and EHEC strains that produce the AE lesion. The locus of enterocyte effacement consists of three segments with genes of known function. Genes associated with the locus of enterocyte effacement encode

Table 12.3 Proteins and genes involved in pathogenesis of EHEC[a]

Genetic locus	Protein description	Gene(s)	Function
Chromosome (locus of enterocyte effacement)	Intimin	*eae*	Adherence
	Tir	*tir*	Intimin receptor
	Secretion proteins	*espA, espB, espD*	Induces signal transduction
	Type III secretion system	*escC, escD, escF, escJ, escN, escR, escS, escT, escU, escV, sepQ, sepZ*	Apparatus for extracellular protein secretion
Phage	Stx	*stx1, stx2, stx2c, stx2d*	Inhibits protein synthesis
Plasmid	EHEC hemolysin	EHEC *hylA*	Disrupts cell membrane permeability
	Catalase-peroxidase	*katP*	

[a]Sources: A. D. O'Brien et al., *Curr. Top. Microbiol. Immunol.* **180:**65–94, 1992; N. T. Perna et al., *Infect. Immun.* **66:**3810–3817, 1998.

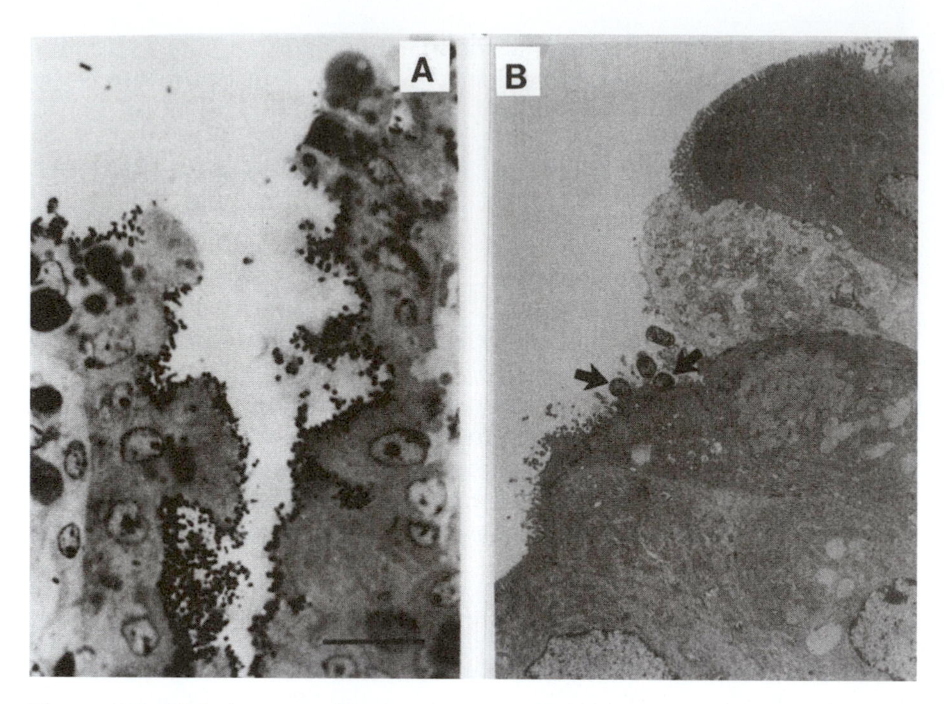

Figure 12.3 AE lesion caused by *E. coli* O157:H7. **(A)** Irregularly shaped cuboidal to low-columnar crypt neck cells with bacteria adherent to the luminal plasma membrane of a gnotobiotic pig (toluidine blue stain). Reproduced from D. H. Francis, J. E. Collins, and J. R. Duimstra, *Infect. Immun.* **51:**953–956, 1986, with permission. **(B)** Transmission electron photomicrograph of rabbit ileum with adherent bacteria (arrow). Reproduced from P. Sherman, R. Soni, and M. Karmali, *Infect. Immun.* **56:**756–761, 1988, with permission. doi:10.1128/9781555817206.ch12.f12.03

proteins involved in attachment and recognition that are known to interact directly with the host.

Intimin is an outer membrane protein encoded by the *eae* (from *E. coli* AE) gene. The protein is produced by EPEC, EHEC, *Hafnia alvei*, and *Citrobacter rodentium*. Intimin is the only described adherence factor of *E. coli* O157:H7 and is important for intestinal colonization in animal models.

The Tir protein is produced by the bacterial cell and moves to the host cell, where it serves as a receptor for intimin, via the type III secretion system. Most bacterial pathogens use existing host cell receptors. The ability of EPEC and EHEC to supply their own receptors may provide additional advantages for initiating the process of adhesion, motility, and signal transduction. The type III secretion system, present in many gram-negative bacterial pathogens, is induced upon contact with the host cells and only exports virulence factors to eukaryotic cells.

The 60-MDa Plasmid (pO157)

E. coli O157:H7 isolates carry a plasmid (pO157) of ~60 MDa (unrelated to the 60-MDa plasmid present in EPEC). Because of the association of the 60-MDa plasmid with EHEC, the plasmid is linked to the pathogenesis of EHEC infections. However, its exact role in virulence has not been determined. The plasmid carries potential virulence genes, including those encoding EHEC hemolysin and catalase-peroxidase. The EHEC catalase-peroxidase is an enzyme that protects the bacterium against oxidative stress, a defense mechanism of mammalian cells during bacterial infection.

Stxs

EHEC produces one or two Stxs. The nomenclature of the Stx family and their important characteristics are listed in Table 12.4. Molecular studies of Stx1 proteins from different *E. coli* strains revealed that Stx1 is either completely identical to the Stx of *S. dysenteriae* type 1 or differs by only one amino acid. Unlike Stx1, toxins of the Stx2 group are not neutralized by serum raised against Stx and do not cross-react with Stx1-specific DNA probes. There is sequence and antigenic variation within toxins of the Stx2 family produced by *E. coli* O157:H7 and other EHEC strains. The subgroups of Stx2 are Stx2, Stx2c, Stx2d, and Stx2e.

Structure of the Stx Family

Stxs are composed of a single enzymatic 32-kDa A subunit in association with a pentamer of 7.7-kDa receptor-binding B subunits. The Stx A subunit can be split by trypsin into an enzymatic A1 fragment (~27 kDa) and a carboxyl-terminal A2 fragment (~4 kDa) that links A1 to the B subunits.

Genetics of Stxs

While all *stx1* operons appear to be identical and are located on the genomes of lysogenic lambdoid bacteriophages, there is considerable heterogeneity in the *stx2* family. Unlike the genes for other Stx2 proteins, which are located on bacteriophages that integrate into the chromosome, the genes encoding Stx of *S. dysenteriae* type 1 and Stx2e are chromosomal. The genes of additional variants of Stx2 have been isolated from EHEC. Finally, Stx and Stx1 production are negatively regulated at the transcriptional level by an iron-Fur protein complex that binds at the *stx1* promoter. This is not affected by temperature, whereas Stx2 production is neither iron nor temperature regulated.

Receptors

All members of the Stx family bind to glycolipids on the eukaryotic cell surface. The alteration of binding specificity between Stx2e and the rest of the Stx family is related to the carbohydrate specificities of receptors. Stx1 binds preferentially to Gb_3 (glycolipid globotriaosylceramide) containing $C_{20:1}$ fatty acid, whereas Stx2c prefers Gb_3 containing $C_{18:1}$ fatty acid.

Table 12.4 Nomenclature and biological characteristics of Stx[a]

		% Amino acid homology to Stx2 subunit:				
Nomenclature	Genetic locus	A	B	Receptor	Activated by intestinal mucus	Disease(s)
Stx	Chromosome	55	57	Gb_3	No	Human diarrhea, HC,[b] HUS
Stx1	Phage	55	57	Gb_3	No	Human diarrhea, HC, HUS
Stx2	Phage	100	100	Gb_3	No	Human diarrhea, HC, HUS
Stx2c	Phage	100	97	Gb_3	No	Human diarrhea, HC, HUS
Stx2d	Phage	99	97	Gb_3	Yes	Human diarrhea, HC, HUS
Stx2e	Chromosome	93	84	Gb_4[c]	No	Pig edema disease

[a]Source: A. R. Melton-Celsa and A. D. O'Brien, p. 121–128, *in* J. B. Kaper and A. D. O'Brien (ed.), Escherichia coli *O157:H7 and Other Shiga Toxin-Producing* E. coli *Strains* (ASM Press, Washington, DC, 1998).

[b]HC, hemorrhagic colitis.

[c]Gb_4, glycolipid globotetraosylceramide.

Mode of Action of the Stxs

Stxs act by inhibiting protein synthesis. While it appears that transfer of the toxin to the Golgi apparatus is essential for intoxication, the mechanism of entry of the A subunit and particularly the role of the B subunit remain unclear. Although the entire toxin is necessary for its toxic effect on whole cells, the A1 subunit can cleave the *N*-glycoside bond in one adenosine position of the 28S ribosomal RNA (rRNA) that comprises 60S ribosomal subunits. This elimination of a single adenine nucleotide inhibits the elongation factor-dependent binding to ribosomes of aminoacyl-bound transfer RNA (tRNA) molecules. Peptide chain elongation is stopped, and overall protein synthesis is suppressed, resulting in cell death.

The Roles of Stxs in Disease

The precise roles of Stxs in mediating colonic disease, HUS, and neurological disorders are not understood. There is no satisfactory animal model for bloody diarrhea (hemorrhagic colitis) or HUS. The severity of the diseases prevents the study of experimental infections in humans. Therefore, our understanding of the role of Stxs in causing disease is obtained from histopathology of diseased human tissues, animal models, and endothelial tissue culture cells. It appears that Stxs contribute to pathogenesis by directly damaging vascular endothelial cells in certain organs, thereby disrupting the homeostatic properties of the cells.

There is a correlation between infection with *E. coli* O157:H7 and development of HUS in humans. Histopathological examination of kidney tissue from HUS patients reveals extensive structural damage in the glomeruli, the basic filtration unit of the kidney. Endotoxin in the presence of Stxs also can activate phagocytic cells to synthesize and release cytokines, superoxide radicals, or proteinases and can amplify endothelial cell damage.

E. coli O157:H7 strains isolated from patients with bloody diarrhea usually produce both Stx1 and Stx2 or Stx2 only; isolates producing only Stx1 are uncommon. Patients infected with *E. coli* O157:H7 producing only Stx2 or Stx2 in combination with Stx1 are more likely to develop serious kidney or circulatory complications than patients infected with EHEC strains producing Stx1 only.

CONCLUSION

The serious nature of hemorrhagic colitis (bloody diarrhea) and HUS caused by *E. coli* O157:H7 places this pathogen in a category apart from other foodborne pathogens that typically cause only mild symptoms. The severity of the illness, combined with its low infectious dose (<100 cells), qualifies *E. coli* O157:H7 to be among the most dangerous of foodborne pathogens. Illness caused by this pathogen has been linked to the consumption of a variety of foods. Recreational water and drinking water are also vehicles of transmission of *E. coli* O157:H7 infection. The pathogenic mechanisms of *E. coli* O157:H7 are not fully understood; however, production of one or more Stxs and AE adherence are important virulence factors. Although other EHEC strains cause disease, *E. coli* O157:H7 is still by far the most important serotype of EHEC in North America. Isolation of non-O157:H7 EHEC strains requires techniques not generally used in clinical laboratories; hence, these bacteria are rarely detected in routine practice. Recognition of non-O157 EHEC strains in foodborne illness requires the identification of serotypes of EHEC other than O157:H7 in persons with bloody diarrhea and/or HUS and preferably in the implicated food. The increased availability in clinical laboratories of techniques such as testing for Stxs or their genes and identification of other virulence markers unique to EHEC may enhance the detection of disease attributable to non-O157 EHEC.

Summary

- Cattle are a major reservoir of *E. coli* O157:H7, with undercooked ground beef being the single most frequently implicated vehicle of transmission.
- Acid tolerance is an important feature of this pathogen.
- EHEC strains other than O157:H7 have been increasingly associated with cases of HUS.
- Production of Stx is an important virulence feature of *E. coli* O157:H7.
- *E. coli* O157:H7 can be readily distinguished from other *E. coli* strains based on biochemical characteristics, such as inability to ferment sorbitol and lack of production of β-glucuronidase.

Suggested reading

Besser, R. E., P. M. Griffin, and L. Slutsker. 1999. *Escherichia coli* O157:H7 gastroenteritis and the hemolytic uremic syndrome: an emerging infectious disease. *Annu. Rev. Med.* **50:**355–367.

Centers for Disease Control and Prevention. 2000. Preliminary FoodNet data on the incidence of foodborne illnesses—selected sites, United States, 1999. *MMWR Morb. Mortal. Wkly. Rep.* **49:**201–205.

Kaper, J. B., J. P. Nataro, and H. L. Mobley. 2004. Pathogenic *Escherichia coli*. *Nat. Rev. Microbiol.* **2:**123–140.

Meng, J., M. P. Doyle, and T. Zhao. 2007. Enterohemorrhagic *Escherichia coli*, p. 249–270. *In* M. P. Doyle and L. R. Beuchat (ed.), *Food Microbiology: Fundamentals and Frontiers*, 3rd ed. ASM Press, Washington, DC.

O'Brien, A. D., V. L. Tesh, A. Donohue-Rolfe, M. P. Jackson, S. Olsnes, K. Sandvig, A. A. Lindberg, and G. T. Keusch. 1992. Shiga toxin: biochemistry, genetics, mode of action, and role in pathogenesis. *Curr. Top. Microbiol. Immunol.* **180:**65–94.

Perna, N. T., G. F. Mayhew, G. Posfai, S. Elliott, M. S. Donnenberg, J. B. Kaper, and F. R. Blattner. 1998. Molecular evolution of a pathogenicity island from enterohemorrhagic *Escherichia coli* O157:H7. *Infect. Immun.* **66:**3810–3817.

Philpott, D., and F. Ebel (ed.). 2003. E. coli *Shiga Toxin Methods and Protocols.* Humana Press, Totowa, NJ.

Questions for critical thought

1. The large outbreak in Germany in 2011 was associated with sprouts contaminated with *E. coli* O104:H4, an STEC similar to *E. coli* O157:H7. However, the numbers of individuals that died and of those that developed HUS were greater than previously associated with outbreaks linked to *E. coli* O157:H7. What, if any, characteristics of *E. coli* O104:H4 made it so virulent?
2. The infectious dose of *E. coli* O157:H7 may be as low as 10 cells. What characteristics of the pathogen contribute to the low infectious dose?
3. *E. coli* O157:H7 produces Stx. Explain how Stxs affect host cells and the role they play in disease.

4. The principal reservoirs for classes of diarrheagenic *E. coli* (EHEC, EPEC, and EHEC) differ. How might this characteristic influence spread of the pathogen and contamination of food?
5. How might acid tolerance contribute to foodborne illness associated with *E. coli* O157:H7?
6. Consider whether the following statements apply to ETEC. Explain your response. (i) *E. coli* O157:H7 is one example of an ETEC strain. (ii) It is a major cause of traveler's diarrhea but can also cause outbreaks in the United States. (iii) ETEC strains may behave more like *Vibrio cholerae* in producing a profuse watery diarrhea without blood or mucus. (iv) ETEC is an important cause of dehydrating diarrhea in infants and children in less-developed countries.
7. On-farm feeding practices can influence shedding of *E. coli* O157:H7 by cattle. How could this information be used to reduce contamination of beef products?
8. Based on the outbreaks described, what measures could be implemented to reduce cases of foodborne illness linked to *E. coli* O157:H7?
9. The *stx* genes associated with *E. coli* O157:H7 are located on bacteriophages. Why hasn't there been spread of the *stx* genes to other bacteria by the bacteriophages?
10. Provide a scenario that may result in cross-contamination of lettuce (or other types of produce) with *E. coli* O157:H7.
11. Why are young children (10 years of age and younger) more likely than adults to develop complications (HUS) when infected with *E. coli* O157:H7?
12. The protein intimin is required for adherence of *E. coli* O157:H7 to epithelial cells. Speculate whether the pathogen would cause disease if it lacked the ability to produce intimin.
13. Based on characteristics of *E. coli* O157:H7, what tests could be performed to distinguish the pathogen from other *E. coli* strains?
14. Speculate as to why produce-linked *E. coli* O157:H7-related outbreaks have increased during the past decade.
15. Based on information contained in this chapter and outside reading, which of the following statements is accurate? (i) EHEC causes bloody diarrhea and is associated with ground beef; its toxin resembles Stx, and EHEC may cause HUS. (ii) EPEC is the oldest known type and is the cause of infant diarrhea in developing countries. (iii) EIEC is commonly called *E. coli* O157:H7, resembles *Shigella*, and is associated with contaminated infant formula. (iv) ETEC causes traveler's diarrhea, which resembles disease caused by *V. cholerae*, and produces labile toxin and/or stable toxin.

13 *Listeria monocytogenes*

Authors' note

Steps in an epidemiological investigation often appear linear. However, the flurry of activity in an investigation often includes interviews, recalls, and laboratory tests that occur all at the same time, and during the investigation the working hypothesis may be changed several times.

LEARNING OBJECTIVES

The information in this chapter will enable the student to:

- use basic biochemical characteristics to identify *Listeria monocytogenes*
- understand what conditions in foods favor *L. monocytogenes* growth
- recognize, from symptoms and time of onset, a case of foodborne illness caused by *L. monocytogenes*
- choose appropriate interventions (heat, preservatives, or formulation) to prevent *L. monocytogenes* growth
- identify environmental sources of *L. monocytogenes*
- understand the role of *L. monocytogenes* toxin(s) and virulence factor(s) in causing foodborne illness

Outbreak

Outbreaks of foodborne illness continue to occur, despite the intervention measures that the food industry continues to develop and the ways in which consumers become better educated about safe food preparation. We all learn a great deal from studying foodborne outbreaks: about the ways in which microorganisms grow and survive and how they cause illness. The time line of foodborne outbreaks, from the moment of contamination until the last case is reported, can be very long. This case study examines important questions raised during an outbreak investigation. How long does it take to detect an outbreak of a sporadic disease? How much evidence does it take to launch a recall? What do you do with 35 million pounds of a recalled food product, like hot dogs?

These are just a few of the questions asked in a listeriosis outbreak that spanned 1998 and 1999, killed 12 people, and made 79 people in 17 states ill. This is a historically important outbreak in terms of numbers, food source, and outcome. The sequence of events is outlined below.

Spring and summer, 1998: A hot-dog-processing plant had an unusually high isolation rate of cells of *Listeria* from the environment.

July 4 weekend, 1998: An overhead air-conditioning unit thought to be the source of *Listeria* cells was cut apart and removed from the plant.

October 1998: The Centers for Disease Control and Prevention (CDC) began investigating four listeriosis cases from Tennessee. A 74-year-old woman who loved hot dogs was dead. Her flu-like illness made her drowsy and headachy and then killed her. Soon, more cases were reported from Ohio,

Outbreak continues on next page

doi:10.1128/9781555817206.ch13

New York, and Connecticut. Officials continued comparing food histories of individuals who were sick and family and friends who were well.

November 1998: A 31-year-old camp counselor died in Ohio. The genetic fingerprints of bacteria isolated from the listeriosis victims in the different states were identical. They must have eaten a common food.

Early December 1998: Hot dogs were statistically implicated as the cause of the outbreak.

Mid-December 1998: Hot dog brands packed at Sara Lee's Bil Mar Foods plant were implicated. With four people dead, the plant stopped shipping product.

22 December 1998: Sara Lee announced the recall of all meat processed at the Bil Mar plant.

23 December 1998: The outbreak strain of *Listeria monocytogenes* was isolated from an unopened package of Bil Mar hot dogs. This meets the U.S. Department of Agriculture (USDA) criterion for a recall.

25 December 1998: Another victim died in upstate New York.

January 1999: The USDA issued its first press release on the recall. It was of no help to a pregnant 27-year-old video store manager. She survived a severe "case of the flu" only to see her premature twin babies delivered stillborn. *Listeria monocytogenes* was isolated from her placenta.

May 1999: The USDA advised immunocompromised consumers not to eat hot dogs or deli meats unless they were thoroughly heated.

This is the story of but one listeria recall. The USDA typically issues about 30 recalls for *L. monocytogenes* contamination each year, and the number of recalls is increasing. Why? As food safety becomes a critical component of food-testing processes for many companies, items are being recalled as precautions or in reaction to specific tests. You have likely noticed this yourself in the news media.

INTRODUCTION

Over the last 25 years, listeriosis has become a major foodborne disease. The outbreak described above shows how quickly a large outbreak can occur. The first outbreak was traced to coleslaw in 1981. Outbreaks were soon linked to cheeses, lunch meats, milk, chicken nuggets, and fish. There are now about 1,662 cases of listeriosis every year, with an estimated 266 deaths. It is estimated that *Listeria* contamination in deli meats alone costs the United States $1.1 billion each year plus the loss of 4,000 quality-adjusted life years, as determined by microbiologists and epidemiologists. It is interesting that listeriosis is primarily foodborne, whereas other microorganisms are transmitted by both food and water.

Many factors contribute to the emergence of listeriosis and other "new" pathogens. These include the following.

- Eliminating the spoilage bacteria that prevent listeriae from growing. *L. monocytogenes* is a relatively poor competitor with other bacteria but thrives in refrigerated conditions under which competing bacteria cannot grow.
- Changing demographics, with more people considered at risk due to old age, immunosuppressants, organ transplants, chemotherapy, etc. Some pathogens, such as *Clostridium botulinum*, can make anyone sick. Others, like *L. monocytogenes*, are opportunistic

Authors' note

An early step in the investigation of foodborne disease is to determine what food(s) was eaten by people who got sick but not by those who did not get sick. These numbers can statistically implicate a food.

pathogens that attack only people whose immune defenses are impaired.

- Changing food production practices, particularly centralization and consolidation. Small institutional kitchens are increasingly consolidated into larger, centralized facilities which make good hygienic practices more challenging.
- Increasing use of refrigeration to preserve foods. *L. monocytogenes* grows better than other organisms in the cold.
- Changing eating habits with increased consumer demand for "fresh," "minimally processed," and "natural" foods that require little cooking or preparation and do not contain preservatives that would prevent listerial growth.
- Changing awareness (have you ever noticed that once you buy a silver Ford Mustang, the road is full of silver Ford Mustangs?) and ability to detect sporadic outbreaks. Computerized databases, DNA fingerprinting, and the Internet have played key roles in enabling us to detect previously unrecognized sporadic outbreaks.

Table 13.1 Symptoms caused by *L. monocytogenes*

Low-grade flu-like infection—not serious, except in pregnant women (who spontaneously abort)
Listeric meningitis—headache, drowsiness, coma
Perinatal infection
Encephalitis
Psychosis
Infectious mononucleosis
Septicemia

The disease listeriosis is not a typical foodborne illness. Listeriosis is sporadic and rare, but it is severe. It can cause meningitis, septicemia, and abortion (Table 13.1). Listeriosis is the third leading cause of death due to foodborne illness. Approximately 15.9% of people who get listeriosis die from it (i.e., the fatality / case ratio is high), and there is a long time lag from when the food is eaten to the onset of illness. *L. monocytogenes*, the causative organism, differs from most other foodborne pathogens. The organism is widely distributed in nature, resistant to adverse environmental conditions, and psychrotrophic and grows in human phagocytes. In addition, *Listeria* survives for long periods in or on food, soil, plants, and hard surfaces. These traits make *Listeria* control difficult for the food industry. It is unclear how *L. monocytogenes* should be viewed. The disease's severity and case fatality rate require that it be controlled. But given the organism's characteristics, it is unrealistic to make all food *Listeria* free.

Authors' note

Once a cell is phagocytized, it can continue growing and move from cell to cell without reentering the bloodstream.

CHARACTERISTICS OF THE ORGANISM

Classification

The Genus *Listeria*

Joseph Lister (Fig. 13.1), the inventor of medical disinfection, had a mouthwash and a genus of bacteria named after him. This genus, *Listeria*, is related to *Clostridium*, *Staphylococcus*, *Streptococcus*, *Lactobacillus*, and *Brochothrix*. *L. monocytogenes* is 1 of 10 species in the genus *Listeria*. The other species are *L. ivanovii*, *L. innocua*, *L. seeligeri*, *L. welshimeri*, *L. murrayi*, *L. denitrificans*, *L. rocourtiae*, *L. marthii*, and *L. grayi*. Within the genus *Listeria*, only *L. monocytogenes* and *L. ivanovii* are pathogens. *L. monocytogenes* is a human pathogen. *L. ivanovii* is primarily an animal pathogen.

Figure 13.1 Joseph Lister. doi:10.1128/9781555817206.ch13.f13.01

Listeria species are differentiated by a few biochemical traits. The biochemical tests that characterize the species are acid production from D-xylose, L-rhamnose, alpha-methyl-D-mannoside, and D-mannitol. The ability to lyse red blood cells differentiates *L. monocytogenes* from nonpathogenic *Listeria* species. Tumbling motility due to peritrichous flagella is another characteristic trait of *L. monocytogenes*, which is a facultative anaerobe. The nonpathogenic species are frequently isolated from the same foods

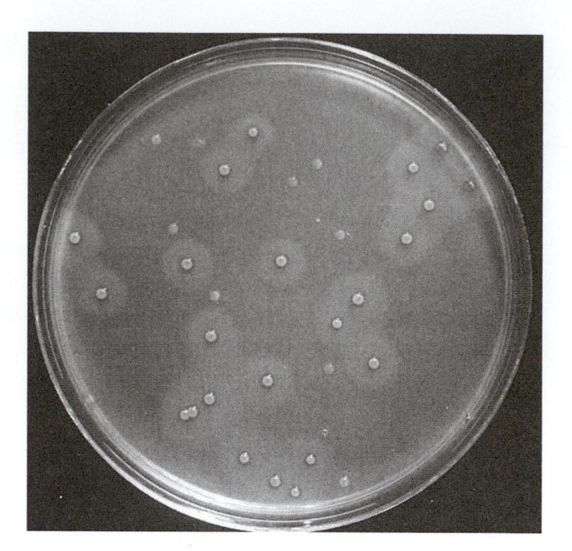

Figure 13.2 *L. monocytogenes* colonies on Oxoid Chromogenic Listeria Agar. Courtesy of Oxoid/Remel. doi:10.1128/9781555817206.ch13.f13.02

and environments as *L. monocytogenes.* "Rapid" tests based on genetic or immunological traits identify listeria isolates to the genus or species level. Special media, such as Oxoid's Chromogenic Listeria Agar (Fig. 13.2), also allow its rapid identification.

Public health officials investigating outbreaks can characterize *L. monocytogenes* isolates below the species level to help identify common sources of the organism. This can be done by genetic fingerprinting or by serotype. There are 13 serotypes of pathogenic *L. monocytogenes*. Ninety-five percent of the organisms isolated from listeriosis victims belong to three serotypes: 1/2a, 1/2b, and 4b. In the United States, the CDC has established a network (PulseNet) of public health laboratories to subtype foodborne pathogenic bacteria by using genetic fingerprinting. PulseNet laboratories use standardized methods for subtyping bacterial DNA by pulsed-field gel electrophoresis (PFGE), which uses alternating currents to increase the resolution of the DNA fingerprints. By comparing PFGE patterns of foodborne pathogens from different locations via the Internet, they can identify outbreaks having a common source.

Susceptibility to Physical and Chemical Agents

L. monocytogenes grows from 0 to 45°C but grows slowly at colder temperatures (Table 13.2). As mentioned above, the increase in cases of listeriosis may be in part due to the fact that we eat more ready-to-eat refrigerated foods and as a society we keep food in refrigeration conditions for longer periods. Freezing does not significantly reduce the size of the bacterial population. Survival and injury during frozen storage depend on the food and the freezing rate. *L. monocytogenes* is killed at temperatures of >50°C.

Table 13.2 Influence of temperature on doubling time of *L. monocytogenes*

Temp (°C)	t_d (hours)[a]
4	43
10	6.6
37	1.1

[a] t_d, doubling time.

L. monocytogenes grows in laboratory media at pH values as low as 4.4. At pH values below 4.3, cells may survive but do not grow. Organic acids such as acetic, citric, and lactic acids at 0.1% can inhibit *L. monocytogenes* growth. The antilisterial activity of these acids is related to their degree of dissociation (see chapter 25). Not all acids are created equal; citric and lactic acids are less harmful at an equivalent pH than acetic acid. HCl is least effective of all.

L. monocytogenes grows best at water activity (a_w) values of ≥0.97. For most strains, the minimum a_w for growth is 0.93, but some strains may grow at a_w values as low as 0.90. The bacterium may survive for long periods at a_w values as low as 0.83. *L. monocytogenes* heat resistance increases as the a_w of the food in which it is heated decreases. This is problematic for food manufacturers which combine low a_w and heat treatments to maintain the safety of foods such as cured meats.

L. monocytogenes grows to high levels in moderate salt concentrations (6.5%). It can even grow in the presence of 10 to 12% sodium chloride. The bacterium survives for long periods in higher salt concentrations. Lowering the temperature increases the survival rate in high-salt environments. Thus, cured meats (like hot dogs, bologna, and ham) are very hospitable environments for listeria growth. For these reasons *Listeria* control in foods is best achieved by multiple hurdles (see chapter 2) coupled with environmental monitoring in processing areas.

Table 13.3 Some foods that have been associated with listeriosis

Cheese
Alfalfa tablets
Fish
Chicken nuggets
Raw milk
Cod roe
Turkey frankfurters
Cook-and-chill chicken
Smoked fish
Deli salads
Deli meat
Hot dogs
Vegetable rennet
Human breast milk
Ice cream
Pork sausage
Homemade sausage
Salted mushrooms

LISTERIOSIS AND SPECIFIC FOODS

Ready-to-Eat Foods

Ready-to-eat foods (such as soft cheeses, frankfurters, delicatessen meats, and poultry products) can pose a high listeriosis risk for susceptible populations. Some foods that have caused listeriosis are listed in Table 13.3. Because 20% of refrigerators have temperatures of >50°F, refrigeration cannot ensure the safety of ready-to-eat foods. In addition, refrigeration offers an environment where *L. monocytogenes* can outcompete mesophilic pathogens.

Milk Products

Raw milk is a source of *L. monocytogenes*. Mexican-style cheese made with unpasteurized milk caused one of the first listeriosis outbreaks in California. Pasteurization is a proven safety intervention. The World Health Organization states, "Pasteurization is a safe process which reduces the number of *L. monocytogenes* organisms in raw milk to levels that do not pose an appreciable risk to human health." U.S. regulatory authorities share this view. However, *L. monocytogenes* grows well in pasteurized milk. This makes postprocess contamination (the contamination of food by environmental bacteria after the food has received its antimicrobial processing) a major concern. *Listeria* cells grow more rapidly in pasteurized milk than in raw milk at 7°C because there are fewer competitors in pasteurized milk. Milk contaminated after pasteurization and refrigerated can support very high listerial populations after 1 week. Temperature abuse can make things even worse. In a recent outbreak, chocolate milk was contaminated in the plant and then held at room temperature (in the summer) for several days. This allowed *L. monocytogenes* populations to reach the very high level of 10^9 colony-forming units (CFU)/ml. The victims told the epidemiologists to check the chocolate milk because "it tasted bad." Nonetheless, in spite of the off-flavor, they drank it.

Cheeses

L. monocytogenes survives cheese manufacturing and ripening because of its temperature hardiness, ability to grow in the cold, and salt tolerance. *L. monocytogenes* is concentrated in the cheese curd during manufacturing. The behavior of listeriae in the curd is influenced by the type of cheese, ranging from growth in feta cheese to significant death in cottage cheese.

Authors' note

On a good day, bacteria can reach a level of 10^9 CFU/ml in a laboratory.

For example, during cheese ripening, *L. monocytogenes* can grow in Camembert cheese, die gradually in Cheddar or Colby cheese, or decrease rapidly during early ripening and then stabilize, as in blue cheese. The reasons for this are unknown. Consumption of soft cheeses by susceptible persons is a risk factor for listeriosis. The CDC recommends that pregnant women and other immunocompromised people avoid eating soft cheeses. For recent outbreaks involving contaminated cheese in Europe, recalling the foods and advising people not to eat the contaminated product were effective control measures, but learning about recalled food products is not always easy. During this outbreak across three countries, several people ate the incriminated cheese products even after the recall was announced; however, numbers were limited to 34 outbreak cases.

Meat and Poultry Products

L. monocytogenes growth in meat and poultry depends on the type of meat, the pH, and the presence of other bacteria. *L. monocytogenes* grows better in poultry than in other meats. Roast beef and summer sausage support the least growth. Contamination of animal muscle tissue can be caused by the presence of *L. monocytogenes* in or on the animal before slaughter or by contamination of the carcass after slaughter. Because *L. monocytogenes* concentrates and multiplies in the kidney, lymph nodes, liver, and spleen, eating organ meat may be more hazardous than eating muscle tissue. *L. monocytogenes* attaches to the surface of raw meats, where it is difficult to remove or kill. It also grows readily in processed meat products, including vacuum-packaged beef, at pH values near 6.0. There is little or no growth around pH 5.0. Ready-to-eat meats that are heated and then cooled in brine before packaging may favor *L. monocytogenes* growth because competitive bacteria are reduced and have a high salt tolerance.

A 1993–1996 USDA monitoring program for *L. monocytogenes* in cooked ready-to-eat meats found it in 0.2 to 5.0% of beef jerky, cooked sausages, salads, and spreads. In sliced ham and sliced luncheon meats the incidence was 5.1 to 81%. This has decreased dramatically over the last 10 years. In Canada, the incidence of *L. monocytogenes* in domestic ready-to-eat, cooked meat products was 24% in 1989 and 1990 but declined to 3% or less during 1991 and 1992. However, in environmental specimens obtained from food-processing plants, the contamination level remained constant from 1989 to 1992. This suggests that once established in a food plant environment, *L. monocytogenes* may persist for years. In fact, strains of *L. monocytogenes* appear to persist within a processing environment and have resistance to some disinfectants, which allows them to survive even better despite normal cleaning and sanitation programs. Scientists continue to evaluate the persistence and tolerance of *L. monocytogenes* strains under different conditions.

Seafoods

L. monocytogenes has been isolated from fresh, frozen, and processed seafood products, including crustaceans, molluscan shellfish, and finfish. Shrimp, smoked mussels, and imitation crabmeat have all been vehicles for illness. Rainbow trout were implicated in two outbreaks. A U.S. Food and Drug Administration (FDA) survey of refrigerated or frozen cooked crabmeat showed contamination levels of 4.1% for domestic products and 8.3% for imported products. Crab and smoked fish were contaminated with *L. monocytogenes* at 7.5 and 13.6%, respectively.

Among the seafoods that may be high-risk foods for listeriosis are molluscs, including mussels, clams, and oysters; raw fish; lightly preserved fish products, including salted, marinated, fermented, and cold-smoked fish; and mildly heat-processed fish products and crustaceans.

Other Foods

A survey of vacuum-packaged processed meat revealed that 53% of the products were contaminated with *L. monocytogenes* and that 4% contained >1,000 CFU/g (a high level for a ready-to-eat food). This corroborates experimental evidence that the growth of *L. monocytogenes* (a facultative anaerobe) is not significantly affected by vacuum packaging. Studies with meat juice, raw chicken, and precooked chicken nuggets revealed that modified atmospheres do not protect against *L. monocytogenes* growth.

Many surviving cells are injured by heating, freezing, or various other treatments. Heat-stressed *L. monocytogenes* may be less pathogenic than nonstressed cells because energy normally devoted to pathogenesis is diverted to dealing with stressors. On nonselective agar (or food), injured cells can repair the damage induced by stress and grow. When injured cells are subjected to additional stress in selective agar (or food), they may not be recovered.

Environmental Sources of *L. monocytogenes*

L. monocytogenes is ubiquitous in the environment. *Listeria* species survive and grow in many water environments: surface water of canals and lakes, ditches, freshwater tributaries, and sewage. *Listeria* organisms have been isolated from alfalfa plants and crops grown on soil treated with sewage sludge. One-half of the radish samples grown in soil inoculated with *L. monocytogenes* were still contaminated 3 months later. *L. monocytogenes* is also present in pasture grasses and grass silages. Decaying plant and fecal materials probably contribute to *L. monocytogenes* in soil. The soil provides a cool, moist environment, and the decaying material provides the nutrients.

L. monocytogenes has been isolated from the feces of healthy animals. Many animal species can get listeriosis. Humans having listeriosis and humans who carry the organism without becoming sick both shed the organism in their feces. Figure 13.3 illustrates how *L. monocytogenes* is spread from the environment to animals and humans and back to the environment.

Food-Processing Plants

L. monocytogenes enters food-processing plants through soil on workers' shoes and clothing and on vehicles. It can also enter the facility with contaminated raw vegetable and animal tissue and with human carriers. The high humidity and nutrient levels of food-processing environments promote listerial growth. *L. monocytogenes* is often detected in moist areas such as floor drains, condensed and stagnant water, floors, residues, and processing equipment. *L. monocytogenes* attaches to surfaces, including stainless steel, glass, and rubber. Perhaps the ways in which these bacterial cells attach and persist in these environments help them to survive. In fact, after six visits to one processing plant in Canada, scientists identified the genetic profiles of strains they named "persistent" for their ability to survive on floor drains and in the plant.

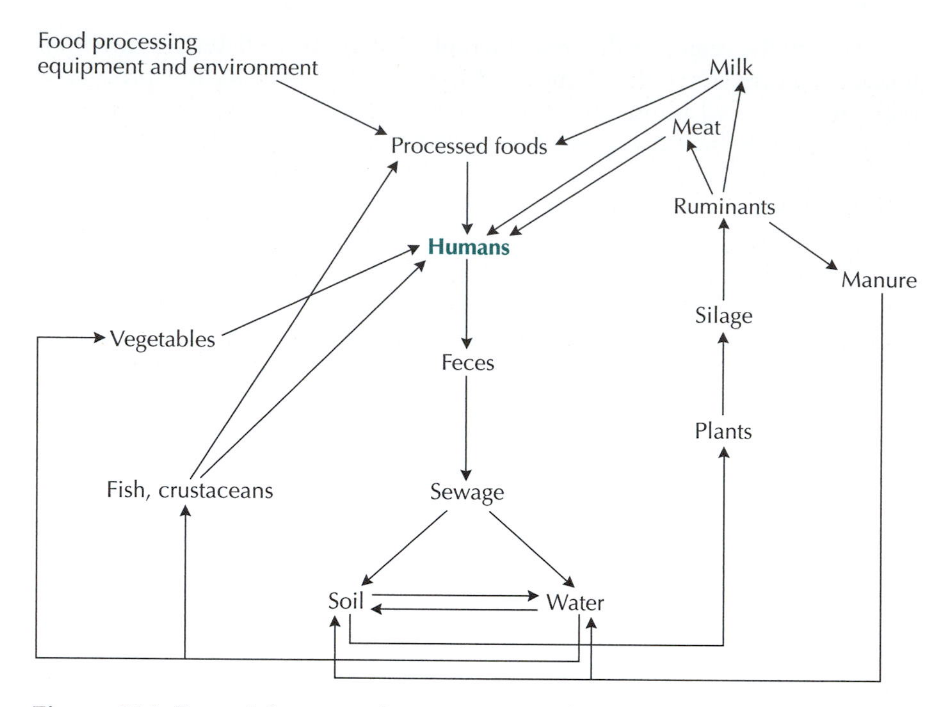

Figure 13.3 Potential routes of transmission of *L. monocytogenes*. Based on B. Swaminathan, p. 383–409, *in* M. P. Doyle, L. R. Beuchat, and T. J. Montville (ed.), *Food Microbiology: Fundamentals and Frontiers*, 2nd ed. (ASM Press, Washington, DC, 2001). doi:10.1128/9781555817206.ch13.f13.03

L. monocytogenes contaminates carcasses from feces during slaughter. A high percentage (11 to 52%) of healthy animals are fecal carriers. *L. monocytogenes* is present in both unclean and clean zones in slaughterhouses, especially on workers' hands. The most heavily contaminated working areas are those for cow dehiding, pig stunning, and hoisting. Studies in turkey and poultry slaughterhouses found *L. monocytogenes* in drip water associated with defeathering, chill water overflow, and recycled cleaning water.

L. monocytogenes does not survive heat processing. It gets into processed foods primarily by postprocess contamination. *L. monocytogenes* is particularly hard to eliminate from food plants because it adheres to food contact surfaces and forms biofilms in hard-to-reach areas such as drains and pipes. Figure 13.4 shows a similar phenomenon in the growth of *Pseudomonas fluorescens* in hard-to-clean sites in equipment and out-of-reach areas like holes that may naturally occur during the manufacturing process or during routine use. This makes proper sanitation difficult. *L. monocytogenes* is often in raw materials used in food plants, so there are ample opportunities for reintroduction of listeriae into processing facilities that have previously eradicated it.

Prevalence and the Regulatory Status of *L. monocytogenes*

L. monocytogenes contamination of food is widespread. It is a major cause of USDA recalls (Fig. 13.5). Contamination levels range from 0% in bakery goods to 16% in some ready-to-eat foods. The rate of contamination of raw foods is much higher, up to 60% in raw chicken. Preserved but not heat-treated fish and meat products are more frequently contaminated than

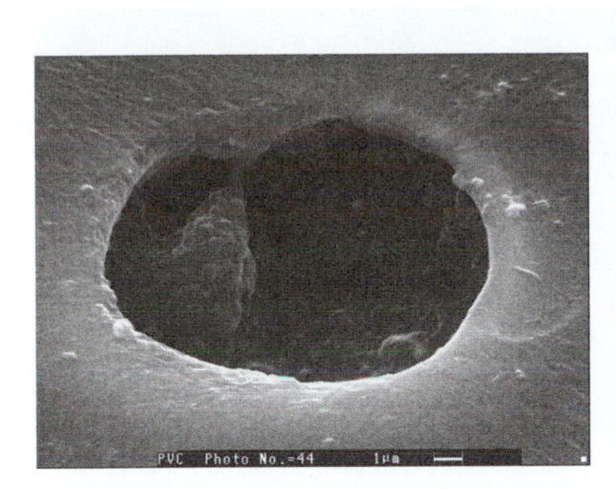

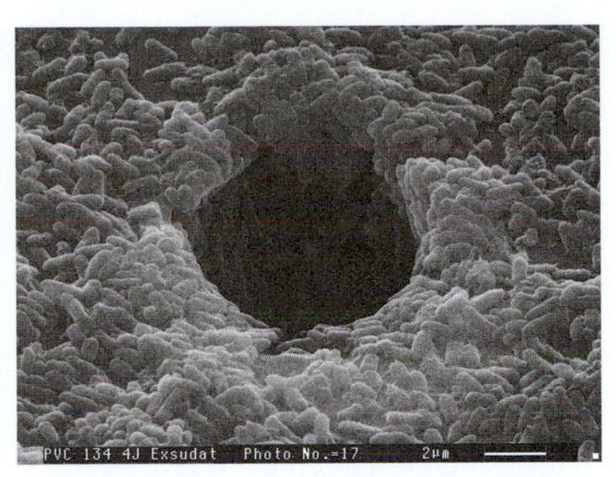

Figure 13.4 (Left) Hole in polyvinyl chloride caused by gas escaping during the polymerization process. Scale bar, 1 µm. **(Right)** Colonization by *Pseudomonas fluorescens* CCL 134 of a hole in a polyvinyl chloride conveyor belt (4-day culture in meat exudate). Imagine that this is *L. monocytogenes* in a biofilm in a cool processing plant environment. Scale bar, 2 µm. Reprinted from B. Carpentier and O. Cerf, *Int. J. Food Microbiol.* **145:**1–8, 2011, with permission. Courtesy of Brigitte Carpentier. doi:10.1128/9781555817206.ch13.f13.04

heat-treated meat foods. *L. monocytogenes* can be isolated from many vegetables, including bean sprouts, cabbage, cucumbers, leafy vegetables, potatoes, prepackaged salads, radishes, salad vegetables, and tomatoes.

There is an international debate about how *L. monocytogenes* should be regulated (Box 13.1). Because *L. monocytogenes* is so common, regulatory agencies in many countries assert that it is impossible to produce *L. monocytogenes*-free foods. They have established "tolerance levels" for *L. monocytogenes*. Products that have caused human listeriosis are placed in a special category and regulated more strictly than foods that have never caused listeriosis. Most European Union countries feel that foods should be free of *L. monocytogenes*, if possible, or have the lowest level possible. Thus, foods intended for susceptible populations must be listeria free. Other foods (except those intended for infants and for special medical purposes)

Authors' note

There is a zero tolerance for L. monocytogenes *in ready-to-eat foods but no detection method that is 100% accurate.*

Figure 13.5 *L. monocytogenes* is a leading cause of recalls. The graph shows the leading causes of USDA recalls in 1997. doi:10.1128/9781555817206.ch13.f13.05

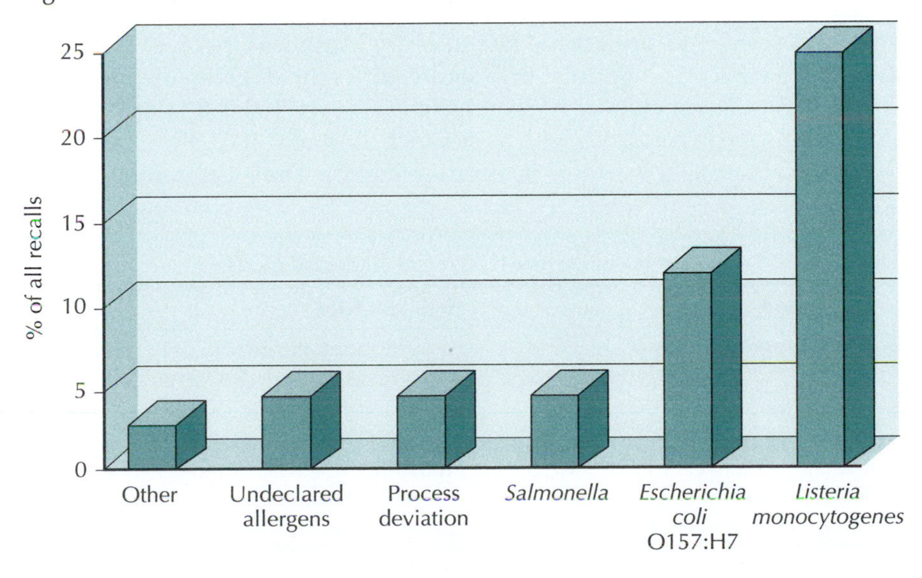

BOX 13.1

Group exercise: industrial and regulatory responses to *Listeria monocytogenes*

Objective

To understand some of the real-world factors involved in setting industrial policies through use of a simulation.

Scenario

You work for MSF Foods, a large cookie company that added a premium pie line 8 months ago. This line has been a huge success, yielding a 4% increase in corporate profits. Marketing has determined that adding cheesecake and cream pie products to the line could fill a new market niche and increase next year's profits by a similar extent.

The product line manager questioned if these products might pose a *Listeria* risk and what kind of end product testing would be required to ensure product safety. (In the cookie line, the company does "test and hold" for *Salmonella*. Should a similar program be instituted for *Listeria* in the pies?) The production side of the company says that a test-and-hold policy would be the kiss of death for this short-shelf-life, refrigerated project. The microbiology department insists that end product testing is the only way to ensure the safety of the product.

A meeting has been called to resolve this issue. Your job is, at the end of the meeting, to write a two-page recommendation to the Vice President of Cookies, Creams, and Cakes.

Your assignment

1. THE PREMEETING: Split into groups and discuss the issues. Decide what additional information is required for you to make a decision. Decide which side of the testing controversy you stand on and will fight for during the meeting.
2. THE MEETING: The group tries to reach a consensus. (Self-explanatory.)
3. THE WORK PRODUCT: Write a two-page memo to the Vice President of Cookies, Creams, and Cakes (who knows a lot about cookies but nothing about microbiology) outlining your recommendation as to what the company should do, and why.

may contain *L. monocytogenes* at <100 CFU/g. In contrast, the United Kingdom and the United States have a "zero tolerance" (in a 25-g sample) for *L. monocytogenes*. Both countries argue that the infective dose must be known before an "acceptable" level for *L. monocytogenes* can be set. The infective dose is unknown, and it may differ for different types of people.

A "citizens' petition" (by which anyone can ask the U.S. government to change its regulations) has proposed a tolerance for *L. monocytogenes* in low-risk foods (Table 13.4). The argument is that if *L. monocytogenes* cannot grow in the food, is not held at temperatures that would support growth (yes, this is redundant with the first condition), and has a low probability of making someone ill, there is no justification for zero tolerance. There are two additional arguments against the U.S. zero-tolerance policy. The first is that the FDA only has legal authority to issue regulations that are based on sound science to protect the health of the American people. However, different parts of the population require different degrees of protection (Table 13.5). The incidence of listeriosis (~0.7 per 100,000 people) is the same in European countries with a tolerance limit of <100 CFU/g as it is in the United States, where there is a zero tolerance. Thus, the zero-tolerance

Table 13.4 Citizens' proposal for low-risk versus high-risk foods

Low-risk food	High-risk food
Does not support growth of *L. monocytogenes* to high numbers	Food supports the growth of *L. monocytogenes* to high numbers
Is not held for extended periods at temperatures that permit growth	Is held for long periods at refrigeration temperatures
Probability of illness is 1 in 100,000 for at-risk populations	Probability of illness is 1 in 6,000 for at-risk populations
Tolerance of <100 CFU/g	Tolerance remains zero

regulation appears to offer no additional health benefit. Secondly, international trade requires harmonization of microbial specifications. When the United States bars the import of foods meeting the European Union's <100-CFU/g tolerance limit, it can be charged with inhibition of free trade through "nontariff trade barriers." However, politically, no one wants to be seen as "in favor of allowing germs in food." The debate over zero tolerance continues. The current dilemma is discussed in Box 13.2.

Table 13.5 Relative risk of listeriosis for different populations

Population	Relative risk
Total	1[a]
Over 70 yr old	3
Pregnant	17
HIV[b] positive	200

[a]Absolute risk, 7 cases per million people.
[b]HIV, human immunodeficiency virus.

Human Carriers

L. monocytogenes can be carried asymptomatically in the feces of many groups. These include healthy people, pregnant women, patients with gastroenteritis, slaughterhouse workers, laboratory workers handling *Listeria*, food handlers, and patients undergoing hemodialysis. *L. monocytogenes* has been isolated from 2 to 6% of fecal samples from healthy people. Patients with listeriosis often excrete high numbers of *L. monocytogenes* organisms. Specimens from 21% of patients had $>10^4$ *L. monocytogenes* organisms/g of feces. The same strain of *L. monocytogenes* was fecally shed by 18% of their housemates. Fecal carriers amplify outbreaks through secondary transmission. (The primary transmission is through the food to the first victim. The secondary transmission is from the feces of the first victim to another person.) This secondary fecal-oral transmission of bacteria that originate in foodborne outbreaks also occurs with other pathogens.

FOODBORNE OUTBREAKS

Foodborne transmission of listeriosis was first documented in 1981 during a Canadian outbreak in Nova Scotia. A case-control study and strain typing proved that food was the source. There were 34 cases among pregnant women and 7 cases among nonpregnant adults over a 6-month period. A case-control study implicated coleslaw. The epidemic strain was later

BOX 13.2

The current dilemma

The United States is currently spending hundreds of millions of dollars to decrease the incidence of listeriosis. But is the money being spent effectively? Is the obsession with listeriosis cost-effective?

It is hard to say. The causative organism is widespread in the environment and difficult to eradicate, yet relatively few foods have been implicated in listeriosis outbreaks. Should all ready-to-eat foods or only those with a historical problem be tested for *Listeria*? There are relatively few cases of listeriosis, and when outbreaks occur, only a tiny percentage of people who eat the product get sick. Although the attack rate is low, the fatality rate is high. Might it be more cost-effective to educate at-risk populations to avoid certain foods rather than test for listeriae? The methods for isolating and identifying *Listeria* leave much to be desired. They have a 10 to 15% false-negative rate; i.e., if 100 samples containing *Listeria* were tested, it would not be detected in 10 to 15 of the samples in which it was present. Is it fair to demand zero tolerance when there is no 100% reliable test? Does it make sense to test products in which the presence of listeriae is unlikely or that have received listericidal treatments? The International Commission on Microbial Specifications for Foods has suggested that these foods need not be tested. The current regulatory emphasis *has* dramatically decreased the isolation of *Listeria* from food, but the number of cases of listeriosis has not had a corresponding decrease. Regulatory agencies, food industry groups, and consumer advocates all agree that the current regulations need to be reexamined, but they disagree on how or why. For the moment, there is emphasis on developing better surveillance for listeriosis so that progress (or lack of progress) can be monitored. Foods that have been associated with listeriosis are under increased scrutiny, but this has not prevented additional recalls. There are efforts to educate at-risk groups, especially pregnant women, who can be reached through their obstetricians, but these need to be more effective. The debate is sure to continue for years.

isolated from an unopened package of coleslaw. Cabbage fertilized with manure from sheep suspected to have had meningitis caused by *Listeria* was the probable source. Harvested cabbage was stored over the winter and spring in an unheated shed. The cold provided a growth advantage for the psychrotrophic *L. monocytogenes*.

Contaminated Mexican-style cheese caused a 1985 listeriosis outbreak in California. There were 142 cases over 8 months. Pregnant women accounted for 93 cases. The remaining 49 were among nonpregnant adults. Most of the victims had a predisposing condition for listeriosis. Almost one-third of the people infected died. However, about 10,000 people consumed the cheese, so the overall attack rate was very low. Inadequate pasteurization of milk and mixing of raw milk with pasteurized milk caused the outbreak.

L. monocytogenes in soft cheese was responsible for a 4-year outbreak of 122 cases in Switzerland. A contaminated pâté caused a 300-case outbreak in the United Kingdom. Recalling the food, advising people not to eat contaminated product, and taking action to prevent *L. monocytogenes* contamination at processing facilities controlled these large outbreaks.

There was a 10-state outbreak of listeriosis between May and November 2000. The *L. monocytogenes* isolates were all serotype 1/2a and had identical PFGE patterns. Eight perinatal and 21 nonperinatal cases were reported. (The perinatal period includes the time before birth to about 1 month after.) Among the 21 nonperinatal-case patients, the median age was 65 years (range, 29 to 92 years); 62% were female. This outbreak resulted in four deaths and three miscarriages or stillbirths. A case-control study implicated deli turkey meat.

CHARACTERISTICS OF DISEASE

L. monocytogenes causes listeriosis in well-defined high-risk groups. These include pregnant women, newborns, and immunocompromised adults. Listeriosis occasionally occurs in otherwise healthy people. In nonpregnant adults, *L. monocytogenes* causes septicemia, meningitis, and meningoencephalitis. The mortality rate is 20 to 25%. Cancer, organ transplants, immunosuppressive therapy, infection with human immunodeficiency virus, and advanced age predispose people to listeriosis. *L. monocytogenes* infection may cause only mild flu-like symptoms in pregnant women but cause stillbirth or abortion of the fetus. While listeriosis outbreaks attract much attention, most cases of human listeriosis are sporadic. Some sporadic cases may be unrecognized common-source outbreaks. Although foods may be implicated in some sporadic cases, the mode of infection is usually unknown. Because they are so hard to investigate, many sporadic cases cannot be associated with food. Incubation times of up to 5 weeks make it hard to get accurate food histories and examine suspect foods. Understanding the epidemiology of sporadic cases would help develop more effective control strategies.

Sporadic listeriosis is a rare disease. In Europe and Canada, passive surveillance suggests that there are 2 to 7 cases per million people. An active-surveillance study in the United States gave a more precise estimate of 7.4 cases per million people. Since 1990, there has been an almost 50% decrease in sporadic listeriosis in the United States. This suggests that current preventive measures are making a difference.

Human listeriosis is poorly understood. Although exposure to *L. monocytogenes* is common, listeriosis is rare. This may be due to high human resistance or low pathogenicity of most strains. Some healthy humans are asymptomatic fecal carriers of *L. monocytogenes*. The risk of clinical disease in carriers is unknown. Although *Listeria* seems to target pregnant women, some pregnant women also have asymptomatic fecal carriage of listeriae and have normal pregnancies. Women who give birth to infected infants may not be infected in later pregnancies.

L. monocytogenes can also cause a second illness, feverish gastroenteritis. The gastroenteritis outbreaks differ from the invasive outbreaks. They affect people with no predisposing risk factors. The infectious dose is higher (1.9×10^5 to 1×10^9 CFU/g or CFU/ml) than that for invasive listeriosis. Finally, the symptoms appear within several hours (18 to 27 h) of exposure, in contrast to the weeks observed for invasive listeriosis. The first case of listeric gastroenteritis was reported in 1994. It was caused by the postprocess contamination of chocolate milk held in an unrefrigerated processing vat. The contaminated milk was then stored unrefrigerated (i.e., incubated) by the buyers, who served it at a large summer gathering; about 75% of those who drank the milk got sick.

INFECTIOUS DOSE

The infectious dose of *L. monocytogenes* depends on the immunological status of the human, the virulence of the microbe, and the food. Studies with human volunteers are impossible (i.e., such studies would be unethical), but in studies with animals, reducing exposure levels reduces clinical disease. There are usually >100 CFU of *L. monocytogenes*/g in foods responsible for outbreaks. However, the frankfurters implicated in the 1998 listeriosis outbreak had <0.3 CFU/g. More epidemiological data are needed to accurately assess the infectious dose.

VIRULENCE FACTORS AND MECHANISMS OF PATHOGENICITY

L. monocytogenes is unique among foodborne pathogens. Other pathogens excrete toxins or multiply in the blood. In contrast, *L. monocytogenes* enters the host's cells, grows inside the cell, and passes directly into nearby cells (Fig. 13.6). Cell-to-cell transmission reduces the bacterium's exposure to antibiotics and circulating antibodies. This membrane-penetrating ability allows *L. monocytogenes* to cross into the brain and placenta.

Pathogenicity of *L. monocytogenes*

There are many tests for studying *L. monocytogenes* pathogenicity, including a fertilized hen egg test, tissue culture assays, and tests using laboratory animals, particularly mice. In these studies, mice are infected by injection into the stomach cavity, intravenously, or by mouth. Virulence is then evaluated by comparing the 50% lethal dose or by counting bacteria in the spleen or liver. Fifty percent lethal dose values range from 10^3 to 10^7 CFU.

When mice are exposed to *L. monocytogenes*, the infection follows a well-defined course, lasting for about 1 week. Within 10 min after intravenous injection, 90% of the bacteria are taken up by the liver and 5 to 10% by the spleen. During the first 6 h, the number of viable listeriae in the liver decreases 10-fold, indicating their rapid destruction. Surviving listeriae then infect susceptible macrophages and multiply. They grow exponentially in the spleen and liver for the next 48 h. If inactivation ensues during the next 3 to 4 days, the host recovers.

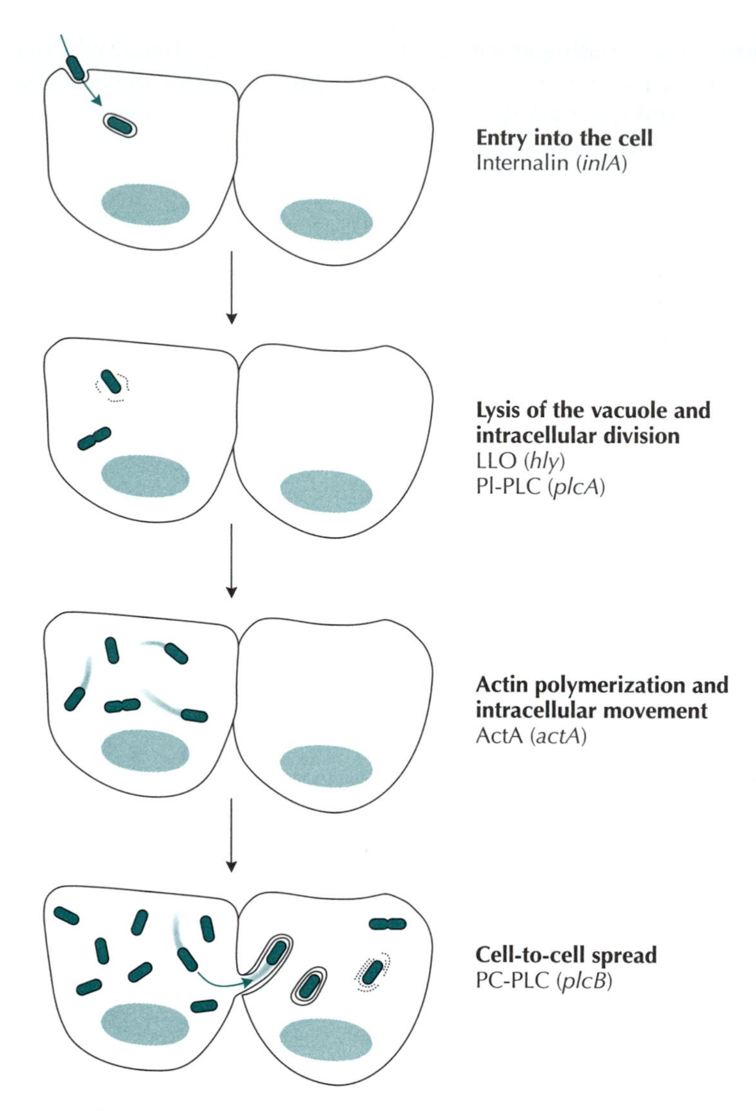

Figure 13.6 Schematic representation of *L. monocytogenes* cell-to-cell spread. Abbreviations: LLO, listeriolysin O; PI-PLC, phosphatidylinositol phospholipase C; PC-PLC, phosphatidylcholine-specific phospholipase C. Redrawn from B. Swaminathan, p. 383–409, *in* M. P. Doyle, L. R. Beuchat, and T. J. Montville (ed.), *Food Microbiology: Fundamentals and Frontiers*, 2nd ed. (ASM Press, Washington, DC, 2001). doi:10.1128/9781555817206.ch13.f13.06

L. monocytogenes enters humans through ingestion, crosses the intestinal barrier, and is internalized by macrophages, where it replicates. The blood then transports the listeriae to lymph nodes. When they reach the liver and the spleen, most listeriae are rapidly killed. The survivors travel by the blood to the brain or the placenta.

Specific Genes Mediate Pathogenicity

L. monocytogenes cells can fight their way into nonphagocytic human cells. Our understanding of how gene regulation influences pathogenicity is in its infancy. A series of *L. monocytogenes* genes gives the instructions for each step of this infectious process. The protein internalin is needed to start the process. Internalin is coded for by the gene *inlA*. This gene is thermoregulated and expressed more readily at 37°C than at 20°C. Mutants lacking this gene cannot form the phagosome that introduces the bacteria to the inside of the mammalian cell. To do real damage, the listeria must be released from the phagosome and liberated into the cytoplasm of the host cell. The *hly* gene is responsible for this. It codes for listeriolysin, an enzyme that breaks open the phagosome. Listeriolysin breaks down red blood cells; its activity is measured as hemolytic activity. *L. monocytogenes* mutants lacking the *hly* gene are not pathogenic. These cells can be entrapped by phagosomes but are not freed into the cytoplasm.

Once the listeria cells are free in the host cytoplasm, the genes that regulate growth and multiplication work in the normal fashion. The listeria cells have the unusual ability to use the host's actin molecules. Actin, together with myosin, is the major component of human muscles. It is in nonmuscle cells at lower concentrations. The *actA* gene in listeria codes for a protein, ActA, which is made at one end of the bacterium. ActA makes the actin propel the cell forward. The force generated by the actin pushes the cell across the infected host's membrane and into the membrane of the adjacent cell. The final gene needed for pathogenicity is *plcB*. This gene codes for a membrane-hydrolyzing enzyme that helps listeriolysin O liberate the bacterium into the cytoplasm of the adjacent human cell.

Summary

- *L. monocytogenes* is a ubiquitous organism that can grow at refrigeration temperatures.
- There are relatively few (5,000/year) cases of listeriosis, but there is a high (~25%) fatality rate among people who get it.
- *L. monocytogenes* is a hardy organism that is relatively resistant to dehydration, low pH, and low a_w.
- Ready-to-eat foods that are preserved by refrigeration pose a special challenge with regard to *L. monocytogenes*.
- Symptomatic and asymptomatic people can shed *L. monocytogenes* in their feces.
- The United States has a zero tolerance for *L. monocytogenes* in ready-to-eat foods, but in Europe there is some tolerance for it in certain foods.
- Outbreaks of listeriosis generally extend over a long period, are caused by refrigerated foods, have high case fatality rates, and involve a disproportionate number of pregnant women.
- *L. monocytogenes* enters the host's cells, grows inside the cell, and passes directly into nearby cells.
- Internalin, listeriolysin, ActA, and a membrane-hydrolyzing enzyme are the four major proteins that govern the pathogenic process.

Suggested reading

Carpentier, B., and O. Cerf. 2011. Review—persistence of *Listeria monocytogenes* in food industry equipment and premises. *Int. J. Food Microbiol.* **145:**1–8.

National Advisory Committee on Microbiological Criteria for Foods. 1991. *Listeria monocytogenes. Int. J. Food Microbiol.* **14:**185–246.

Ryser, E. T., and E. H. Marth. 1991. Listeria, *Listeriosis, and Food Safety.* Marcel Dekker, New York, NY.

Swaminathan, B., D. Cabanes, W. Zhang, and P. Cossart. 2007. *Listeria monocytogenes,* p. 457–491. *In* M. P. Doyle and L. R. Beuchat (ed.), *Food Microbiology: Fundamentals and Frontiers,* 3rd ed. ASM Press, Washington, DC.

Tompkin, R. B. 2002. Control of *Listeria monocytogenes* in the food-processing environment. *J. Food Prot.* **65:**709–725.

Questions for critical thought

1. What reasons might explain our lack of awareness of *L. monocytogenes* prior to 1980?
2. What are the pros and cons of a tolerance versus zero tolerance for *L. monocytogenes* in ready-to-eat food?
3. Make up an outbreak of listeriosis. Include food, setting, population, symptoms, the chain of events that caused the outbreak, and how it might have been prevented.
4. Diagram the life of *L. monocytogenes* as an intracellular pathogen, noting important gene products.
5. What characteristics make *L. monocytogenes* a "successful" foodborne pathogen?
6. Why do listeriae grow more rapidly in pasteurized milk than in raw milk?
7. Write down one thing you do not understand about *L. monocytogenes.* Discuss it with a classmate.
8. Choose one of the "more questions than answers" issues, and write a page about it.
9. Use the CDC, USDA, or FDA website to find a recent listeria outbreak or recall. How do its characteristics fit with those described in the text?
10. Write a really good question about *Listeria* that might be used on your exam.

More questions than answers

1. Are all *L. monocytogenes* strains equally pathogenic?
2. Are all of them capable of causing outbreaks?
3. What is the infectious dose of *L. monocytogenes* for nonpregnant adults, persons with immune system deficiencies, and pregnant women?
4. Do the infectious doses vary significantly among strains?
5. Is the expression of virulence genes influenced by the food the listeriae are in?
6. Do all strains cause feverish gastroenteritis in humans?
7. Are some *L. monocytogenes* strains better able to form biofilms and persist in the environment?
8. Is it realistic to expect all ready-to-eat foods to be free of *Listeria*?
9. How can we ensure that ready-to-eat salads are not contaminated with *L. monocytogenes*?
10. How can we prevent *Listeria* contamination of other ready-to-eat foods such as frankfurters, soft cheeses, and smoked fish?
11. Considering all that you know about bacterial growth, why might *L. monocytogenes* cells be "persistent" compared to other types of bacteria? (Remember that it is cold in most food-processing plants.)

14 *Salmonella* Species

LEARNING OBJECTIVES

The information in this chapter will enable the student to:

- use basic biochemical characteristics to identify *Salmonella*
- understand what conditions in foods favor *Salmonella* growth
- recognize, from symptoms and time of onset, a case of foodborne illness caused by *Salmonella*
- choose appropriate interventions (heat, preservatives, and formulation) to prevent the growth of *Salmonella*
- identify environmental sources of the organism
- understand the roles of *Salmonella* toxins and virulence factors in causing foodborne illness

Outbreak

The convenience of frozen food entrées is especially appealing when time is limited and you may not feel like preparing a meal from scratch. A particularly delicious frozen meal is cheesy chicken and rice. Although the frozen dinners can be cooked in conventional ovens, most consumers prefer the ease of heating such meals in a microwave oven. Cheesy chicken and rice frozen entrées that were not cooked properly resulted in a total of 44 cases of salmonellosis during the spring and summer months of 2010. Approximately 40% of the patients were hospitalized; no deaths were reported. Testing of unopened packages of the cheesy chicken and rice frozen entrées revealed that they were contaminated with *Salmonella enterica* serovar Chester having genetic fingerprints indistinguishable from the outbreak pattern. Heating meals in microwave ovens may not heat foods thoroughly, since microwave ovens vary in strength and tend to cook foods unevenly. Foods should reach an internal temperature of 165°F to ensure that they are cooked properly.

INTRODUCTION

National epidemiological registries continue to underscore the importance of *Salmonella* spp. as leading causes of foodborne bacterial illnesses in humans. Incidents of foodborne salmonellosis tend to dwarf those associated with most other foodborne pathogens (Table 14.1). It is noteworthy that the problem of human salmonellosis from the consumption of contaminated foods generally remains on the increase worldwide. Notwithstanding recent improvements in procedures for the epidemiological investigation

doi:10.1128/9781555817206.ch14

Table 14.1 Examples of major foodborne outbreaks of salmonellosis worldwide since 1981

Yr	Location(s)	Vehicle	Serovar(s)	No. of cases[a]	No. of deaths
1981	The Netherlands	Salad base	Indiana	600[b]	0
1981	Scotland	Raw milk	Typhimurium PT 204	654	2
1984	Canada	Cheddar cheese	Typhimurium PT 10	2,700	0
1984	France, England	Liver pâté	Goldcoast	756	0
1984	International	Aspic glaze	Enteritidis PT 4	766	2
1984	United States	Salad bars	Typhimurium	751	0
1985	United States	Pasteurized milk	Typhimurium	16,284	7
1987	People's Republic of China	Egg drink	Typhimurium	1,113	NS[c]
1987	Norway	Chocolate	Typhimurium	361	0
1988	Japan	Cuttlefish	Champaign	330	0
1988	Japan	Cooked eggs	*Salmonella* spp.	10,476	NS
1991	United States, Canada	Cantaloupes	Poona	>400	NS
1991	Germany	Fruit soup	Enteritidis	600	NS
1993	France	Mayonnaise	Enteritidis	751	0
1993	United States	Tomatoes	Montevideo	100	0
1993	Germany	Paprika chips	Saint-Paul, Javiana, Rubislaw	>670	0
1994	Finland, Sweden	Alfalfa sprouts	Bovismorbificans	492	0
1994	United States	Ice cream	Enteritidis PT 8	740	0
1995	United States	Orange juice	Hartford, Gaminara	62	0
1995	United States	Restaurant foods	Newport	850	0
1996	United States	Alfalfa sprouts	Montevideo, Meleagrides	481	1
1997	United States	Stuffed ham	Heidelberg	746	1
1998	United States	Toasted oat cereal	Agona	209	0
1998	Canada	Cheddar cheese	Enteritidis PT 8	700	0
1999	Japan	Dried squid	*Salmonella* spp.	>453	0
1999	Australia	Orange juice	Typhimurium	427	NS
1999	Japan	Cuttlefish chips	Oranienburg	≥1,500	NS
1999	Canada	Alfalfa sprouts	Paratyphi B (Java)	>53	NS
1999	Japan	Peanut sauce	Enteritidis PT 1	644	NS
1999	United States, Canada	Orange juice	Muenchen	>220	0
2000	United States	Orange juice	Enteritidis	>74	0
2000	United States	Mung bean sprouts	Enteritidis	>45	0
2000	Canada	Bakery goods	Enteritidis	15	0
2004	Great Britain	Lettuce	Newport	>350	0
2004	United States	Roma tomatoes	Braenderup	125	0
			Javiana	390	0
			Typhimurium	27	0
			Anatum	5	0
			Thompson	4	0
			Muenchen	4	0
2005	Great Britain	Imported shell eggs	Enteritidis	68	0

Table 14.1 *(continued)*

Yr	Location(s)	Vehicle	Serovar(s)	No. of cases[a]	No. of deaths
2005	Canada	Roast beef	*Salmonella* spp.	155	0
2006	Great Britain	Chocolate	Montevideo	>46	0
2006	United States	Peanut butter	Tennessee	>288	0
2007	United States	Veggie Booty brand snack	Wandsworth	65	0
2008	United States	Cantaloupes	Litchfield	51	0
2009	United States	Alfalfa sprouts	Saintpaul	235	0
2010	United States	Ground red and black pepper	Montevideo	272	0
2011	United States	Ground turkey	Heidelberg	111	0

[a]Confirmed cases unless stated otherwise.
[b]Estimated number of cases.
[c]NS, not specified.

of foodborne outbreaks in many countries, the global increases in foodborne salmonellosis are considered real. They are not a result of enhanced surveillance programs and/or greater resourcefulness of medical and public health officials. Events such as the ongoing pandemic of egg-borne *Salmonella enterica* serovar Enteritidis infection clearly have an impact on current disease statistics. Major outbreaks of foodborne salmonellosis in the last few decades are of interest because they underline the multiplicity of foods and *Salmonella* serovars that have been implicated in human illness. In 1974, temperature abuse of egg-containing potato salad served at an outdoor barbecue led to an estimated 3,400 human cases of *S. enterica* serovar Newport infection in which cross-contamination of the salad by an infected food handler was suspected. The large Swedish outbreak of serovar Enteritidis phage type (PT) 4 infection in 1977 was attributed to the consumption of a mayonnaise dressing in a school cafeteria. In 1984, Canada experienced its largest outbreak of foodborne salmonellosis. It was attributed to the consumption of Cheddar cheese manufactured from heat-treated and pasteurized milk, and it resulted in >2,700 confirmed cases of serovar Typhimurium PT 10 infection. Manual override of a flow diversion valve reportedly led to the entry of raw milk into vats of thermized and pasteurized cheese milk. The following year, the largest outbreak of foodborne salmonellosis in the United States occurred, involving 16,284 confirmed cases of illness. Although the cause of this outbreak was never determined, a cross-connection between raw and pasteurized milk lines was likely the cause. A large outbreak of salmonellosis affecting >10,000 Japanese consumers was attributed to a cooked egg dish. One of the largest (>650 cases in Ontario alone) sprout-related salmonellosis outbreaks in the world occurred in 2005 in Canada and was associated with consumption of contaminated mung bean sprouts. In 1993, paprika imported from South America was incriminated as the contaminated ingredient in the manufacture of potato chips distributed in Germany. A major outbreak of foodborne salmonellosis that occurred in the United States involved ice cream contaminated with serovar Enteritidis. The transportation of ice cream mix in an unsanitized truck that had previously carried raw eggs was the source of contamination. Despite the general perception about the involvement of meat and egg products as the primary sources of human *Salmonella* infections, many outbreaks in recent years have been associated with tomatoes, orange juice, and vegetable sprouts.

CHARACTERISTICS OF THE ORGANISM

In the early 19th century, clinical pathologists in France first documented the association of human intestinal ulceration with a contagious agent. The disease was later identified as typhoid fever. During the first quarter of the 20th century, great advances occurred in the serological detection of somatic (O) and flagellar (H) antigens within the *Salmonella* group. An antigenic scheme for the classification of salmonellae was later developed and now includes >2,500 serovars.

Biochemical Identification

Salmonella spp. are facultatively anaerobic gram-negative rod-shaped bacteria belonging to the family *Enterobacteriaceae*. Although members of the genus are motile via peritrichous flagella (flagella distributed uniformly over the cell surface), nonflagellated variants, such as *S. enterica* serovars Pullorum and Gallinarum, and nonmotile strains resulting from dysfunctional flagella do occur. Salmonellae are chemoorganotrophic (able to utilize a wide range of organic substrates), with an ability to metabolize nutrients by both respiratory and fermentative pathways. The bacteria grow optimally at 37°C and catabolize D-glucose and other carbohydrates with the production of acid and gas. Salmonellae are oxidase negative and catalase negative and grow on citrate as a sole carbon source. They generally produce hydrogen sulfide, decarboxylate lysine, and ornithine and do not hydrolyze urea. Many of these traits have formed the basis for the presumptive biochemical identification of *Salmonella* isolates. Accordingly, a typical *Salmonella* isolate would produce acid and gas from glucose in triple sugar iron (TSI) agar medium and would not utilize lactose or sucrose in TSI or in differential plating media, such as brilliant green, xylose-lysine-deoxycholate, and Hektoen enteric agars. Additionally, typical salmonellae readily produce an alkaline reaction from the decarboxylation of lysine to cadaverine in lysine ion agar, generate hydrogen sulfide gas in TSI and lysine ion media, and fail to hydrolyze urea. The dynamics of genetic variability (the result of bacterial mutations and conjugative intra- and intergeneric exchange of plasmids encoding determinant biochemical traits) continue to reduce the proportion of "typical" *Salmonella* biotypes. *Salmonella* utilization of lactose and sucrose is plasmid mediated, and the occurrence of *lac*$^+$ and/or *suc*$^+$ biotypes in clinical isolates and food materials is of public health concern. These isolates could escape detection on the disaccharide-dependent plating media that are commonly used in hospital and food industry laboratories. Bismuth sulfite agar remains a medium of choice for isolating salmonellae because, in addition to its high level of selectivity, it responds effectively to the production of extremely low levels of hydrogen sulfide gas. The diagnostic hurdles created by the changing patterns of disaccharide utilization by *Salmonella* are being further complicated by the increasing occurrence of biotypes that cannot decarboxylate lysine, that possess urease activity, that produce indole, and that readily grow in the presence of KCN (potassium cyanide). The variability in biochemical traits is leading to the replacement of traditional testing protocols with molecular technologies targeted at the identification of stable genes and/or their products that are unique to the genus *Salmonella*.

Taxonomy and Nomenclature

According to the World Health Organization (WHO) Collaborating Centre for Reference and Research on *Salmonella* (Institut Pasteur, Paris, France), *S. enterica* and *Salmonella bongori* currently include 2,443 and 20 serovars, respectively (Table 14.2). The Centers for Disease Control and Prevention have adopted as their official nomenclature the following scheme. The genus *Salmonella* contains two species, each of which contains multiple serovars (Table 14.2). The two species are *S. enterica*, the type species, and *S. bongori*, which was formerly subspecies V. *S. enterica* is divided into six subspecies, which are referred to by a roman numeral and a name (I, *S. enterica* subsp. *enterica*; II, *S. enterica* subsp. *salamae*; IIIa, *S. enterica* subsp. *arizonae*; IIIb, *S. enterica* subsp. *diarizonae*; IV, *S. enterica* subsp. *houtenae*; and VI, *S. enterica* subsp. *indica*). *S. enterica* subspecies are differentiated biochemically and by genomic relatedness. *Salmonella* serovars that have been linked to recent cases of foodborne illness include *S. enterica* serovars Enteritidis, Typhimurium, Newport, and Stanley.

Serological Identification

The aim of serological testing procedures is to determine the complete antigenic formulas of individual *Salmonella* isolates. Antigenic determinants include somatic (O) lipopolysaccharides (LPS) on the external surface of the bacterial outer membrane, flagellar (H) antigens associated with the peritrichous flagella, and the capsular (Vi [virulence]) antigen, which occurs only in serovars Typhi, Paratyphi C, and Dublin. The heat-stable somatic antigens are classified as major or minor antigens. The former category consists of antigens, such as the somatic factors O:4 and O:3, which are specific determinants for the somatic groups B and E, respectively. In contrast, minor somatic antigenic components, such as O:12, are nondiscriminatory, as evidenced by their presence in different somatic groups. Smooth variants are strains with well-developed serotypic LPS that readily agglutinate with specific antibodies, whereas rough variants exhibit incomplete LPS antigens, resulting in weak or no agglutination with *Salmonella* somatic antibodies. Flagellar (H) antigens are heat-labile proteins, and individual *Salmonella* strains may produce one (monophasic) or two (diphasic) sets of flagellar antigens. Although serovars such as Dublin produce a single set of flagellar antigens, most serovars can elaborate either of two sets of antigens, i.e., phase 1 and phase 2 antigens. Capsular (K) antigens, commonly encountered in members of the family *Enterobacteriaceae*, are limited to the Vi antigen in the genus *Salmonella*.

Table 14.2 The genus *Salmonella*[a]

***Salmonella* species and subspecies (no.)**	**No. of serovars**
S. enterica subsp. *enterica* (I)	1,454
S. enterica subsp. *salamae* (II)	489
S. enterica subsp. *arizonae* (IIIa)	94
S. enterica subsp. *diarizonae* (IIIb)	324
S. enterica subsp. *houtenae* (IV)	70
S. enterica subsp. *indica* (VI)	12
S. bongori (V)	20
Total	2,463

[a]Source: J.-Y. D'Aoust, p. 1233–1299, *in* B. M. Lund, A. C. Baird-Parker, and G. W. Gould (ed.), *The Microbiological Safety and Quality of Food* (Aspen Publishers Inc., Gaithersburg, MD, 2000).

We present serovar Infantis (6,7:r:1,5 [see below]) as an example. Commercially available polyvalent somatic antisera each consist of a mixture of antibodies specific for a limited number of major antigens; e.g., polyvalent B (poly-B) antiserum (Difco Laboratories, Detroit, Michigan) recognizes somatic groups C_1, C_2, F, G, and H. Following a positive agglutination with poly-B antiserum, single-group antisera representing the five somatic groups included in the poly-B reagent would be used to define the serogroup of the isolate. The test isolate would react with the C_1 group antiserum, indicating that antigens 6 and 7 are present. Flagellar antigens would then be determined by broth agglutination reactions using poly-H antisera or the Spicer-Edwards series of antisera. In the former assay, a positive agglutination reaction with one of the five polyvalent antisera (poly-A to -E; Difco) would lead to testing with single-factor antisera to specifically identify the phase 1 and/or phase 2 flagellar antigens present. Agglutination in poly-C flagellar antiserum and subsequent reaction of the isolate with single-group H antisera would confirm the presence of the r antigen (phase 1). The empirical antigenic formula of the isolate would then be 6,7:r. Phase reversal in semisolid agar supplemented with r antiserum would immobilize phase 1 salmonellae at or near the point of inoculation, thereby facilitating the recovery of phase 2 cells from the edge of the zone of migration. Serological testing of phase 2 cells with poly-E and 1-complex antisera would confirm the presence of the flagellar 1 factor. Confirmation of the flagellar 5 antigen with single-factor antiserum would yield the final antigenic formula 6,7:r:1,5, which corresponds to serovar Infantis. A similar analytical approach would be used with the Spicer-Edwards poly-H antisera, in which the identification of flagellar antigens would arise from the pattern of agglutination reactions among the four Spicer-Edwards antisera and with three additional polyvalent antisera, including the L, 1, and e,n complexes.

Physiology

Growth and Survival

Temperature. *Salmonella* spp. are resilient and can adapt to extreme environmental conditions. Some *Salmonella* strains can grow at elevated temperatures (54°C), and others exhibit psychrotrophic properties (they are able to grow in foods stored at 2 to 4°C) (Table 14.3). Moreover, preconditioning of cells to low temperatures can greatly increase the growth and survival of salmonellae in refrigerated food products. Studies of the maximum temperatures for growth of *Salmonella* spp. in foods are generally lacking. Notwithstanding an early report of growth of salmonellae in inoculated custard and chicken à la king at 45.6°C, more recent evidence indicates that prolonged exposure of mesophilic strains to thermal stress conditions results in mutants of serovar Typhimurium capable of growth at 54°C.

Salmonella can survive for extended periods in foods stored at freezing and room temperatures. Several factors can influence the survival of salmonellae during frozen storage of foods: (i) the composition of the freezing matrix, (ii) the kinetics of the freezing process, (iii) the physiological state of the foodborne salmonellae, and (iv) serovar-specific responses. The viability of salmonellae in dry foods stored at 25°C decreases with increasing storage temperature and with increasing moisture content.

Many factors may enhance the heat resistance of *Salmonella* and other foodborne bacterial pathogens in food ingredients and finished products. The heat resistance of *Salmonella* spp. increases as the water activity (a_w) of

Table 14.3 Physiological limits for the growth of *Salmonella* spp. in foods and bacteriological media

Parameter	Limits (time to double in no.) Minimum	Maximum	Product	Serovar(s)
Temp (°C)	2 (24 h)		Minced beef[a]	Typhimurium
	2 (2 days)		Minced chicken[b]	Typhimurium
	4 (≤10 days)		Shell eggs[b]	Enteritidis
		54.0[c]	Agar medium	Typhimurium
pH	3.99[d]		Tomatoes	Infantis
	4.05[e]		Liquid medium	Anatum, Tennessee, Senftenberg
		9.5	Egg washwater[b]	Typhimurium
a_w	0.93[f]		Rehydrated dried soup[b]	Oranienburg

[a]Naturally contaminated.
[b]Artificially contaminated.
[c]Mutants selected to grow at elevated temperature.
[d]Growth within 24 h at 22°C.
[e]Acidified with HCl or citric acid; growth within 24 h at 30°C.
[f]Growth within 3 days at 30°C.

the heating matrix (food) decreases. The type of solutes used to alter the a_w of the heating menstruum can influence the level of acquired heat resistance. Other important features associated with this adaptive response include the greater heat resistance of salmonellae grown in nutritionally rich versus minimal media, of stationary-phase versus logarithmic-phase cells, and of salmonellae previously stored in a dry environment. The ability of *Salmonella* to acquire greater heat resistance following exposure to sublethal temperatures is equally important. This adaptive response has potentially serious implications for the safety of thermal processes that expose or maintain food products at marginally lethal temperatures. Changes in the fatty acid composition of cell membranes in heat-stressed *Salmonella* provide a greater proportion of saturated membrane phospholipids. This results in reduced fluidity of the bacterial cell membrane and an associated increase in membrane resistance to heat damage. The likelihood that other protective cellular functions are triggered by heat shock stimuli cannot be ruled out.

The following scenario involving *Salmonella* contamination of a chocolate confectionery product demonstrates the complex interactions of food, temperature, and organism. The survival of salmonellae in dry-roasted cocoa beans can lead to contamination of in-line and finished products. Thermal inactivation of salmonellae in molten chocolate is difficult. This is because the time-temperature conditions that are required to eliminate the pathogen in this sucrose-containing product with low a_w would likely result in an organoleptically (off-taste and off-odor) unacceptable product. The problem is further compounded by the ability of salmonellae to survive for many years in the finished product when it is stored at room temperature. Clearly, effective decontamination of raw cocoa beans and stringent in-plant control measures to prevent cross-contamination of in-line products are of great importance in this food industry.

pH. The physiological adaptability of *Salmonella* spp. is further demonstrated by their ability to grow at pH values ranging from 4.5 to 9.5, with an optimum pH for growth of 6.5 to 7.5 (Table 14.3). Of the various organic and inorganic acids used in product acidification, propionic and acetic acids are more bactericidal than the common food-associated lactic and citric acids. Interestingly, the antibacterial action of organic acids decreases with increasing length of the fatty acid chain. Early research on the ability of *Salmonella* to grow in acidic environments revealed that wild-type strains preconditioned on pH gradient plates could grow in liquid and solid media at considerably lower pH values than the starting wild-type strains. This raises concerns regarding the safety of fermented foods, such as cured sausages and fermented raw milk products. The starter culture-dependent acidification of fermented foods could provide a favorable environment for the growth of salmonellae in the product to a state of increased acid tolerance. The growth and/or enhanced survival of salmonellae during the fermentative process would result in a contaminated ready-to-eat product. Ultimately, this may promote survival of *Salmonella* in the host.

Brief exposure of serovar Typhimurium to mild acid environments of pH 5.5 to 6.0 (preshock) followed by exposure of the adapted cells to a pH of ≤4.5 (acid shock) triggers a complex acid tolerance response (ATR). This permits the survival of the microorganism under extreme acid environments (pH 3.0 to 4.0). The response translates into an induced synthesis of 43 acid shock and outer membrane proteins, a reduced growth rate, and pH homeostasis. It is demonstrated by the bacterial maintenance of internal pH values of 7.0 to 7.1 and 5.0 to 5.5 upon sequential exposure of cells to external pH values of 5.0 and 3.3, respectively. Aside from the log-phase ATR, *Salmonella* also possesses a stationary-phase ATR that provides greater acid resistance than the log-phase ATR. This is induced at pHs of <5.5 and functions maximally at pH 4.3. This stationary-phase ATR induces the synthesis of only 15 shock proteins. The induction of the remaining acid-protective mechanism associated with stationary-phase *Salmonella* is independent of the external pH and dependent on the alternative sigma factor (σ^s) encoded by the *rpoS* locus. The mechanism enhances the ability of stationary-phase cells to survive under hostile environmental conditions. Thus, three possibly overlapping cellular systems confer acid tolerance on *Salmonella* spp. These are (i) the pH-dependent, *rpoS*-independent log-phase ATR; (ii) the pH-dependent, *rpoS*-independent stationary-phase ATR; and (iii) the pH-independent, *rpoS*-dependent stationary-phase acid resistance. These systems likely operate in the acidic environments that prevail in fermented and in acidified foods and in phagocytic cells of the infected host.

Acid stress can also enhance bacterial resistance to other adverse environmental conditions. The growth of serovar Typhimurium at pH 5.8 increases thermal resistance at 50°C, enhances tolerance to high osmotic stress (2.5 M NaCl), and produces greater surface hydrophobicity and an increased resistance to the antibacterial lactoperoxidase system and surface-active agents, such as crystal violet and polymyxin B.

Osmolarity. High salt concentrations have long been recognized for their ability to extend the shelf life of foods by inhibiting microbial growth. Foods with a_w values of <0.93 do not support the growth of salmonellae. Although *Salmonella* is generally inhibited in the presence of 3 to 4% NaCl, bacterial salt tolerance increases with increasing temperature in the range of 10 to 30°C.

The pH, salt concentration, and temperature of the microenvironment can exert profound effects on the growth kinetics of *Salmonella* spp. It is now accepted that *Salmonella* spp. have the ability to grow under acidic (pH < 5.0) conditions or in environments of high salinity (>2% NaCl) with increasing temperature. Interestingly, the presence of salt in acidified foods can reduce the antibacterial action of organic acids in which low concentrations of NaCl or KCl stimulate the growth of serovar Enteritidis in broth medium acidified to pH 5.19 with acetic acid.

Modified atmosphere. Concerns about the ability of *Salmonella* to survive under extremes of pH, temperature, and salinity are further heightened by the widespread refrigerated storage of foods packaged under vacuum or modified atmosphere to prolong shelf life. Gaseous mixtures consisting of 60 to 80% (vol/vol) CO_2 with various proportions of N_2 and/or O_2 can inhibit the growth of aerobic spoilage microorganisms, such as *Pseudomonas* spp., without promoting the growth of *Salmonella* spp. The safety of modified-atmosphere- and vacuum-packaged foods that contain high levels of salt is coming under question, since anaerobic conditions may enhance *Salmonella* salt tolerance.

RESERVOIRS

Salmonella spp. will continue to be significant human pathogens in the global food supply for several reasons, including their presence in the environment; intensive farming practices used in the meat, fish, and shellfish industries (which promote spread among animals); and the recycling of slaughterhouse by-products into animal feeds. Poultry meat and eggs are predominant reservoirs of *Salmonella* spp. in many countries (Fig. 14.1). This overshadows the importance of other animal meats, such as pork, beef, and mutton, as potential vehicles of infection. To address the problem of *Salmonella* in meat products, the U.S. Department of Agriculture (USDA) Food Safety Inspection Service requires the meat and poultry industries to implement Hazard Analysis and Critical Control Point (HACCP) plans in all of their plants. The Food Safety Inspection Service conducts *Salmonella* testing to verify that the implemented HACCP are helping to control *Salmonella* on finished products. The rate of *Salmonella* contamination has been reduced for all meat animals since the implementation of HACCP.

The continuing pandemic of human serovar Enteritidis infections associated with the eating of raw or lightly cooked shell eggs and egg-containing

Authors' note

Several government agencies are responsible for controlling salmonellosis. The Food and Drug Administration (FDA) inspects imported foods and milk pasteurization plants, promotes better food preparation techniques in restaurants and food-processing plants, and regulates the sale of turtles. The USDA monitors the health of food animals, inspects egg pasteurization plants, and is responsible for the quality of slaughtered and processed meat. The U.S. Environmental Protection Agency (EPA) regulates and monitors the safety of our drinking water supplies.

Figure 14.1 Consumption of improperly cooked chicken can result in salmonellosis. Consumers must pay attention to the food whether eating at home or at a restaurant to be certain that it is cooked properly; in the case of chicken, consumers should make sure it is not pink or raw in the center. Salmonellosis is the second leading cause of bacterial foodborne illness in the United States. doi:10.1128/9781555817206.ch14.f14.01

products further emphasizes the importance of poultry as vehicles of human salmonellosis. The need for sustained and stringent bacteriological control of poultry husbandry practices is urgent. This egg-related pandemic is of particular concern, because the problem arises from transovarian transmission of serovar Enteritidis into the interior of the egg prior to shell deposition. The viability of these internalized serovar Enteritidis organisms remains unaffected by egg surface sanitizing practices. Increased efforts to intervene and reduce the prevalence of serovar Enteritidis have led to significant reductions in egg-borne outbreaks in the United Kingdom and the United States.

Rapid depletion of wild stocks of fish and shellfish in recent years has greatly increased the importance of the international aquaculture industry. The feeding of raw meat scraps and offal, feces potentially contaminated with typhoid and paratyphoid salmonellae, and animal feeds that may harbor *Salmonella* to reared species is common in developing countries. Accordingly, aquaculture farmers are relying heavily on antibiotics applied at subtherapeutic levels to safeguard the vigor of farmed fish and shellfish. The use of antibiotics creates a serious public health concern, since these antimicrobial agents are the mainstay treatment for systemic salmonellosis in humans.

Fruits and vegetables have gained notoriety in recent years as vehicles of human salmonellosis. The situation has developed from the increased global export of fresh and dehydrated fruits and vegetables from countries that enjoy tropical and subtropical climates. The prevailing hygienic conditions during the production, harvesting, and distribution of products in these countries do not always meet minimum standards and may facilitate product contamination. Operational changes in favor of field irrigation with treated effluents, washing of fruits and vegetables with disinfected water, education of local workers on the hygienic handling of fresh produce, and greater protection of products from environmental contamination during all phases of handling would greatly enhance the safety of fresh fruits and vegetables. These recommendations must be more widely adopted, since recent large outbreaks of salmonellosis have been linked to the consumption of contaminated lettuce and bean sprouts.

CHARACTERISTICS OF DISEASE

Symptoms and Treatment

Human *Salmonella* infections can lead to several clinical conditions, including enteric (typhoid) fever, uncomplicated enterocolitis, and systemic infections by nontyphoid microorganisms (Fig. 14.2). Enteric fever is a serious human disease associated with the typhoid and paratyphoid strains. Symptoms of enteric fever appear after a period of incubation ranging from 7 to 28 days and may include diarrhea, prolonged and spiking fever, abdominal pain, headache, and prostration. Diagnosis of the disease relies on isolation of the infective agent from blood or urine samples in the early stages of the disease or from stools after the onset of clinical symptoms. An asymptomatic chronic carrier state commonly follows the acute phase of enteric fever. The treatment of enteric fever is based on supportive therapy and/or the use of chloramphenicol, ampicillin, or trimethoprim-sulfamethoxazole to eliminate the systemic infection. Marked global increases in the resistance of typhoid and paratyphoid organisms to these antibacterial drugs in the last decade have limited their effectiveness in human therapy.

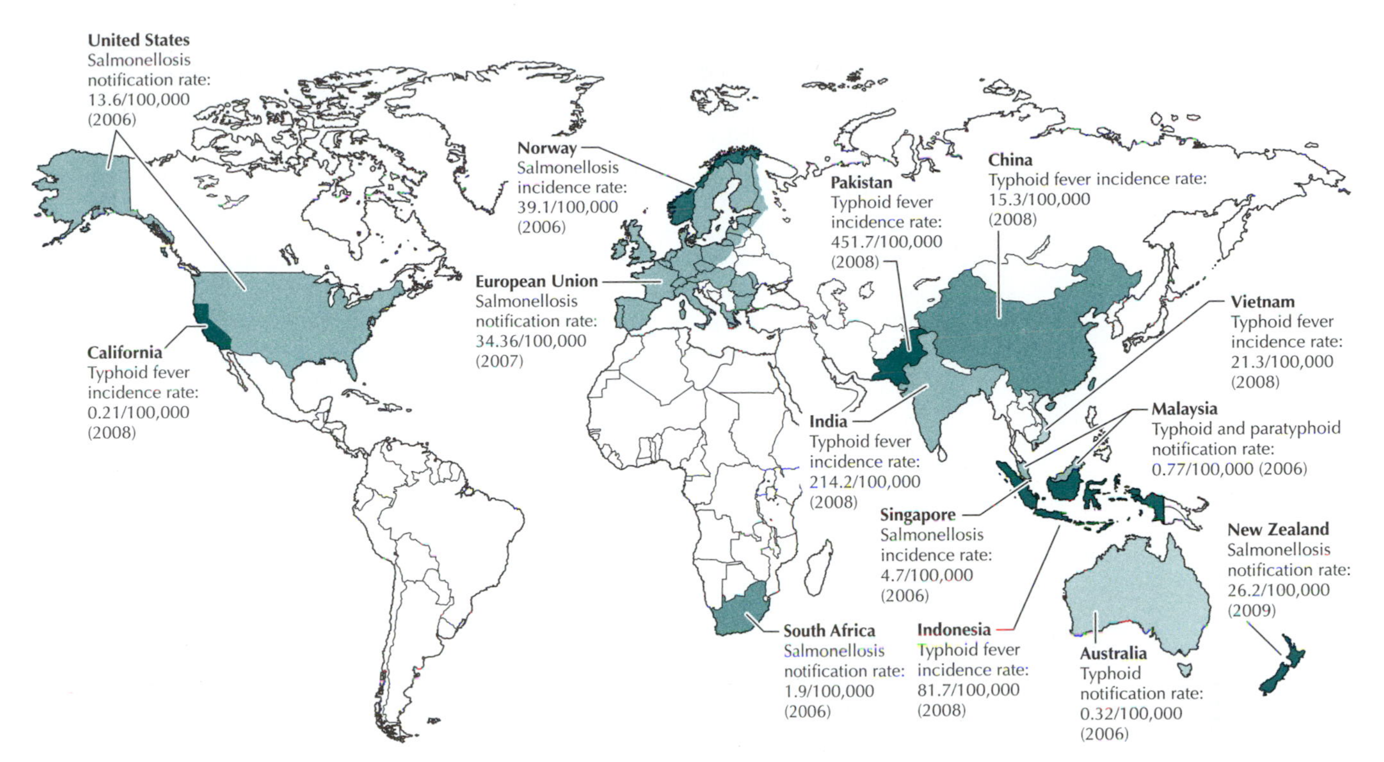

Figure 14.2 Incidences of salmonellosis and typhoid fever in different parts of the world, per 100,000 population. The absence of potable water and proper sewage treatment can result in increased incidences of salmonellosis and typhoid fever. Incidences of both are typically higher in developing countries. Regardless, cases of salmonellosis are underreported, since many sick individuals do not seek medical attention. doi:10.1128/9781555817206.ch14.f14.02

Human infections with nontyphoid *Salmonella* strains commonly result in symptoms 8 to 72 h after ingestion of the invasive pathogen. The illness is usually self-limiting, and remission of the characteristic nonbloody diarrheal stools and abdominal pain usually occurs within 5 days of the onset of symptoms. The successful treatment of uncomplicated cases may require only supportive therapy, such as fluid and electrolyte replacement. Antibiotics are not used in such episodes, because they prolong the carrier state. Asymptomatic persistence of salmonellae in the gut is probably due to antibiotic-dependent repression of the native gut microflora. Human infections with nontyphoid strains can also degenerate into systemic infections and precipitate various chronic conditions.

Salmonella can induce chronic conditions, such as aseptic reactive arthritis, Reiter's syndrome, and ankylosing spondylitis. Bacterial prerequisites for the onset of these chronic diseases include the ability of the bacterial strain to infect mucosal surfaces, the presence of outer membrane LPS, and the ability to invade host cells.

Preventative Measures

The global impact of typhoid and paratyphoid salmonellae on human health led to the early development of parenteral vaccines consisting of heat-, alcohol-, or acetone-killed cells. The phenol- and heat-killed typhoid vaccine is widely used. However, this vaccine can cause adverse reactions, including fever, headache, pain, and swelling at the site of injection. The pandemic of poultry- and egg-borne serovar Enteritidis infections that continues to afflict consumers underscores the potential benefit of vaccines for human and veterinary applications. Live attenuated vaccines continue to generate great research interest because such preparations induce strong and durable immune responses in vaccinees.

Antibiotic Resistance

Current global increases in the numbers of antibiotic-resistant *Salmonella* serovars in humans and farm animals are alarming. The major public health concern is that *Salmonella* will become resistant to antibiotics used in human medicine, thereby greatly reducing therapeutic options and threatening the lives of infected individuals. The liberal administration of antimicrobial agents in hospitals and other treatment centers contributes significantly to the emergence of many resistant strains. The emergence of antimicrobial resistance in bacterial pathogens, many of which can be traced to food-producing animals, has generated much controversy and public media attention.

Because evolutionary processes offer genetic and phenotypic variability for resistance genes, there will always be some level of background resistance in bacterial populations. However, without selective pressures, resistance levels are low, and so are the chances for therapy failures. Resistance in *Salmonella* to a single antibiotic was first noted in the early 1960s. The appearance of serovar Typhimurium definitive type 104 (DT104) in the late 1980s raised major concerns because of its multiple resistances against ampicillin, chloramphenicol, streptomycin, sulfonamides, and tetracycline. Some strains are also resistant to gentamicin, trimethoprim, fluoroquinolones, and other antibiotics. Fluoroquinolones belong to a class of newer antimicrobial agents with a broad spectrum of activity, high efficiency, and widespread application in human and veterinary medicine. An increase in the resistance of DT104 strains to fluoroquinolones has led many European countries to ban the nonhuman use of fluoroquinolones.

The genetic determinants for *Salmonella* virulence and antibiotic resistance can occur on the same plasmid. This increases the disease potential of *Salmonella* DT104. In the next few decades (or perhaps sooner), we may well witness the international scientific community feverishly trying to develop new drugs to replace those currently in use, since they will no longer be effective against the multiple-antibiotic-resistant *Salmonella*.

No studies have conclusively linked the subtherapeutic use of antibiotics in agriculture with the development of resistance. The obvious need for data led to the joint development in 1996 of the National Antimicrobial Resistance Monitoring System by the FDA's Center for Veterinary Medicine, the Centers for Disease Control and Prevention, and the USDA. The purpose of this program is to provide baseline data so that changes in the antimicrobial resistance and susceptibility of bacteria can be identified. Additionally, in June 2000 the WHO released a set of global principles aimed at reducing the risks related to the use of antimicrobials in food animals. An Institute of Food Technologists report (2006) provides new recommendations designed to fill knowledge gaps associated with the implications of antimicrobial resistance for the food system.

INFECTIOUS DOSE

Newborns, infants, the elderly, and immunocompromised individuals are more susceptible to *Salmonella* infections than healthy adults. The incompletely developed immune system in newborns and infants, the frequently

Table 14.4 Human infectious doses of *Salmonella*[a]

Food or ingredient	Serovar(s)	Infectious dose (CFU)[b]
Eggnog	Meleagridis	10^6–10^7
	Anatum	10^5–10^7
Goat cheese	Zanzibar	10^5–10^{11}
Carmine dye	Cubana	10^4
Imitation ice cream	Typhimurium	10^4
Chocolate	Eastbourne	10^2
Hamburger	Newport	10^1–10^2
Cheddar cheese	Heidelberg	10^2
Chocolate	Napoli	10^1–10^2
Cheddar cheese	Typhimurium	10^0–10^1
Chocolate	Typhimurium	$\leq 10^1$
Paprika potato chips	Saint-Paul, Javiana, Rubislaw	$\leq 4.5 \times 10^1$
Alfalfa sprouts	Newport	$\leq 4.6 \times 10^2$
Ice cream	Enteritidis	$\leq 2.8 \times 10^1$

[a]Adapted from J.-Y. D'Aoust, *Int. J. Food Microbiol.* **24:**11–31, 1994.
[b]CFU, colony-forming units.

weak and/or delayed immunological responses in the elderly and debilitated persons, and the generally low gastric acid production in infants and seniors facilitate the intestinal colonization and systemic spread of salmonellae in these populations.

The ingestion of only a few *Salmonella* cells can be infectious (Table 14.4). Determinant factors in salmonellosis are not limited to the immunological heterogeneity within human populations and the virulence of infecting strains but may also include the chemical composition of contaminated food. A common denominator of foods associated with low infectious doses is a high fat content (e.g., cocoa butter in chocolate, milk fat in cheese, and animal fat in meat). *Salmonella* organisms may become entrapped within hydrophobic lipid micelles and be protected against the bactericidal action of gastric acidity. Food producers, processors, and distributors need to be reminded that low levels of salmonellae in a finished food product could lead to serious public health consequences. This could undermine the reputation and economic viability of the incriminated food manufacturer.

PATHOGENICITY AND VIRULENCE FACTORS

Specific and Nonspecific Human Responses

The presence of viable salmonellae in the human intestinal tract confirms the successful evasion by ingested organisms of nonspecific host defenses. Antibacterial lactoperoxidase in saliva, gastric acidity, mucoid secretions from intestinal cells, intestinal peristalsis, and sloughing of luminal epithelial cells synergistically inhibit bacterial colonization of the intestine. In addition to these hurdles to bacterial infection, the antibacterial action of phagocytic cells coupled with the immune responses mounts a formidable defense against the systemic spread of *Salmonella*. The human diarrheagenic response to foodborne salmonellosis results from the migration of the pathogen from the oral cavity to intestinal tissues and mesenteric lymph follicles (enterocolitis). The failure of host defense systems to hold the invasive *Salmonella* in check can degenerate into septicemia and other chronic clinical conditions.

Attachment and Invasion

The establishment of a human *Salmonella* infection depends on the ability of the bacterium to attach (colonization) and enter (invasion) intestinal cells. Salmonellae must successfully compete with the indigenous gut microflora for suitable attachment sites on the intestinal wall. Upon contact with epithelial cells, *Salmonella* produces proteinaceous appendages on its surface. After bacterial attachment, signaling between the pathogen and the host cell results in *Salmonella* invasion of intestinal cells. Histologically, *Salmonella* invasion is characterized by membrane ruffling of epithelial cells and programmed death (apoptosis) of epithelial and phagocytic cells. Diarrhea associated with salmonellae now appears to be in response to bacterial invasion of intestinal cells rather than the action of a putative enterotoxin.

Several sets of genes are involved in the invasion process. The *inv* pathogenicity island is a multigenic locus consisting of 30 genes. Many of these genes encode the synthesis of enzymes and transcriptional activators responsible for the regulation, expression, and translocation of important effectors to the surfaces of host cells. Genes in the pathogenicity island are useful targets for polymerase chain reaction (PCR)-based detection of *Salmonella* in foods.

Growth and Survival within Host Cells

In contrast to several bacterial pathogens, such as *Yersinia*, *Shigella*, and enteroinvasive *Escherichia coli*, which replicate within the cytoplasm of host cells, salmonellae are confined to endocytotic vacuoles in which bacterial replication begins within hours following internalization. The infected vacuoles move from the apical to the basal pole of the host cell, where *Salmonella* organisms are released into the tissue. A bacterial surface mechanism that facilitates the migration of salmonellae deeper into layers of tissue upon their release has been tentatively identified.

The systemic migration of *Salmonella* exposes it to phagocytosis and to the antibacterial conditions in the cytoplasm of the host defense cells. Survival of the bacterial cell within the hostile confines of the phagocytic cells determines the host's fate. Whether the host develops enteric fever from *Salmonella* infection is determined partly by the genetics of the host and the salmonellae. Intracellular pathogens have developed different strategies to survive within phagocytic cells. These include (i) escaping the phagosomes, (ii) inhibiting acidification of the phagosomes, (iii) preventing phagosome-lysosome fusion, and (iv) withstanding the toxic environment of the phagolysosome. Although there is evidence to support the ability of *Salmonella* to prevent acidification of the phagosomes, fusion with the lysosome, or maturation of the phagolysosome, *Salmonella* is a hardy bacterium that can survive and replicate within the bactericidal milieu of the acidified phagolysosome.

Virulence Plasmids

Virulence plasmids are large DNA structures that replicate in synchrony with the bacterial chromosome. These plasmids contain many virulence loci ranging from 30 to 60 megadaltons in size and occur with a frequency of one or two copies per chromosome. The presence of virulence plasmids within the genus *Salmonella* is limited and has been confirmed in serovars Typhimurium, Dublin, Gallinarum-Pullorum, Enteritidis, Choleraesuis, and Abortusovis. The highly infectious serovar Typhi does not carry a virulence plasmid. Although limited in its distribution in nature, the plasmid is self-transmissible. Gene products from the transcription of this plasmid aid systemic spread and infection of tissues other than the intestine but not *Salmonella* adhesion to and invasion of epithelial cells. More specifically, virulence plasmids may enable carrier strains to rapidly multiply within host cells and overwhelm host defense mechanisms. These plasmids also confer on the salmonellae the ability to induce lysis of macrophages and elicit inflammatory response and enteritis in their animal hosts. The *Salmonella* virulence plasmids contain highly conserved nucleotide sequences and exhibit functional homology in which plasmid transfer from a wild-type serovar to another, plasmid-cured serovar restores the virulence of the recipient strain. The *Salmonella* virulence plasmid also contains a fimbrial operon that encodes an adhesin involved in colonization of the small intestine.

Other Virulence Factors

Siderophores are yet another part of the *Salmonella* virulence weaponry. These elements retrieve essential iron from host tissues to drive key cellular functions, such as the electron transport chain and enzymes associated with iron cofactors. To this end, *Salmonella* must compete with host transferrin, lactoferrin, and ferritin ligands for available iron. For example, transferrin scavenges tissue fluids for Fe^{3+} ions to form Fe^{3+}-transferrin complexes that bind to surface host cell receptors. Upon internalization, the complexes dissociate and the released Fe^{3+} is complexed with ferritin for intracellular storage. Siderophores do not appear to be the only mechanism by which *Salmonella* can acquire iron from its host.

Diarrheagenic enterotoxin may be a *Salmonella* virulence factor. The release of toxin into the cytoplasm of infected host cells precipitates an activation of adenyl cyclase localized in the epithelial cell membrane and a marked increase in the cytoplasmic concentration of cyclic adenosine monophosphate (AMP) in host cells. The simultaneous fluid exsorption into the intestine results from a net secretion of Cl^{2+} ions and depressed Na^+ absorption at the level of the intestinal villi. Enterotoxigenicity is a virulence phenotype of *Salmonella*, including serovar Typhi, which is expressed within hours following bacterial contact with the targeted host cells.

In addition to enterotoxin, *Salmonella* strains generally elaborate a thermolabile cytotoxic protein, which is localized in the bacterial outer membrane. Hostile environments, such as acidic pH and elevated (42°C) temperature, cause release of the toxin, possibly as a result of induced bacterial lysis. The virulence attribute of cytotoxin stems from its inhibition of protein synthesis and its lysis of host cells, thereby promoting the dissemination of viable salmonellae into host tissues.

Three additional virulence determinants located within or on the external surface of the *Salmonella* outer membrane are worth mentioning. The capsular polysaccharide Vi antigen occurs in most strains of serovar Typhi, in a few strains of serovar Paratyphi C, and, rarely, in serovar Dublin. The length of LPS that protrudes from the bacterial outer membrane not only defines the rough (short-LPS) and smooth (LPS) phenotypes but also plays an important role in preventing attack by the host immune system. Porins are outer membrane proteins that function as transmembrane (outer membrane) channels in regulating the influx of nutrients, antibiotics, and other small molecular species. Low osmolarity, low nutrient availability, and low temperature can regulate the expression of these genes.

Summary

- Intensive animal husbandry practices in meat and poultry production and processing industries continue to make raw poultry and meats principal vehicles of human foodborne salmonellosis.
- Unless changes in agricultural and aquacultural practices are implemented, the prevalence of human foodborne salmonellosis will continue.
- Salmonellae are classified as serovars of two species.
- The emergence of multiple-antibiotic-resistant strains may be linked to the use of antibiotics in agriculture.
- Salmonellae quickly adapt to environments of high salinity and low pH, creating a problem with respect to control in food matrices.
- Virulence genes are located on the chromosome in pathogenicity islands and in large virulence plasmids.
- Epidemiological evidence suggests a low infective dose.
- *Salmonella* invasion proteins may activate apoptosis in epithelial and phagocytic cell types.

Suggested reading

Bellido-Blasco, J. B., A. Arnedo-Pena, E. Cordero-Cutillas, M. Canós-Cabedo, C. Herrero-Carot, and L. Safont-Adsuara. 2002. The effect of alcoholic beverages on the occurrence of a *Salmonella* food-borne outbreak. *Epidemiology* **13:**228–230.

Brenner, F. W., R. G. Villar, F. J. Angulo, R. Tauxe, and B. Swaminathan. 2000. *Salmonella* nomenclature. *J. Clin. Microbiol.* **38:**2465–2467.

D'Aoust, J.-Y. 1991. Pathogenicity of foodborne *Salmonella. Int. J. Food Microbiol.* **12:**17–40.

D'Aoust, J.-Y. 1994. *Salmonella* and the international food trade. *Int. J. Food Microbiol.* **24:**11–31.

D'Aoust, J.-Y. 2000. *Salmonella,* p. 1233–1299. *In* B. M. Lund, A. C. Baird-Parker, and G. W. Gould (ed.), *The Microbiological Safety and Quality of Food*. Aspen Publishers Inc., Gaithersburg, MD.

D'Aoust, J.-Y., and J. Maurer. 2007. *Salmonella* species, p. 187–236. *In* M. P. Doyle and L. R. Beuchat (ed.), *Food Microbiology: Fundamentals and Frontiers*, 3rd ed. ASM Press, Washington, DC.

Pui, C. F., W. C. Wong, L. C. Chai, R. Tunung, P. Jeyaletchumi, M. S. Noor Hidayah, A. Ubong, M. G. Farinazleen, Y. K. Cheah, and R. Son. 2011. *Salmonella*: a foodborne pathogen. *Int. Food Res. J.* **18:**465–473.

Questions for critical thought

1. A large outbreak of salmonellosis was linked to consumption of peanut butter. Peanut butter has an a_w of approximately 0.70. Would you expect *Salmonella* to grow, die, or survive under such a condition?
2. The typing scheme for *Salmonella* is based on two antigens. What is meant by the nomenclature 6,7:r:1,5?

3. There are many *Salmonella* serovars but only two species. What are the *Salmonella* species and subspecies? What serovars are most commonly associated with outbreaks in the United States?
4. Serovar Enteritidis has been difficult to control in poultry. What unique characteristic(s) of this pathogen contributes to the control problem?
5. *Salmonella* DT104 is resistant to multiple antibiotics. Within the bacterial cell, where are the genes located, and why is that important?
6. Some *Salmonella* strains produce capsule, and capsule antigens are unique to some serovar Typhi strains. What role may capsule play in survival of the organism in a host or in a food-processing facility?
7. How might the ATR contribute to bacterial survival in food and subsequently in the host?
8. How would you determine whether a particular *Salmonella* isolate was resistant or sensitive to a given antibiotic? Could you determine whether the resistance was transient? How?
9. How could motility and antibodies against the flagella be used for detection of *Salmonella*?
10. To prevent foodborne illness associated with large meals, such as holiday dinners, what food-handling practices should be employed?
11. Individuals on antibiotic therapy are more prone to infection with *Salmonella*. Explain why.
12. Would individuals infected with a strain of *Salmonella* lacking the gene for the diarrheagenic enterotoxin experience diarrhea? Explain your answer.
13. Based on the impacts of pH, temperature, salinity, and a_w on the growth and survival of *Salmonella*, develop a hurdle technology to prevent the growth of *Salmonella* in the food of your choice.
14. According to a report by Bellido-Blasco et al. (see "Suggested reading"), consumption of alcohol can mitigate or prevent foodborne illness linked to the consumption of *Salmonella*-contaminated food. Should the Centers for Disease Control and Prevention recommend that individuals who consume raw foods (e.g., oysters, sushi, and seed sprouts) consider consuming alcohol as a preventative measure? Explain your answer; base it on societal issues and science.

15

Shigella Species

LEARNING OBJECTIVES

The information in this chapter will enable the student to:

- use basic biochemical characteristics to identify *Shigella*
- understand what conditions in foods favor *Shigella* growth
- recognize, from symptoms and time of onset, a case of foodborne illness caused by *Shigella*
- choose appropriate interventions (heat, preservatives, and formulation) to prevent the growth of *Shigella*
- identify environmental sources of the organism
- understand the roles of *Shigella* toxins and virulence factors in causing foodborne illness

Outbreak

Maybe it is a good thing that airlines no longer serve meals on most flights. An estimated 300 to 1,500 airline passengers became infected with *Shigella sonnei* following consumption of contaminated carrots. In the 2004 outbreak, 47 passengers had culture-confirmed cases of shigellosis. Luckily, the onset time was not rapid and passengers did not suffer diarrhea in-flight. The epidemiological investigation determined that carrots were the only ingredient common to salads served in economy class. The carrots were likely contaminated before they were shipped to the caterer, since no one worker could be linked to preparation of all carrots and all salads.

Individuals recovering from shigellosis can shed the pathogen for months in their feces. In short, food handlers can transmit *Shigella* through their feces without even showing symptoms. During a 2-month period, 52 individuals that had all eaten at the same cafeteria developed shigellosis. One of the food handlers had traveled to Morocco shortly before the outbreak started. Seven individuals had stool samples positive for the same strain as was seen in Morocco during the same time. Washing hands is especially important after going to the bathroom and before food preparation.

In a 2-month period during the summer of 1998, seven outbreaks of *Shigella sonnei* infection associated with eating fresh parsley occurred. Fresh parsley is often used as a garnish or sprinkled on top of a meal prior to serving to add taste and color. Hundreds of people became ill, with most experiencing severe short-term diarrhea. Most of the patients had eaten at restaurants that served chopped, uncooked parsley. The causative agent was identified as *S. sonnei*, and based on pulsed-field gel electrophoresis, epidemiological traceback, and data from other investigations, one farm in Mexico was implicated as the source of the contaminated parsley. As a

doi:10.1128/9781555817206.ch15

result of these outbreaks, changes in food-handling practices were proposed, eliminating practices such as chopping and holding large quantities of parsley and chopping of parsley at room temperature. *Shigella* is spread through contaminated water used for irrigation and postharvest processing of fresh produce. Transmission occurs through the fecal-oral route, with as few as 10 to 100 organisms capable of causing infection. Humans and other primates are the only reservoirs for *S. sonnei*.

INTRODUCTION

Bacillary dysentery, or shigellosis, is caused by *Shigella* species. Dysentery was the term used by Hippocrates to describe an illness characterized by frequent passage of stools containing blood and mucus accompanied by painful abdominal cramps. Perhaps one of the greatest historical impacts of this disease has been its powerful influence on military operations. Protracted military campaigns and sieges have almost always spawned epidemics of dysentery, causing large numbers of military and civilian casualties. With a low infectious dose required to cause disease coupled with oral transmission by fecally contaminated food and water, it is not surprising that dysentery caused by *Shigella* spp. follows in the wake of many natural (earthquakes, floods, and famine) and man-made (war) disasters. Apart from these special circumstances, shigellosis remains an important disease in developed and developing countries.

During the past two decades, several large outbreaks of shigellosis (listed below) have been linked to the consumption of contaminated food. Disease is caused by ingestion of these contaminated foods, and in some instances it subsequently leads to rapid dissemination through contaminated feces from infected individuals.

1989 and 1994: Shigellosis aboard cruise ships. In October 1989, 14% of passengers and 3% of crew members aboard a cruise ship reported having gastrointestinal symptoms. A multiple-antibiotic-resistant strain of *Shigella flexneri* 2a was isolated from several ill passengers and members of the crew. The vehicle of the outbreak was German potato salad. Contamination was introduced by infected food handlers, initially in the country where the food was originally prepared and subsequently by a member of the galley crew on the cruise ship. Another outbreak of shigellosis occurred in August 1994 on the cruise ship S.S. *Viking Serenade*. Thirty-seven percent (586) of the passengers and 4% (24) of the crew reported having diarrhea, and one death occurred. *S. flexneri* 2a was isolated from patients, and the suspected vehicle was spring onions.

1990: Operation Desert Shield. Diarrheal diseases during a military operation can be a major factor in reducing troop readiness. Enteric pathogens were isolated from 214 U.S. soldiers in Operation Desert Shield, and among those soldiers, 113 were diagnosed as having shigellosis; *S. sonnei* was the most prevalent species isolated. Shigellosis accounted for more time lost from military duties and was responsible for more severe morbidity than enterotoxigenic *Escherichia coli*, the most common enteric pathogen isolated from U.S. troops in Saudi Arabia. The suspected vehicle was contaminated fresh vegetables, specifically, lettuce. Twelve heads of lettuce were tested, and enteric pathogens were isolated from all of them.

1991: Moose soup in Alaska. In September 1991 in Galena, Alaska, 25 people who participated in a gathering of local residents contracted shigellosis associated with eating homemade moose soup. One of five women who made the soup reported having gastroenteritis while preparing it. *S. sonnei* was isolated from a hospitalized patient.

1994: Contaminated produce. Lettuce and green onions have been implicated in illness. An outbreak in Norway of 110 culture-confirmed cases of shigellosis caused by *S. sonnei* was reported in 1994. Iceberg lettuce from Spain, served in a salad bar, was suspected as the source of the outbreak in Norway and was likely responsible for increases in shigellosis in other European countries, including the United Kingdom and Sweden. *S. sonnei* was isolated from patients from several northwest European countries but was not isolated from any foods. Epidemiological evidence indicated that imported lettuce was the vehicle of these outbreaks. An outbreak of *S. flexneri* serotype 6 (mannitol-negative) infection occurred in the Midwest in 1994. Although not confirmed, the suspected vehicle was Mexican green onions (scallions, or spring onions). Seventeen cases of shigellosis were contracted at a church potluck meal in Indiana, 29 cases were contracted at an anniversary reception in Indiana, and 26 culture-confirmed mannitol-negative *S. flexneri* or *Shigella* sp. cases were reported to the Illinois State Department of Health. *S. flexneri* serotype 6 was also isolated from patients in Missouri, Minnesota, Wisconsin, Michigan, and Kentucky. Green onions were implicated as the vehicle of infection. An infected worker most likely contaminated the onions at the time of harvest or packing.

2000: Five-layer bean dip. An outbreak of shigellosis associated with contaminated five-layer (bean, salsa, guacamole, nacho cheese, and sour cream) party dip occurred in three West Coast states. The causative agent, *S. sonnei*, was isolated from at least 30 patients. The pathogen was isolated from only one layer (cheese) of the dip and was initially detected by a polymerase chain reaction (PCR) assay targeting shigellae. *Shigella* was isolated subsequently by enrichment, followed by plating on selective agar.

2001: Tomatoes. An outbreak involving more than 880 people was linked, based on epidemiological information, to the use by several restaurants of overripe and bruised tomatoes purchased from a single distributor. The strain was isolated from ill and symptomatic workers, but they reported not being ill prior to the tomatoes' arrival.

One of the striking features regarding foodborne outbreaks of shigellosis is that contamination of foods usually occurs not at the processing plant but, rather, through an infected food handler. As is evident from the examples above and in Table 15.1, these incidents can be anything from contamination of foods by infected food handlers at small-town gatherings and picnics to large-scale outbreaks, such as those on cruise ships and at institutions.

Classification and Biochemical Characteristics

There are four species in the genus *Shigella*, serologically grouped (41 serotypes) based on their somatic (O) antigens: *Shigella dysenteriae* (group A), *S. flexneri* (group B), *Shigella boydii* (group C), and *S. sonnei* (group D). As members of the family *Enterobacteriaceae*, they are genetically almost identical to

Table 15.1 Examples of foodborne outbreaks caused by *Shigella* spp.

Yr	Location	Source of contamination[a]	Species
1986	Texas	Shredded lettuce	*S. sonnei*
1987	Rainbow Family gathering	Food handlers	*S. sonnei*
1988–1989	Monroe, New York	Multiple sources	*S. sonnei*
1988	Outdoor music festival, Michigan	Food handlers	*S. sonnei*
1988	Commercial airline	Cold sandwiches	*S. sonnei*
1989	Cruise ship	Potato salad	*S. flexneri*
1990	Operation Desert Shield (U.S. troops)	Fresh produce	*Shigella* spp.
1991	Alaska	Moose soup	*S. sonnei*
1992–1993	Operation Restore Hope, Somalia (U.S. troops)		*Shigella* spp.
1994	Europe	Shredded lettuce from Spain	*S. sonnei*
1994	Midwest	Green onions	*S. flexneri*
1994	Cruise ship		*S. flexneri*
1998	Various U.S. locations	Fresh parsley	*S. sonnei*
2000	West Coast	Bean dip	*S. sonnei*
2001	New York	Tomato	*S. flexneri*
2004	Commercial airline	Carrots	*S. sonnei*
2010	Chicago		*S. sonnei*
2011	Belgium	Food handler	*S. sonnei*

[a]The source of contamination is listed when known.

the escherichiae and closely related to the salmonellae. *Shigella* spp. are nonmotile, oxidase-negative, gram-negative rods. An important biochemical characteristic that distinguishes these bacteria from other enteric bacteria is their inability to ferment lactose; however, some strains of *S. sonnei* may ferment lactose slowly or utilize citric acid as a sole carbon source. They do not produce H_2S, except for *S. flexneri* 6 and *S. boydii* serotypes 13 and 14, and do not produce gas from glucose. *Shigella* spp. are inhibited by potassium cyanide and do not synthesize lysine decarboxylase. Enteroinvasive *E. coli* (EIEC) has pathogenic and biochemical properties similar to those of *Shigella* spp. These similarities pose a problem in distinguishing these pathogens. For example, EIEC is nonmotile and is unable to ferment lactose. Some serotypes of EIEC also have O antigens identical to those of *Shigella*.

Shigella spp. are not particular in their growth requirements and are routinely cultivated in the laboratory on artificial medium. Cultures of *Shigella* are easily isolated and grown from analytical samples, including water and clinical specimens. In the latter case, *Shigella* spp. are present in fecal specimens in large numbers (10^3 to 10^9 per g of stool) during the acute phase of infection, and therefore, identification is readily accomplished using culture media, biochemical analysis, and serological typing. Shigellae are shed by, and continue to be detected from, convalescent patients (10^2 to 10^3 per g of stool) for weeks or longer after the initial infection. Isolation of *Shigella* at this stage of infection is more difficult because a selective enrichment broth for shigellae is not available, and therefore, shigellae can be outgrown by the resident bacterial fecal flora.

Isolation of *Shigella* spp. from foods is not as easy as from other sources. Foods have many different physical attributes that may affect the recovery of shigellae. These factors include composition, such as the fat content of the food; physical parameters, such as pH and salt; and the natural microbial flora of the food. In the last case, other microbes in a sample may overgrow shigellae during culture in broth media. The amount of time from the clinical report of a suspected outbreak to the analysis of the food samples can be considerable, thus lessening the chances of identifying the causative agent. The physiological state of shigellae present in the food is a contributing factor in the successful recovery of the pathogen. *Shigella* spp. may be present in low numbers or in a poor physiological state in suspect food samples. Under these conditions, special enrichment procedures are required for successful isolation and detection of shigellae.

Shigella in Foods

Shigella spp. are not associated with any specific foods. Common foods that have been implicated in outbreaks caused by shigellae include potato salad, chicken, tossed salad, and shellfish. Establishments where contaminated foods have been served include the home, restaurants, camps, picnics, schools, airlines, sorority houses, and military mess halls. In many cases, the source (food) was not identified. From 1983 to 1987, 2,397 foodborne outbreaks representing 54,453 cases were reported to the Centers for Disease Control and Prevention (CDC). In only 38% of the cases was the source of the etiological agent identified. Whereas epidemiological methods may strongly imply a common food source, *Shigella* spp. are not often recovered from foods and identified by using standard bacteriological methods. Also, since shigellae are not commonly associated with any particular food, routine testing of foods to identify these pathogens is not usually performed.

The traditional approach to address the problem of microbially contaminated foods in the processing plant is to inspect the final product. There are several drawbacks to this approach. Current bacteriological methods are often time-consuming and laborious. An alternative to end product testing is the Hazard Analysis and Critical Control Point (HACCP) system. The HACCP system identifies certain points of the processing system that may be most vulnerable to microbial contamination and chemical and physical hazards.

In contrast, establishing specific critical control points for preventing *Shigella* contamination of foods is not always suitable for the HACCP concept. The pathogen is usually introduced into the food supply by an infected person, such as a food handler with poor personal hygiene. In some cases, this may occur at the manufacturing site, but more likely it happens at a point between the processing plant and the consumer. Another factor is the fact that foods, such as vegetables (lettuce is a good example), can be contaminated at the site of collection and shipped directly to market. Although the HACCP system is a method for controlling food safety and preventing foodborne outbreaks, pathogens such as *Shigella* that are not indigenous to, but rather introduced into, foods are most likely to go undetected.

Survival and Growth in Foods

Depending upon growth conditions, *Shigella* spp. can survive in media with a pH range of 2 to 3 for several hours. However, shigellae do not usually survive well in low-pH foods or in stool samples. Studies using citrus juices (orange and lemon), carbonated beverages, and wine revealed that shigellae were recovered after 1 to 6 days. In neutral-pH foods, such as butter or margarine, shigellae were recovered after 100 days when stored frozen or at 6°C.

Shigella can survive a temperature range of −20°C to room temperature. However, shigellae survive longer in foods stored frozen or at refrigeration temperature than in those stored at room temperature. In foods such as salads containing mayonnaise and some cheese products, *Shigella* has survived for 13 to 92 days. *Shigella* can survive for an extended period on dry surfaces and in foods such as frozen shrimp, ice cream, and minced pork. Growth of *Shigella* is impeded in the presence of 3.8 to 5.2% NaCl at pH 4.8 to 5.0, in 300 to 700 mg of $NaNO_2$/liter, and in 0.5 to 1.5 mg of sodium hypochlorite (NaClO)/liter of water at 4°C. *Shigella* is sensitive to ionizing radiation, with a reduction of 10^7 colony-forming units (CFU)/g at 3 kilograys.

CHARACTERISTICS OF DISEASE

Shigellosis is differentiated from diseases caused by most other foodborne pathogens described in this book by at least two important characteristics: (i) the production of bloody diarrhea or dysentery and (ii) the low infectious dose. Dysentery involves bloody diarrhea, but the passage of bloody mucoid stools is accompanied by severe abdominal and rectal pain, cramps, and fever. While abdominal pain and diarrhea are experienced by nearly all patients with shigellosis, fever occurs in about one-third and gross blood in the stools occurs in ~40% of cases. The clinical features of shigellosis range from a mild watery diarrhea to severe dysentery. The dysentery stage caused by *Shigella* spp. may or may not be preceded by watery diarrhea. During the dysentery stage, there is extensive bacterial colonization of the colon and invasion of the cells of the colon. As the infection progresses, dead cells of the mucosal surface slough off. This leads to the presence of blood, pus, and mucus in the stools.

Shigella is a serious pathogen that causes disease in otherwise healthy individuals (Box 15.1). The greatest frequency of illness is among children <6 years of age. The incubation period for shigellosis is 1 to 7 days, but the symptoms usually begin within 3 days. The severity of illness differs depending on the strain involved; however, regardless of the severity of the illness, shigellosis is self-limiting. If left untreated, clinical illness usually persists for 1 to 2 weeks (although it may last as long as 1 month), and the patient recovers. Since the infection is self-limited in normally healthy patients and full recovery occurs without the use of antibiotics, drug therapy is usually not indicated. However, the antibiotic of choice for treatment of shigellosis is trimethoprim-sulfamethoxazole. Complications arising from the disease include severe dehydration, intestinal perforation, septicemia, seizures, hemolytic-uremic syndrome, and Reiter's syndrome.

BOX 15.1

Shigella is easily transmitted

Shigella is easily passed from person to person by food. Outbreaks are often associated with poor hygiene, especially improper hand washing after use of the bathroom. The pathogen multiplies rapidly in food at room temperature. *Shigella* produces a powerful toxin, Shiga toxin, once in the host. Shigellosis is an infection, not an intoxication. That is, vegetative cells must be ingested, after which the cells multiply and cause illness. As few as 100 cells may cause illness.

FOODBORNE OUTBREAKS

Although the number of reported foodborne outbreaks of shigellosis in the United States has declined recently, shigellosis continues to be a major public health concern. Surveillance data suggest that approximately 450,000 cases of shigellosis occur yearly in the United States. *Shigella sonnei* was associated with 72% of those cases. This made it the third leading cause of foodborne outbreaks by bacterial pathogens. Worldwide, the World Health Organization (WHO) estimates that *Shigella* spp. are responsible for 164.7 million cases of shigellosis annually in developing countries and 1.5 million cases in developed countries. The CDC, WHO, and other international agencies have instituted surveillance systems to monitor foodborne outbreaks caused by *Shigella*.

Humans are the natural reservoir of *Shigella*. Human-to-human transmission of *Shigella* is through the fecal-oral route. Most cases of shigellosis result from the ingestion of fecally contaminated food or water. With foods, the major cause of contamination is poor personal hygiene of food handlers. From infected carriers, shigellae are spread by several routes, including food, fingers, feces, and flies. The highest incidence of shigellosis occurs during the warmer months of the year. Improper storage of contaminated foods is the second most common factor contributing to foodborne outbreaks of shigellosis. Other contributing factors are inadequate cooking, contaminated equipment, and food obtained from unsafe sources. To reduce the spread of shigellosis, infected patients should be monitored until stool samples are negative for *Shigella*. The low infectious dose of *Shigella* underlies the high rate of transmission. As few as 100 cells of *Shigella* can cause illness, facilitating person-to-person spread, as well as foodborne and waterborne outbreaks of diarrhea.

Shigellosis can be widespread in institutional settings, such as prisons, mental hospitals, and nursing homes, where crowding and/or insufficiently hygienic conditions create an environment for direct fecal-oral contamination. The occurrence of disasters that destroy the sanitary waste treatment and water purification infrastructure is often associated with large outbreaks of shigellosis.

VIRULENCE FACTORS

The clinical symptoms of shigellosis can be directly attributed to the hallmarks of *Shigella* virulence: the ability to induce diarrhea, invade epithelial cells of the intestine, multiply intracellularly, and spread from cell to cell. The production of enterotoxins by the bacteria while they are in the small bowel probably causes the diarrhea that precedes dysentery. The ability of *Shigella* to invade epithelial cells and move from cell to cell is regulated by an array of genes. A bacterium's mechanism for regulating expression of the genes involved in virulence is important for pathogenicity. *S. dysenteriae* produces a thermolabile toxin, designated Shiga toxin, that is involved in the pathogenesis of *Shigella* diarrhea. Strains that are invasive and produce the toxin cause the most severe infections.

The growth temperature is an important factor in controlling virulence. Virulent strains of *Shigella* are invasive when grown at 37°C but noninvasive when grown at 30°C. This strategy ensures that the organism conserves energy by synthesizing virulence products only when it is in the host.

Genetic Regulation

Given the complexity of the interactions between host and pathogen, it is not surprising that *Shigella* virulence requires several genes. These include both chromosomal and plasmid-carried genes. The large plasmid has an indispensable role in invasion by *S. sonnei* and *S. flexneri*. Other *Shigella* spp., as well as strains of EIEC, contain similar plasmids, which are functionally interchangeable and show significant degrees of DNA relatedness. The plasmids of *Shigella* and EIEC are probably derived from a common ancestor.

The *ipa* group of genes encode invasion plasmid antigens, the main antigens detected with sera from convalescent patients and experimentally challenged monkeys. These genes are required for the invasion of mammalian cells. Ultimately, the proteins form a complex on the bacterial cell surface and are responsible for transducing the signal leading to entry of *Shigella* into the host cells via bacterium-directed phagocytosis. The products of the *ipa* genes have also been postulated to be the contact hemolysin responsible for lysis of the phagocytic vacuole minutes after entry of the bacterium into the host cell. The ability of *S. flexneri* to induce programmed cell death in infected macrophages is an additional property assigned to IpaB.

In contrast to the genes of the virulence plasmid that are responsible for the invasion of mammalian tissues, most of the chromosomal loci associated with *Shigella* virulence are involved in regulation or survival within the host. Although *Shigella* and *E. coli* are very closely related at the genetic level, there are significant differences beyond the presence of the virulence plasmid in *Shigella*. In addition to extra genes in the *Shigella* chromosome, there are genes present in the closely related *E. coli* that are missing from the chromosome of *Shigella*; genetic divergence accounts for differences in the virulence and pathogenicity of *Shigella* and *E. coli*.

Authors' note

Genes that encode virulence are highly regulated by temperature. At typical ambient temperatures (below 30°C, or 86°F) at which foods would be expected to be held, virulence gene expression is suppressed. Once the organism is ingested and exposed to body temperature (37°C, or 98.6°F), virulence genes are activated and the organism is now capable of causing severe illness.

CONCLUSIONS

Although foodborne infections due to *Shigella* spp. may not be as frequent as those caused by other foodborne pathogens, they have the potential for explosive spread due to the extremely low infectious dose that can cause overt clinical disease. In addition, cases of bacillary dysentery frequently require medical attention (even hospitalization), resulting in time lost from work, as the severity and duration of symptoms can be incapacitating. There is no effective vaccine against dysentery caused by *Shigella*. These features, coupled with the wide geographical distribution of the strains and the sensitivity of the human population to *Shigella* infection, make *Shigella* a formidable public health threat.

Summary

- The infective dose may be as low as 100 cells.
- Shigellosis is self-limiting.
- Humans are the natural reservoir of *Shigella*.
- *Shigella* is spread through the fecal-oral route; therefore, hand washing is one of the most effective control measures.
- Expression of virulence genes is regulated by temperature.
- *Shigella* shares many virulence genes with *E. coli*.
- Genes responsible for virulence are located on a virulence plasmid.

Suggested reading

Lampel, K. A., and A. T. Maurelli. 2007. *Shigella* species, p. 323–342. *In* M. P. Doyle and L. R. Beuchat (ed.), *Food Microbiology: Fundamentals and Frontiers*, 3rd ed. ASM Press, Washington, DC.

Niyogi, S. K. 2005. Shigellosis. *J. Clin. Microbiol.* **43:**133–143.

Parsot, C., and P. J. Sansonetti. 1996. Invasion and the pathogenesis of *Shigella* infections. *Curr. Top. Microbiol. Immunol.* **209:**25–42.

Questions for critical thought

1. How many *Shigella* species are considered pathogenic to humans? Is *Shigella* considered a zoonotic microbe?
2. Outbreaks of shigellosis are more commonly associated with day care and other institutional settings rather than linked to food; speculate as to why.
3. Shigellosis is distinguished from diseases caused by other foodborne pathogens by at least two characteristics. What are the two key characteristics?
4. Why is the use of antibiotics to treat shigellosis controversial?
5. Are there any long-term consequences to a *Shigella* infection?
6. What is meant by the term infectious dose? Compared to other foodborne pathogens, does *Shigella* have a low infectious dose?
7. How does temperature play a role in *Shigella* pathogenesis?
8. An outbreak of 52 shigellosis cases in Belgium was linked to one food handler. What steps did the public health department follow to determine a linkage of the outbreak with the food handler?
9. Once *Shigella* has invaded epithelial cells of the host, what unique function can the pathogen perform, and how does that influence pathogenesis?
10. What is the most likely route by which food becomes contaminated with *Shigella*?
11. The enterohemorrhagic organism *E. coli* O157:H7 also produces Shiga toxin and illness present as diarrhea, bloody diarrhea, and cramps. How would you determine whether a person was infected with *Shigella* and not *E. coli* O157:H7?
12. What characteristics of the pathogen with respect to reservoir and shedding pattern may exacerbate its spread and the subsequent outbreak of shigellosis?
13. Many cases of shigellosis are linked to consumption of fresh produce. What measures could be implemented to reduce the risk of shigellosis from consumption of fresh vegetables?
14. This is an extra-hard question. The Food and Drug Administration (FDA) approved the use of a *Listeria*-specific bacteriophage preparation on ready-to-eat meat and poultry products as a means to control the foodborne pathogen *Listeria monocytogenes*. Provide a brief narrative to address each point listed below with respect to the use of phages to control *Shigella* associated with a food of your choice. The points to be addressed are phage source, specificity of phages, phage titer, allergenicity, immunogenicity, method of treating food, determining efficacy of phage treatment, benefit compared to existing methods, and potential pitfalls associated with utilization of phages as a natural decontamination system.

16 *Staphylococcus aureus*

LEARNING OBJECTIVES

The information in this chapter will enable the student to:

- use basic biochemical characteristics to identify *Staphylococcus aureus*
- understand what conditions in foods favor *S. aureus* growth
- recognize, from symptoms and time of onset, a case of *S. aureus* food poisoning
- choose appropriate interventions (heat, preservatives, or formulation) to prevent the growth of *S. aureus*
- identify sources of *S. aureus*
- understand the role of staphylococcal enterotoxins in causing foodborne illness

Outbreak

My fear of flying is related to bad food, not hijackings, crashes, or boring in-flight movies. As a college student, I boarded a flight in Tokyo, Japan. It would stop in Anchorage, Alaska, for a fresh crew and (as it turned out, not so fresh) food. The first leg of the trip was uneventful. An hour out of Alaska, we ate peanuts and soda. Breakfast would break the boredom of the flight—in a way I could not have imagined. Breakfast smelled good. It was an omelet with ham and cheese, served nice and hot. But before the service cart reached the back of the plane, there were rumblings in the front. About 30 min after eating, a passenger rushed to the rear of the plane, ignored the attendant's admonition that the seat belt sign was on, and vomited on a blue-haired woman before finally reaching the toilet. Although the air was calm, most of the passengers started to turn green. The trickle of passengers trying to reach the toilet turned into a stream. The flow crested 2.5 h after breakfast. The first man to become ill was lucky. He got a toilet. There were only four bathrooms for the remaining 250 people with vomiting and diarrhea! A burly first-class passenger blocked the way to the front of the cabin. In first class, the passenger-to-toilet ratio was 20 to 2, and first-class food had not even made them sick! About 80% of the passengers had diarrhea, 70% had vomiting, and 85% had nausea. I will spare you the details of what happened when they ran out of "discomfort bags." Fortunately, the crew was served a "microbiologically insensitive" meal. Only one of them got sick. She will never pass up steak in favor of an omelet again.

After we landed, two men in lab coats told me that I would be exempt from the customs line if I would "do them a favor." My rectal swabs tested positive for *Staphylococcus aureus* toxin and the organism. They found staphylococcal toxin and

Outbreak continues on next page

doi:10.1128/9781555817206.ch16

the organism in the cheese omelets and the ham, too. Other investigators visited the kitchen. They discovered that the meal had been prepared 14 h before the flight and had been held at room temperature. It was then held at inadequate refrigeration during the flight and microwaved to heat just before serving. When the public health investigators went to shake hands with the head chef, they saw a pus-filled boil on his hand. They cultured it, and guess what they found—*S. aureus*.

An alternate food poisoning scenario: "flesh-eating" bacteria

Stories of methicillin-resistant *Staphylococcus aureus* are usually reported on the front page of tabloid newspapers. Although they are usually hospital-acquired infections, they can also be caused by food. A family bought shredded barbecued pork at a deli, heated it in their microwave, and ate it. Three to four hours later, three of them became ill with vomiting and nausea. They had eaten no other meal together. Their children who did not eat the pork did not become sick. Two of the adults were taken to the hospital, treated, and released. Different and sometimes multiple isolates of *S. aureus* were isolated from their stools, the pork, and nasal swabs from three food handlers at the deli. The strain related to the outbreak was methicillin-resistant *Staphylococcus aureus* and produced staphylococcal enterotoxin C.

CHARACTERISTICS OF THE ORGANISM

Historical Aspects and General Considerations

The link between staphylococci and foodborne illness was made in 1914, when researchers found that drinking contaminated milk caused vomiting and diarrhea. In 1930, G. M. Dack and coworkers voluntarily consumed fluids from cultures of "a yellow hemolytic *Staphylococcus*" isolated from contaminated cake. They became ill with "vomiting, abdominal cramps and diarrhea."

S. aureus is well characterized. It excretes a variety of compounds. Many of these, including the staphylococcal enterotoxins, are virulence factors. The staphylococcal toxins cause at least two human diseases, toxic shock syndrome (TSS) and staphylococcal food poisoning.

S. aureus is now recognized as the main agent of staphylococcal food poisoning, a common cause of gastroenteritis. Unlike many other forms of gastroenteritis, staphylococcal food poisoning usually is not caused by eating live bacteria. One gets it by eating staphylococcal toxins that have already been made in the contaminated food. This form of food poisoning is known as "intoxication" or "poisoning" because it does not require that the bacteria infect the victim. Indeed, outbreaks have been caused by foods in which the organism was killed but the toxin remained. Staphylococcal toxin is unique because it is not destroyed by heating, even by canning. This is illustrated by a large outbreak that occurred with canned mushrooms. *S. aureus* grew and produced toxin in the mushrooms before they were canned. The canned mushrooms were processed at 121°C for 2.4 min (see chapter 2) to produce a "commercially sterile" product. However, people who ate the mushrooms became ill with typical symptoms of staphylococcal food poisoning. Staphylococcal enterotoxin was isolated from the food.

Sources of Staphylococcal Food Contamination

People are the main reservoir of *S. aureus*. Humans are "natural" carriers and spread staphylococci to other people and to food. In humans, the nose interior is the main colonization site. *S. aureus* also occurs on the skin. *S. aureus* spreads by direct contact, by skin fragments, or through respiratory droplets when people cough or sneeze. Most staphylococcal food poisoning is traced to food contaminated by humans during preparation. In addition to contamination by food handlers, meat grinders, knives, storage utensils, cutting blocks, and saw blades may also introduce *S. aureus* into food. Conditions often associated with outbreaks of staphylococcal illness are inadequate refrigeration, preparing foods too far in advance, poor personal hygiene, inadequate cooking or heating of food, and prolonged use of warming plates when serving foods.

Animals are also *S. aureus* sources. For example, mastitis (infection of cow teats) is a serious problem for the dairy industry. It is often caused by *S. aureus*. Bovine mastitis is the single most costly agricultural disease in the United States. Mastitis is also a public health concern because the bacteria can contaminate milk and dairy products. This can be controlled through strict hygiene of automated milking machines, milk handlers, and facilities.

The source of staphylococcal food contamination for a specific outbreak is not always known. Regardless of its source, *S. aureus* is present in many foods (Table 16.1). The staphylococcus levels are usually low initially: <100 colony-forming units (CFU)/g. However, they can grow to high levels, $>10^6$ CFU/g, and cause staphylococcal food poisoning under favorable conditions.

Resistance to Adverse Environmental Conditions

S. aureus's unique resistance to inhibitors helps it grow under conditions in which other pathogens cannot. Thus, foods with high salt or low water activity are often implicated in outbreaks. If fact, under aerobic conditions, *S. aureus* can grow at water activities as low as 0.86, making it the most osmotolerant foodborne pathogen. Under anaerobic conditions, cells do not grow below a water activity of 0.90. Foods with low water activity include cured meats, puddings, and sauces.

Table 16.1 Prevalence of *S. aureus* in several common food products[a]

Product	No. of samples tested	% Positive for *S. aureus*	*S. aureus* content (CFU/g)[b]
Ground beef	74	57	≥100
	1,830	8	≥1,000
	1,090	9	>100
Big game	112	46	<10
Pork sausage	67	25	≥100
Ground turkey	50	6	≤10
	75	80	≤3.4
Salmon steaks	86	2	≤3.6
Oysters	59	10	≤3.6
Blue crab meat	896	52	≤3
Peeled shrimp	1,468	27	≤3
Lobster tail	1,315	24	≤3
Assorted cream pies	465	1	≤25
Tuna pot pies	1,290	2	≤10
Delicatessen salads	517	12	≤3

[a]Adapted from L. M. Jablonski and G. A. Bohach, p. 411–434, *in* M. P. Doyle, L. R. Beuchat, and T. J. Montville (ed.), *Food Microbiology: Fundamentals and Frontiers*, 2nd ed. (ASM Press, Washington, DC, 2001).

[b]Determined by either direct plate count or most-probable-number technique.

Staphylococci have an efficient osmoprotectant system for growth at low water activity. Several compounds accumulate in the cell or enhance its growth under osmotic stress. Glycine betaine is the most important osmoprotectant. To various degrees, other compounds, including L-proline, proline betaine, choline, and taurine, also act as compatible solutes. Proline and glycine betaine accumulate to very high levels in *S. aureus* in response to low water activity. This lowers the intracellular water activity to match the external water activity. However, the transport of these compounds requires energy, diverting it from other cellular purposes. Under extreme conditions of growth, the organism may not produce toxin.

Because there are so many human and animal reservoirs, controlling staphylococcal food poisoning is especially challenging. *S. aureus* persists in sites such as the mucosal surface because it can bind to and be internalized by many different cell types. Once bound, *S. aureus* triggers a series of host-specific changes that are similar to those induced by other intracellular pathogens. These include phagosome formation, protein tyrosine kinase activation, and changes in cell morphology, which are explained more fully for the case of *Listeria monocytogenes* in chapter 13.

FOODBORNE OUTBREAKS

Incidence of Staphylococcal Food Poisoning

There is little reason to report staphylococcal food poisoning, because people usually recover in 24 to 48 h and have no reason to go to the doctor. Although there is national surveillance for staphylococcal food poisoning, it is not an officially reportable disease. Only 1 to 5% of all staphylococcal food poisoning cases in the United States are reported. Most of these are highly publicized outbreaks. Sporadic cases in the home are usually unreported. Staphylococcal food poisoning accounts for about 14% of the total outbreaks of foodborne illness within the United States. There are about 25 major outbreaks of staphylococcal food poisoning annually in the United States. Occurrence of staphylococcal food poisoning is seasonal: most cases are in the late summer, when temperatures are warm and food is stored improperly. A second peak occurs in November and December, presumably due to the mishandling of holiday leftovers. Although international reporting is also poor, staphylococcal food poisoning is a leading cause of foodborne illness worldwide. In one study, 40% of foodborne gastroenteritis outbreaks in Hungary were due to staphylococcal food poisoning. The percentage is slightly lower in Japan, at 20 to 25%. In Great Britain, meat or poultry products caused 75% of the 359 staphylococcal food poisoning cases reported between 1969 and 1990.

A Typical Large Staphylococcal Food Poisoning Outbreak

The Food and Drug Administration (FDA) reported an outbreak having many typical elements of staphylococcal food poisoning. The type of food involved, means of contamination, inadequate food-handling measures, and symptoms all pointed to *S. aureus*. The outbreak was traced to one meal fed to 5,824 schoolchildren at 16 sites in Texas. A total of 1,364 children developed typical staphylococcal food poisoning. Investigations revealed that 95% of the ill children had eaten chicken salad which contained *S. aureus* at high levels.

The meal was prepared in a central kitchen the day before. Frozen chickens were boiled for 3 h. After being cooked, the chickens were deboned, cooled to room temperature, ground into small pieces, placed into 30.5-cm-deep pans, and stored overnight in a walk-in refrigerator at 5.5 to 7°C. The following morning, the other salad ingredients were added and the mixture was blended. The food was placed in containers and trucked to the schools between 9:30 a.m. and 10:30 a.m. It was kept at room temperature until served between 11:30 a.m. and noon.

The chicken was probably contaminated after cooking when it was deboned. The storage of the warm chicken in the deep pans prevented rapid cooling. It provided a good environment for staphylococcal growth and toxin production. Holding the food in warm classrooms gave an additional opportunity for growth. The screening of food handlers to identify *S. aureus* carriers, cooling the chicken more rapidly, and refrigerating the salad after preparation could have prevented the incident.

CHARACTERISTICS OF DISEASE

Staphylococcal food poisoning is a self-limiting illness causing emesis (vomiting) after an unusually short time of onset, as little as 30 min after eating (Box 16.1). However, vomiting is not the only symptom. Likewise, many patients with staphylococcal food poisoning do not vomit. Nausea, cramps, diarrhea, headaches, and/or prostration are other common symptoms. In a summary of clinical symptoms involving 2,992 patients diagnosed with staphylococcal food poisoning, 82% complained of vomiting, 74% felt nauseated, 68% had diarrhea, and 64% exhibited abdominal pain. In all cases of diarrhea, vomiting was always present. The lack of fever is consistent with the illness being caused by a toxin, not an infection.

Authors' note

The period between the consumption of tainted food and onset of illness can be an important clue in foodborne illness.

Symptoms usually develop within 6 h after eating. According to one report, 75% of the victims had symptoms of staphylococcal food poisoning within 6 to 10 h after eating. The average incubation period is 4.4 h, although illness can appear in ~30 min. Death due to staphylococcal food poisoning is rare. The fatality rate ranges from 0.03% for the general public to 4.4% for more susceptible populations, such as children and the elderly. Approximately 10% of patients with confirmed staphylococcal food poisoning seek medical help. Treatment is usually minimal, although fluids are given when diarrhea and vomiting are severe.

BOX 16.1

Things are not always what they seem

You would immediately suspect *S. aureus* intoxication if you were presented with a scenario in which more than 25 people became ill with vomiting and nausea within 30 min of eating at a buffet. Such was the case in an outbreak that originated in a restaurant in Ohio in 2000. However, toxin was not detected in the food or the victims' feces. Statistical analysis implicated the salad. Testing of the salad and a victim's vomitus revealed the presence of methomyl, an organophosphate choline esterase inhibitor. Methomyl was used as "fly bait" distributed around the restaurant by hand or by spoon. The spoon was missing. In a separate case of food-related poisoning, three people who ate roti (an Indian dish) died. The restaurant stored methomyl in an unlabeled tin can. These outbreaks illustrate two important lessons: food utensils should never be used to distribute chemicals, and chemicals and insecticides should be clearly labeled and stored in a locked cabinet.

TOXIC DOSE

Toxin Dose Required

The best data on minimum toxic dose are from analyses of food recovered from outbreaks. Staphylococcal toxins are quite potent, and 1 ng (10^{-9} g) of staphylococcal enterotoxin per g of food can cause illness. However, the staphylococcal enterotoxin level in outbreak foods is relatively large, ranging from 1 to 5 μg of ingested toxin. *S. aureus* levels of $>10^5$ cells/g of food may produce the level of enterotoxin necessary to cause illness. Other studies suggest that 10^5 to 10^8 cells/g is the typical range, although lower cell populations are sometimes implicated.

Many factors contribute to the severity and likelihood of getting staphylococcal food poisoning. These include individual susceptibility to the toxin and how much food was eaten. The toxin type may also be important. Although enterotoxin A causes more outbreaks, staphylococcal enterotoxin B produces more severe symptoms. Forty-six percent of individuals exposed to staphylococcal enterotoxin B had symptoms severe enough to be hospitalized. But only 5% of individuals exposed to enterotoxin A required hospitalization. This may reflect different levels of toxin expression, as enterotoxin B is generally produced at higher levels than enterotoxin A.

The staphylococcal enterotoxin acts at the viscera (gut). The emetic response results from a stimulation of its neural receptors. These transmit impulses through the nerves, ultimately stimulating the brain's vomiting center. Different levels of ingested enterotoxins are required to make human volunteers or monkeys vomit. Monkeys are generally less susceptible than humans. Human volunteers require 20 to 25 μg of staphylococcal enterotoxin B (0.4 μg/kg of body weight) to induce vomiting. In rhesus monkeys, the 50% emetic dose is about 1 μg/kg.

MICROBIOLOGY, TOXINS, AND PATHOGENICITY

Nomenclature, Characteristics, and Distribution of Enterotoxin-Producing Staphylococci

The term staphylococci informally describes a group of small spherical, gram-positive bacteria. Their cells have a diameter ranging from 0.5 to 1.5 μm (Fig. 16.1). They are catalase positive (i.e., they have enzymes that break down hydrogen peroxide). Staphylococci have typical cell walls containing peptidoglycan and teichoic acids. Some characteristics that differentiate *S. aureus* from other staphylococcal species are summarized in Table 16.2.

Bergey's Manual of Determinative Bacteriology (the "dictionary" for the classification of bacteria) puts staphylococci in the family *Micrococcaceae*. This family includes the genera *Micrococcus*, *Staphylococcus*, and *Planococcus*. The genus *Staphylococcus* is further subdivided into more than 23 species and subspecies. Many of these contaminate food. Staphylococci that form black colonies on Baird-Parker agar (Fig. 16.2) are suspected of being *S. aureus*.

Identification as *S. aureus* is usually confirmed by the coagulase test. (Coagulase is an enzyme highly associated with toxin production.) In a positive coagulase test, the culture filtrate clots the serum reagent. Since clotting can be subjective, only samples showing a +4 reaction (Fig. 16.3) are considered positive. However, several other species of *Staphylococcus*, including both coagulase-negative and coagulase-positive isolates, can

Authors' note

Having a medium named after oneself is a high honor. The photo in Fig. 16.2 is of a container of Baird-Parker agar autographed by Tony Baird-Parker.

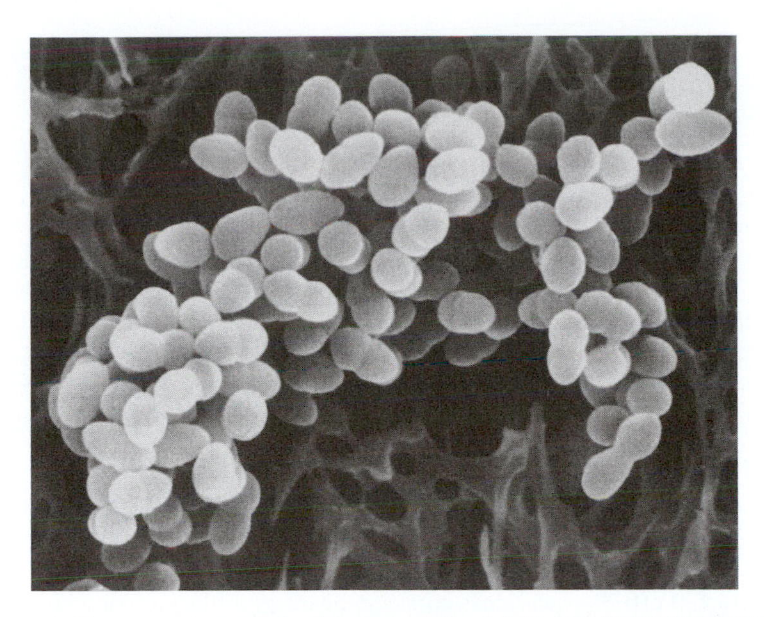

Figure 16.1 Electron micrograph of *S. aureus* cells. doi:10.1128/9781555817206.ch16.f16.01

produce staphylococcal enterotoxins. The FDA has noted that neither the coagulase test, the thermostable nuclease test, nor carbohydrate fermentation patterns are reliable tools for differentiating toxic from nontoxic strains. Nonetheless, nearly all staphylococcal food poisoning is attributed to *S. aureus*. Better methods have recently been developed to identify specific enterotoxin-producing *S. aureus* strains in outbreaks. Three genetic techniques—pulsed-field gel electrophoresis, randomly amplified polymorphic DNA, and polymerase chain reaction—have traced the pathogen to manually handled dairy products, nasal cavities, and manually handled vegetables. While the FDA and U.S. Department of Agriculture (USDA) have been slow to adopt these methods for the initial identification of *S. aureus*, they are widely used in the investigation of outbreaks.

Table 16.2 General characteristics of selected species of *Staphylococcus*[a]

	Presence of phenotype in species[b]					
Characteristic	***S. aureus***	***S. chromogenes***	***S. hyicus***	***S. intermedius***	***S. epidermidis***	***S. saprophyticus***
Coagulase	+	−	+	+	−	−
Thermostable nuclease	+	−	+	+	±	−
Clumping factor	+	−	−	+	−	−
Yellow pigment	+	+	−	−	−	±
Hemolytic activity	+	−	−	+	±	−
Phosphatase	+	+	+	+	±	−
Lysostaphin	Sensitive	Sensitive	Sensitive	Sensitive	Slightly sensitive	ND
Hyaluronidase	+	−	+	−	±	ND
Mannitol fermentation	+	±	−	±	−	±
Novobiocin resistance	−	−	−	−	−	+

[a]Reprinted from L. M. Jablonski and G. A. Bohach, p. 411–434, *in* M. P. Doyle, L. R. Beuchat, and T. J. Montville (ed.), *Food Microbiology: Fundamentals and Frontiers*, 2nd ed. (ASM Press, Washington, DC, 2001).

[b]ND, not determined; +, present; −, absent; ±, variable.

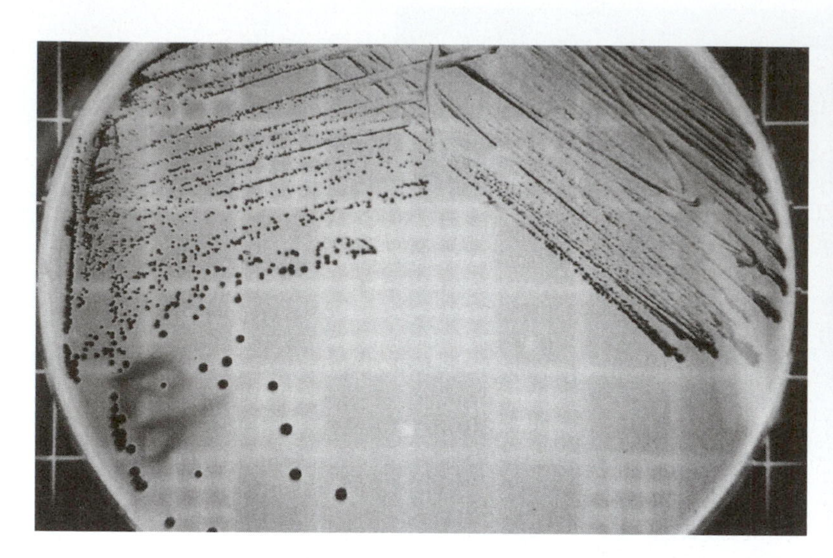

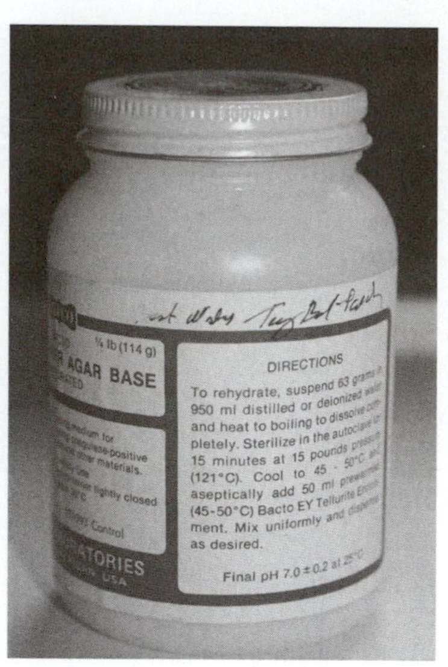

Figure 16.2 **(Left)** *Staphylococcus aureus* appears as a black colony surrounded by an opaque halo when plated on Baird-Parker agar; **(right)** a bottle of Baird-Parker agar autographed by Tony Baird-Parker (photo by Ilene Levine). doi:10.1128/9781555817206.ch16.f16.02

Introduction to and Nomenclature of the Staphylococcal Enterotoxins

Current Classification Scheme Based on Antigenicity

S. aureus makes multiple toxins that can be differentiated by immunological tests. Staphylococcal enterotoxins are named by letter in the order of their discovery. Staphylococcal enterotoxin A, enterotoxin B, enterotoxin C, enterotoxin D, and enterotoxin E are the major types. Protein sequencing and recombinant DNA methods have yielded the primary sequences of the enterotoxins (Fig. 16.4). They have a high degree of homology (similarity of amino acid sequence). Staphylococcal enterotoxins G, H, J, and I are more recent discoveries. An exotoxin produced by the *S. aureus* associated with TSS was initially called staphylococcal enterotoxin F. When it was discovered that this enterotoxin did not cause vomiting in animals (as the other staphylococcal enterotoxins do), its name was changed to toxic shock syndrome toxin 1 (Box 16.2).

Figure 16.3 Coagulase reaction for *S. aureus*. The light color represents liquid, and the dark color represents clotted matter. doi:10.1128/9781555817206.ch16.f16.03

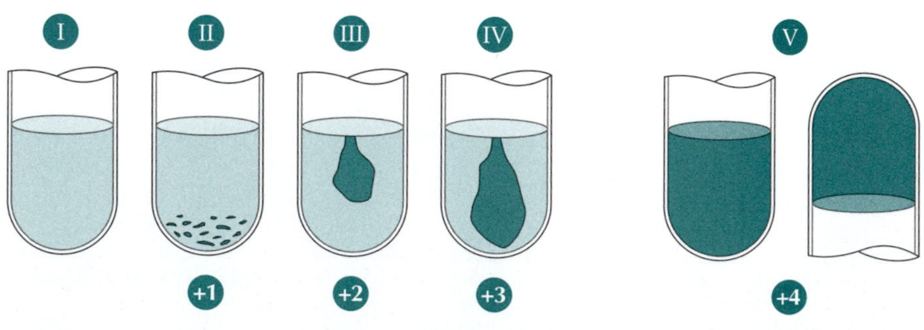

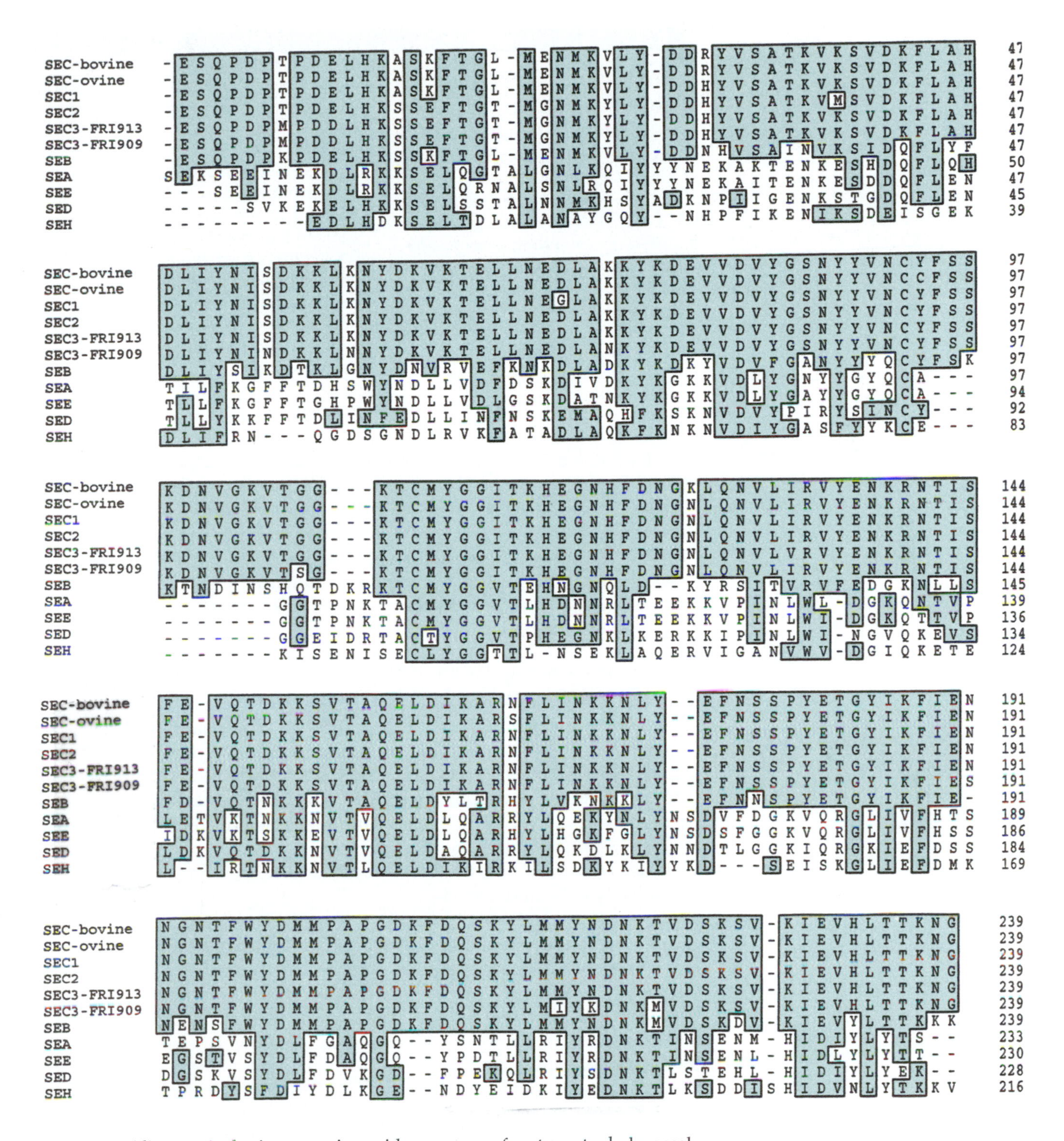

Figure 16.4 Alignment of primary amino acid sequences of mature staphylococcal enterotoxin (SE) proteins. Also shown are residue sequence numbers (on the right) and dashes to indicate gaps in the sequences made by alignment. Sequence alignment and output were done with the PileUp and PrettyPlot programs, respectively. Reprinted from L. M. Jablonski and G. A. Bohach, p. 353–375, *in* M. P. Doyle, L. R. Beuchat, and T. J. Montville (ed.), *Food Microbiology: Fundamentals and Frontiers* (ASM Press, Washington, DC, 1997). doi:10.1128/9781555817206.ch16.f16.04

BOX 16.2

Toxic shock syndrome: changes in microbial ecology give rise to a new threat from an old pathogen

TSS is not caused by food but is caused by *S. aureus*. In women, TSS causes vomiting, diarrhea, renal problems, central nervous system symptoms, and peeling of skin from the hands and feet. It can be fatal. TSS emerged in the mid-1970s when superabsorbent tampons were introduced into the highly competitive feminine-hygiene market. The superabsorbency led to extended periods of usage. This led to extended time for growth of and toxin production by *S. aureus* in a moist, warm, nutrient-rich environment. Toxic shock syndrome toxin 1 causes 75% of the cases, but staphylococcal enterotoxin B (the toxin that causes food poisoning) is responsible for 25% of the cases. The federal government now regulates tampon absorbency and advises menstruating women to change their tampons frequently.

Each enterotoxin type has enough antigenic distinctness to be differentiated from other toxins using antibodies. However, some cross-reactivity can occur. The level of cross-reactivity generally correlates with similar amino acid sequences. The type C enterotoxin subtypes and their molecular variants are cross-reactive, as are enterotoxins A and E, the two major serological types with the greatest sequence similarity. The two most distantly related enterotoxins recognized by a common antibody are staphylococcal enterotoxins A and D.

Staphylococcal Regulation of Staphylococcal Enterotoxin Expression

General Considerations

Staphylococcal enterotoxins are produced in low quantities during the exponential growth phase. The amount made is strain dependent. Staphylococcal enterotoxins B and C are produced in the highest quantities, up to 350 μg/ml. Staphylococcal enterotoxins A, D, and E are easily detectable by gel diffusion assays, which detect as little as 100 ng of enterotoxin per ml of culture. Strains that produce very low levels of toxins require more sensitive detection methods. Of these, staphylococcal enterotoxins D and J are most likely to be undetected.

Molecular Regulation of Staphylococcal Enterotoxin Production

Three genes that regulate expression of *S. aureus* virulence factors are *agr* (accessory gene regulator), *sar* (staphylococcal accessory regulator), and *sae* (*S. aureus* exoprotein expression). The best characterized is *agr*. Mutations in *agr* decrease toxin expression. Gene regulation by *agr* can be transcriptional or translational. Not all staphylococcal toxins are regulated by *agr*. Staphylococcal enterotoxin A expression is not affected by *agr* mutations. Since at least 15 genes are under *agr* control, it is considered a "global regulator."

Regulation of Staphylococcal Enterotoxin Gene Expression

The properties of *agr* explain several aspects of enterotoxin production and, in many cases, explain what were once known as "glucose effects." For example, *agr* expression coincides with expression of enterotoxins B and C during growth. All have their highest expression during late exponential and postexponential growth. Staphylococcal enterotoxin A, which is

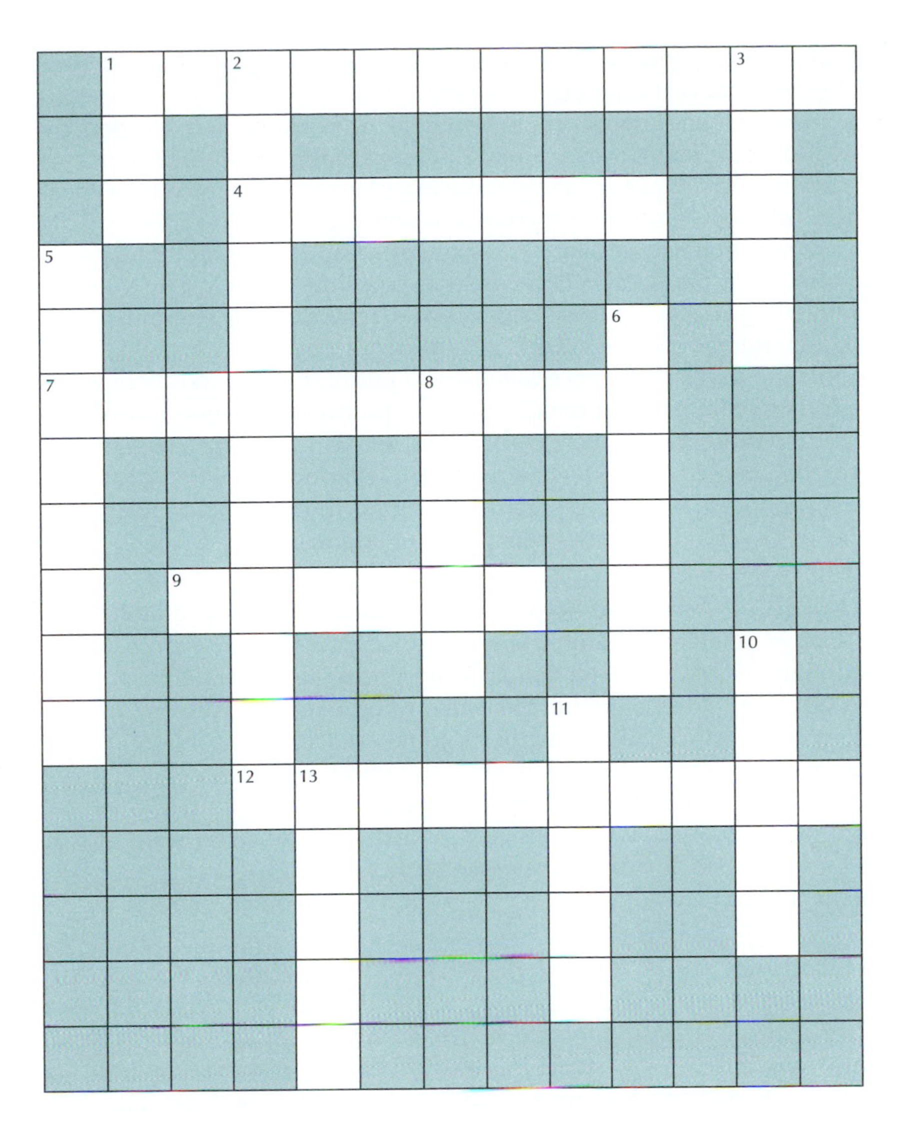

Across

1. Basis of classification for staphylococcal enterotoxins
4. Usual type of treatment for staphylococcal food poisoning
7. Type of diffusion assay used to identify staphylococcal enterotoxin
9. Cases of staphylococcal food poisoning peak in this season
12. Most cases of staphylococcal food poisoning can be attributed to poor ________

Down

1. A global regulator that controls toxin production
2. Staphylococcal enterotoxin is stable to elevated ________ that kill the bacterium
3. *Staphylococcus aureus* produces a heat-stable ________
5. One symptom of staphylococcal food poisoning
6. Main environmental reservoir of *Staphylococcus aureus*
8. *Staphylococcus aureus* can grow at ________ water activities than other foodborne pathogens
10. Toxic syndrome that is not caused by foodborne *Staphylococcus aureus*
11. Symptoms of staphylococcal food poisoning have a ______ onset
13. The most common contributing factor associated with staphylococcal food poisoning is temperature ______

not regulated by *agr*, is produced earlier. Furthermore, the production of several staphylococcal enterotoxins is inhibited by growth in media containing glucose; this affects *agr*. Enterotoxin C expression is affected by glucose in two different ways. First, glucose metabolism indirectly influences enterotoxin C production by reducing pH: *agr* is expressed the most at neutral pH, but growth in media containing glucose lowers pH. This directly reduces *agr* expression. Expression of *sec* and other *agr* target genes is also affected. Glucose also reduces *sec* expression in strains lacking *agr*. This suggests a second glucose-dependent mechanism for stopping toxin production, independent of *agr* and not influenced by pH.

In many organisms, a process known as autoinduction coordinates the expression of proteins. In autoinduction, the production of one compound induces the production of others in the same organism. The *agr* locus is at the center of the *S. aureus* autoinduction response. Components of the *agr* operon comprise a quorum sensing apparatus analogous to the signal transduction pathway of bacterial two-component (receiver and response) systems.

Authors' note

Quorum sensing is the process by which bacteria determine the size of their population and whether it is big enough to act in a certain fashion. This is analogous to legislative bodies that require a quorum (certain number of people) to do business. Signal transduction is the mechanism by which bacteria "sense" some external factor and transduce it to an internal response.

Other Molecular Aspects of Staphylococcal Enterotoxin Expression

Many factors selectively inhibit enterotoxin expression. Their effects on regulation and signal transduction are only beginning to be defined. *agr* is not the only signal transduction mechanism for *S. aureus*. Glucose's negative effect is not entirely due to higher acid levels. Cultures containing glucose produce less enterotoxin even when the acid is neutralized.

S. aureus is osmotolerant. When it grows in foods with low water activity, less enterotoxin is produced. In experiments with enterotoxin C-producing strains, levels of *sec* messenger RNA (mRNA) and enterotoxin C protein are both reduced in response to high NaCl concentrations.

Toxin Structures

Staphylococcal enterotoxins are single polypeptides of 25 to 28 kilodaltons. Most are neutral or basic proteins that have no net charge at pH values ranging from 7 to 8.6 (i.e., their isoelectric points fall in this range). All staphylococcal enterotoxins are monomeric (single-unit) proteins. They are made as larger precursors with a signal peptide that is cut off as they are excreted from the cell.

The three-dimensional shapes of all staphylococcal enterotoxins are similar. Staphylococcal enterotoxins A, B, C, and E have a lower α-helix content (<10%) than β-pleated sheet/β-turn structure content (approximately 60 to 85%). It is the three-dimensional structure that gives staphylococcal enterotoxins their remarkable heat stability.

Staphylococcal enterotoxins are very stable. Toxicity and antigenicity are not completely destroyed by boiling or even canning. The temperatures needed to inactivate staphylococcal toxins are much higher than those needed to kill *S. aureus* cells. In many cases of staphylococcal food poisoning, no live bacteria are found in the food.

Staphylococcal Enterotoxin Antigenic Properties

Each enterotoxin type has enough antigenic distinctness to be differentiated from other toxins using antibodies. However, some cross-reactivity can occur. The level of cross-reactivity generally correlates with similar amino acid sequences. The type C enterotoxin subtypes and their molecular variants are cross-reactive, as are enterotoxins A and E, the two major

BOX 16.3

S. aureus enterotoxin

The characterization of a staphylococcal isolate as an enterotoxin producer is an important task. There are many *Staphylococcus* species in the environment that pose no threat to the safety of foods. These need to be differentiated from the ones that pose a real hazard. The first step in the isolation of *S. aureus* is plating on Baird-Parker medium. Presumptive *S. aureus* colonies are very distinctive. They are black, with a halo of precipitation against a clearing zone. Direct tests for enterotoxin are difficult. So suspect colonies are subcultured in broth, and the supernatants are tested for coagulase and/or thermostable nuclease activity. The coagulase test determines the culture's ability to coagulate a special type of rabbit serum. In a strong positive reaction, the tube can be inverted without the clot falling out. Weaker reactions need to be confirmed by another method, usually a thermostable nuclease. This test uses a color change to show the ability to break down DNA. 3M Petrifilm *S. aureus* Count Plates cleverly combine the reactions of Baird-Parker medium with the thermostable nuclease reaction on a single piece of Petrifilm. However, both coagulase and thermostable nuclease activities correlate with, but do not prove, enterotoxin production. The official method for detecting enterotoxin is the microslide diffusion assay. This method is tedious and complex. The microslide apparatus is constructed by placing a drilled Plexiglas template on a slab of diffusion agar supported by a microscope slide. Antitoxin(s) (usually A and B [AntiA and AntiB in the figure] since these are the most common) are placed in the center well. Positive controls of purified enterotoxin are placed in wells 3 (EntA) and 5 (EntB). Test samples are placed in wells 2 and 4. Thus, the reaction of example 1 has immunoprecipitation lines between the antitoxin and the control enterotoxins, but not the sample. Toxin is absent. When the sample contains toxin, its immunodiffusion line crosses the immunodiffusion line caused by the toxin standard. Thus, in example 2, enterotoxin B (but not A) is present in test sample 4. There are a variety of other toxin tests which come in kit form and are based on enzyme-linked immunosorbent assays. They are simpler to run and easier to interpret. Hopefully, the acceptance of enzyme-linked immunoassays for staphylococcal enterotoxins will make detection easier.

The figure and explanations for the immunodiffusion assay are derived from the FDA's Bacteriological Analytical Manual.

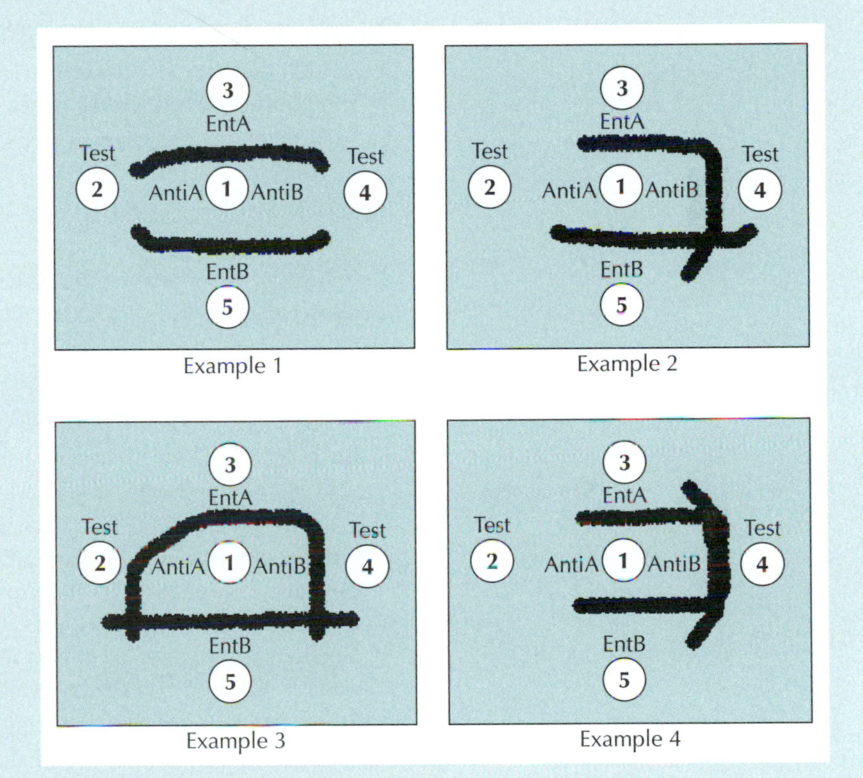

doi:10.1128/9781555817206.ch16.fBox16.03

serological types with the greatest sequence similarity. The two most distantly related enterotoxins recognized by a common antibody are staphylococcal enterotoxins A and D. Box 16.3 illustrates how antigenicity can be used as a toxin detection system.

Summary

- *S. aureus* is commonly associated with humans.
- Poor sanitation causes many outbreaks.

- Symptoms of vomiting and diarrhea develop in 0.5 to 6 h after eating.
- *S. aureus* makes several antigenically distinct enterotoxins.
- The cells are heat sensitive, but the toxins are heat resistant.
- *S. aureus* is the most osmotolerant foodborne pathogen.
- There is considerable knowledge of enterotoxin genetics and mechanism of action.

Suggested reading

Dinges, M. M., P. M. Orwin, and P. M. Schievery. 2000. Exotoxins of *Staphylococcus aureus*. *Clin. Microbiol. Rev.* **13:**16–34.

Seok, K. S., and G. A. Bohach. 2007. *Staphylococcus aureus*, p. 493–518. *In* M. P. Doyle and L. R. Beuchat (ed.), *Food Microbiology: Fundamentals and Frontiers*, 3rd ed. ASM Press, Washington, DC.

Questions for critical thought

1. Which factors contributed to the outbreak described at the beginning of the chapter? How could it have been prevented?
2. Why did the passengers get sick even though the food was properly heated?
3. What would *you* do if you were served undercooked chicken on a transatlantic flight?
4. What is the difference between enterotoxins and exotoxins? Name an exotoxin (from another chapter) that is not an enterotoxin.
5. Why do food microbiologists sometimes refer to staphylococcal food poisoning as a "double bucket" syndrome?
6. What is signal transduction? Why do bacteria need it?
7. Both botulinal toxin and staphylococcal toxin are proteins, but the former is very heat sensitive and the latter very heat resistant. Why?
8. You work for Chubby Chuck's Chickenry (CCC), which produces (among other chicken specialties) about 1,000 pounds of chicken salad per day. It is packed in 10-pound tubs and distributed under refrigeration to restaurants and delis. Chubby Chuck heard about the terrible Texas outbreak and tells you it is your job to make sure this never happens at CCC. What do you tell Chuck?
9. Being the diligent plant manager of CCC, you test products for staphylococcal enterotoxins. On Friday, you get the results of Thursday's testing. Two batches of chicken salad have tested positive for *Staphylococcus* enterotoxin A. Batch 200498-6 was packaged and shipped out. Batch 200491-6 was batched into 10-gallon tubs which are still in the cold storage room of the plant. What do you do? What (and when) do you tell Chubby Chuck?
10. Interpret the results of the microslide immunodiffusion assays shown in Box 16.3, examples 3 and 4. (Remember that the test preparations in wells 2 and 4 can contain EntA, EntB, neither, or both.)
11. Why can *S. aureus* grow to a lower water activity aerobically than anaerobically? (Hint: consult chapter 2.)

17 *Vibrio* Species

LEARNING OBJECTIVES

The information in this chapter will enable the student to:

- use basic biochemical characteristics to identify *Vibrio*
- understand what conditions in foods favor *Vibrio* growth
- recognize, from symptoms and time of onset, a case of foodborne illness caused by *Vibrio*
- choose appropriate interventions (heat, preservatives, and formulation) to prevent the growth of *Vibrio*
- identify environmental sources of the organism
- understand the role of *Vibrio* toxins and virulence factors in causing foodborne illness

Outbreak

According to some seafood aficionados, the only way to consume certain types of seafood is raw. That can lead to unintended outcomes: diarrhea, stomachaches, and vomiting. In 2006, 177 individuals in three states became sick with *Vibrio parahaemolyticus* following consumption of contaminated shellfish, most likely clams and oysters. The illness is self-limiting, with diarrhea, vomiting, and chills lasting 1 to 3 days. Since vibrios multiply rapidly, even low levels of *V. parahaemolyticus* in harvested products can rapidly increase to infectious levels if the products are not rapidly refrigerated after harvest and maintained at proper temperatures during transport, processing, and storage (i.e., <50°F).

Raw oysters are consumed for their taste and because of the old wives' tale that they act as an aphrodisiac. People rarely consider the hazards associated with the consumption of raw shellfish. During a 2-month period (July and August) in 1997, one of the largest outbreaks of culture-confirmed *Vibrio parahaemolyticus* infection occurred in North America. At least 209 persons became ill with gastroenteritis from eating raw oysters. Symptoms included diarrhea, abdominal cramps, nausea, and vomiting. Although symptoms of gastroenteritis may attract the attention of others, they would certainly not be considered attractive. Oysters sampled from the beds implicated in the outbreak yielded <200 *V. parahaemolyticus* colony-forming units (CFU)/g of oyster meat, suggesting that human illness may occur at levels of bacteria lower than the current action level. The United States and Canada permit the sale of oysters if there are $<10^4$ CFU of *V. parahaemolyticus* per g of oyster meat.

doi:10.1128/9781555817206.ch17

INTRODUCTION

Over 20 *Vibrio* species have now been described. At least 12 are capable of causing infection in humans, although with the exception of *Vibrio cholerae* and *V. parahaemolyticus*, little is known about the virulence mechanisms they employ. Of the 12 pathogens, 8 are directly food associated, and these are the subject of this chapter. (*Vibrio carchariae* and *Vibrio damsela* infections appear to result solely from wound infections, and the routes of infection for *Vibrio metschnikovii* and *Vibrio cincinnatiensis* are unclear.)

One of the most consistent aspects of vibrio infections is a recent history of seafood consumption. Vibrios, the predominant bacterial genus in estuarine waters, are associated with a great variety of seafoods. Approximately 40 to 60% of finfish and shellfish at supermarkets may contain *Vibrio* spp., with *V. parahaemolyticus* and *Vibrio alginolyticus* being the most commonly isolated. Vibrios are most frequently isolated from molluscan shellfish during the summer months. The estimated annual incidence of *Vibrio* infections increased 78% from its baseline in 1996 to 1998 to 2006. Aside from additional measures to reduce contamination of seafood, consumers should be informed that they are at risk for *Vibrio* infections when they consume raw seafood.

CHARACTERISTICS OF THE ORGANISM

Many different enrichment broths for the isolation of vibrios have been described, although alkaline peptone water remains the most commonly used. These are frequently coupled with thiosulfate-citrate-bile salts-sucrose (TCBS) agar or other plating media. In order to distinguish vibrios from the *Enterobacteriaceae*, sucrose-positive colonies are subjected to the oxidase test (used to detect the presence of cytochrome *c*); however, erroneous results may arise when colonies are obtained directly from TCBS agar. Therefore, sucrose-positive colonies on TCBS should be subcultured by heavy inoculation onto a nonselective medium, such as blood agar, and allowed to grow for 5 to 8 h before being tested for oxidase activity.

Epidemiology

The numbers of vibrios in both surface waters and shellfish correlate with seasonality, generally being greater during the warm-weather months between April and October (Fig. 17.1). Similarly, vibrios are more commonly isolated from the warmer waters of the Gulf and East coasts than from those of the West and Pacific Northwest. Seasonality is most notable for *Vibrio vulnificus* and *V. parahaemolyticus* infections, whereas infections with some vibrios, such as *Vibrio fluvialis*, occur throughout the year.

CHARACTERISTICS OF DISEASE

There is considerable variation in the severities of the various *Vibrio*-associated diseases. The outcomes of infections depend on an individual's underlying health (e.g., patients with chronic liver disease are more likely to succumb to severe illness). An exception to this generalization is *V. cholerae* O1/O139, which can readily make noncompromised individuals sick. Most cases are associated with a recent history of seafood consumption. Because vibrios are part of the normal estuarine microflora caused by fecal contamination, *Vibrio* infections cannot be controlled through shellfish sanitation programs. It is therefore essential that raw seafood be kept cold to prevent significant bacterial growth and that it be properly cooked before consumption.

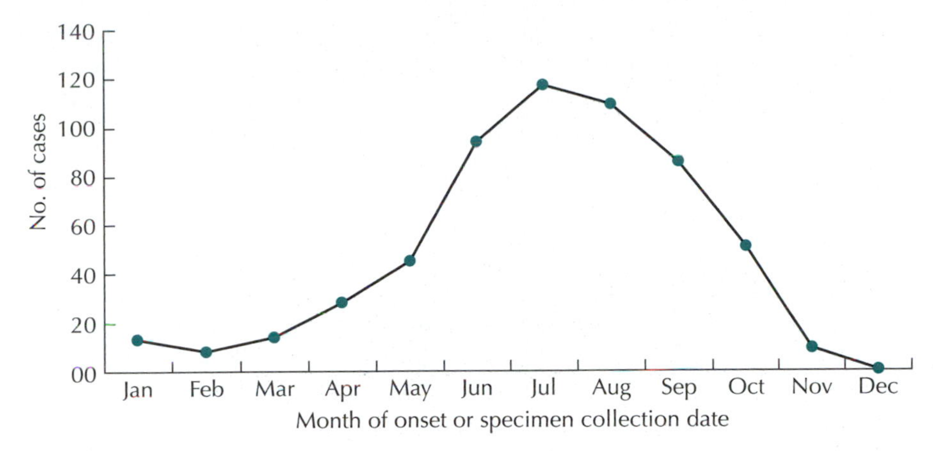

Figure 17.1 The number of cases of *Vibrio* infection typically peaks during the summer, with the highest frequency occurring in July. Redrawn from http://www.cdc.gov/nationalsurveillance/PDFs/Jackson_Vibrio_CSTE2008_FINAL.pdf. doi:10.1128/9781555817206.ch17.f17.01

Symptoms of foodborne illness for the two most common *Vibrio* species differ and are worth noting. Symptoms of *V. vulnificus* foodborne illness generally include fever, chills, and nausea. Diarrhea is generally not experienced. Mortality occurs in 40 to 60% of cases. The most common symptoms of *V. parahaemolyticus* infection are diarrhea, abdominal cramps, nausea, and vomiting. The mortality rate is extremely low.

SUSCEPTIBILITY TO PHYSICAL AND CHEMICAL TREATMENTS

With the exception of *V. cholerae* and *V. parahaemolyticus*, relatively little is known about the effects of preservation methods on inactivation of vibrios. Generally, vibrios are sensitive to cold, and seafoods can be protective for vibrios at refrigeration temperatures. Vibrios have survived for extended periods in products stored at refrigeration temperatures and after prolonged frozen storage. Thermal processing is a very effective means of reducing populations of *Vibrio* in foods. The U.S. Food and Drug Administration recommends steaming shellstock oysters, clams, and mussels for 4 to 9 min; frying shucked oysters for 10 min at 375°F; or baking oysters for 10 min at 450°F. Thorough heating of shellfish to an internal temperature of at least 60°C for several minutes should kill pathogenic vibrios. Irradiation and high hydrostatic pressure also reduce the numbers of *Vibrio* organisms. High hydrostatic pressure has an additional advantage in that the process perfectly shucks the oyster. A large variety of dried spices, the oils of several herbs, tomato sauce, and several organic acids exhibit bactericidal activities. The process of depuration, in which filter-feeding bivalves are purified of certain bacteria by pumping bacterium-free water through the bivalve tissues, removes *Salmonella* and *Escherichia coli* but not *Vibrio*.

V. cholerae

V. cholerae O1 causes cholera, one of the few foodborne illnesses with *epidemic* (many people in one area become sick at the same time) and *pandemic* (similar to an epidemic but over a larger area) potential. Important distinctions within the species are made on the basis of production of cholera enterotoxin (cholera toxin [CT]), serogroup, and potential for epidemic spread. Not all *V. cholerae* strains of the O1 serogroup produce CT; however, *V. cholerae* O1 CT-producing strains have long been associated with epidemic and pandemic cholera. Nearly 200 serogroups of *V. cholerae* have been described. Serotyping of *V. cholerae* strains is based on the lipopolysaccharide (LPS), also known as the somatic (O) antigen; H antigens are not useful in serotyping.

Isolation and Identification

Suspected *V. cholerae* isolates can be cultured on a standard series of biochemical media used for identification of *Enterobacteriaceae* and *Vibrionaceae*. Both conventional tube tests and commercially available enteric identification systems work well.

Nucleic acid probes are not routinely used for the identification of *V. cholerae* due to the ease of identifying this species by conventional methods. DNA probes are useful in distinguishing CT-producing strains of *V. cholerae* through the detection of CT (*ctx*) genes. Restriction fragment length polymorphism analysis has been useful in epidemiological studies of *V. cholerae*.

Reservoirs

V. cholerae is part of the normal, free-living bacterial flora in estuarine areas. Non-O1/non-O139 strains are much more commonly isolated from the environment than are O1 strains, even in epidemic settings in which fecal contamination of the environment might be expected. Periodic introduction of such environmental isolates into the human population through the ingestion of uncooked or undercooked shellfish appears to be responsible for localized outbreaks along the U.S. Gulf Coast and in Australia. The persistence of *V. cholerae* within the environment may be facilitated by its ability to assume survival forms, including a viable but nonculturable (VBNC) state. In this dormant state, the cells are reduced in size and become ovoid. Although VBNC vibrios are not culturable with nonselective enrichment broth or plates, studies have demonstrated that VBNC *V. cholerae* O1 organisms injected into ligated rabbit ileal loops or ingested by volunteers have yielded culturable *V. cholerae* O1 in intestinal contents or stool specimens, respectively.

Long-term carriage of *V. cholerae* in humans is extremely rare and is not important in the transmission of disease. However, even after the cessation of symptoms, patients who have not been treated with antibiotics may continue to excrete vibrios for 1 to 2 weeks. Asymptomatic carriers are most commonly identified among household members of persons with acute illness: in various studies, the rate of asymptomatic carriage in this group has ranged from 4 to almost 22%.

Foodborne Outbreaks

The critical role of water in the transmission of cholera has been recognized for more than a century. In 1854, the London physician and epidemiologist John Snow determined that illness was associated with consumption of water derived from the Thames River at a point below major sewage inflows. The range of food items implicated in the transmission of cholera includes crabs, shrimp, raw fish, mussels, cockles, squid, oysters, clams, rice, raw pork, millet gruel, cooked rice, food bought from street vendors, frozen coconut milk, and raw vegetables and fruit. One shared characteristic of the foods is their neutral or nearly neutral pH. When foodborne outbreaks occur and a number of different foods are suspected, foods with an acid pH can be eliminated. Food acts to buffer *V. cholerae* O1 against killing by gastric acid. In the United States, crabs, shrimp, and oysters have been the most frequently implicated vehicles of *V. cholerae* illness.

Characteristics of Disease

The explosive, dehydrating diarrhea characteristic of cholera actually occurs in only a minority of persons infected with CT-producing *V. cholerae* O1/O139. Most infections with *V. cholerae* O1 are mild or even asymptomatic. The incubation period of cholera can range from several hours to 5 days and is dependent in part on the *inoculum* size (the number of cells ingested). The onset of illness may be sudden, with watery diarrhea, or there can be loss of appetite, abdominal pain, and simple diarrhea. Initially, the stool is brown with fecal matter, but once diarrhea starts, it becomes a pale gray color with a slightly fishy odor. Mucus in the stool gives the characteristic "rice water" appearance. Vomiting can occur a few hours after the onset of diarrhea. In healthy North American volunteers, doses of 10^{11} CFU of *V. cholerae* were required to cause diarrhea when the inoculum was given in buffered saline (pH 7.2). However, a lower number of bacteria caused illness in volunteers when 10^6 vibrios were given, with food (fish and rice) acting as a buffer.

Virulence Mechanisms

Infection due to *V. cholerae* O1/O139 begins with the ingestion of food or water contaminated with the pathogen. After passage through the acid barrier of the stomach, vibrios colonize the small intestine using one or more adherence factors. Production of CT (and possibly other toxins) disrupts ion transport by intestinal epithelial cells. This leads to the severe diarrhea characteristic of cholera with excessive loss of water and electrolytes. Cholera halotoxin is composed of five identical B subunits and a single A subunit; the individual subunits are not sufficient to cause secretogenic activity (an increase in secretion of cellular components) in animals or intact cell culture systems.

V. cholerae produces a variety of extracellular products that are harmful to eukaryotic cells. In addition to the well-characterized toxins (soluble hemagglutinin and hemolysins), *V. cholerae* can produce a number of other toxic factors, but the responsible proteins and genes have not yet been purified or cloned. A number of other virulence factors are involved in illness, including colonization factors, flagella, LPS, and polysaccharide capsule.

V. mimicus

Before 1981, *Vibrio mimicus* was known as sucrose-negative *V. cholerae* non-O1. This species was determined to be a distinctly different species on the basis of biochemical reactions and DNA hybridization studies, and the name *mimicus* was given because of its similarity to *V. cholerae*. This organism is isolated mainly from cases of gastroenteritis but can also cause ear infections. The reservoir of *V. mimicus* is the aquatic environment. Besides being present free in the water, *V. mimicus* was also isolated from the roots of aquatic plants, from sediments, and from plankton at levels of up to 6×10^4 CFU per 100 g of plankton.

Foodborne Outbreaks

Gastroenteritis due to *V. mimicus* has been associated only with consumption of seafood. In the United States, consumption of raw oysters is the main cause of illness due to this species. In Japan, consumption of raw fish has resulted in at least two outbreaks involving *V. mimicus* of serogroup O41.

Characteristics of Disease

Disease due to *V. mimicus* is characterized by diarrhea, nausea, vomiting, and abdominal cramps in most patients. In some individuals, fever, headache, and bloody diarrhea also occur. There are no volunteer or epidemiological data to enable estimation of an infectious dose for *V. mimicus*. There is no particularly susceptible population, other than people who eat raw oysters.

Virulence Factors

V. mimicus does not produce unique enterotoxins, but many strains produce toxins that were first described for other *Vibrio* species, including CT, thermostable direct hemolysin (TDH), Zot, and a heat-stable enterotoxin apparently identical to the NAG-ST produced by *V. cholerae* non-O1/non-O139 strains. There is little information about potential intestinal colonization factors of *V. mimicus*.

V. parahaemolyticus

Along with *V. cholerae*, *V. parahaemolyticus* is the best described of the pathogenic vibrios, with numerous studies reported since the first description of its involvement in a major outbreak of food poisoning in 1950. Between 1973 and 1998, a total of 40 outbreaks of *V. parahaemolyticus* infection in the United States were reported to the Centers for Disease Control and Prevention, with >1,000 persons involved. In the United States the percentage of *Vibrio* infections attributed to *V. parahaemolyticus* continues to increase, from 54% in 2005 to 58% in 2009.

Classification

V. parahaemolyticus is serotyped according to both its somatic (O) and capsular polysaccharide (K) antigens. Currently, 12 O (LPS) antigens and 59 K antigens are recognized. Although many environmental and some clinical isolates are not typeable by the K antigen, most clinical strains can be classified based on their O types. A special consideration in the taxonomy of *V. parahaemolyticus* is the ability of certain strains to produce a hemolysin, called TDH or the Kanagawa hemolysin, which is linked to virulence in the species.

Reservoirs

V. parahaemolyticus occurs naturally in estuarine waters throughout the world and is easily isolated from coastal waters of the United States, as well as from sediment, suspended particles, plankton, and a variety of fish and shellfish. The last source includes at least 30 different species, among them clams, oysters, lobsters, scallops, shrimp, and crabs. A high percentage of seafood samples tested positive for the species. *V. parahaemolyticus* counts are season dependent; samples analyzed in January and February are often free of *V. parahaemolyticus*.

Foodborne Outbreaks

Gastroenteritis from *V. parahaemolyticus* is almost exclusively associated with seafood that is consumed raw, inadequately cooked, or cooked but recontaminated. In Japan, *V. parahaemolyticus* is a major cause of foodborne illness. Approximately 70% of all bacterial foodborne illnesses in the 1960s were the result of this pathogen. Seafood, including fish, crabs, shrimps, lobsters, and oysters, is primarily associated with U.S. outbreaks. The first major outbreak (with 320 persons ill) in the United States occurred in Maryland in 1971, a result of eating improperly steamed crabs. Subsequent outbreaks occurred throughout U.S. coastal regions and Hawaii. The largest U.S. outbreak of *V. parahaemolyticus* infection occurred during the summer of 1978 and affected 1,133 of 1,700 persons attending a dinner in Port Allen, Louisiana.

Characteristics of Disease

V. parahaemolyticus has generation times of 8 to 9 min at 37°C and 12 to 18 min in seafood. Hence, *V. parahaemolyticus* has the ability to grow rapidly, both in vitro and in vivo, contributing to the infectious dose required for illness. Symptoms may begin 4 to >30 h after the ingestion of contaminated food, with a mean onset time of 23.6 h. Primary symptoms include diarrhea and abdominal cramps, along with nausea, vomiting, and fever. The symptoms subside in 3 to 5 days in most individuals. Approximately 10^5 to 10^7 CFU is required for illness. The numbers of *V. parahaemolyticus* organisms present in fish and shellfish are usually no greater than 10^4 Kanagawa phenotype-positive cells, suggesting that temperature abuse of contaminated food occurs prior to consumption. The temperature abuse permits the growth of the pathogen to levels that result in illness.

Virulence Mechanisms

Although the epidemiological linkage between virulence for humans and the ability of *V. parahaemolyticus* isolates to produce the Kanagawa hemolysin has long been established, the molecular mechanisms by which this factor can cause diarrhea have only recently been elucidated. *V. parahaemolyticus* possesses at least three hemolytic components—a thermolabile hemolysin gene (*tlh*), a TDH gene (*tdh*), and a TDH-related gene (*trh*)—linked to disease.

While much is understood regarding the toxins of *V. parahaemolyticus*, little is known of the adherence process. This is an essential step in the pathogenesis of most enteropathogens (pathogens that infect the intestines). Several adhesive factors have been proposed, including the outer membrane, lateral flagella, pili, and a mannose-resistant, cell-associated hemagglutinin; however, the importance of any of these factors in human disease is unknown.

Authors' note

Determining whether strains produce Kanagawa hemolysin or TDH requires the use of a special blood agar called Wagatsuma agar. With the explosion of polymerase chain reaction (PCR)-based methods, the presence of the gene (tdh) *that encodes the toxin can now be easily determined.*

V. vulnificus

V. vulnificus is the most serious of the pathogenic vibrios in the United States. In Florida, *V. vulnificus* is the leading cause of reported deaths due to foodborne illness. Approximately 86% of people experiencing *V. vulnificus* foodborne illness require hospitalization. Among the population at risk for infection by this bacterium, primary septicemia cases resulting from raw oyster consumption typically have fatality rates of 60%. This is the highest death rate for any foodborne disease agent in the United States. The Centers for Disease Control and Prevention estimates that there are

~50 cases of foodborne *V. vulnificus* infection annually. The bacterium can produce wound infections in addition to gastroenteritis and primary septicemias. Wound infections carry a 20 to 25% fatality rate and are also seawater and/or shellfish associated. Surgery is usually required to clean the infected tissue. In some cases, amputation of the infected limb is required.

Authors' note

Raw oysters have been enjoyed for centuries. However, food microbiologists know that nearly all cases of Vibrio parahaemolyticus *and* V. vulnificus *infection result from consumption of raw oysters. A fresh oyster is not necessarily uncontaminated.*

Classification

The phenotypic traits of this species have been fully described in several studies. The isolation of *V. vulnificus* from blood samples is straightforward, as the bacterium grows readily on TCBS, MacConkey, and blood agars. Isolation from the environment is much more difficult. Vibrios comprise 50% or more of estuarine bacterial populations, and most have not been characterized. Considerable variation exists in the phenotypic traits of *V. vulnificus*, including lactose and sucrose fermentation, considered among the most important characteristics in identifying the species.

Susceptibility to Control Methods

V. vulnificus is susceptible to freezing, low-temperature pasteurization, high hydrostatic pressure, and ionizing radiation. The pathogen can be killed through exposure to horseradish-based sauces. Application of these sauces to raw oysters will not make the oysters safe to eat, since they do not kill bacteria within the oysters.

Reservoirs

V. vulnificus is widespread in estuarine environments, from the Gulf, Atlantic, and Pacific coasts of the United States to locations around the world. The presence of *V. vulnificus* in water is not associated with the presence of fecal coliforms. Water temperature does have an impact on the presence and levels of the pathogen in water. *V. vulnificus* is seldom isolated from water or oysters when water temperatures are low. This was thought to result from cold-induced death during cold-weather months. However, it may be related to a cold-induced VBNC state, in which the cells remain viable but are no longer culturable on the routine media normally employed for their isolation.

Foodborne Outbreaks

There is no report of more than one person developing *V. vulnificus* infection following consumption of the same lot of oysters. Raw oysters from the same lot, or even the same serving, may have very different levels of the pathogen. In fact, two oysters taken from the same estuarine location may have vastly different populations of *V. vulnificus*. In most cases *V. vulnificus* illness occurs between April and October, when the water is warmer. Nearly all cases of *V. vulnificus* infection result from consumption of raw oysters, and most of these infections result in primary septicemias.

Characteristics of Disease

Discussion here is limited to the primary foodborne form of infection caused by *V. vulnificus*. Symptoms vary considerably among cases, with onset times ranging from 7 h to several days, with a median of 26 h. Symptoms include fever, chills, nausea, and hypotension. Surprisingly, symptoms typical of gastroenteritis—abdominal pain, vomiting, and diarrhea—are not common. In severe cases, survival depends to a great extent directly on prompt antibiotic administration. The infectious dose of *V. vulnificus* is not known. *V. vulnificus* is susceptible to most antibiotics.

Virulence Mechanisms

The polysaccharide capsule, LPS, and a large number of extracellular compounds contribute to virulence. The polysaccharide capsule, which is produced by nearly all strains of *V. vulnificus*, is essential to the bacterium's ability to initiate infection. Elevated levels of serum iron appear to be essential for *V. vulnificus* to multiply in the human host. Avirulent strains produce "translucent" acapsular colonies (Fig. 17.2). Symptoms which occur during *V. vulnificus* septicemia, including fever, tissue edema, hemorrhage, and especially the significant hypotension, are those classically associated with endotoxic shock caused by gram-negative bacteria.

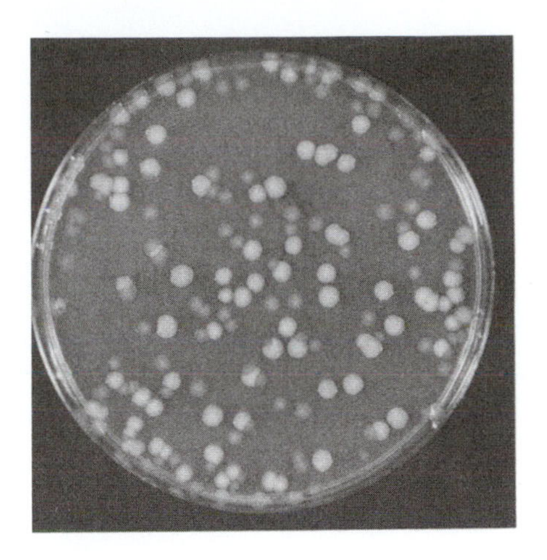

Figure 17.2 Opaque (encapsulated) and translucent (acapsular) colonies of *V. vulnificus*. Virulent strains are generally encapsulated, whereas nonencapsulated cells are usually avirulent. Reprinted from L. M. Simpson, V. K. White, S. F. Zane, and J. D. Oliver, *Infect. Immun.* **55**:269–272, 1987, with permission. doi:10.1128/9781555817206.ch17.f17.02

V. fluvialis, V. furnissii, V. hollisae, and *V. alginolyticus*

V. fluvialis has been isolated frequently from brackish and marine waters and sediments in the United States, as well as other countries. It also has been isolated from fish and shellfish from the Pacific Northwest and Gulf Coast. *Vibrio furnissii* has been isolated from river and estuarine waters, marine molluscs, and crustaceans throughout the world. The distribution of *Vibrio hollisae* is not well documented, although it is likely a marine species. It appears that the bacterium prefers warm waters. *V. alginolyticus* is often found in high numbers in seawater and seafood obtained throughout the world. It is easily isolated from fish, clams, crabs, oysters, mussels, and shrimp, as well as water. Human infections caused by vibrios such as *V. fluvialis, V. furnissii, V. hollisae,* and *V. alginolyticus* are less common and usually less severe than those caused by the pathogens discussed above, although deaths have been reported. Consumption of contaminated seafood is usually the source of infection.

Summary

- Illness associated with *V. cholerae* is not common in the United States.
- *V. vulnificus* and *V. parahaemolyticus* are most commonly associated with foodborne illness.
- *V. vulnificus* is solely responsible for 95% of all deaths due to seafoodborne infections in the United States.
- Numbers of vibrios in water are usually largest during the warm-weather months between April and October.
- Proper thermal processing of seafood kills *Vibrio* species.

Suggested reading

Karaolis, D. K. R., J. A. Johnson, C. C. Bailey, E. C. Boedeker, J. B. Kaper, and P. R. Reeves. 1998. A *Vibrio cholerae* pathogenicity island associated with epidemic and pandemic strains. *Proc. Natl. Acad. Sci. USA* **95**:3134–3139.

Linkous, D. A., and J. D. Oliver. 1999. Pathogenesis of *Vibrio vulnificus*. *FEMS Microbiol. Lett.* **174**:207–214.

Oliver, J. D., and J. B. Kaper. 2007. *Vibrio* species, p. 343–380. *In* M. P. Doyle and L. R. Beuchat (ed.), *Food Microbiology: Fundamentals and Frontiers*, 3rd ed. ASM Press, Washington, DC.

Strom, M. S., and R. N. Paranjpye. 2000. Epidemiology and pathogenesis of *Vibrio vulnificus*. *Microbes Infect.* **2**:177–188.

Tilton, R. C., and R. W. Ryan. 1987. Clinical and ecological characteristics of *Vibrio vulnificus* in the northeastern United States. *Diagn. Microbiol. Infect. Dis.* **6**:109–117.

Questions for critical thought

1. Which characteristics can be used to distinguish *V. vulnificus* from other *Vibrio* species?
2. Which characteristics typically differentiate *Vibrio vulnificus* infection from *Vibrio parahaemolyticus* infection?
3. *Vibrio vulnificus* infection can be contracted by eating raw or undercooked seafood or by a wound inflicted during oyster shucking, and it can result in death when the patient becomes septicemic. Which of the aforementioned would also apply to *V. parahaemolyticus*?
4. *Vibrio* may enter a VBNC state in water. What are key characteristics of this state, and how is this state associated with survival of the organism?
5. Why would treatment of oysters with Tabasco sauce or other similar agents be ineffective in eliminating all *Vibrio* spp. from an oyster?
6. Sanitation programs are often of limited value in the control of vibrios associated with seafood. Why?
7. Explain how depuration of shellfish would aid in mitigating illness associated with the consumption of raw or minimally cooked shellfish.
8. Which *Vibrio* species is associated with cutaneous infection? Which group of people is most at risk? Are cutaneous infections a significant health risk?
9. What do the common virulence characteristics of *V. parahaemolyticus* and *V. vulnificus* include? How do these contribute to pathogenesis?
10. Upon review of a patient's symptoms, how could you presumptively state whether the individual was ill from *V. vulnificus* or *V. parahaemolyticus*?
11. Outbreaks of *V. cholerae* typically occur after large natural disasters (hurricanes, earthquakes, or floods). Explain why such outbreaks typically follow natural disasters. Why do outbreaks associated with *V. vulnificus* and *V. parahaemolyticus* not occur with great frequency after natural disasters?
12. There are many preservation methods that could be used to ensure the safety of seafood. Which methods could you use to ensure the safety of fresh seafood? Suggest advantages and disadvantages of those methods.

18 *Yersinia enterocolitica*

LEARNING OBJECTIVES

The information in this chapter will enable the student to:

- use basic biochemical characteristics to identify *Yersinia*
- understand what conditions in foods favor *Yersinia* growth
- recognize, from symptoms and time of onset, a case of foodborne illness caused by *Yersinia*
- choose appropriate interventions (heat, preservatives, and formulation) to prevent the growth of *Yersinia*
- identify environmental sources of the organism
- understand the role of *Yersinia* toxins and virulence factors in causing foodborne illness

Outbreak

In the United States, Norway, Canada, Japan, and Europe, *Yersinia enterocolitica* serotype O:3 is the dominant cause of yersiniosis. But nothing is cast in stone when it comes to microbes and foodborne illness. In March 2011, 21 cases of yersiniosis were linked to consumption of bagged salad mix contaminated with *Y. enterocolitica* O:9. No deaths occurred. Cases of yersiniosis are often linked to consumption of contaminated pork products.

Brawn (mmm, good!) is a traditional Christmastime favorite in Norway prepared by layering pork meat (precooked head muscles), veal, lard, and spices in a mold. The brawn mold is cooked to an assumed core temperature of 74°C before the temperature is reduced, and the meat is maintained at a temperature of 70°C for at least 30 min. After removal from the mold, the brawn is packed for sale sliced or whole. In 2006, 11 patients developed yersiniosis; 4 were hospitalized and 2 died. Pigs may carry *Yersinia* in their nasal cavities and act as a reservoir of the organism. The brawn contains a significant amount of lard, which may actually protect the organism. The product may have been contaminated during handling, with cross-contamination of each piece occurring during slicing. A cooking thermometer, which is relatively inexpensive, should have been used and may have saved lives.

One of the largest outbreaks of foodborne disease caused by *Yersinia enterocolitica* occurred between June and August of 1982. At least 172 *Y. enterocolitica* culture-positive cases were identified, and 41% occurred among children <5 years of age. Consumption of pasteurized milk was epidemiologically implicated as the mode of transmission of *Y. enterocolitica*. Although *Y. enterocolitica* was not isolated from the milk, the same serotype as found in the outbreak was isolated from a

doi:10.1128/9781555817206.ch18

milk crate on a hog farm where the outdated milk from the implicated processing plant was fed to hogs. The milk crates, if reused without proper sanitizing, might have resulted in cross-contamination of the outside of milk cartons destined for consumers. The symptoms of *Yersinia* enteritis, severe abdominal pain and fever, mimic those of appendicitis. Indeed, 17 patients underwent unnecessary appendectomies (removal of the appendix).

INTRODUCTION

Y. enterocolitica first emerged as a human pathogen during the 1930s. *Y. enterocolitica* shows between 10 and 30% DNA homology with other genera in the family *Enterobacteriaceae* and is ~50% related to *Yersinia pseudotuberculosis* and *Yersinia pestis* (the cause of plague). The last two species show >90% DNA homology, and genetic analysis has revealed that *Y. pestis* is a clone of *Y. pseudotuberculosis* that evolved some 1,500 to 20,000 years ago, shortly before the first known pandemics of human plague.

Characteristics of the Organism

The genus *Yersinia* has 18 species and is classified within the family *Enterobacteriaceae*. As with other members of this family, yersiniae are gram-negative, oxidase-negative, rod-shaped, facultative anaerobes that ferment glucose. The genus includes three well-characterized pathogens of mammals, one pathogen of fish, and several other species whose etiologic role in disease is uncertain. The four known pathogenic species are *Y. pestis*, the causative agent of bubonic and pneumonic plague (the Black Death); *Y. pseudotuberculosis*, a rodent pathogen that occasionally causes disease in humans; *Yersinia ruckeri*, a cause of enteric disease in salmonids and other freshwater fish; and *Y. enterocolitica*, a versatile intestinal pathogen and the most prevalent *Yersinia* species among humans.

Y. pestis is transmitted to its host by the bites of fleas or respiratory aerosols, whereas *Y. pseudotuberculosis* and *Y. enterocolitica* are foodborne pathogens. Nevertheless, these three species share a number of essential virulence determinants that enable them to overcome host defenses. Analogs of these virulence determinants occur in several other enterobacteria, such as enteropathogenic and enterohemorrhagic *Escherichia coli* and *Salmonella* and *Shigella* species, as well as in various pathogens of animals (e.g., *Pseudomonas aeruginosa* and *Bordetella* species) and plants (e.g., *Erwinia amylovora*, *Xanthomonas campestris*, and *Pseudomonas syringae*). This provides evidence for the horizontal transfer of virulence genes among diverse bacterial pathogens. In addition, yersiniae may have acquired a number of human genes that enable them to undermine key aspects of the physiological response to infection.

Classification

Y. enterocolitica is a heterogeneous species that is divisible into a large number of subgroups, largely according to biochemical activity and lipopolysaccharide (LPS) O antigens (Table 18.1). Biotyping is based on the ability of *Y. enterocolitica* to metabolize selected organic substrates and provides a convenient means to subdivide the species into subtypes of clinical and epidemiological significance (Table 18.1). Most pathogenic strains of humans and domestic animals occur within biovars 1B, 2, 3, 4, and 5. By contrast, *Y. enterocolitica* strains of biovar 1A are commonly obtained from terrestri-

Table 18.1 Biotyping scheme for *Y. enterocolitica*

Test	Reaction of biovar[a]:					
	1A	1B	2	3	4	5
Lipase (Tween hydrolysis)	+	+	−	−	−	−
Esculin hydrolysis	D	−	−	−	−	−
Indole production	+	+	(+)	−	−	−
D-Xylose fermentation	+	+	+	+	−	D
Voges-Proskauer reaction	+	+	+	+	+	(+)
Trehalose fermentation	+	+	+	+	+	−
Nitrate reduction	+	+	+	+	+	−
Pyrazinamidase	+	−	−	−	−	−
β-D-Glucosidase	+	−	−	−	−	−
Proline peptidase	D	−	−	−	−	−

[a]+, positive; (+), delayed positive; −, negative; D, different reactions.

al and freshwater ecosystems. For this reason, they are often referred to as environmental strains, although some of them may be responsible for intestinal infections. Not all isolates of *Y. enterocolitica* obtained from soil, water, or unprocessed foods can be assigned to a biovar. These strains lack the characteristic virulence determinants of biovars 1B though 5 (see below) and may represent novel nonpathogenic subtypes or even new *Yersinia* species.

Authors' note

Identification of Y. enterocolitica *is time-consuming, since commercial rapid test kits are not available.*

Serotyping of *Y. enterocolitica*, based on LPS surface O antigens, coincides to some extent with biovar typing and provides a useful additional tool to subdivide this species in a way that relates to its pathological significance. Serogroup O:3 is the variety most frequently isolated from humans. Almost all of these isolates belong to biovar 4. Other serogroups commonly obtained from humans include O:9 (biovar 2) and O:5,27 (biovar 2 or 3), particularly in northern Europe. The most frequent *Y. enterocolitica* biovar obtained from human clinical samples worldwide is biovar 4. Biovar 1B bacteria are usually isolated from patients in the United States and are referred to as American strains. They have caused several foodborne outbreaks of yersiniosis in the United States.

At least 18 flagellar (H) antigens of *Y. enterocolitica*, designated by lowercase letters (a,b; b,c; b,c,e,f,k; m; etc.), have also been identified. There is some overlap between the H antigens of *Y. enterocolitica* and those of related species, but complete O and H serotyping is seldom done.

Other schemes for subtyping *Yersinia* species include bacteriophage typing, multienzyme electrophoresis, and the demonstration of restriction fragment length polymorphisms of chromosomal and plasmid DNA. These techniques can be used to facilitate epidemiological investigations of outbreaks or to trace the sources of sporadic infections.

Susceptibility and Tolerance

Y. enterocolitica is unusual because it can grow at temperatures below 4°C. The doubling time at the optimum growth temperature (ca. 28 to 30°C) is ~34 min, which increases to 1 h at 22°C, 5 h at 7°C, and ~40 h at 1°C. *Y. enterocolitica* readily withstands freezing and can survive in frozen foods for extended periods even after repeated freezing and thawing. *Y. enterocolitica* generally survives better at room temperature and refrigeration temperature than at intermediate temperatures. *Y. enterocolitica* persists longer in cooked foods than in raw foods, probably due to increased availability of nutrients in cooked foods. Also, the presence of other psychrotrophic (growing between 5 and 35°C) bacteria, including nonpathogenic strains of *Y. enterocolitica*, in unprocessed food may restrict bacterial growth. The number of viable *Y. enterocolitica* organisms may increase more than a millionfold on cooked beef or pork within 24 h at 25°C or within 10 days at 7°C. Growth is slower on raw beef and pork. *Y. enterocolitica* can grow at refrigeration temperature in vacuum-packed meat, boiled eggs, boiled fish, pasteurized liquid eggs, pasteurized whole milk, cottage cheese, and tofu (soybean curd). Growth also occurs in refrigerated seafoods, such as oysters, raw shrimp, and cooked crabmeat, but at a lower rate than in pork or beef. Yersiniae can also persist for extended periods in refrigerated vegetables and cottage cheese.

Y. enterocolitica and *Y. pseudotuberculosis* can grow over a pH range of approximately 4 to 10, with an optimum pH of ca. 7.6. They tolerate alkaline conditions extremely well, but their acid tolerance is less apparent and depends on the acidulant used, the environmental temperature, the composition of the medium, and the growth phase of the bacteria. The acid tolerance of *Y. enterocolitica* is enhanced by the production of urease, which hydrolyzes urea to release ammonia and elevates the cytoplasmic pH.

Y. enterocolitica and *Y. pseudotuberculosis* are susceptible to heat and are easily killed by pasteurization at 71.8°C for 18 s or 62.8°C for 30 min. Exposure of surface-contaminated meat to hot water (80°C) for 10 to 20 s reduced bacterial viability by at least 99.9%. *Y. enterocolitica* is also easily killed by ionizing and ultraviolet (UV) irradiation and by sodium nitrate and nitrite added to food, although it is relatively resistant to these salts in solution. The pathogen can also tolerate NaCl at concentrations of up to 5%. *Y. enterocolitica* is susceptible to organic acids, such as lactic and acetic acids, and to chlorine. However, some resistance to chlorine occurs among yersiniae grown under conditions that are similar to natural aquatic environments.

CHARACTERISTICS OF INFECTION

Infections with *Y. enterocolitica* typically manifest as nonspecific, self-limiting diarrhea, but they may lead to a variety of autoimmune diseases (Table 18.2). The risk of these diseases is determined partly by host factors, in particular, age and immune status. *Y. enterocolitica* enters the gastrointestinal tract after ingestion of contaminated food or water. The median infectious dose for humans is not known, but it likely exceeds 10^4 colony-forming units (CFU). Stomach acid is a significant barrier to infection with *Y. enterocolitica*, and in individuals with decreased levels of stomach acid, the infectious dose may be lower.

Most symptomatic infections with *Y. enterocolitica* occur in children, especially in those <5 years of age. In these patients, yersiniosis causes diarrhea, often accompanied by low fever and abdominal pain. The diarrhea varies from watery to mucoid. The illness typically lasts from a few days to 3 weeks, although some patients develop diarrhea that may persist for several months.

In children older than 5 years and adolescents, acute yersiniosis often causes pain in the abdomen that is mistaken for appendicitis. This is referred to as pseudoappendicular syndrome. Some children who experience this also have fever but little or no diarrhea. The pseudoappendicular syndrome happens more frequently with the more virulent strains of

Table 18.2 Clinical symptoms and diseases associated with *Y. enterocolitica* infections

Common symptoms and diseases
Diarrhea (gastroenteritis), especially in young children
Enterocolitis
Pseudoappendicular syndrome due to terminal ileitis; acute mesenteric lymphadenitis
Pharyngitis
Postinfection autoimmune sequelae
Arthritis, especially associated with HLA-B27
Erythema nodosum
Uveitis, associated with HLA-B27
Glomerulonephritis (uncommon)
Myocarditis (uncommon)
Thyroiditis (uncommon)
Less common symptoms and diseases
Septicemia
Visceral abscesses, e.g., in liver, spleen, or lungs
Skin infection (pustules, wound infection)
Pneumonia
Endocarditis
Osteomyelitis
Peritonitis
Meningitis
Intussusception
Eye infection (conjunctivitis)

Y. enterocolitica, notably strains of biovar 1B. *Y. enterocolitica* is rarely found in patients with true appendicitis.

Although *Y. enterocolitica* is seldom isolated outside the intestine, there appears to be no tissue in which it cannot grow. Factors that may make an individual more susceptible to *Yersinia* bacteremia include decreased immune function, malnutrition, chronic kidney disease, liver disease, alcoholism, diabetes, and acute and chronic iron overload states. Spread of the organism in the body can lead to various diseases, including abscesses, catheter-associated infections, heart disease, and meningitis. *Yersinia* bacteremia has a fatality rate between 30 and 60%.

Although most individuals with yersiniosis recover without long-term complications, infections with *Y. enterocolitica* are noteworthy for the large variety of immunological complications, such as reactive arthritis, carditis, and thyroiditis, which follow acute infection. Of these, reactive arthritis is the most widely recognized.

RESERVOIRS

Infections with *Yersinia* species are *zoonoses* (diseases transmitted from animals to humans). The subgroups of *Y. enterocolitica* that commonly occur in humans also occur in domestic animals, whereas those which are infrequent in humans generally reside in wild rodents. *Y. enterocolitica* can survive in many environments and has been isolated from the intestinal tracts of many different mammalian species, as well as from birds, frogs, fish, flies, fleas, crabs, and oysters.

Foods that are frequently positive for *Y. enterocolitica* include pork, beef, lamb, poultry, and dairy products (notably milk, cream, and ice cream). *Y. enterocolitica* is also commonly found in terrestrial and freshwater systems, including soil, vegetation, lakes, rivers, wells, and streams. It can survive for extended periods in soil, vegetation, streams, lakes, wells, and spring water, particularly at low environmental temperatures. Many environmental isolates of *Y. enterocolitica* lack markers of bacterial virulence and may not be a risk to human or animal health.

Although *Y. enterocolitica* has been recovered from a variety of wild and domesticated animals, pigs are the only animal species from which *Y. enterocolitica* of biovar 4, serogroup O:3 (the variety most commonly associated with human disease), has been isolated with any degree of frequency. Pigs may also carry *Y. enterocolitica* of serogroups O:9 and O:5,27, particularly in regions where human infections with these varieties are common. In countries with a high incidence of human yersiniosis, *Y. enterocolitica* is commonly isolated from pigs at slaughterhouses. The tissue most frequently culture positive at slaughter is the tonsils. This appears to be the preferred site of *Y. enterocolitica* infection in pigs. Yersiniae have also been isolated from feces, gut tissue, and the tongue, cecum, and rectum. *Y. enterocolitica* is seldom isolated from meat offered for retail sale. However, standard methods of bacterial isolation and detection may underestimate the true incidence of contamination. Further evidence that pigs are a significant reservoir of human infections is based on epidemiological studies linking consumption of raw or undercooked pork with yersiniosis. Infection also occurs after handling contaminated pig intestines while preparing chitterlings.

Food animals are seldom infected with biovar 1B strains of *Y. enterocolitica*, the reservoir of which remains unknown. The relatively low incidence of human yersiniosis caused by these strains, despite their high virulence, suggests limited contact between their reservoir and humans. Yersiniae of this biovar are pathogens of rodents; therefore, rats or mice are likely the natural reservoir of these strains.

FOODBORNE OUTBREAKS

Considering the widespread occurrence of *Y. enterocolitica* in nature and its abilities to colonize food animals, to persist within animals and the environment, and to grow at refrigeration temperature, outbreaks of yersiniosis are surprisingly uncommon. Most foodborne outbreaks in which a source was identified have been traced to milk (Table 18.3). *Y. enterocolitica* is easily destroyed by pasteurization. Therefore, infection results from the consumption of raw milk or milk that is contaminated after pasteurization. During the mid-1970s, two outbreaks of yersiniosis caused by *Y. enterocolitica* O:5,27 occurred among 138 Canadian schoolchildren who had consumed raw milk, but the organism was not recovered from the suspected source. In 1976, serogroup O:8 *Y. enterocolitica* was responsible for an outbreak in New York State which affected 217 people, 38 of whom were culture positive. The source of infection was chocolate milk, which evidently became contaminated after pasteurization.

Table 18.3 Selected foodborne outbreaks of infection with *Y. enterocolitica*

Location	Yr	Mo	No. of cases	Serogroup	Source(s)
Canada	1976	April	138	O:5,27	Raw milk?[a]
New York	1976	September	38	O:8	Chocolate-flavored milk
Japan	1980	April	1,051	O:3	Milk
New York	1981	July	159	O:8	Powdered milk, chow mein
Washington	1981	December	50	O:8	Tofu and spring water
Pennsylvania	1982	February	16	O:8	Bean sprouts and well water
Southern United States	1982	June	172	O:13a,13b	Milk?
Hungary	1983	December	8	O:3	Pork cheese (sausage)
Georgia (United States)	1989	November	15	O:3	Pork chitterlings
Northeastern United States	1995	October	10	O:8	Pasteurized milk?
Norway	2006	February	11	O:9	RTE[b] pork
Norway	2011	March	21	O:9	RTE salad mix

[a] ?, bacteria were not isolated from the suspected source.
[b] RTE, ready to eat.

In 1981, an outbreak of infection with *Y. enterocolitica* O:8 affected 35% of 455 individuals at a diet camp in New York State. Seven patients were hospitalized as a result of infection, five of whom underwent appendectomies. The source of the infection was reconstituted powdered milk and/or chow mein. An infected food handler probably contaminated the food during preparation.

In 1989, an outbreak of infection with *Y. enterocolitica* O:3 affected 15 infants and children in metropolitan Atlanta, Georgia. In this instance, bacteria were transmitted from raw chitterlings to the affected children on the hands of food handlers. Other foods that have been responsible for outbreaks of yersiniosis include pork cheese (a type of sausage prepared from chitterlings), bean sprouts, and tofu. In the outbreaks associated with bean sprouts and tofu, contaminated well or spring water was the probable source of yersiniae. Water was also the likely source of infection in a case of *Y. enterocolitica* bacteremia in a 75-year-old man in New York State and a small family outbreak in Ontario, Canada. Several outbreaks of presumed foodborne infection with *Y. enterocolitica* O:3 have occurred in the United Kingdom and Japan, but in most cases, the source of these outbreaks was not identified.

MECHANISMS OF PATHOGENICITY

Y. enterocolitica is an invasive *enteric* (intestinal) pathogen whose virulence determinants have been the subject of intensive investigation. Not all strains of *Y. enterocolitica* are equally virulent. *Y. enterocolitica* strains of biovars 1B, 2, 3, 4, and 5 possess many interactive virulence determinants.

Pathological Changes

Examination of surgical specimens from patients with yersiniosis shows that *Y. enterocolitica* is an invasive pathogen. The distal ileum (part of the intestine) is the main infection site. Human volunteers cannot be used to study yersiniosis, since they may develop other diseases. Thus, most information regarding the pathogenesis of yersiniosis has been obtained from animal models, in particular mice and rabbits. Studies of experimentally infected rabbits and pigs show that after penetrating the epithelium of the intestine, the pathogens can spread throughout the body. If the bacteria reach the lymph nodes, they can enter the bloodstream and can disseminate to any organ. However, they preferentially localize in the liver and spleen.

VIRULENCE DETERMINANTS

Chromosomal Determinants of Virulence

The ability of *Y. enterocolitica* to enter mammalian cells is associated with invasin, an outer membrane protein encoded by the chromosomal *inv* gene. Invasins are related to intimin, an essential virulence determinant of enteropathogenic and enterohemorrhagic strains of *E. coli*, which require the protein in order to produce the distinctive attaching-effacing lesions that characterize infection with these bacteria.

The amino terminus of invasin is inserted in the bacterial outer membrane, while the carboxyl terminus is exposed on the surface. The carboxyl terminus binds the host cell integrins. The internalization process is controlled entirely by the host cell, because dead bacteria and even latex particles coated with invasin are internalized. Although DNA sequences the same as that of *inv* occur in all *Yersinia* species (except *Y. ruckeri*), the gene is functional only in *Y. pseudotuberculosis* and the classical pathogenic biovars (1B through 5) of *Y. enterocolitica*. This suggests that invasin plays a key role in virulence.

Most strains of *Y. enterocolitica* secrete a heat-stable enterotoxin, known as Yst (or Yst-a), which is active in infant mice. The contribution of Yst to diarrhea associated with yersiniosis is uncertain. Some strains of *Y. enterocolitica* can produce Yst or the other enterotoxins over a wide range of temperatures, from 4 to 37°C. Since these toxins are relatively acid stable, they could resist inactivation by stomach acid. If they were ingested preformed in food, they could cause foodborne illness. In artificially inoculated foods, however, these toxins are produced mainly at 25°C during the stationary phase of bacterial growth. The storage conditions required for their production in food generally result in severe spoilage, making the ingestion of preformed Yst unlikely.

Under laboratory conditions, *Y. enterocolitica* is motile when grown at 25°C but not when grown at 37°C. The expression of the genes encoding both flagellin and invasin appears to be regulated at the level of transcription. Mutants of *Y. enterocolitica* that are defective in the expression of invasin are extremely motile. Flagellar proteins are made in individuals with yersiniosis, based on assays for the detection of flagellar proteins. Motility does not appear to contribute to the virulence of *Y. enterocolitica* in mice.

Other Virulence Determinants

The observation that patients suffering from iron overload have increased susceptibility to severe infections with *Y. enterocolitica* suggests that the availability of iron in tissues may determine the outcome of yersiniosis. Some isolates of *Y. enterocolitica* are *hemolytic* (they lyse red blood cells) due to the production of phospholipase A. A strain of *Y. enterocolitica* in which the *yplA* gene encoding this enzyme was deleted had decreased virulence for inoculated mice. YplA is secreted by *Y. enterocolitica* via the same export system as used for flagellar proteins.

In *Y. enterocolitica,* acid tolerance is associated with the production of urease. This enzyme catalyzes the release of ammonia from urea and enables the bacteria to resist pHs as low as 2.5. Urease also contributes to the survival of *Y. enterocolitica* in host tissues, but the mechanism by which this occurs is not known. The urease of *Y. enterocolitica* is also unusual in that it displays optimal activity at pH 3.5 to 4.5, suggesting a physiological role in protecting the bacterium from acid.

All fully virulent, highly invasive strains of *Y. enterocolitica* carry a ca. 70-kb plasmid, called pYV (for plasmid for *Yersinia* virulence), which is found in *Y. enterocolitica, Y. pestis,* and *Y. pseudotuberculosis*. pYV permits the bacteria to resist phagocytosis (ingestion by specific host cells) and lysis. Thus, the pathogen can grow within tissues. When pYV-negative strains of *Y. enterocolitica* are incubated with host epithelial cells or phagocytes, they penetrate the cells in large numbers without causing damage. By contrast, pYV-positive bacteria generally remain outside host cells (i.e., resist phagocytosis).

Enteropathogenic yersiniae use a special secretory pathway (the type III secretory pathway) to inject proteins into the cytosol of eukaryotic cells. This facilitates pathogenesis. The transport of *Yersinia* outer membrane proteins from the bacterial cytoplasm into the host cell cytosol may occur in one step from bacteria that are closely bound to the host cell.

Pathogenesis of *Yersinia*-Induced Autoimmunity

Following an infection with pYV-positive *Y. enterocolitica,* a small number of patients develop *autoimmune* (the person's own immune system starts to attack them) arthritis. A similar condition can also occur after infections with *Campylobacter, Salmonella, Shigella,* or *Chlamydia* species. The pathogenesis of reactive arthritis is poorly understood. Men and women are affected equally. Arthritis typically follows the onset of diarrhea or the pseudoappendicular syndrome by 1 to 2 weeks, with a range of 1 to 38 days. The joints most commonly involved are the knees, ankles, toes, tarsal joints, fingers, wrists, and elbows. The duration of arthritis is typically <3 months.

Many other autoimmune complications have been linked to *Y. enterocolitica* infection, including various thyroid disorders, autoimmune thyroiditis, and Graves' disease (hyperthyroidism). The last is an immune system disorder caused by a person producing antibodies to the thyrotropin receptor. The main link between *Y. enterocolitica* and thyroid diseases is that patients with these disorders frequently have elevated levels of antibodies to *Y. enterocolitica* O:3.

Other autoimmune complications of yersiniosis, including Reiter's syndrome, uveitis, acute proliferative glomerulonephritis, collagenous colitis, and rheumatic-like carditis, have been reported, mostly from Scandinavian countries. Yersiniosis has also been linked to various thyroid disorders, including nontoxic goiter and Hashimoto's thyroiditis, although the causative role of yersiniae in these conditions is uncertain. In Japan, *Y. pseudotuberculosis* has been linked to Kawasaki's disease.

Summary

- *Y. enterocolitica* expresses factors, such as urease, flagella, and smooth LPS, which facilitate its passage through the stomach and the mucus layer of the small intestine.

- Once *Y. enterocolitica* begins to replicate in the intestine at 37°C, LPS becomes rough, exposing proteins on the bacterial surface that are involved in attachment and invasion.
- The higher infectivity of *Y. enterocolitica* when grown at ambient temperature compared with its infectivity when grown at 37°C may account for the small number of reports of human-to-human transmission of yersiniosis.
- The well-defined life cycle of *Y. enterocolitica* with its distinctive temperature-induced phases is reminiscent of the flea-rat-flea cycle of *Y. pestis*.
- Outbreaks of yersiniosis are surprisingly uncommon, considering the ubiquity of *Y. enterocolitica* in nature and its abilities to colonize food animals, to persist within animals and the environment, and to proliferate at refrigeration temperature.

Suggested reading

Andersen, J. K., R. Sorensen, and M. Glensbjerg. 1991. Aspects of the epidemiology of *Yersinia enterocolitica*: a review. *Int. J. Food Microbiol.* **13:**231–237.

Bottone, E. J. 1997. *Yersinia enterocolitica*: the charisma continues. *Clin. Microbiol. Rev.* **10:**257–276.

Burnens, A. P., A. Frey, and J. Nicolet. 1996. Association between clinical presentation, biogroups and virulence attributes of *Yersinia enterocolitica* strains in human diarrhoeal disease. *Epidemiol. Infect.* **116:**27–34.

Dube, P. 2009. Interaction of *Yersinia* with the gut: mechanism of pathogenesis and immune evasion. *Curr. Top. Microbiol. Immunol.* **337:**61–91.

Robins-Browne, R. M. 2007. *Yersinia enterocolitica*, p. 293–322. *In* M. P. Doyle and L. R. Beuchat (ed.), *Food Microbiology: Fundamentals and Frontiers*, 3rd ed. ASM Press, Washington, DC.

Questions for critical thought

1. Very few cases of foodborne illness, compared to those associated with *Salmonella* or *Campylobacter*, linked to *Y. enterocolitica* occur each year in the United States. Based on this information, should regulatory agencies still be concerned with contamination of the food supply with *Y. enterocolitica*?
2. Contrast *Y. enterocolitica* with the other foodborne pathogen that is capable of growth at refrigeration temperatures. Which is of greater concern to pregnant women? Which has a higher mortality rate?
3. Why is *Y. enterocolitica* considered unusual among pathogenic enterobacteria? Explain.
4. What domestic animal is considered a reservoir for *Y. enterocolitica*? How was this conclusion drawn?
5. Which group is most likely to develop pseudoappendicular syndrome? What is pseudoappendicular syndrome?
6. Characterizing the pathogenesis of an organism is important for developing strategies to prevent and treat infection. Why is it that human volunteers cannot be used to study *Y. enterocolitica* infection?
7. Which four key factors are associated with the virulence of *Y. enterocolitica*?
8. Infection with *Y. enterocolitica* can result in other diseases. What diseases can develop, and how does infection with *Y. enterocolitica* play a role in those diseases?

9. Identify unique characteristics of *Y. enterocolitica* that would facilitate identification of the pathogen.
10. In reference to the large *Y. enterocolitica* outbreak linked to consumption of pasteurized milk, what measure(s) should have been implemented to prevent such an outbreak?
11. *Y. enterocolitica* is motile when grown at 25°C but not when grown at 37°C. This seems counter to properties required for an enteric pathogen. Provide several reasons why the organism would prefer motility at 25°C to motility at 37°C.
12. Yersiniosis is seldom treated with antibiotics. Most strains of *Y. enterocolitica* show high antibiotic susceptibility. However, they are resistant to β-lactam antibiotics. Indicate why *Y. enterocolitica* is resistant to β-lactam antibiotics.
13. Select two of the autoimmune diseases from Table 18.2 and use your investigative skills to determine the characteristics of each disease.
14. Virulent strains of *Yersinia* carry the plasmid for *Yersinia* virulence (pYV). What properties that enhance virulence does the plasmid impart to strains? Outline a set of experiments that could be done to prove the role of pYV in virulence.
15. You work for the Centers for Disease Control and Prevention (CDC), and there is a large budget cut. The director wants you to eliminate one foodborne pathogen from the top-five list for which data are collected in an effort to save money. You decide to remove *Y. enterocolitica* from the list. Write a one-page justification for your decision based on your knowledge of other pathogens discussed in this book.
16. Develop a list of outbreaks and, if available, why the outbreaks occurred. Based on that information for each outbreak, propose three measures that could be implemented to ensure the safety of the foods involved.

Other Microbes Important in Food

19

Lactic Acid Bacteria and Their Fermentation Products

LEARNING OBJECTIVES

The information in this chapter will help the student to:

- understand the biochemical basis of food fermentations
- appreciate the similarities and differences among fermentations of dairy, vegetable, and meat products
- relate specific bacteria to specific fermentations
- outline the process for making fermented foods
- understand the benefits of using fermentation as a food processing method

INTRODUCTION

"Fermentation" is a word with many meanings. Pasteur used it to describe life in the absence of oxygen. A more formal biochemical definition of fermentation is the process that bacteria use to make energy from carbohydrates in the absence of oxygen. In common usage, *fermentation* also describes any biological process (for example, fermentations that make vinegar, antibiotics, monosodium glutamate, amino acids, citric acid, etc.) whether oxygen is present or not. With a few exceptions, food fermentations are bioprocesses that change food properties while the bacteria generate energy in the absence of oxygen. These changes go far beyond acid production. Fermentations add value to foods by producing flavor compounds and carbonation, altering texture, and increasing nutrient bioavailability (Fig. 19.1 and Table 19.1).

This chapter simplifies fermentation biochemistry for students who have not had a biochemistry course. These biochemical principles are applied to the fermentation of dairy, vegetable, and meat products.

THE BIOCHEMICAL FOUNDATION OF FOOD FERMENTATION

Energy is released when a compound is oxidized (i.e., an electron is "lost") (think back to Chemistry 101). Fermentation products such as alcohol are more oxidized than the starting product (usually a sugar), but they are not completely oxidized. For example, sugar can be partially oxidized to alcohol. The alcohol can be further oxidized when burned, releasing additional energy.

When one compound is oxidized and loses an electron, that electron must "go" somewhere. It goes to an electron acceptor, which becomes reduced. This can be expressed as follows:

Reduced electron donor (compound A) → Oxidized compound A

Oxidized electron acceptor (compound B) → Reduced compound B

doi:10.1128/9781555817206.ch19

Figure 19.1 The bread, sauerkraut, meat, and cheese in this sandwich, as well as the pickles, are all fermentation products. Source: iStockphoto; © Charles Brutlag. doi:10.1128/9781555817206.ch19.f19.01

When oxygen is the electron acceptor in the oxidation of sugars, the following reactions occur:

$$\text{Sugar } (C_6H_{12}O_6) \rightarrow CO_2$$
$$O_2 \rightarrow H_2O$$
$$\text{Sugar} + O_2 \rightarrow CO_2 + H_2O \text{ (+ energy)}$$

This complete oxidation gives the cell a lot of energy. This energy is stored, by mechanisms more appropriately discussed in a biochemistry textbook, in the form of phosphorylated compounds such as adenosine 5′-triphosphate (ATP). In the example above, the complete oxidation of glucose produces 34 ATP molecules. This energy can be transferred to other compounds by moving the phosphoryl group. (Biochemists call this "phosphoryl group transfer.")

In fermented foods, there is no oxygen to serve as an electron acceptor, so part of the sugar must serve as the electron acceptor. This part of the sugar cannot be further oxidized to generate more energy. This incomplete oxidation does not yield very much energy (only 1 or 2 molecules of

Table 19.1 Uses of fermentation

Uses of fermentation
To preserve vegetables and fruits
To develop characteristic sensory properties, i.e., flavor, aroma, and texture
To destroy naturally occurring toxins and undesirable components in raw materials
To improve digestibility, especially of some legumes
To enrich products with desired microbial metabolites, e.g., L-(+)-lactic acid or amino acids
To create new products for new markets
To increase dietary value

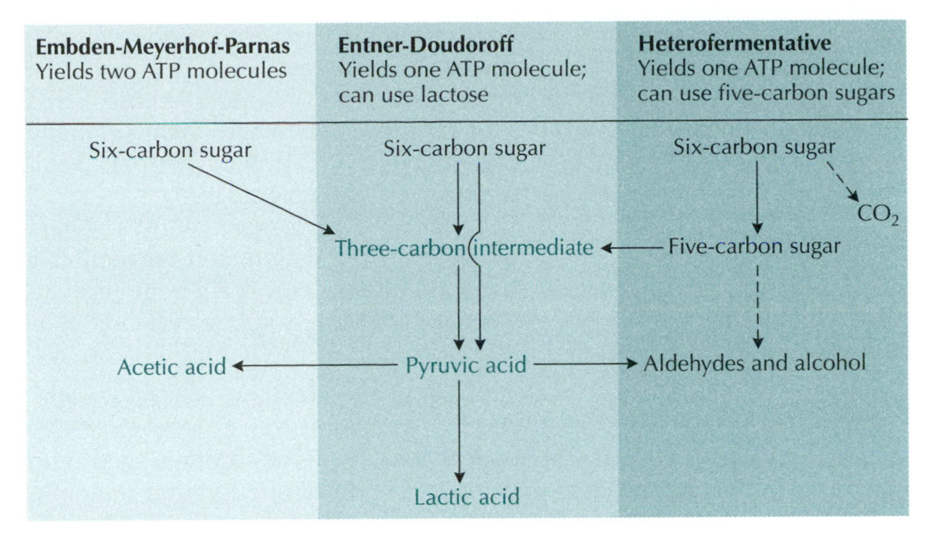

Figure 19.2 Simplified catabolic pathways used in fermented foods. doi:10.1128/9781555817206.ch19.f19.02

ATP from 1 molecule of glucose), but in the land of the blind, the one-eyed man is king. The incompletely oxidized fermentation products such as ethanol, acetic acid, and lactic acid are sometimes oxidized by other "salvaging" organisms. Ethanol is oxidized to acetic acid with the generation of one ATP molecule by *Acetobacter aceti*. *Propionibacterium freudenreichii* subsp. *shermanii* makes one ATP molecule by oxidizing lactic acid to propionic acid.

Catabolic Pathways

Figure 2.13 shows the detailed, step-by-step biochemical pathways used by fermentative bacteria. The more generalized Fig. 19.2 allows the pathways to be understood with less biochemical detail. The Embden-Meyerhof-Parnas (EMP) pathway from glucose to pyruvic acid may be the most important catabolic pathway. One molecule of glucose is converted to two molecules of pyruvic acid. At this point, the oxidation-reduction reactions are not balanced. They can be balanced by oxidizing pyruvate to compounds such as lactic acid or ethanol. Nonfermentative organisms (like us) completely oxidize the pyruvic acid to carbon dioxide and water, using oxygen as the terminal electron acceptor. In terms of ATP generated per mole of glucose used (i.e., two), the EMP pathway is the "best" of the fermentative pathways. Since its only product is lactic acid, organisms that use it are called "homolactic."

The Entner-Doudoroff and heterofermentative pathways yield only one ATP molecule per mole of glucose. The Entner-Doudoroff pathway is important in dairy fermentations. Lactose, or "milk sugar," is a disaccharide broken down to glucose and galactose during fermentation. Milk contains 4.5 to 5.0% lactose and no other free carbohydrates. The EMP pathway cannot metabolize the galactose portion of lactose. Organisms that use the EMP excrete galactose, wasting half of their energy source. The "wasted" galactose can be taken up and metabolized by organisms using the Entner-Doudoroff pathway, but these organisms produce only one three-carbon intermediate and therefore one ATP molecule.

The heterofermentative pathway releases a carbon dioxide molecule, makes one three-carbon intermediate, and generates one ATP molecule. However, it allows organisms to use five-carbon sugars (pentoses) that other organisms cannot use. Thus, in environments where there are many pentoses but no hexoses (six-carbon sugars that can yield two ATP molecules), heterofermentative organisms have a big advantage.

Many factors determine which pathway an organism uses. These include the genes for specific enzymes, their regulation, and the available sugars. Bacteria in the genera *Lactococcus* and *Pediococcus* are homofermentative, producing only lactic acid. Some *Lactobacillus* species are homofermentative, some are heterofermentative, and some can use both pathways.

Genetics of Lactic Acid Bacteria

Genes in lactic acid bacteria (LAB) can be located on the chromosomes or on insertions to the chromosome (introns), move from one location to another (by transposable elements), or be located on extrachromosomal (not on the chromosome) pieces of DNA. These extrachromosomal pieces of DNA are called plasmids. A plasmid location for important genes such as lactose metabolism can be problematic. Plasmids are *independently* replicating extrachromosomal circular DNA molecules. They (and the characteristics they code for) can be lost when plasmid replication is not synchronized with bacterial division. Plasmids are of particular importance in lactococci, in which they encode characteristics essential for dairy fermentations, including lactose metabolism, proteinase activity, oligopeptide transport, bacteriophage resistance, exopolysaccharide production, and citric acid utilization.

Research on the genetics of LAB initially focused on plasmids and natural gene transfer in lactococci, but it has advanced to the use of genomic studies (i.e., those dealing with the sequence of the whole genome) to characterize important phenotypes. LAB genomes (chromosomes) are relatively small, with 2,000 to 3,000 protein-encoding genes on genomes that range in size from 1.8 to 3.4 Mbp. (A mega base pair equals 1,000,000 base pairs.) The large range (almost twofold) of genomic sizes suggests that the evolution of LAB species has entailed gene loss, duplication, and acquisition at various rates. The changes are important on an evolutionary basis. As LAB evolved into nutrient-rich environments, the genes for biosynthetic pathways were lost and the genes for degradative pathways (e.g., for transport of proteases) were acquired. This genetic fluidity can be seen when LAB genomes are compared. When orthologs (genes derived from a common ancestral gene) of LAB were examined, most of the proteins that they code for could be determined. However, the functions of ~10% could be described only vaguely and the functions for another 10% of the genome are totally unknown. Genomics yields considerable information but reveals the extent of our ignorance.

DAIRY FERMENTATIONS

Many microbes are used to make fermented milk products (Table 19.2). These are shown in Fig. 19.3. Note that *Lactococcus* spp. normally form chains in liquid media rather than the clusters when grown on solid media. *Pediococcus* spp. are often found as tetrads, while *Leuconostoc* spp. have a single division plane and form chains. The main bacteria used for acid production are the homofermentative LAB. Heterofermentative LAB contribute flavor

Table 19.2 Microorganisms involved in the manufacture of cheeses and fermented milks

Product	Principal acid producer	Intentionally introduced secondary microbiota
Cheeses		
Colby, Cheddar, cottage, cream	*Lactococcus lactis* subsp. *cremoris* or *lactis*	None
Gouda, Edam, Havarti	*Lactococcus lactis* subsp. *cremoris* or *lactis*	*Leuconostoc* spp., *Lactococcus lactis* subsp. *lactis*
Brick, Limburger	*Lactococcus lactis* subsp. *cremoris* or *lactis*	*Geotrichum candidum, Brevibacterium linens, Micrococcus* spp.
Camembert	*Lactococcus lactis* subsp. *cremoris* or *lactis*	*Penicillium camemberti*, sometimes *Brevibacterium linens*
Blue	*Lactococcus lactis* subsp. *cremoris* or *lactis*	*Lactococcus lactis* subsp. *lactis, Penicillium roqueforti*
Mozzarella, provolone, Romano, Parmesan	*Streptococcus thermophilus, Lactobacillus delbrueckii* subsp. *bulgaricus, Lactobacillus helveticus*	None; animal lipases added to Romano for piquant or rancid flavor
Swiss	*Streptococcus thermophilus, Lactobacillus helveticus, Lactobacillus delbrueckii* subsp. *bulgaricus*	*Propionibacterium freudenreichii* subsp. *shermanii*
Fermented milks		
Yogurt	*Streptococcus thermophilus, Lactobacillus delbrueckii* subsp. *bulgaricus*	None
Buttermilk	*Lactococcus lactis* subsp. *cremoris* or *lactis*	*Leuconostoc* spp., *Lactococcus lactis* subsp. *lactis*
Sour cream	*Lactococcus lactis* subsp. *cremoris* or *lactis*	None

compounds as described below. The homo- and heterofermentative bacteria are often paired to give the flavor and texture characteristic of a given cheese. Paired cultures often acidify faster than single cultures.

Pasteurization kills most of the natural LAB and other bacteria present in raw milk. This facilitates the addition of LAB "starter cultures" that do not have to compete with large numbers of naturally occurring milk bacteria. Starter cultures allow manufacturers to control the rate and extent of acid development in the fermented food. This results in greater process control and a more consistent product.

Other bacteria, referred to as the secondary microbiota, are added to some fermented products to influence flavor and texture. They typically comprise 10 to 20% of the total starter culture. *Leuconostoc* species and strains of *Lactococcus lactis* subsp. *lactis* that metabolize citric acid produce aroma compounds and carbon dioxide in cultured buttermilk and cheeses such as Gouda, Edam, blue, and Havarti. Heterofermentative lactobacilli (*Lactobacillus brevis, Lactobacillus fermentum*, and *Lactobacillus kefiri*) are part of the varied microbiota (including several yeast species) that produce ethanol, carbon dioxide, and lactic acid in more exotic cultured milks such as kefir and koumiss. This ethanol is not very volatile and does not contribute much flavor. However, when esterified to make volatile compounds, it has a major impact on taste. *Propionibacterium freudenreichii* subsp. *shermanii* is added to make Swiss-type cheeses. Its scavenging metabolism converts L-lactic acid to propionic acid, acetic acid, and carbon dioxide. The carbon dioxide forms the "eyes" (holes) in Swiss cheese. Propionibacteria also ferment citric acid to glutamic acid, a natural flavor enhancer. *Penicillium roqueforti* is a secondary culture for the production of blue cheese, while *Penicillium camemberti* is used to make Camembert cheese.

A generic scheme for making cheese is shown in Fig. 19.4. It is important to note that only high-quality milk should be used in the fermentation.

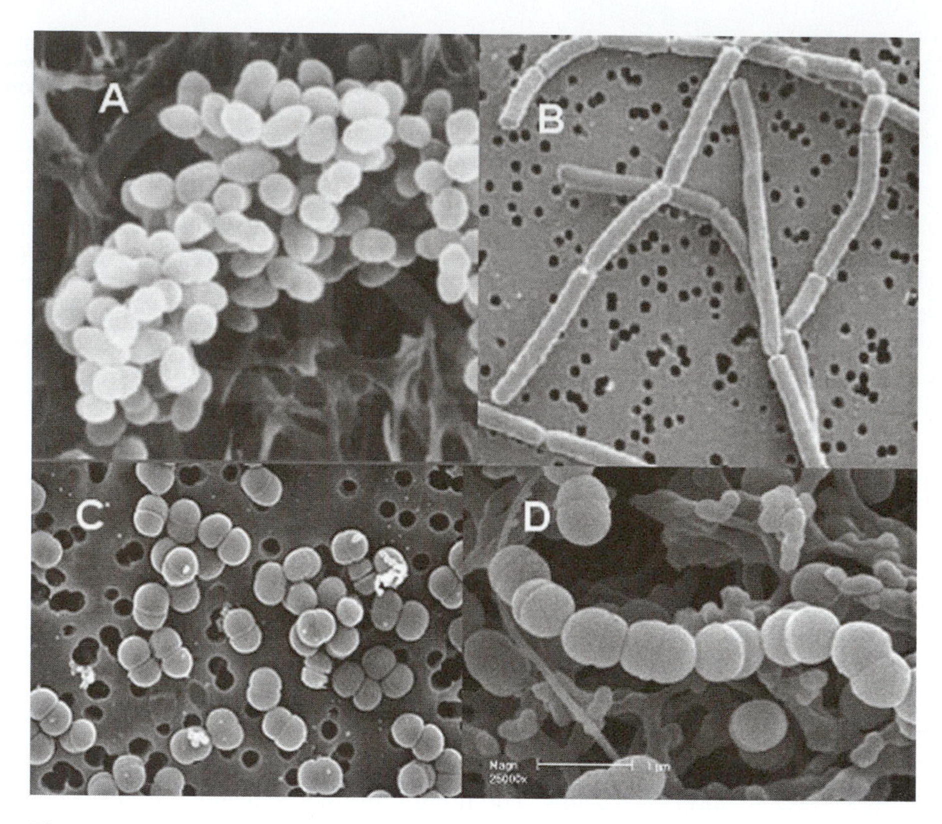

Figure 19.3 LAB associated with food fermentations. **(A)** *Lactococcus*; **(B)** *Lactobacillus*; **(C)** *Pediococcus*; **(D)** *Leuconostoc*. (B and C) Courtesy of the U.S. Department of Energy Joint Genome Institute; (D) reprinted from G. Kaletunç, J. Lee, H. Alpas, and F. Bozoglu, *Appl. Environ. Microbiol.* **70:**1116–1122, 2004, with permission from the American Society for Microbiology. doi:10.1128/9781555817206.ch19.f19.03

The removal of sediment, by a process called clarification, is important for both aesthetic and safety reasons. Pathogens like *Listeria monocytogenes* can be inside the leukocytes shed in the milk. Clarification removes these leukocytes. Other ingredients such as salt, seasoning, or even additional milk solids are added prior to heating, which kills any remaining bacteria. Homogenization stabilizes the product to prevent fat separation during the fermentation and to ensure an even distribution of all the product ingredients. In the actual fermentation step, the starter cultures rapidly (within 4 to 8 h) lower the pH to <5.3 in cheese and <4.6 in fermented milk products. At this pH, only acid-tolerant bacteria can grow. After the fermentation, the product may be packaged, aged, or subjected to secondary fermentations for flavor development.

Yogurt has traditionally been made using *Streptococcus thermophilus* and *Lactobacillus bulgaricus*. *Lactobacillus acidophilus* and *Bifidobacterium* spp. are often added due to their popularity as probiotics. The first step in yogurt manufacture is to concentrate the milk by 25% using a vacuum dehydrator. Milk solids (5%, wt/wt) are then added and the mixture is heated to 90°C for 30 to 90 min. After the mixture is cooled to 45°C, the starter culture is added at 2% (vol/vol) and the mixture is incubated for 3 to 5 h. The final product has a titratable acidity of 0.8 to 0.9% and about 10^9 organisms/g. These may die off during cold storage to a population of 10^6 organisms/g, the minimum required to make a "live and active culture" claim for the yogurt.

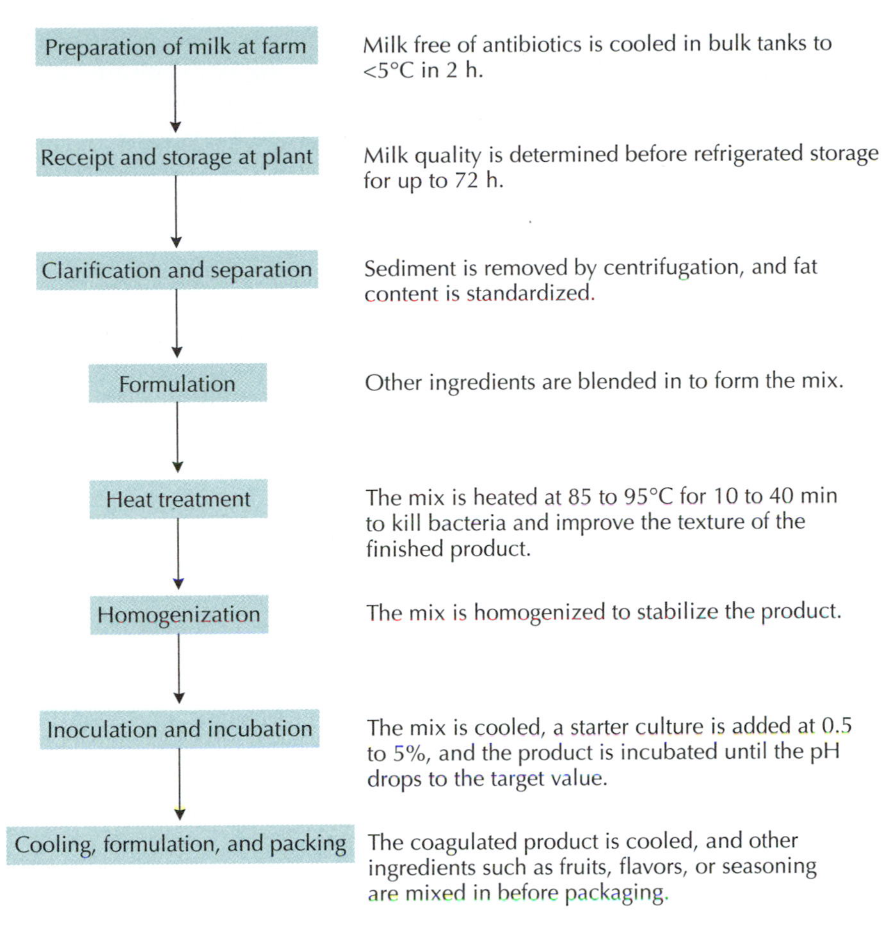

Figure 19.4 Generic scheme for making fermented dairy products. doi:10.1128/9781555817206.ch19.f19.04

Starter Cultures

The key to the commercial success of fermented milk products is consistent and predictable rates of acid production (Table 19.3). The addition of defined bacterial cultures, i.e., starter cultures, to initiate the fermentation guarantees predictable acidification. Acidity has profound effects on moisture control, protein retention, mineral loss, protein hydration, and interactions between protein molecules. These, in turn, determine the product's sensory characteristics. This consistency can be undermined if bacteriophages attack the starter cultures. The adulteration of milk by antibiotics can also wreak havoc with starter cultures. In countries where milk is routinely screened for antibiotics, this is no longer a big issue.

Table 19.3 Benefits of using starter cultures in manufacture of fermented food

Rapid acid production
Decreased rate of contamination
Production of flavor compounds
More consistent product
Greater process predictability

The flavor of cultured milk products is determined by microorganisms used as starter cultures and the secondary microbiota. With cultured milks and some cheeses, such as mozzarella, cream, and cottage cheeses, the short time from processing to consumption (1 day to 4 weeks) is not long enough for secondary organisms to generate "ripened" flavors. In fact, cottage cheese can be made by a "direct set" method, in which acid is added directly to the milk, rather than fermenting it.

Production of Aroma Compounds

Lactic acid is the main end product in dairy fermentations. It contributes the acid taste, but not the aroma. The main aromas and flavors of fermented milks come from acetic acid, acetaldehyde, and diacetyl. In yogurt, these are formed by *Streptococcus thermophilus* and *Lactobacillus delbrueckii* subsp. *bulgaricus*. In buttermilk and some cheeses, *Leuconostoc* species and citrate-utilizing *Lactococcus lactis* subsp. *lactis* produce the aroma compounds.

Milk contains 0.15 to 0.2% citric acid. Because citric acid is converted to pyruvic acid without the need to regenerate nicotinamide adenine dinucleotide (NAD), virtually all of it can be converted to diacetyl, which has a buttery flavor. Citric acid is metabolized by *Leuconostoc* species, citrate-metabolizing *Lactococcus lactis* subsp. *lactis*, and facultative heterofermentative lactobacilli to make diacetyl, acetic acid, and carbon dioxide. The carbon dioxide causes the holes (eyes) in Gouda and Edam cheeses and the effervescent quality of buttermilk.

Several metabolic pathways form acetaldehyde. The enzyme threonine aldolase cleaves threonine to glycine and acetaldehyde. Acetaldehyde is also formed by citric acid-metabolizing *Lactococcus lactis* subsp. *lactis*. Fermented milks develop a "yogurt" or "green apple" flavor defect when the ratio of diacetyl to acetaldehyde is high. Excessive acetaldehyde in yogurt is caused by prolonged fermentation and is associated with high acid content. The use of *Leuconostoc* that converts acetaldehyde to ethanol prevents these excessive acetaldehyde levels. Rapid cooling and refrigerated storage also reduce off-flavor production.

Proteolysis is required for the growth of many LAB. They require anywhere from 6 to 16 amino acids to grow. The protein casein comprises 80% of milk protein, but there are few free amino acids in milk. This makes the ability to break casein down to amino acids a very important trait for LAB. The casein can be hydrolyzed by extracellular proteases, by proteases in the cell membrane, or by intracellular proteases.

Proteolysis also generates flavors in ripened cheeses. The free amino acids and peptides produced by the extracellular proteases can have positive or negative effects in the cheese. A major negative effect of proteolysis is bitterness. This is caused by the breakdown of casein into hydrophobic peptides that are 3 to 27 amino acids long. Other bacterial peptidases can hydrolyze these to nonbitter peptides and amino acids. Therefore, the accumulation of bitter peptides depends on their relative rates of production and destruction.

VEGETABLE FERMENTATIONS

The ancients fermented vegetables to retain their nutritive value. Many different microorganisms are used for this purpose. LAB and yeasts are preferentially used in the West (Europe and America). Molds are used to ferment many Eastern foods. LAB are used extensively for biopreservation

Table 19.4 Steps in a typical vegetable fermentation

1. Select vegetables that are sound, undamaged, uniformly sized, and at the proper ripeness.
2. Pretreat, for example, by peeling, blanching, or cooking.
3. Place whole, pierced, shredded, or sliced vegetables in fermentation vessels. These can hold from 100 liters to 100 tons. Starter cultures can also be added.
4. Completely cover the vegetables with brine. Seal the fermentation vessels to exclude oxygen and ensure anaerobic conditions.
5. Let the fermentation take its natural course. The fermentation time depends on the temperature, the type of product, and the bacteria present.
6. Distribute the final fermented product fresh, unpackaged, packaged, or pasteurized.

of vegetables. The market for pickled vegetable products is $2 billion in the United States alone.

Lactic acid fermentation of vegetables was practiced by the Chinese in prehistoric times. The oldest written evidence dates from the first century C.E., when Plinius described the preservation of white cabbage in earthen vessels. This causes a lactic acid fermentation that turns cabbage into sauerkraut. The development of heat sterilization and refrigeration systems has made fermentations less important as a preservation method in industrialized countries. Nonetheless, in developing countries, fermentation is still a critical means of food preservation.

Vegetable fermentations involve complex microbiological, biochemical, chemical, and physical reactions. Fermentations are also influenced by many external factors. These are classified into four groups: technological factors, ingredients, raw material quality, and native microbiota (i.e., organisms that are naturally associated with the vegetable).

Table 19.4 shows a generic process for manufacturing fermented vegetables. It is important to start with sound vegetables. Bruised or damaged vegetables can have high levels of natural microbiotas. Uniform sizing facilitates processing and packaging. Blanching or cooking may be required to inactivate enzymes or degas the vegetables. The placement of the vegetables in a fermentation vessel is straightforward. Starter cultures are not usually added because they cannot compete with the vegetables' natural microbiota. Anaerobic conditions (i.e., the exclusion of oxygen) are required since this is biochemically a fermentation. Anaerobiosis is easily achieved by covering the fermentation vessel. Any residual dissolved oxygen is removed by the respiration of the vegetables and any aerobic bacteria present. The fermentation proceeds at whatever speed is dictated by the natural microbiota. The finished product can be distributed in many forms. Because the botanical, physical, and chemical properties of various vegetables differ, the process described above must be tailored to specific products.

Ingredients and Additives Used during Fermentations

Salt is added to fermentations for many reasons. Salty flavors are essential to many fermented foods. The amount of salt used depends on the vegetable and on consumer preference. In sauerkraut production, salt enhances fluid release from shredded cabbage. It also helps create anaerobic conditions in fermentation vessels. In addition, salt has a selective effect on the vegetables' natural microbiota. Increasing amounts of salt favor the growth of LAB and inhibit the undesirable bacteria and fungi. In sauerkraut fermentations, heterofermentative LAB are favored by the low salt concentration (~1%) and are greatly inhibited at 3%. Higher salt concentrations favor homofermentative species, accelerating the fermentation. Salt content below 0.8% often results in undesirable fermentation as well as in soft sauerkraut.

Other food additives are used in specific fermented vegetables and are regulated by law in most countries. The addition of ascorbic acid to sauerkraut prevents gray or brown discoloration. Citric acid or sulfur dioxide is also used for this purpose. Sorbic acid prevents the growth of yeasts, molds, and other microbes.

Sauerkraut Fermentation

The overall microbial population, as well as the population of LAB, changes during the course of a vegetable fermentation. The fermentation of cabbage, for example, has four stages.

1. Fermentation starts as soon as chopped cabbage is placed into vessels. In commercial processing, these vessels can contain 100 tons of chopped cabbage. Salt is added to an equilibrium concentration of 2%. The cabbage is tightly packed to form a bowl at the top of the vessel. To create an anaerobic environment, the cabbage is covered with a plastic sheet which is then filled with water. Under the anaerobic conditions, the number of strictly aerobic bacteria decreases, while facultatively anaerobic enterobacteria grow for the first 2 or 3 days. During this period, dissolved oxygen is consumed by the microorganisms and by plant respiration. The formation of lactic, acetic, formic, and succinic acids lowers the pH. Carbon dioxide may generate foam.
2. During the second stage of fermentation, non-LAB are suppressed by anaerobic conditions and are overgrown by the LAB. Cabbage typically contains 4 to 5% sugar, with about half glucose and half fructose. The fermentation of these sugars to lactic acid is initiated by heterofermentative LAB, namely, *Leuconostoc mesenteroides*, followed by *Lactobacillus brevis*. This succession of microorganisms is complete after 3 to 6 days, and the lactic acid concentration increases to ~1%.
3. The third stage of fermentation is dominated by homofermentative LAB. Their growth is favored by the complete lack of oxygen, low pH, and high salt content. The homofermentative LAB increase the total acid content to 1.5 to 2.0%. Most sauerkraut is pasteurized when it reaches pH 3.8 to 4.1.
4. The fourth stage of fermentation is optional. For economic reasons, sauerkraut may be stored in the fermentation vessel for up to a year. During the storage period, *Lactobacillus brevis* and some heterofermentative species metabolize the pentoses released by the breakdown of cell walls. The acid content may increase to 2.5%.

Pickle Fermentation

Pickles are made from cucumbers. Although more than 40 flavor compounds have been isolated from pickles, most of them originate from the cucumber. Indeed, pickles made by direct acidification outsell those made by fermentation. Dill pickles provide the vegetable component sold in the restaurants that feed the fast food nation (unless they hold the pickles and hold the lettuce).

Commercial cucumber fermentations are conducted outdoors in vessels that contain 8 to 10,000 gal of product. Starter cultures are rarely used.

After the cucumbers are covered with brine, the vessel is sealed with a wooden headboard. Cucumbers contain glucose and fructose at about 1% each; these undergo a homofermentation, typically by naturally occurring *Lactobacillus plantarum*. The final pH of ~3.7 is well below the regulatory cutoff for high-acid foods (i.e., pH 4.6). This level of acid gives a 5-log reduction of pathogens and ensures product safety. Like sauerkraut, pickles can be stored in their vessels for up to a year.

"Sweet" pickles are fermented in a similar fashion using small cucumbers. After the fermentation, the cucumbers are washed and then soaked in 25 to 30% sugar.

The carbon dioxide produced by cucumber respiration or excessive growth of heterofermentative *Leuconostoc mesenteroides* can accumulate in the pickles and cause "floaters," also known as "bloaters." Bloater formation is a serious problem in cucumber fermentation. Damage is worse with larger cucumbers, at higher fermentation temperatures, and when the dissolved carbon dioxide concentration in the brine is high. The brine's carbon dioxide content can be reduced by limiting growth of heterofermentative LAB and removing carbon dioxide from the brine by purging with nitrogen or other gases.

MEAT FERMENTATIONS

Fermented meats are not as popular as fermented vegetable or dairy products but are still a major class of fermented food worth billions of dollars. The natural microbiota of meat is gram-negative, aerobic putrefying spoilage bacteria. Fermentation by LAB prevents spoilage and turns the raw meat into a totally different product. The most common types of fermented meats are dry and semidry fermented sausages.

The U.S. Department of Agriculture's Food Safety Inspection Service requires that shelf-stable sausages contain nitrite and curing agents, be fermented to a pH of <5.0, and have a moisture/protein ratio of $\leq 3.1:1.0$. The production of such a sausage is straightforward (Table 19.5). Grinding the meat reduces its particle size. It is then blended with nitrites, curing agents, spices, a fermentable carbohydrate, and sometimes glucono-delta-lactone. The nitrite inhibits *Clostridium botulinum*, contributes to the cured meat taste, and converts myoglobin to nitrosomyoglobin (the pink color of cured meat). Fermentable sugars are added because meat does not contain much fermentable carbohydrate. Glucono-delta-lactone is converted to gluconic acid. This is converted by LAB to lactic and acetic acids and accelerates acidification. The starter cultures (10^7 colony-forming units [CFU]/g) are also added at this point. The mixture is stuffed into oxygen-impermeable casing to ensure an anaerobic fermentation. The sausage links are incubated

Table 19.5 Steps in the manufacture of fermented sausages

1. Reduce particle size of high-quality meat by grinding.
2. Add curing salts (nitrite at <150 parts per million), 3% glucose as a fermentable sugar, spices, and starter culture at $\sim 10^7$ CFU/g.
3. Blend the ingredients.
4. Vacuum stuff the meat into the casing.
5. Incubate (also referred to as "ripen") the sausages.
6. Heat to inactivate the starter culture and pathogens.
7. Age (dry) the sausages.

to promote microbial growth. The starter cultures decrease the pH from 5.6 to 4.8 within 8 h. The lactic acid coagulates the protein. This facilitates drying. The finished sausage is sometimes given a heat treatment.

There are several ways to start a meat fermentation. The native bacteria of lactobacilli, coagulase-negative staphylococci, and micrococci and yeasts can be used. Unfortunately, because their presence on a given lot of meat is variable, the use of indigenous bacteria makes process control difficult. A *slightly* more sophisticated approach is using some of the last batch to inoculate the next batch. Defined starter cultures are relatively new to the meat industry and represent the third way to start a fermentation. Attempts to use the native meat LAB were unsuccessful because they do not tolerate the freeze-drying process used to make starter cultures. The next step in the starter culture progression was the use of freeze-dried *Pediococcus acidilactici*, an organism not normally associated with meat. However, it is homofermentative, can tolerate 6% salt, and is not proteolytic or lipolytic. It also has the advantage of growing at 43 to 50°C, a temperature that precludes the growth of pathogens. However, freeze-dried cultures were hard to dissolve and had long lag times, so in the 1980s, manufacturers started using frozen cultures. When inoculated into the meat at 10^7 CFU/g, they shorten the lag time and acidify very rapidly.

Summary

- Fermentation is an incomplete oxidation of sugars in the absence of oxygen using an internal organic compound as an electron acceptor.
- "Fermentation" can also mean "bioprocess."
- Fermentative metabolism yields only one or two ATP molecules per mole of glucose, much less than oxidative metabolism.
- Homofermentative bacteria make only lactic acid.
- Heterofermentative bacteria make lactic acid, ethanol, carbon dioxide, and acetic acid.
- Homofermentative bacteria are used to make acid, whereas heterofermentative bacteria contribute flavor.
- Starter cultures increase the speed and consistency of fermentations.
- LAB have several methods for transferring genes.
- Plasmids carry the genes for many traits important to fermentations.
- Vegetable fermentations rely on indigenous microbiotas.

Suggested reading

Breidt, F., Jr., R. F. McFeeters, and I. Díaz-Muñiz. 2007. Fermented vegetables, p. 783–794. *In* M. P. Doyle and L. R. Beuchat (ed.), *Food Microbiology: Fundamentals and Frontiers*, 3rd ed. ASM Press, Washington, DC.

Fox, P. E. (ed.). 1993. *Cheese: Chemistry, Physics, and Microbiology*, vol. 1 and 2. Chapman and Hall, Ltd., London, England.

Johnson, M. E., and M. E. Steele. 2007. Fermented dairy products, p. 767–782. *In* M. P. Doyle and L. R. Beuchat (ed.), *Food Microbiology: Fundamentals and Frontiers*, 3rd ed. ASM Press, Washington, DC.

Ricke, C. S., I. Z. Diaz, and J. T. Keeton. 2007. Fermented meat, poultry, and fish products, p. 795–816. *In* M. P. Doyle and L. R. Beuchat (ed.), *Food Microbiology: Fundamentals and Frontiers*, 3rd ed. ASM Press, Washington, DC.

Salminen, S., and A. von Wright (ed.). 1998. *Lactic Acid Bacteria.* Marcel Dekker, Inc., New York, NY.

Questions for critical thought

1. How is the saying "In the land of the blind, the one-eyed man is king" relevant to energy production by fermentative metabolism? Can you think of another folk saying that captures the same concept?
2. If you were a *Lactobacillus* species able to use pentoses and hexoses and landed in an environment with both, which would you use? Why?
3. What are three general usages of the word "fermentation"?
4. During cheese manufacture, why does pH have profound effects on moisture control, retention of coagulants, and hydration of proteins?
5. Why are commercial starter cultures rarely used for pickle fermentation? If you wanted to revolutionize the pickle industry by using starter cultures, what would you need to do?
6. The year is 2020, and you are an associate professor at Big State University. The Peanut Producers and Packers want to develop a fermented peanut product that will add value and increase the demand for peanuts. The Peanut Producers and Packers will allocate $300,000 to this project and have e-mailed several universities requesting one-page research proposals. You would like the money. Write the proposal.
7. Why is the oxidation of ethanol and lactic acid referred to as "scavenging metabolism"?
8. Identify the microorganisms associated with each component of the sandwich shown in Fig. 19.1.

20

Yeast-Based and Other Fermentations

LEARNING OBJECTIVES

The information in this chapter will help the student to:

- appreciate the role of yeast in several fermentations
- relate the steps of bread making to its fermentation and final characteristics
- describe the biochemical basis of beer and wine fermentations
- identify the similarities between beer and wine making
- understand the roles of different ingredients and process steps in the production of beer
- describe the differences and similarities in the production of red and white wines
- put the evolution of vinegar processing technologies into a historical context
- describe the role of fermentation in the production of cocoa, coffee, and indigenous fermentations

INTRODUCTION

Enthusiasm is the yeast that makes your hopes shine to the stars.

HENRY FORD

Yeasts raise, ferment, carbonate, and otherwise transform a wide variety of agricultural products. This chapter examines these abilities of yeast in the context of bread, beer, and wine. Bread uses the carbon dioxide produced by the yeast to leaven, or raise, it. This fermentation takes some time, so those in a hurry (e.g., the Israelites during their exodus from Egypt) eat unleavened bread. The production of beer is more complex and involves several steps. Enzymes must be made, and they must break down starch to fermentable carbohydrates before the carbohydrates can be fermented to alcohol. Since wine grapes contain simple sugars that are easily fermented, wine making is in some ways simpler than beer making, but in other ways, due to the complexity of grapes and their chemistry, it is more complex. Vinegar production is in theory very easy, since it involves only a single oxidation step. In reality, getting enough oxygen into the fermenting liquid can be quite challenging. Finally, the microbiology of some non-Western fermented food is briefly discussed.

doi:10.1128/9781555817206.ch20

FERMENTATIONS THAT USE YEAST

Bread

He who has no bread has no authority.

TURKISH PROVERB

As ancient societies became more sophisticated, their cereal food progressed from porridges and gruels to unleavened flatbreads and, finally, to breads leavened with yeast. This occurred as early as 2700 B.C.E., as evidenced by archeological remains of Egyptian baking ovens (Fig. 20.1). In those early times, bakeries were often attached to breweries so that the yeast by-product of brewing (*Saccharomyces cerevisiae*) could be used in bread making. Today, specialized strains of yeast are used for each purpose.

The first step in bread making is to mix flour, sugar (as a fermentable carbohydrate), fat (for texture), salt, and other ingredients. The yeast *Saccharomyces cerevisiae* (baker's yeast) is added in the form of dried powder, block, or cream at 1 to 6% on a weight basis per weight of flour. The yeast's main role in bread production is to produce the carbon dioxide that makes the bread rise. However, the yeast also produces amylases that break down starch to the more fermentable glucose. Water is added, and the dough is kneaded so that gluten protein in the flour stretches and the dough forms a viscoelastic (i.e., both viscous and

Figure 20.1 Scenes of Egyptian bread making from an excavation by Georg Steindorff. Note the kneading of the bread in the first two rows and the filling of conical bread molds in the next two rows. The finished bread is removed from the molds in the bottom panel. The scribes may be early food microbiologists documenting the process. doi:10.1128/9781555817206.ch20.f20.01

Like many other microbes, yeasts can be "good" or "bad."
Reprinted with permission from CartoonStock.

"I'm afraid you have a yeast infection."

elastic) mass. The bread is fermented or "proofed" at 28 to 32°C for several hours. The dough is then remixed (or "punched" in home usage) to evenly distribute the gas cells, portioned into loaves, and fermented again so that the loaves double in volume. During the baking at around 200°C for ~30 min, carbon dioxide expands to increase the bread volume by ~40%. The gas is captured by the protein as it "sets" by heat denaturation.

For sourdough breads, a stable mixture of heterofermentative lactic acid bacteria (LAB) is also added, at levels of 10^7 to 10^9 colony-forming units (CFU)/g. The *Lactobacillus* species, primarily *Lactobacillus sanfranciscensis*, add a characteristic "bite" to the bread, improve texture, and prevent spoilage. When sourdough breads are made by artisans, the inoculum is usually made using some of the last batch. The sequential transfer of the cultures can propagate the starter culture for decades.

Beer

Beer is proof that God loves us and wants us to be happy.

Attributed to Benjamin Franklin

Authors' note

This statement applies to those of us who are at least 21 years old. God's love is manifest to minors as ice cream.

Beer production, using methods remarkably similar to those we use today, has been traced to ancient times (Box 20.1). Beer making has become a large industry, producing more than 100 million gallons per day worldwide. In 2010, $101 billion worth of beer was sold in the United States alone. The essential ingredients of beer are hops (for flavor and antimicrobial activity), yeast (to produce alcohol and carbon dioxide), water (for obvious reasons), and malt (to provide the fermentable carbohydrate). German purity laws used to forbid the use of any other ingredients. Other countries (including the United States) allow less expensive grains, including corn, to be used as "adjuncts" to augment or replace the malt as the source of the fermentable carbohydrates.

The brewing process consists of malting, mashing, wort boiling, fermentation, and postfermentation treatments (Fig. 20.2). The yeast cannot ferment the starch in the barley, so it is "malted" and mashed. During malting, the grain is steeped in water for 24 to 48 h to induce germination. This breaks down the cell wall and protein matrix that contains the starch granules and liberates amylases. During mashing, the amylases break the starch into fermentable carbohydrates. At the end of this process, the liquid malt is almost ready to ferment. Herbs, usually hops, are added at this point for flavor and antimicrobial activity. Different hops give different beers their distinctive flavors. The mixture of hops and liquid malt is boiled and then cooled to ~20°C before the yeast is added.

Different species of yeast are used to produce different types of beer. *Saccharomyces cerevisiae,* which grows on top of the fermentation mix, is used for ales. *Saccharomyces carlsbergensis,* which settles to the bottom, is used for lagers. In both cases, the inoculum level is quite high, $\sim 10^7$/ml, and increases eightfold over the course of the fermentation; i.e., growth consists of three doublings. The yeast from 1 fermentation is used to inoculate the next batch for 10 to 15 fermentations. Then, a new defined inoculum from a standard source is introduced. This prevents genetic drift which would change the beer-making qualities of the yeast. The yeast uses the various sugars in the wort sequentially, with simple sugars being fermented first and maltotriose later. Ethanol production continues after the yeast stops growing and flavor compounds, such as aldehydes and higher alcohols, are produced.

The postfermentation processing of beer includes aging to remove the "green" flavors caused by diacetyl and acetaldehyde. These compounds are present in low concentrations but have a low flavor threshold (i.e., humans can taste them at very low concentrations) and cause off-flavors. Aging in casks, storage in the presence of yeast at 15°C, and avoiding exposure to oxygen help remove these flavor defects. Most beers are filtered until clear, although the cloudiness of boutique beers is part of their charm. Filter aids, such as cellulose fiber or pumice, enhance the removal of the yeast by filtration. Membrane filtration can be used as a means of "cold pasteurization." This retains the beer's "draft" taste and avoids the energy expense of heat pasteurization. For heat pasteurization, the beer is held for 5 to 30 min at 60°C to kill any yeast which remains after filtration. The pasteurized beer is then dispensed into sanitized bottles. Draft

BOX 20.1

The Hymn to Ninkasi shows that the coproduction of bread and beer predates the 19th century B.C.E.

The Hymn to Ninkasi	Contemporary Process
Borne of the flowing water, Tenderly cared for by the Ninhursag, Borne of the flowing water, Tenderly cared for by the Ninhursag,	**Bread**
You are the one who handles the dough [and] with a big shovel, Mixing in a pit, the bappir with sweet aromatics, Ninkasi, you are the one who handles the dough [and] with a big shovel, Mixing in a pit, the bappir with [date]-honey,	Preparing dough for baking
You are the one who bakes the bappir in the big oven, Puts in order the piles of hulled grains, Ninkasi, you are the one who bakes the bappir in the big oven, Puts in order the piles of hulled grains,	Baking the bread
You are the one who waters the malt set on the ground, The noble dogs keep away even the potentates, Ninkasi, you are the one who waters the malt set on the ground, The noble dogs keep away even the potentates,	**Beer**
You are the one who soaks the malt in a jar, The waves rise, the waves fall. Ninkasi, you are the one who soaks the malt in a jar, The waves rise, the waves fall.	Malting
You are the one who spreads the cooked mash on large reed mats, Coolness overcomes, Ninkasi, you are the one who spreads the cooked mash on large reed mats, Coolness overcomes,	Mashing
You are the one who holds with both hands the great sweet wort, Brewing [it] with honey [and] wine (You the sweet wort to the vessel) Ninkasi, (...) (You the sweet wort to the vessel)	Making the wort
The filtering vat, which makes a pleasant sound, You place appropriately on a large collector vat. Ninkasi, the filtering vat, which makes a pleasant sound, You place appropriately on a large collector vat.	Brewing
When you pour out the filtered beer of the collector vat, It is [like] the onrush of Tigris and Euphrates. Ninkasi, you are the one who pours out the filtered beer of the collector vat,	Filtering
It is [like] the onrush of Tigris and Euphrates.	Bottling

Translation by Miguel Civil

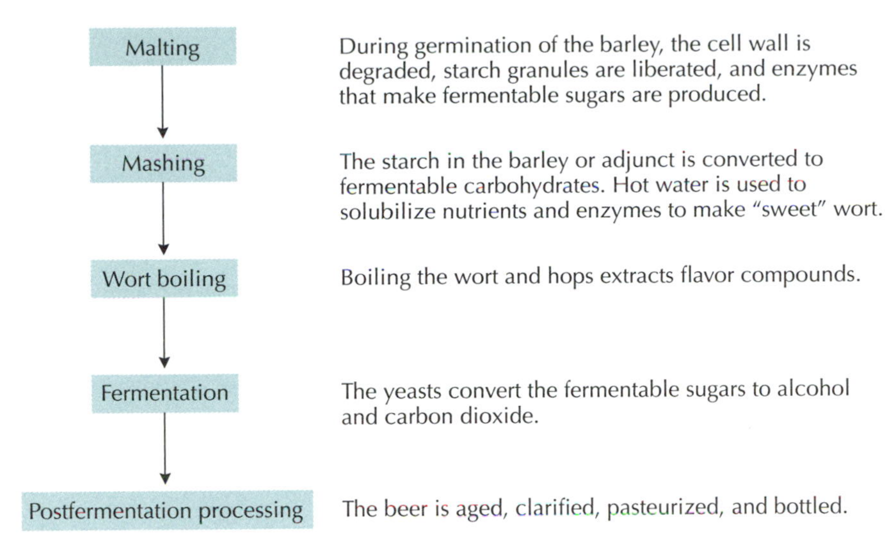

Figure 20.2 Schematic for making beer. Adapted from I. Campbell, p. 735–745, *in* M. P. Doyle, L. R. Beuchat, and T. J. Montville (ed.), *Food Microbiology: Fundamentals and Frontiers*, 2nd ed. (ASM Press, Washington, DC, 2001). doi:10.1128/9781555817206.ch20.f20.02

beer, which is dispensed into kegs, is not heated and must be refrigerated during shipping and storage.

Wine

> *No nation is drunken where wine is cheap; and none sober, where the dearness of wine substitutes ardent spirits as the common beverage. It is, in truth, the only antidote to the bane of whiskey.*
>
> THOMAS JEFFERSON

Wine making also started in ancient times, probably when the natural yeasts contaminated grape juice and fermented it to an alcoholic beverage. It was not until the late 19th century that the French microbiologist Louis Pasteur discovered the role of microbes in this fermentation. He also discovered the role of bacteria in wine spoilage. Pasteurization was invented to kill these bacteria in wine. Only later was it applied to milk. Together, France Italy, Spain, and the United States produce more than 15 million tons of wine per year.

Figure 20.3 outlines the wine-making process. Many important aspects of wine making come before the actual fermentation. The most obvious is the grapes. Different varieties of grape are used for white and red wines. Riesling, chardonnay, and sauvignon blanc are among the grape varieties used for white wines. Merlot, pinot noir, and shiraz can be used to make red wines. The sugar/acid ratios, climate, soil, vine age, and other factors contribute to the flavor of the grape. The flavor of the grape, of course, is transmitted to the wine. The crushing and pretreatments are different for red and white wines. For white wines, the juice is drained away from the skins immediately after the grapes are crushed. The juice is then clarified and sent to the fermentation tank. For red wines, the juice and skins go directly to the fermentor. The skins float, and the juice is pumped over them to extract the purple and red anthocyanins and phenolics that give red

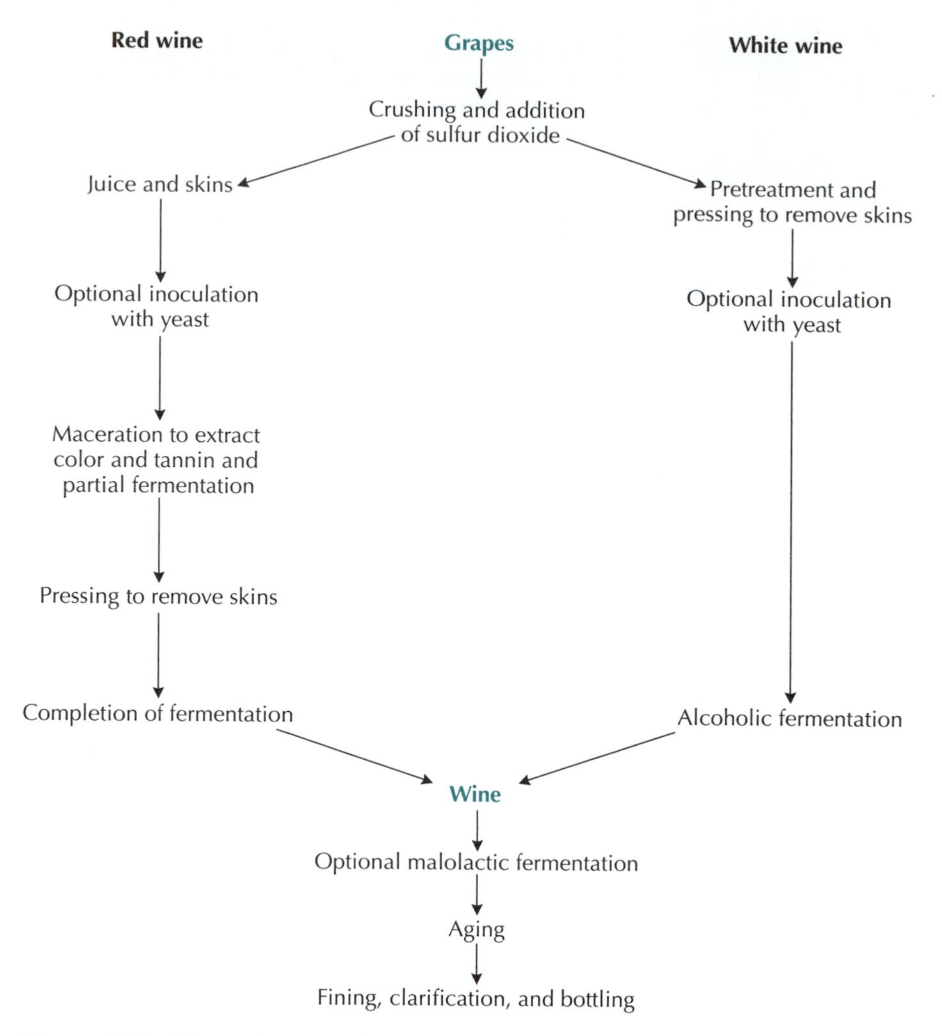

Figure 20.3 Schematic for making wines. Seven hundred eighty-four million gallons of wine was consumed in the United States in 2010. Figure adapted from G. H. Fleet, p. 747–772, *in* M. P. Doyle, L. R. Beuchat, and T. J. Montville (ed.), *Food Microbiology: Fundamentals and Frontiers*, 2nd ed. (ASM Press, Washington, DC, 2001). doi:10.1128/9781555817206.ch20.f20.03

wines their color, astringency, and, possibly, their health benefit. After the fermentation has started, the liquid is pressed through the skins into a fermentation tank.

Grapes can be fermented in barrels, but large stainless steel tanks are more commonly used for mass-produced wines. To inoculate or not to inoculate grape juice with a defined inoculum of yeast is a subject of ongoing debate. As with other fermentations, using a defined inoculum gives a more consistent product and a faster process. *S. cerevisiae* or *Saccharomyces bayanus* is used at levels of 10^6 to 10^7 CFU/g. However, *S. cerevisiae* does not naturally occur at high levels on grapes. If the wine is not inoculated and native yeasts are used, the process is less predictable. The wine can be extraordinarily good or very bad. These native yeasts can come from the grapes or the equipment in the winery. The native grape yeasts that cause wine fermentations are mostly *Kloeckera* and *Hansenia* species, with some *Candida*, *Pichia*, and *Hansenula* organisms. Although

Across

5. Provides fermentable carbohydrate
6. Fermented product that is made in a single oxidation step
7. Virus that attacks bacteria
10. Staple food not normally considered fermented although fermentation is usually involved in making it
11. Used to make beer, wine, and bread
12. Enzymes that break down starch to fermentable carbohydrates
14. Discoverer of the role of bacteria in wine and spoilage
15. Type of bread made without yeast

Down

1. Type of fermentation that decreases acidity of wine
2. Breads made using lactic acid bacteria
3. Type of fermentation that produces only lactic acid
4. Method of killing yeast after the fermentation is complete
7. In days of old, they were connected to bakeries
8. Shorthand for "lactic acid bacteria"
9. Carbon dioxide gas does this during baking
13. Provides flavor and antimicrobial activity in beer

these yeasts occur in low numbers (~10 to 1,000 CFU/g) on immature fruit, the population increases to 10^4 to 10^6 CFU/g as the grapes ripen.

The fermentation of white wines usually takes 1 to 2 weeks at 10 to 18°C. The relatively low temperature retains important volatile flavors. Red wines are fermented for a shorter period (~7 days) at a higher temperature (20 to 30°C) to extract the red color. As with the growth of other microorganisms, yeast growth is influenced by intrinsic and extrinsic factors of the grape juice. These include the pH, oxygen concentration, sugar concentration, temperature, additional ingredients, and other indigenous members of the microbiota. The fermentation ends when glucose and fructose are completely utilized. In addition to ethanol and carbon dioxide, the fermentation can produce glycerol (which smoothens the taste and imparts viscosity) and higher-alcohol esters and aldehydes as flavor compounds. The production of succinic and acetic acids causes wine defects.

Malic acid occurs naturally in some grapes and can give wine a very sour taste. In fact, it is used to manufacture SweeTarts and Jolly Ranchers. Malic acid is a by-product of the tricarboxylic acid cycle and generally decreases as the fruit ripens. Wines with too much malic acid are bitter. The malolactic fermentation is used to decrease the acidity of these wines. This secondary fermentation starts 1 to 3 weeks after the alcohol fermentation is finished. It is carried out by *Oenococcus oeni* (formerly classified as a leuconostoc), which decarboxylates malic acid to lactic acid.

$$\underset{\text{Malic acid}}{HOOC\text{-}C(OH)\text{-}CH_2\text{-}COOH} \rightarrow \underset{\text{Lactic acid}}{HOOC\text{-}C(OH)\text{-}CH_3} + \underset{\text{Carbon dioxide}}{CO_2}$$

In addition to creating a mellower flavor, the malolactic fermentation increases the pH by 0.3 to 0.5 unit.

The postfermentation processes for wines are straightforward. Red wines are stored in oak barrels for 1 to 2 years for flavor development. Some of the flavor comes from the barrels, which may be made with smoked woods, resins, or other flavor-inducing mechanisms. White wines are not usually aged. Potassium sorbate at 100 to 200 μg/ml can be added to red or white wine to control yeast spoilage. The wine is then filtered to clarity and bottled.

VINEGAR FERMENTATION

Men are like wine—some turn to vinegar, but the best improve with age.

POPE JOHN XXIII

Vinegar fermentation is almost as ancient as that of wine. It had to come after the discovery of wine, since vinegar is made from wine (or, these days, from other alcoholic substrates). Vinegar was undoubtedly discovered accidentally when *Gluconobacter* or *Acetobacter* spp. contaminated wine and turned it to vinegar. This turned out to be not such a bad thing, since the vinegar could then be used to preserve other foods.

Vinegar is made by oxidizing ethanol to acetic acid:

$$CH_3CH_2OH + O_2 \rightarrow CH_3COOH + H_2O$$

This occurs in a two-step bioprocess. The first step in vinegar production is the alcoholic fermentation by yeast to produce alcohol. Ethanol from wine or hard cider is frequently used, but vinegar can be made by the fermentation of almost any fruit or starchy material. In the second step, the ethanol is oxidized to acetic acid. Since the *Gluconobacter oxydans* used in making vinegar is a strict aerobe, and the ethanol must be oxidized to acetic acid, oxygen transfer is the rate-limiting step in the process. The oldest processors simply left wine in wooden barrels exposed to the air, inoculated it with *Gluconobacter*, and waited. In several weeks, or perhaps months, the wine became vinegar. The "trickling" fermentor described below radically reduced this time to a day or two by increasing the oxygen transfer rate. The trickling fermentor is a large wooden box with slatted sides to allow oxygenation. The box is filled with wood shavings, on which the *Gluconobacter* grows as biofilms. A rotating sprayer at the top of the chamber sprays the alcoholic liquid on the top of the chips, and it trickles down over the immobilized *Gluconobacter*, being converted to acetic acid in the process. The large surface area of the bacteria on the wood shavings exposed to oxygen makes this method very efficient. Its demise occurred not due to technical limitations but because the craftsmen who could build trickling fermentors retired or died. The trickling fermentor is probably the first bioreactor using immobilized-cell technology. The most modern method of vinegar production is the submerged culture reactor (Fig. 20.4). By pumping oxygen through these very large fermentors, at rates equal to the volume of the fermentor, large quantities of vinegar can be made in a very short time.

Figure 20.4 A large-scale fermentor used in the modern production of vinegar. doi:10.1128/9781555817206.ch20.f20.04

COCOA AND COFFEE FERMENTATIONS

Cocoa

Cocoa? Cocoa! Damn miserable puny stuff, fit for kittens and unwashed boys. Did Shakespeare drink cocoa?

Shirley Jackson

Good cocoa flavor is the result of a complex fermentation involving microbial succession. Cocoa is made from the plant *Theobroma cacao*. It is native to South America and grows within 15° north and south of the equator. The cacao "beans" or seeds are contained in pods about the size of a coconut (Fig. 20.5). The pods are filled with mucilage (a thick gooey substance)

Figure 20.5 A cocoa pod. doi:10.1128/9781555817206.ch20.f20.05

and seeds, much like a pumpkin. Without the fermentation, the cacao seeds are bitter and astringent. To overcome this, the seeds and mucilage are removed from the pods, put into heaps or containers, and fermented for 2 to 8 days. The material is turned periodically to introduce oxygen. The acetic acid and heat generated by the fermentation keep the seeds from germinating.

The mucilage contains sucrose and other sugars at a level of 10 to 12%. The mucilage also contains citric acid, which contributes to a pH of 3.4 to 4.0. The fermentative microbes come from the environment, the plants, and the workers. During the first phase of the fermentation, which lasts 2 or 3 days, the yeasts are the most important microbes. They produce alcohols and aldehydes and create anaerobic conditions that favor the growth of LAB to levels as high as 10^7 CFU/ml. This phase lasts up to 3 days. The LAB produce large amounts of hydrolytic enzymes, invertases, glycosidases, and proteases. The microbial population is primarily homofermentative but contains some heterofermentative LAB. The turning of the fermenting mass introduces air that favors acetic acid bacteria in the final stage of the fermentation. Levels of *Gluconobacter* species and *Acetobacter* species can reach 10^6 CFU/g. The fermentation provides flavor precursors that are converted to chocolate flavors during drying and roasting.

There are two main methods of fermenting the cocoa. In Ghana and other small-scale cocoa-producing countries, the seeds are put into heaps on the ground, covered with banana leaves, and turned by hand (Fig. 20.6). All of this is very labor-intensive but produces very-high-quality seed. These seeds yield the very best chocolate. The second fermentation method, favored on large plantations and estates, uses large boxes instead of heaps to contain the fermenting mass. These boxes range from $1 \times 1 \times 1$ m to $7 \times 5 \times 1$ m. These are frequently grouped or arranged in tiers to make turning

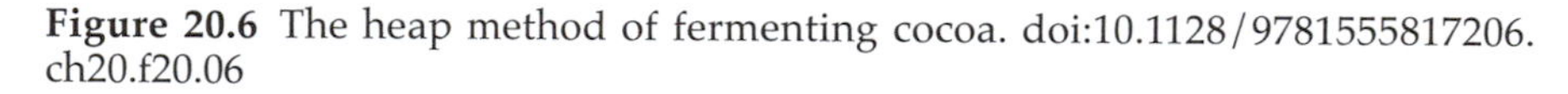

Figure 20.6 The heap method of fermenting cocoa. doi:10.1128/9781555817206.ch20.f20.06

easier. In both methods, a liquid fermentate called "sweats" is produced and is drained away from the beans. This makes drying easier. The larger scale of the box fermentation makes the seed fermentation quicker and cheaper than that of heap fermentation, but the box fermentation yields a lower quality. What chocolate manufacturers would like, of course, is a fermentation that is quick and cheap, like the box fermentation, but that yields a high-quality seed like the heap fermentation. This will require considerable research on the microbial ecology of both systems.

The development of good cocoa flavor is still very much an art. The complexity of the microbial succession prevents the use of starter cultures. The flavors are also dictated by the genetic potential of the plant. While a good fermentation cannot upgrade an inferior product, a bad fermentation can ruin a superior cocoa.

Coffee

Coffee, the finest organic suspension ever devised

CAPTAIN KATHRYN JANEWAY, *Star Trek: Voyager*

Coffee is made from several species of *Coffea*, including *Coffea arabica*, *C. robusta*, and *C. canephora*, which are grown in South America, Central America, Hawaii, Ethiopia, and India. It takes about a year from the time a tree flowers until it bears ripe "coffee cherries." Coffee cherries, which are much smaller than cocoa pods, contain coffee beans surrounded by fleshy mucilage. The mucilage must be removed to liberate the beans. This can be done mechanically, by enzymes, by dehydration, or by fermentation.

The fermentation plays a much less critical role in coffee production than in cocoa production. Its main role is to liberate the beans. The fermentation contributes little to bean flavor or quality. When the beans are fermented, they are first mechanically depulped. The beans, covered with residual mucilage, are submerged in tanks of water. Native yeast, molds, LAB, and gram-negative bacteria ferment the mucilage to water-soluble products that can then be washed away. Since the main component of mucilage is pectin, the fermentation is dominated by a pectinolytic population. During the fermentation, which takes 12 to 60 h, the pH of the beans drops to ~3.7 from the pH of 5.4 to 6.4 characteristic of the native bean. The beans are then subjected to 10 to 25 days of sun drying characterized by an ill-defined microbial succession. The beans can then be roasted and ground.

FERMENTED FOODS OF NON-WESTERN SOCIETIES

There are dozens of fermented foods that play important roles throughout Asia, Africa, and Latin America (Box 20.2). Fermented foods play very important roles in countries that lack refrigeration, energy-intensive methods of food preservation, or the rudimentary technology required for an indigenous food industry. The fermentation may be used for preservation (as in the acidification of kimchi or gari), to improve nutritional value or health (as in the case of tempeh or kefir), or simply to create a pleasurable effect (as in the case of sake).

Authors' note

It is hard to find the right "politically correct" adjective since many "traditional," "indigenous," and "non-Western" societies are more adept in the use of fermented foods than is mainstream American society.

BOX 20.2

Pulque, the enduring drink of the Mexican high desert scrub

Pulque is the nectar of the Aztecs. Its consumption dates back to the Mesoamerican period (the 1500s). It is still consumed by the older generation of campesinos (farmers and farm workers) in a small area of south-central Mexico, but folks of college age prefer beer. If you do want to drink pulque, you have to know where to look for it. It is available in different town markets on different days of the week. You just have to know.

Pulque is an acidic (pH 4), alcoholic (3 to 6%), viscous drink. It is made by indigenous bacteria that ferment the *aguamiel* (sap) of the agave cactus. LAB make the acid, *Zymomonas* (not yeast) makes the alcohol, and *Leuconostoc* generates viscosity. Pulque contains more than 20 species of bacteria. No one knows exactly which species are responsible for the fermentation or are environmental contaminants. Pulque is considered a very healthy drink. It is a good source of vitamin B_1 and may have prebiotic (stimulates the growth of probiotic), probiotic (see chapter 26), and antidiabetic properties. It also contains calories and is a source of liquid, no small thing in the arid Mexican highlands.

Pulque can taste good, but in a way that is hard to describe. The best description I can come up with is acidic, alcoholic, viscous Alka-Seltzer. But I consider this a favorable description. Pulque that is starting to go bad burns the back of the throat and explodes in the stomach. Pulque can completely spoil in a day due to the action of *Acetobacter* spp., putrefaction from the action of proteases, and production of many off-flavors. It was noted in the 1500s that no dead animal smelled as bad as spoiled pulque. Because one never knows if a given bottle of pulque will be good or bad, not many people buy it or drink it. Its "commercialization" is limited to its sale in used 2-liter soda bottles, or by the cup on market day. The absence of defined starter cultures, varying climatic conditions, differences in the agave plants, and the idiosyncratic practices of individual rancheros (farmers) make every batch unique. This lack of consistency limits pulque to an artisanal beverage. One might think (naively, as I did) that all these problems could be solved by the use of starter cultures. However, there are many equally important variables. Where is the agave plant grown? What is the soil like? How old is the plant when it is tapped? Sap production continues for 3 months; when was *this* sap obtained? What is the population of naturally occurring bacteria? What are the time and temperature of the fermentation? Is the fermentation vessel a ceramic pot or a plastic jug? Finally, there is no agreement as to what constitutes "good" pulque.

Extra-credit homework assignment: The year is 2020. You have been hired by a major alcoholic beverage company to commercialize pulque as its next big-hit niche drink. Your boss asks for a one-page outline of your proposed pulque project. Write it and turn it in to your instructor.

Author Thomas Montville and a Mexican colleague drawing samples from an agave plant. doi:10.1128/9781555817206.ch20.fBox20.02

In Table 20.1, these fermentations have been loosely organized according to their main microbiotas. The substrates for these fermentations are usually local crops that are starchy or have some naturally occurring fermentable carbohydrate. In almost all cases, native microbiotas serve as the inocula, although some fermentations, like tempeh and sake, are inoculated.

Table 20.1 Other regionally important food fermentations[a]

Product	Organism(s)	Description	Product
Fermentations based on yeasts and molds			
Soy sauce	*Aspergillus oryzae*, *Aspergillus sojae*, yeast, pediococci	Soy and roasted wheat are fermented at 35–40°C for 2–4 mo with 12–19% salt, pH 4.6–4.8, and pasteurized at 70–80°C.	Flavoring agent
Miso	*A. oryzae* and secondary fermentation of yeasts and pediococci	Rice and/or soybean paste is fermented for 1 wk–1 yr at 25–30°C with 4–14% salt.	Soup base
Sake	Yeast, mold, and LAB	Rice is fermented by a mold called koji to liberate the sugar from the starch. This is then fermented by yeast to produce alcohol. Some types of sake also undergo a lactic acid fermentation.	Alcoholic beverage
Sufu	*Mucor* species, *Rhizopus chinensis*, *Actinomucor elegans*	Curd of soybean milk is pressed, cubed, boiled in brine, and inoculated. Then it is aged for 1–12 mo.	Condiment
Tempeh	Molds and bacteria	Soybeans are soaked and inoculated with *Rhizopus oligosporus*. Incubation is for 24 h at 37°C. Bacteria are sometimes added as a prefermentation to lower pH or produce flavors.	Mold-penetrated and -covered cake yields a high-quality protein that can be deep-fried or used in soup.
Lao-chao	*Rhizopus oryzae*, *R. chinensis*, *Endomycopsis* spp.	Inoculated rice is incubated for 1 or 2 days at room temperature to yield a juicy, sweet, slightly alcoholic product.	Dessert
Oncom	*Neurospora intermedia*, *R. oligosporus*	Deoiled peanut cakes are inoculated and incubated at 25–30°C for 1 or 2 days. The surface becomes covered with colored conidia or spores.	Fermented peanut press cake
Fermentations based on LAB			
Gari	LAB and yeast	Cassava is peeled, washed, and grated and then fermented for 2 days as water is pressed out.	Staple in stews or eaten with cold sweet water
Kimchi	LAB	Natural fermentation of cabbage or radish taproot and garlic. A variety of seasonings can also be added.	Korean staple as sour carbonated vegetable
Poi	Initially lactobacilli and lactococci, followed by yeast	Corns of taro plants are ground, hydrated, and fermented by the native microbiota for 1–3 days.	Eaten as main or side dish
Fermentations based on other bacteria			
Kefir	Bacteria and yeast	Cow or goat milk is inoculated with kefir grains. The kefir grains contain a variety of yeasts and bacteria, proteins, lipids, and carbohydrates. The lactose is fermented overnight at room temperature to yield a sour, carbonated, and slightly alcoholic beverage.	Beverage
Nattō	*Bacillus subtilis*	Whole soybeans are fermented at 25–40°C for 12 h–2 wk.	Eaten as food

[a]Adapted from L. R. Beuchat, p. 701–720, *in* M. P. Doyle, L. R. Beuchat, and T. J. Montville (ed.), *Food Microbiology: Fundamentals and Frontiers*, 2nd ed. (ASM Press, Washington, DC, 2001).

The frequent use of molds and fungi, usually associated with spoilage, may strike you as "unnatural." But it is no more unnatural than the use of LAB, which spoil meat, to initiate dairy fermentations. LAB are important in many indigenous vegetable fermentations in which their acidity imparts a distinctive taste and serves as a preservative (Fig. 20.7). Many traditional fermentations use mixed, often uncharacterized, microbiotas.

Figure 20.7 One of the authors (Thomas Montville) enjoying a typical Korean meal composed largely of fermented foods. doi:10.1128/9781555817206.ch20.f20.07

Summary

- Beer and bread use yeast fermentations to produce carbon dioxide and ethanol.
- The production of beer involves malting to liberate hydrolytic enzymes, mashing to liberate fermentable carbohydrates, the actual fermentation, clarification, pasteurization, and bottling.
- The production of wine is technically simpler than that of beer, since the fermentable sugars are naturally present in the grapes.
- A secondary fermentation by *Oenococcus oeni* can be used to remove malic acid from wines.
- Vinegar production is an oxidative fermentation which has used barrels, trickling wooden boxes, and large fermentors to overcome oxygenation as the rate-limiting step.
- Mucilage is fermented in coffee and cocoa fermentations to liberate the beans for further processing.
- Fermentation is much more important to cocoa quality than to coffee quality.
- Indigenous fermentations use the substrates in local crops to preserve them and/or make new products with improved nutritional characteristics.

Suggested reading

Campbell, I. 2007. Beer, p. 851–862. *In* M. P. Doyle and L. R. Beuchat (ed.), *Food Microbiology: Fundamentals and Frontiers*, 3rd ed. ASM Press, Washington, DC.

Fleet, G. H. 2007. Wine, p. 863–890. *In* M. P. Doyle and L. R. Beuchat (ed.), *Food Microbiology: Fundamentals and Frontiers*, 3rd ed. ASM Press, Washington, DC.

Nout, M. J. R., P. K. Sarkar, and L. R. Beuchat. 2007. Indigenous fermented foods, p. 817–835. *In* M. P. Doyle and L. R. Beuchat (ed.), *Food Microbiology: Fundamentals and Frontiers*, 3rd ed. ASM Press, Washington, DC.

Thompson, S. S., K. B. Miller, and A. S. Lopez. 2007. Cocoa and coffee, p. 837–850. *In* M. P. Doyle and L. R. Beuchat (ed.), *Food Microbiology: Fundamentals and Frontiers*, 3rd ed. ASM Press, Washington, DC.

Questions for critical thought

1. Why are different types of yeast used in the fermentation of bread and beer?
2. Why is mashing required in the production of beer but not wine?
3. What is an adjunct? Why is it used? Compare this to the use of adjunct professors at colleges and universities.
4. Why are fresh, "defined" inocula used to start beer fermentations after using the "prior batch" inoculation method for the 10 to 15 prior batches?
5. Why is refrigerated storage required for draft beer but not bottled beer?
6. Research the differences between beer, ale, and stout. Briefly describe these. (If over 21, which do you prefer? Why?)
7. Trivia question: Benjamin Franklin's out-of-wedlock son, William, shared his father's love of beer and wine. As governor of New Jersey, he signed the charter of which colonial college?
8. What are the pros and cons of using defined inocula in wine fermentations? If you used all of your savings to buy a winery, which would you use?
9. What is the terminal electron acceptor in the fermentation of wine to vinegar?
10. What common role does fermentation play in cocoa and coffee production?
11. Choose one of the indigenous fermentations shown in Table 20.1 and research it to write a one-page report which describes its history, microbiology, and culinary uses.
12. Use dimensional analysis and simplifying assumptions to estimate how many 1-liter bottles of wine can be made from the 15 million tons of wine produced annually.
13. What is the mechanism by which vinegar preserves food?

21 Spoilage Organisms

LEARNING OBJECTIVES

The information in this chapter will enable the student to:

- gain knowledge about microorganisms responsible for spoilage of a wide range of food products
- discuss intrinsic mechanisms that inhibit spoilage of food
- understand the impact of food processing on microbial spoilage
- identify the sources of microorganisms responsible for spoilage of specific products
- discuss procedures that can be implemented to minimize contamination of raw materials

INTRODUCTION

The phone rings, and your friend from out of town asks if she can drop by for a visit. You go so far as to suggest that you will make dinner. You open the refrigerator to gather the necessary ingredients to make a delicious meal. Upon opening a jar of spaghetti sauce, you notice a greenish white fuzzy mass on the surface of the sauce. The luncheon meat that you remove to make antipasto has a green sheen and a strange odor, and it feels slimy. For a salad, you take out lettuce, but it has turned brown, and the peppers have large dark spots. In order to gather your thoughts, you decide to sit down and have a glass of milk. When pouring the milk, you notice that it is lumpy and a bit smelly. You start to realize that you should have paid more attention in the food microbiology course that you attended as an undergraduate.

A product (meat, dairy, fruit, or seafood) is considered spoiled if sensory changes make it unacceptable to the consumer. Products stored under the same conditions spoil at different rates depending on product composition and microflora (Table 21.1). Factors associated with food spoilage include color defects or changes in texture, the development of off-flavors or off-odors, slime, or any other characteristic that makes the food undesirable for consumption. While enzymatic activity within a food contributes to changes during storage, *organoleptically* detectable (i.e., detectable through a change in odor or color) spoilage is generally a result of decomposition and the formation of metabolites resulting from microbial growth. This chapter discusses spoilage of many foods, from meat to dairy products to produce. The types of microorganisms involved, conditions conducive to spoilage, and defects associated with spoilage are covered.

doi:10.1128/9781555817206.ch21

Table 21.1 Typical shelf lives of food products found in many college students' refrigerators[a]

Food product	Typical shelf life	Food product	Typical shelf life
Dairy		**Drinks**	
Milk	7 days	Juice	7–10 days
Butter	1–3 mo	Beer	6 wk
Cream cheese	2 wk	Wine	1 wk, uncorked
Yogurt	7–14 days	**Condiments**	
Eggs		Ketchup	5 mo
Fresh	5 wk	Mustard	8 mo
Liquid egg	3–10 days	Mayonnaise	3 mo
Meat		Hummus	1 wk
Ground beef	2 days	Pickles	6 mo
Ground turkey	2 days	Chocolate syrup	8 mo
Hot dogs	2 wk	Peanut butter	4 mo
Bacon	1 wk	Jelly	5 mo
Chicken nuggets	2 days		

[a]Based on http://www.fda.gov/downloads/Food/ResourcesForYou/HealthEducators/ucm109315.pdf.

MEAT, POULTRY, AND SEAFOOD PRODUCTS

Origin of the Microflora in Meat

The levels of bacteria in muscle tissues of healthy live animals are extremely low. High numbers of bacteria are present on the hide, hair, and hooves of red-meat animals, as well as in the gastrointestinal tract. Microorganisms on the hide include bacteria, such as *Staphylococcus*, *Micrococcus*, and *Pseudomonas* species, and fungi, such as yeasts and molds, which are normally associated with the skin microflora, as well as species contributed by fecal material and soil. The numbers and composition of this microflora are influenced by environmental conditions.

The majority of bacteria on a dressed red-meat carcass originate from the hide. During hide removal, bacteria are carried from the hide onto the underlying tissue with the initial incision. Unlike those of cattle and sheep, the skin of hogs is usually not removed but is scalded and left on the carcass. Recontamination can also occur during dehairing due to the presence of debris in dehairing machines. Contamination can occur if the intestinal tract is pierced or if fecal material is introduced from the rectum during the removal of the abdominal contents. Handling can result in cross-contamination of other carcasses. In addition to the hide and viscera, the processing environment, such as floors, walls, contact surfaces, knives, and workers' hands, can be a source of contamination in red meats.

Origin of the Microflora in Poultry

The skin, feathers, and feet of poultry harbor microorganisms resident on the skin, as well as from litter and feces. Although present on the skin, psychrotrophic bacteria, consisting primarily of *Acinetobacter* and *Moraxella*, are primarily associated with the feathers. Contamination and cross-contamination with fecal material may occur during transportation of birds from growing houses to slaughter facilities and during processing, when the birds are hung and bled. Following processing, carcasses are chilled rapidly to limit the growth of microbes. Slush ice, continuous-immersion, spray, air, and carbon dioxide chilling systems have been utilized or proposed for this purpose. Potable water should be used in these systems, since bacteria present in untreated water can contribute to spoilage.

Origins of Microfloras in Finfish

The numbers and compositions of microfloras on finfish are influenced by the environment from which the fish are taken, the season, and the conditions of harvesting, handling, and processing. Water temperature has a significant influence on the initial number and types of bacteria on the surfaces of the fish. Higher numbers of bacteria are generally present on fish from warm subtropical or tropical waters than on fish from colder waters. Fish taken from temperate waters harbor predominantly psychrotrophic bacteria, while mesophilic bacteria predominate on fish taken from tropical areas. Bacteria from the genera *Acinetobacter, Aeromonas, Cytophaga, Flavobacterium, Moraxella, Pseudomonas, Shewanella,* and *Vibrio* dominate on fish and shellfish taken from temperate waters, while *Bacillus*, coryneforms, and *Micrococcus* frequently predominate on fish taken from subtropical and tropical waters. The initial microflora on fish is influenced by the method of harvesting. Trawled fish generally have higher microbial levels than those that are line caught. In trawling, the dragging of fish and debris along the ocean bottom stirs up mud that contaminates the fish. In addition, the compaction of fish in trawling nets may cause expression of intestinal contents, with subsequent contamination of the fish surface. A delay in chilling fish also enhances the possibility of rapid microbial growth. In some fish in the families Scombridae and Scomberesocidae (tuna, mackerel, and skipjack), this can lead to the generation of toxic levels of histamine by *Morganella* (*Proteus*) *morganii* and related gram-negative bacteria that produce histidine decarboxylase. *Photobacterium phosphoreum* may produce histamine during low-temperature storage of these fish.

Origins of Microfloras in Shellfish

Unlike other crustacean shellfish (lobsters, crabs, or crayfish) that are kept alive until they are heat processed, shrimp die soon after harvesting. Decomposition begins soon after death and involves bacteria on the shrimp surface that originate from the marine environment or from contamination during handling and washing. Molluscan shellfish (oysters, clams, scallops, and mussels) are stationary filter feeders, and thus, their microfloras depend greatly on the quality of the water in which they reside, the quality of the wash water, and other factors. Bacteria, including *Pseudomonas* spp., *Shewanella putrefaciens, Acinetobacter*, and *Moraxella*, that spoil finfish also cause spoilage of shellfish.

Bacterial Attachment to Food Surfaces

Spoilage of meat, poultry, and seafood generally occurs as a result of the growth of bacteria that have colonized muscle surfaces. The first stage in colonization and growth involves the attachment of microbial cells to the muscle surface. Bacterial attachment to muscle surfaces involves two stages. The first is a loose, reversible sorption that may be related to van der Waals forces or other physicochemical factors. One of the factors that influence attachment at this point is the population of bacteria in the water film. The second stage consists of an irreversible attachment to surfaces involving the production of an extracellular polysaccharide layer known as a *glycocalyx*.

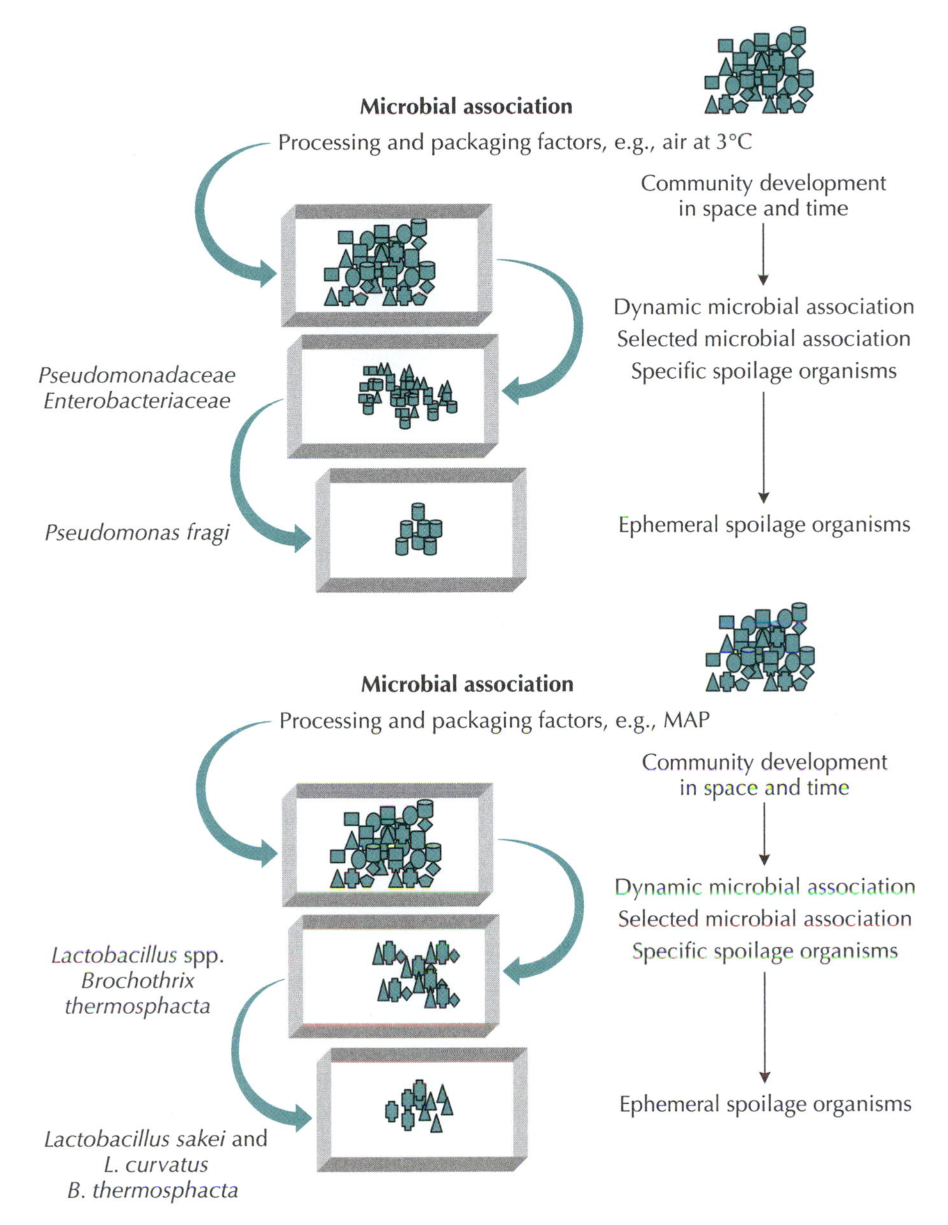

Figure 21.1 Microbial progression in muscle foods under various processing and packaging conditions. Redrawn from G.-J. E. Nychas, D. L. Marshall, and J. N. Sofos, p. 112, *in* M. P. Doyle and L. R. Beuchat (ed.), *Food Microbiology: Fundamentals and Frontiers*, 3rd ed. (ASM Press, Washington, DC, 2007). doi:10.1128/9781555817206.ch21.f21.01

Many factors can influence bacterial attachment, including surface characteristics, growth phase, temperature, and the motility of the bacteria. The development of microbial association with muscle foods is illustrated in Fig. 21.1.

Microbial Progression during Storage

The initial microflora of muscle foods is highly variable. It comes from the microorganisms resident in and on the live animal; environmental sources, such as vegetation, water, and soil; ingredients used in meat products; workers' hands; and contact surfaces in processing facilities. Most perishable meat, poultry, and seafood products are stored at refrigeration temperatures to prolong their shelf life. As microbial growth occurs during storage, the composition of the microflora is altered so that it is dominated by a few, or often a single, microbial species, usually of the genera *Pseudomonas, Lactobacillus, Moraxella,* and *Acinetobacter* or the species *Brochothrix thermosphacta* (Fig. 21.1). *Pseudomonas* species are able to compete successfully on aerobically stored refrigerated muscle foods, since they have a competitive growth rate, but *Moraxella* and *Acinetobacter* species are less capable of competing at refrigeration temperatures and lower pH. The growth of the aerobic spoilage microflora is suppressed during storage under vacuum and modified atmospheres. Under these conditions, lactic acid bacteria are favored because of their growth rate, their fermentative metabolism, and their ability to grow at the pH range of meat. At higher pHs, *B. thermosphacta* and *S. putrefaciens* may grow and contribute to spoilage. The water activity (a_w) of some types of processed meats is lowered by dehydration or the addition of solutes, such as salt or sugar. When low a_w restricts the growth of bacteria, growth of fungi may occur. Microbial growth does not occur on products with a_w values of <0.60.

Muscle Tissue as a Growth Medium

The composition (percent adipose, lean, and carbohydrate) of the meat influences the microbial growth and type of spoilage. Research suggests that spoilage defects in meat become evident when the number of spoilage bacteria on the surface reaches 10^7 colony-forming units (CFU)/cm^2. During aerobic spoilage, off-odors are first detected when levels reach 10^7 CFU/cm^2. When numbers reach 10^8 CFU/cm^2, the muscle tissue surface begins to feel tacky, the first stage in slime formation.

Composition and Spoilage of Red Meats

The a_w of red-meat lean muscle tissue is 0.99, with a corresponding water content of 74 to 80%. The protein content may vary from 15 to 22% on a wet-weight basis. The lipid contents of intact red meats vary from 2.5 to 37%, and the carbohydrate compositions range from 0 to 1.2%. Glycolysis leads to the accumulation of lactic acid, and as a result, the pH of the muscle tissue decreases. The spoilage of meats stored at ambient temperature results from the growth of mesophiles, predominantly *Clostridium perfringens* and members of the family *Enterobacteriaceae*. Spoilage deep within muscle tissue, known as "sours" or "bone taint," has been attributed to a slow cooling of carcasses, resulting in the growth of anaerobic mesophiles thought to be already present in the muscle tissues. If the surfaces of whole carcasses or fresh meat cuts become dry, bacterial growth may be restricted and fungal spoilage may occur. *Thamnidium, Mucor,* and *Rhizopus* may produce a whiskery, airy, or cottony gray to black growth on beef due to the presence of mycelia. Black spot has been attributed to the growth of *Cladosporidium,* white spot has been attributed to the growth of *Sporotrichum* and *Chrysosporium,* and green patches have been attributed to *Penicillium.* Molds do not grow on beef held at temperatures below −5°C.

Composition and Spoilage of Poultry Muscle

The mechanism of microbial spoilage of poultry muscle is similar to that of red meat. Spoilage is generally restricted to the outer surfaces of the skin and cuts and has been characterized by off-odors and sliminess, as well as various types of discoloration. Skin may provide a barrier to the introduction of spoilage microorganisms to the underlying muscle tissue. Although the pH of breast muscle (pH 5.7 to 5.9) differs from that of leg muscle (pH 6.4 to 6.7), the organisms that cause spoilage are similar and include *Pseudomonas*, *Aeromonas*, and *S. putrefaciens*. For poultry carcasses packaged in oxygen-impermeable films, spoilage may be caused by *Shewanella*, *B. thermosphacta*, and atypical lactobacilli. The ability to make sulfide compounds, such as hydrogen sulfide, dimethyl sulfide, and methyl mercaptan, makes *S. putrefaciens* an important component of the spoilage microflora.

Composition and Spoilage of Finfish

The internal muscle tissue of a healthy live fish is generally sterile. Bacteria are present on the outer slime layer of the skin and gill surfaces and in the intestines in the case of feeding fish. The compositions of fish muscle are highly variable among species and may fluctuate widely, depending upon size, season, fishing grounds, and diet. The average composition of nonfatty fish, such as cod, has been characterized as 18% protein and <1% lipid, whereas in fatty fish, such as herring, the lipid content may range from 1 to 30%, with the water content varying so that fat and water constitute ~80% of the muscle tissue. As with other muscle foods, the spoilage microflora of fresh ice-stored fish consists largely of *Pseudomonas* spp. *S. putrefaciens* may also contribute to the spoilage of seafood, and *Acinetobacter* and *Moraxella* may constitute a smaller portion of the spoilage microbes. The spoilage characteristics of fresh fish can be divided into four stages. Stage I occurs from 0 to 6 days after death and involves shifting of bacterial populations without odor. Stage II occurs from 7 to 10 days after death, with bacterial growth becoming apparent and the development of a slightly fishy odor. Stage III occurs from 11 to 14 days after death and is characterized by rapid bacterial growth, a sour and fishy odor, and the start of slime formation on the skin. Stage IV occurs >14 days after death, at which point bacterial numbers are stationary, proteolysis begins, the skin is extremely slimy, and the odor is offensive.

Composition and Spoilage of Shellfish

Crustacean and molluscan shellfish generally contain larger amounts of free amino acids than finfish. Trimethylamine oxide is present in crustacean shellfish, with the exception of cephalopods, scallops, and cockles, and is absent in molluscan tissue. Crustaceans possess potent cathepsin-like enzymes, which rapidly break down proteins, leading to tissue softening and the development of volatile off-odors. Removal of the head after harvest can extend shelf life by eliminating an organ that releases degradative enzymes. Some crustacean meats (primarily shrimp) suffer from a visual defect known as black spot melanosis, which is due to polyphenol oxidase activity and not to microbial action.

Molluscan shellfish contain a lower total nitrogen concentration in their flesh than do finfish or crustacean shellfish and much more carbohydrate, mostly in the form of glycogen. As a result, the spoilage pattern of molluscan shellfish differs from that of other seafoods and is generally fermentative, with the pH of tissues declining as spoilage progresses, yielding a predominance of lactobacilli and streptococci.

Factors Influencing Spoilage

Proteolytic and Lipolytic Activities

Although *Pseudomonas* and other aerobic spoilage bacteria are able to produce proteolytic enzymes, their production is delayed until the late logarithmic phase of growth. Proteolysis occurs only in populations of $>10^8$ CFU/cm^2, when spoilage is well advanced and the bacteria are approaching their maximum cell density.

Oxidative rancidity of fat occurs when unsaturated fatty acids react with oxygen from the storage environment. Stable compounds, such as aldehydes, ketones, and short-chain fatty acids, are produced, resulting in the eventual development of rancid flavors and odors. Autoxidation, independent of microbial activity, occurs in muscle foods stored in aerobic environments. The rate is influenced by the proportion of unsaturated fatty acids in the fat. Generally, lipase production is restricted while carbohydrate substrates in muscle tissue are being utilized; it is unlikely that microbial lipolytic activity would occur until glucose on the muscle surface is depleted. At this point, amino acids would also be degraded, and the resulting spoilage characteristics would possibly mask the effects of rancidity.

Spoilage of Adipose Tissue

Adipose tissue consists predominantly of insoluble fat, which cannot be used for microbial growth until it is broken down and emulsified. While the spoilage processes and growth rates of spoilage bacteria are similar for adipose and muscle tissues, the low level of carbohydrates in adipose tissue means that spoilage odors are detected when lower numbers of bacteria are present and most, if not all, available glucose is depleted. This occurs when populations exceed 10^6 CFU/cm^2. The growth rates of some psychrotrophic microorganisms, such as *Hafnia alvei*, *Serratia liquefaciens*, and *Lactobacillus plantarum*, are higher on fat than on lean beef and pork tissues. Spoilage bacteria, such as *S. putrefaciens*, may also grow. In practice, spoilage of fat before lean tissue is unlikely, given the presence of muscle tissue fluids in vacuum-packaged cuts and the restriction of bacterial growth due to the drying of carcass surfaces.

Spoilage under Anaerobic Conditions

Lactic acid bacteria dominate the spoilage microflora of muscle foods when oxygen is excluded from the storage environment. If the pH of the muscle tissue is high or residual amounts of oxygen are present, other microorganisms, such as *B. thermosphacta* and *S. putrefaciens*, may cause spoilage. The growth rate of bacteria under anaerobic conditions is considerably lower than that under aerobic conditions. In addition, the maximum cell density achieved under anaerobic conditions ($\sim10^8$ CFU/cm^2) is considerably less than that achieved under aerobic conditions ($>10^9$ CFU/cm^2). The sour, acid, cheesy odors or cheesy and dairy flavors that develop in muscle tissue under anaerobic conditions can be attributed, at least in part, to the accumulation of short-chain fatty acids and amines. The presence of *B. thermosphacta* in significant numbers causes more rapid spoilage than that of lactic acid bacteria.

Other Meat Characteristics Associated with Spoilage

Animals subjected to excessive stress or exercise before slaughter can have depleted levels of muscle glycogen. This results in a condition known as dark, firm, and dry. Spoilage of dark, firm, and dry meat occurs more quickly than spoilage of meat with a normal pH. Rapid spoilage is caused by the absence of glucose, and hence lactic acid, in the tissues. Pale, soft, exudative muscle tissue is a condition that occurs in pork and turkey, and to a lesser extent in beef, in which sugar utilization decreases the muscle pH to its ultimate level while the muscle temperature is still high. There is debate over whether pale, soft, exudative meats spoil more slowly than meat with a normal pH.

Comminuted Products

The limited shelf life of comminuted (ground and blended) muscle foods has been attributed to (i) a higher initial microbial load due to use of a lower-quality product for grinding, (ii) contamination during processing, and (iii) the effects of the comminution of the muscle tissue. On the surfaces of aerobically stored comminuted products, *Pseudomonas*, *Acinetobacter*, and *Moraxella* species make up the main microflora, while in the interior, due to the limited availability of oxygen, lactic acid bacteria are dominant. Occasional contaminants, such as *Aeromonas* spp. or members of the family *Enterobacteriaceae*, occur more often on comminuted products than on intact tissue.

Cooked Products

Cooking muscle foods destroys vegetative bacterial cells, although endospores may survive. In perishable, cooked, uncured meats, spoilage is caused by bacteria that survive heat processing or by postprocessing contaminants. Microorganisms responsible for spoilage of these products include psychrotrophic micrococci, streptococci, lactobacilli, and *B. thermosphacta*.

Processed Products

Microbial spoilage of processed products depends on the nature of the product and the ingredients used. Spoilage of processed meats may be characterized as slimy spoilage, souring, or greening. *Slimy spoilage*, which is usually confined to the outside surfaces of product casings, results from the growth of some yeasts; two genera of lactic acid bacteria, *Lactobacillus* and *Enterococcus*; and *B. thermosphacta* (Fig. 21.2). *Souring* occurs, typically beneath casings, when bacteria, such as lactobacilli, enterococci, and *B. thermosphacta*, utilize lactose and other sugars to produce acids. While *greening* of fresh meats may be the result of hydrogen sulfide production by certain bacteria, greening of cured meat products may also develop in the presence of hydrogen peroxide, which may form on the surfaces of vacuum- or modified-atmosphere-packaged meats when they are exposed to air. *Lactobacillus viridescens* is the most common cause of this type of greening, but species of *Streptococcus* and *Leuconostoc* may also produce the defect. The presence of these bacteria is often a result of poor sanitation.

Cured meats are spoiled by microorganisms that tolerate low a_w, such as lactobacilli or micrococci. If sucrose is added to cured meats, a slimy dextran layer may form due to the activity of *Leuconostoc* spp. or other bacteria, such as *L. viridescens*. *B. thermosphacta* may also be involved in the spoilage of these products. If dried meats are properly prepared and stored, their low a_w renders them microbiologically stable. To restrict the growth of some species of microbes that can grow at low a_w, the water content of dried meat products must be sufficiently reduced. Microbial spoilage of

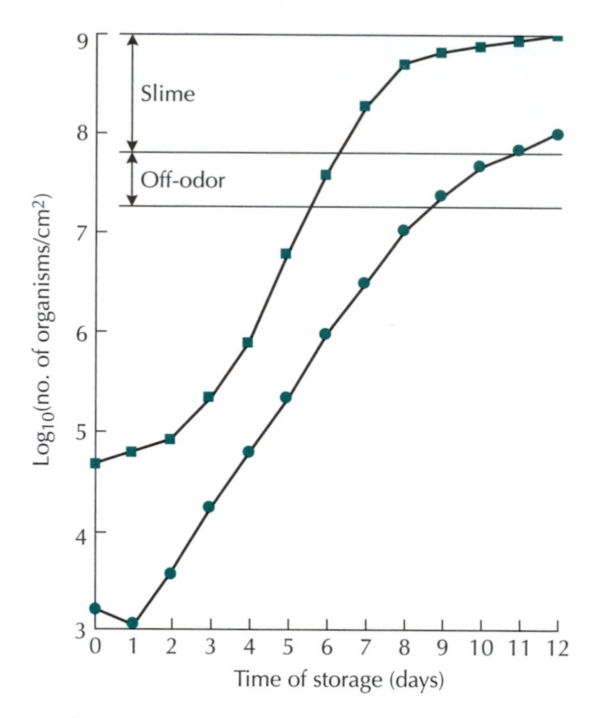

Figure 21.2 Profile for development of slime and off-odor on processed meats stored at 5°C. doi:10.1128/9781555817206.ch21.f21.02

these products should not occur unless exposure to high relative humidity or other high-moisture conditions results in an uptake of moisture. The addition of antifungal agents may retard the growth of microbes in these products.

Control of Spoilage of Muscle Foods

Modifying intrinsic characteristics of products or extrinsic characteristics of the storage environment can control the growth of spoilage microorganisms on muscle foods. The shelf life of processed meats can be extended by processing procedures and ingredients that either prevent the growth of spoilage microorganisms or select for a less offensive spoilage microflora. For fresh meats, the extension of shelf life has become especially important as the food industry moves toward the centralization of processing activities, with products distributed to more distant domestic and international markets. Specific methods to prevent spoilage and prolong the shelf life of muscle foods include effective good manufacturing practices at lower levels for products entering storage and distribution; rinsing by water, immersion, and spray systems to remove physical and microbial contaminants from carcasses; and incorporation of antimicrobial compounds, such as chlorine and organic acids, in wash water. The use of lactic, acetic, propionic, and citric acids has been investigated, although lactic and acetic acids are most often used to reduce microbial levels on carcass surfaces. The effectiveness of organic acids is influenced by factors such as the type of acid, concentration, temperature, and point of application in processing.

Temperature is the most important environmental parameter influencing the growth of microorganisms in muscle foods. As the temperature is decreased below the optimum for the growth of microorganisms, generation

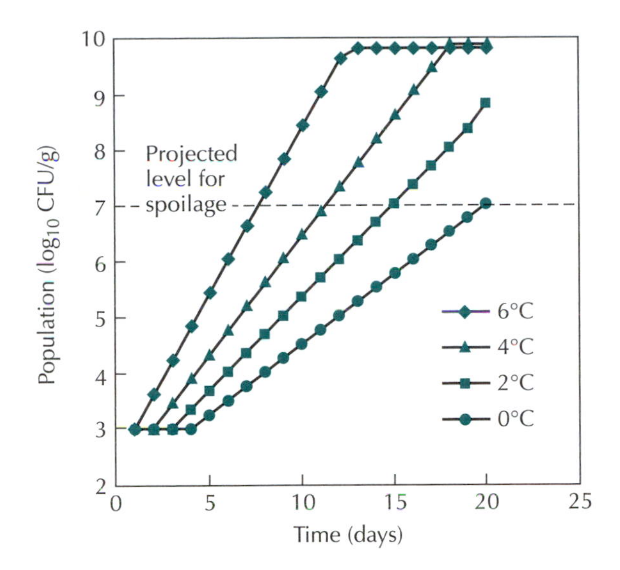

Figure 21.3 Predicted populations of total aerobic bacteria in ground beef as affected by storage temperature. Redrawn from T. C. Jackson, D. L. Marshall, G. R. Acuff, and J. S. Dickson, p. 91–109, *in* M. P. Doyle, L. R. Beuchat, and T. J. Montville (ed.), *Food Microbiology: Fundamentals and Frontiers*, 2nd ed. (ASM Press, Washington, DC, 2001). doi:10.1128/9781555817206.ch21.f21.03

times and lag times are extended, and growth is therefore slowed (Fig. 21.3). Under aerobic storage conditions, the comparatively high growth rate of *Pseudomonas* spp. allows successful competition with mesophiles and other psychrotrophs at temperatures below 20°C. Likewise, under anaerobic conditions at temperatures below 20°C, the rapid growth rate of psychrotrophic lactobacilli allows successful competition with other psychrotrophic spoilage microorganisms. Storage of foods at temperatures at which mesophiles can grow allows these microorganisms to contribute to spoilage. Higher storage temperatures may also allow the growth of psychrotrophic spoilage microorganisms that would otherwise be restricted by the pH or lactic acid concentration of muscle tissue at reduced temperatures. Likewise, the minimum temperature for the growth of psychrotrophic spoilage microorganisms may be increased by other factors that influence growth, e.g., pH, a_w, or oxidation-reduction potential, or by inhibitory agents, such as food additives or increased carbon dioxide.

The shelf life of muscle foods can be extended by storage under vacuum or modified atmospheres. Modified-atmosphere packaging (MAP) involves the storage of products under a high-oxygen barrier film with a headspace containing gas with a composition different from that of air. Typically, this headspace contains elevated amounts of carbon dioxide, and it may also contain nitrogen and oxygen in various proportions. Much of the extended shelf life of such products results from a modification of the spoilage microflora from an aerobic psychrotrophic population, consisting of bacteria such as *Pseudomonas*, *Moraxella*, and *Acinetobacter*, to one consisting predominantly of lactic acid bacteria and *B. thermosphacta* (Fig. 21.4). Carbon dioxide is commonly used in modified-atmosphere environments due to its bacteriostatic activity. Carbon dioxide increases the lag phases and generation times of many microorganisms. Oxygen is occasionally included in modified atmospheres used to package fresh red meats to maintain the

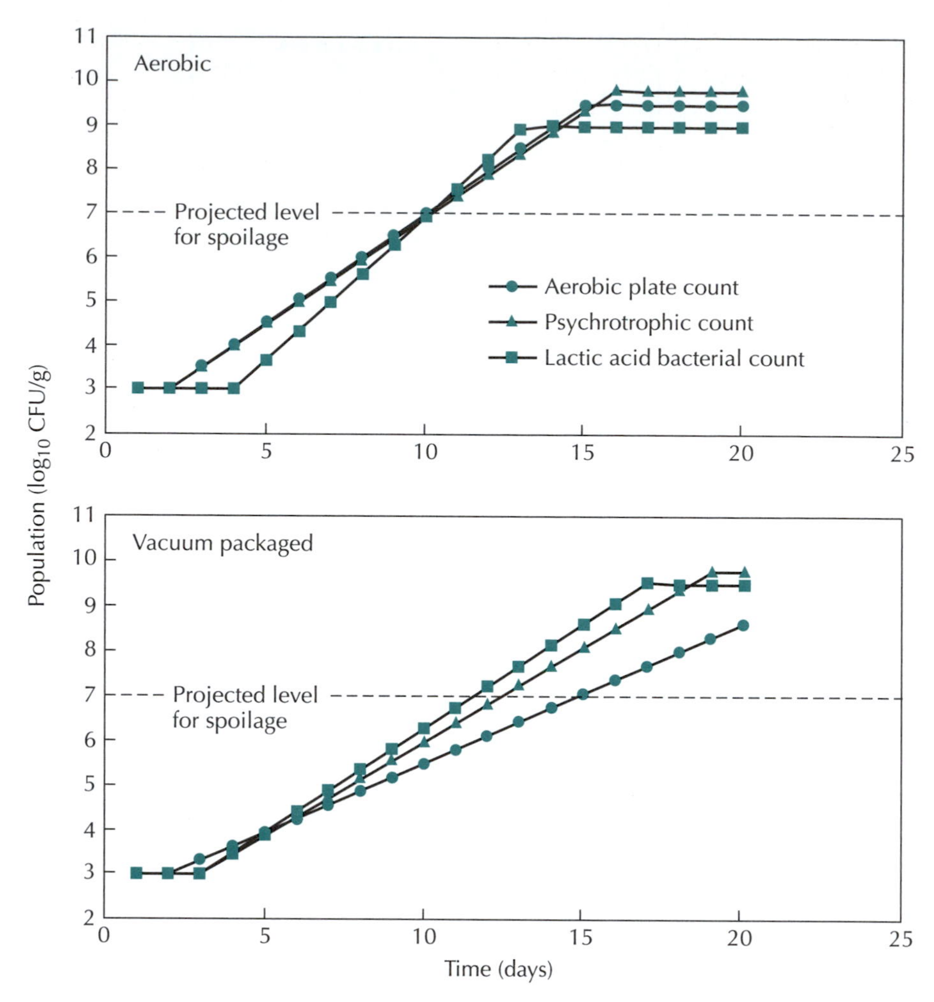

Figure 21.4 Populations of total aerobic, psychrotrophic, and lactic acid bacteria in aerobically packaged and vacuum-packaged ground pork stored at 0°C. The growth curves are based on the derived values for lag phase duration and generation time. Redrawn from T. C. Jackson, D. L. Marshall, G. R. Acuff, and J. S. Dickson, p. 91–109, *in* M. P. Doyle, L. R. Beuchat, and T. J. Montville (ed.), *Food Microbiology: Fundamentals and Frontiers*, 2nd ed. (ASM Press, Washington, DC, 2001). doi:10.1128/9781555817206.ch21.f21.04

oxymyoglobin responsible for the fresh red, or "bloomed," color desired by the consumer. However, if significant amounts of oxygen are present, aerobic microorganisms and *B. thermosphacta* may grow, resulting in spoilage. Nitrogen may also be included to displace oxygen and delay oxidative rancidity and to inhibit the growth of aerobic microorganisms.

Irradiation of meats enhances microbiological safety and quality by significantly reducing populations of pathogenic and spoilage bacteria. At present, relatively low doses of irradiation, i.e., <5 kilograys (kGy), are receiving the most interest. The doses of radiation required to inactivate 90% of the vegetative cells of pathogenic bacteria vary with processing conditions (temperature and presence or absence of air) but typically range from a low of <0.2 kGy for *Aeromonas* and *Campylobacter* to as high as 0.77 kGy for *Listeria* and *Salmonella*. Although there is some variation among species, the doses required to inactivate 90% of the vegetative cells of spoilage bacteria typically fall in the range of 0.2 to 0.8 kGy. In contrast, the dose required for 90% inactivation of *Clostridium* spores is as high as 3.5 kGy. Low-dose irradiation (<5 kGy) is effective in extending the shelf life of fresh meats. Because of significant reductions in both pathogenic and spoilage bacteria, low-dose irradiation has been suggested as a pasteurization process for fresh meats. One of the major drawbacks to irradiation is consumer hesitation to purchase irradiated foods.

BOX 21.1

Is it really spoiled?

Milk is spoiled when it starts to look like yogurt. Yogurt is spoiled when it starts to look like cottage cheese. Cottage cheese is spoiled when it starts to look like regular cheese. Regular cheese is nothing but spoiled milk anyway and cannot get any more spoiled than it is already.

MILK AND DAIRY PRODUCTS

This section focuses on spoilage defects in dairy products that result from microbial growth. Modern dairy processing utilizes pasteurization, heat sterilization, fermentation, dehydration, refrigeration, and freezing as preservation treatments. Combining a preservation method with component separation processes (i.e., churning, filtration, centrifugation, and coagulation) results in an assortment of dairy foods having vastly different tastes and textures and a complex variety of spoilage microfloras. Spoilage of dairy foods is manifested as off-flavors and -odors and changes in texture and appearance (Box 21.1). Some defects of dairy products caused by microorganisms are listed in Tables 21.2 and 21.3.

Milk and Dairy Products as Growth Media

Milk

Milk is a good growth medium for many microorganisms because of its high water content, near-neutral pH, and variety of available nutrients. Although milk is a good growth medium, the addition of yeast extract or protein hydrolysates often increases growth rates, suggesting that milk is not an ideal growth medium. Even though milk has a high fat content, few spoilage microorganisms utilize it as a carbon or energy source. This is because the fat is in the form of globules surrounded by a protective membrane composed of glycoproteins, lipoproteins, and phospholipids. Moreover, many microorganisms cannot utilize lactose and therefore must rely on proteolysis or lipolysis to obtain carbon and energy. Finally, freshly collected raw milk contains various microbial growth inhibitors.

The major nutritional components of milk are lactose, fat, protein, minerals, and various nonprotein nitrogenous compounds. Carbon sources in milk include lactose, protein, and fat. The citrate in milk can be utilized

Table 21.2 Some defects of fluid milk that result from microbial growth[a]

Defect	Associated microorganisms	Type of enzyme	Metabolic product(s)
Bitterness	Psychrotrophic bacteria, *B. cereus*	Protease Peptidase	Bitter peptides
Rancidity	Psychrotrophic bacteria	Lipase	Free fatty acids
Fruity taste	Psychrotrophic bacteria	Esterase	Ethyl esters
Coagulation	*Bacillus* spp.	Protease	Casein destabilization
Sourness	Lactic acid bacteria	Glycolytic	Lactic and acetic acids
Malty taste	Lactic acid bacteria	Oxidase	3-Methylbutanal
Ropiness	Lactic acid bacteria	Polymerase	Exopolysaccharides

[a]Adapted from J. F. Frank, p. 142–155, *in* M. P. Doyle and L. R. Beuchat (ed.), *Food Microbiology: Fundamentals and Frontiers*, 3rd ed. (ASM Press, Washington, DC, 2007).

Table 21.3 Common types of spoilage in dairy products

Spoilage type	Organisms involved	Sign(s) of spoilage
Souring	*Lactobacillus* spp. *Streptococcus* spp.	Sour milk Curd formation
Sweet curding	*Bacillus* spp. *Proteus* spp.	Alkaline pH Curd formation
Gas production	*Clostridium* spp. Coliform bacteria	Explosion of slimy milk
Ropiness	*Alcaligenes* spp. *Klebsiella* spp. *Enterobacter* spp.	Stringy or slimy milk
Red rot	*Serratia marcescens*	Red coloration
Gray rot	*Clostridium* spp.	Gray coloration
Dairy mold	*Penicillium* spp. *Geotrichum* spp.	Moldy appearance

by many microorganisms but is not present in amounts sufficient to support significant growth. Sufficient glucose is present in milk to allow the initiation of growth by some microorganisms, but for fermentative microorganisms to continue growth, they must have the transport system and hydrolytic enzymes for lactose utilization. Other spoilage microorganisms may oxidize lactose to lactobionic acid. The lactose in milk is present in large quantities and is more than sufficient to support extensive microbial growth.

There are primarily two types of proteins in milk, caseins and whey proteins. Caseins are present in the form of highly hydrated micelles and are readily susceptible to proteolysis. Whey proteins (β-lactoglobulin, α-lactalbumin, serum albumin, and immunoglobulins) remain soluble in milk after the precipitation of casein. They are less susceptible to microbial proteolysis than caseins. Milk also contains nonprotein nitrogenous compounds, such as urea, peptides, and amino acids, that are readily available for microbial utilization, but these compounds are present in insufficient quantity to support the extensive growth required for spoilage.

The major microbial inhibitors in raw milk are lactoferrin and the lactoperoxidase system. Natural inhibitors of less importance include lysozyme, specific immunoglobulins, folate, and vitamin B_{12} binding systems. Lactoferrin, a glycoprotein, acts as an antimicrobial agent by binding iron. Psychrotrophic aerobes that commonly spoil refrigerated milk are inhibited by lactoferrin, but the presence of citrate in cow's milk limits its effectiveness, as the citrate competes with lactoferrin for binding the iron. The most effective natural microbial inhibitor in cow's milk is the lactoperoxidase system. Lactoperoxidase catalyzes the oxidation of thiocyanate and the simultaneous reduction of hydrogen peroxide, resulting in the accumulation of hypothiocyanite. Lactic acid bacteria, coliforms, and various pathogens are inhibited by this system.

The minimum heat treatment required for fluid milk to be sold for consumption in the United States is 72°C for 15 s, though most processors use slightly higher temperatures and longer holding times. Pasteurization affects the growth rate of spoilage microorganisms by destroying inhibitor systems. More severe heat treatments affect microbial growth by increasing available nitrogen through protein hydrolysis and by the liberation of inhibitory sulfhydryl compounds. Lactoperoxidase is only partially inactivated by normal pasteurization treatments.

Dairy Products

Dairy products are very different growth environments than fluid milk because they have nutrients removed or concentrated and may have a lower pH or a_w. Yogurt is essentially acidified milk and therefore provides a nutrient-rich low-pH environment. Cheeses are less acidified than yogurt, but they have added salt and less water, resulting in lower a_w. In addition, the solid nature of cheeses limits the mobility of spoilage microorganisms. Liquid milk concentrates, such as evaporated skim milk, do not have sufficiently low a_w to inhibit spoilage and must be canned or refrigerated for preservation. Milk-derived powders have sufficiently low a_w to completely inhibit microbial growth. Butter is a water-in-oil emulsion, so microorganisms are trapped within serum droplets. If butter is salted, the mean salt content of the water droplets is 6 to 8%, sufficient to inhibit gram-negative spoilage organisms that could grow during refrigeration. Unsalted butter is usually made from acidified cream and relies on low pH and refrigeration for preservation.

Psychrotrophic Spoilage

The preservation of fluid milk relies on effective sanitation, pasteurization, timely marketing, and refrigeration. Immediately following collection, raw milk is rapidly cooled and is refrigerated until it is consumed (obviously, during pasteurization the milk is heated). There is often enough time between milk collection and consumption for psychrotrophic bacteria to grow. Many of the flavor defects detected by milk drinkers result from this growth. Pasteurized milk is expected to have a shelf life of 14 to 20 days, so contamination of the contents of a container with even one rapidly growing psychrotrophic microorganism can lead to spoilage.

Psychrotrophic Bacteria in Milk

Psychrotrophic bacteria that spoil raw and pasteurized milk are primarily aerobic gram-negative rods in the family *Pseudomonadaceae*, with occasional representatives from the family *Neisseriaceae* and the genera *Flavobacterium* and *Alcaligenes*. Representatives of other genera, including *Bacillus*, *Micrococcus*, *Aerococcus*, and *Staphylococcus*, and the family *Enterobacteriaceae* may be present in raw milk and may be psychrotrophic, but they are usually outgrown by the gram-negative obligate aerobes (especially *Pseudomonas* spp.) when milk is held at its typical storage temperature of 3 to 7°C. The psychrotrophic spoilage microflora of milk is generally proteolytic, with many isolates able to produce extracellular lipases, phospholipase, and other hydrolytic enzymes but unable to utilize lactose. The bacterium most often associated with flavor defects in refrigerated milk is *Pseudomonas fluorescens*, with *Pseudomonas fragi*, *Pseudomonas putida*, and *Pseudomonas lundensis* also commonly encountered. Psychrotrophic bacteria commonly found in raw milk are inactivated by pasteurization.

Sources of psychrotrophic bacteria in milk. Soil, water, animals, and plant material constitute the natural habitat of psychrotrophic bacteria found in milk. Plant materials may contain $>10^8$ psychrotrophs per g, and farm water usually contains low levels of psychrotrophic microorganisms. Its use to clean and rinse milking equipment provides a direct means for their entry into milk. Psychrotrophic bacteria isolated from water are often very active producers of extracellular enzymes and grow rapidly at refrigeration temperatures. Milking equipment, utensils, and storage tanks are the major source of psychrotrophic contamination of raw milk. Pasteurized milk products become contaminated with psychrotrophic bacteria by exposure to contaminated equipment or air. Although the levels of psychrotrophic bacteria in air are generally quite low, only one viable cell per container can spoil the product.

Growth characteristics and product defects. Generation times (in milk) of the most rapidly growing psychrotrophic *Pseudomonas* spp. isolated from raw milk are 8 to 12 h at 3°C and 5.5 to 10.5 h at 3 to 5°C. These growth rates are sufficient to cause spoilage within 5 days if the milk initially contains only one cell per milliliter. However, most psychrotrophic pseudomonads present in raw milk grow much more slowly, causing refrigerated milk to spoil in 10 to 20 days. Defects of fluid milk associated with the growth of psychrotrophic bacteria are caused by the extracellular enzymes they make. Sufficient enzyme to cause defects is usually present when the population of psychrotrophs reaches 10^6 to 10^7 CFU/ml. Bitter, putrid flavors and coagulation result from proteolysis. Lipolysis leads to the formation of rancid and fruity flavors.

Proteases and their contribution to spoilage. *P. fluorescens* and other psychrotrophs that may be present in raw milk generally produce protease during the late exponential and stationary phases of growth. The temperature for optimum protease production by psychrotrophic *Pseudomonas* spp. is lower than the temperature for optimum growth. Relatively large amounts of protease are produced at temperatures as low as 5°C, and production is inhibited in milk held at 2°C. The effects of calcium and iron ions on protease production by *Pseudomonas* spp. are relevant to dairy spoilage. Ionic calcium is required for protease synthesis. Iron, which may be at a growth-limiting concentration in milk, represses protease production by *Pseudomonas* spp. when added to milk.

The properties of the proteinases produced by *P. fluorescens* that are most relevant to dairy product spoilage include temperature optima from 30 to 45°C with significant activity at 4°C and a pH optimum that is near neutral or alkaline, and all are metalloenzymes containing either Zn^{2+} or Ca^{2+}. Perhaps the most important characteristic of these enzymes in regard to dairy product spoilage is their extreme heat stability: they retain significant activity after ultrahigh-temperature (UHT) milk processing.

Proteases of psychrotrophic bacteria cause product defects either at the time they are produced in the product or as a result of the enzyme surviving a heating process. Degradation of casein in milk by enzymes produced by psychrotrophs results in the liberation of bitter peptides. Bitterness is a common off-flavor in pasteurized milk that has been subject to postpasteurization contamination with psychrotrophic bacteria. Continued proteolysis results in putrid off-flavors associated with low-molecular-weight degradation products, such as ammonia, amines, and sulfides. Bitterness in UHT-processed (commercially sterile) milk develops when sufficient psychrotrophic bacterial growth occurs in raw milk (estimated at 10^5 to

10^7 CFU/ml), leaving behind residual enzymes after heat treatment. Low-level protease activity in UHT-processed milk can also result in coagulation or sediment formation.

Lipases and their contribution to spoilage. Lipases of psychrotrophic pseudomonads, like proteases, are produced in the late logarithmic or stationary phase of growth. As with protease, optimal synthesis of lipase generally occurs below the optimum temperature for growth. Milk is an excellent medium for lipase production by pseudomonads. Ionic calcium is required for lipase activity, as the activity inhibited by ethylenediaminetetraacetic acid (EDTA) is reversed by the addition of Ca^{2+}. Supplementation of milk with iron delays the onset of lipase production by *P. fluorescens* in raw milk.

Temperatures for optimal activity of lipases produced by *Pseudomonas* spp. range from 22 to 70°C, with most between 30 and 40°C. The optimal pH for activity is from 7.0 to 9.0, with most lipases having optima between 7.5 and 8.5. *P. fluorescens* lipase in milk is active at refrigeration temperatures and has significant activity at subfreezing temperatures and low a_w. The heat stability of lipases from psychrotrophic pseudomonads is similar to that of their proteases. Low-temperature inactivation is markedly affected by milk components, with milk salts increasing susceptibility and casein increasing stability.

The triglycerides in raw milk are present in globules that are protected from enzymatic degradation by a membrane. Milk becomes susceptible to lipolysis if this membrane is disrupted by excessive shear force (from pumping, agitation, freezing, etc.). Raw milk contains components that degrade the fat globule membrane and cause rancidity in raw and pasteurized milk. This process is independent of microorganisms. The rancid flavor and odor resulting from lipase action are usually due to the liberation of C_4 to C_8 fatty acids. Fatty acids of higher molecular weight produce a "soapy" flavor. Low levels of unsaturated fatty acids liberated by enzymatic activity may be oxidized to ketones and aldehydes to produce an oxidized, or "cardboardy," off-flavor. *P. fragi* produces a fruity off-flavor in milk by esterifying free fatty acids with ethanol. Lipolytic off-flavors in UHT-processed products generally take several weeks to months to develop due to the small amounts of lipase present or appear in products made from raw milk with excessively high populations of psychrotrophs (>10^6 CFU/ml).

A rancid defect in butter may result from the growth of lipolytic microorganisms, residual heat-stable microbial lipase, or milk lipase activity in the raw milk. The typical odor of rancid butter is associated with low-molecular-weight fatty acids (C_4 to C_8). Microbial lipases in butter are active even if the product is stored at −10°C. Growth of psychrotrophic bacteria in butter occurs only if the product is made from sweet rather than ripened (sour) cream. Sweet cream butter is preserved by salt and refrigeration. Butter is a water-in-fat emulsion, so moisture and salt do not equilibrate during storage. If salt and moisture are not evenly distributed in the product during manufacture, then lipolytic psychrotrophs will grow in pockets of high a_w present in the product.

Cheese is more susceptible to defects caused by bacterial lipases than to those caused by proteases. This is because most proteases are concentrated along with the fat in the curd. The acidic environment of most cheeses limits, but may not eliminate, lipase activity. Some cheeses, such as Camembert and Brie, increase in pH to near neutrality during ripening. More acidic cheeses, e.g., Cheddar, are susceptible if cured for several months or if large amounts of lipase are present. Psychrotrophic bacteria do not grow in cured cheeses because of the low pH and salt content, but they may grow in high-moisture fresh products, such as cottage cheese.

Control of Product Defects Associated with Psychrotrophic Bacteria

Preventing defects caused by psychrotrophic bacteria in raw milk involves limiting contamination levels, rapid cooling immediately after milking, and maintenance of cold storage temperatures. Rapid cooling of milk after collection is important. Fresh milk from the cow enters the farm storage tank (e.g., a bulk tank) at 30 to 37°C. Sanitary standards in the United States require raw milk to be cooled to 7°C within 2 h after milking, but most farm systems achieve rapid cooling to <4°C. Since milk is often picked up from the farm every 48 h, three additional milkings will be added to the previously collected milk. The second milking should not warm the previously collected milk to >10°C.

Preventing contamination of pasteurized dairy products with psychrotrophic bacteria is primarily a matter of equipment cleaning and sanitation, although airborne psychrotrophs may also limit product shelf life. Even when filling equipment is effectively cleaned and sanitized, it can still become a source of psychrotrophic microorganisms, which accumulate during the normal hours of continuous use.

Spoilage by Fermentative Nonsporeformers

Fluid milk, cheese, and cultured milks are the major dairy products susceptible to spoilage by non-spore-forming fermentative bacteria. Non-spore-forming bacteria responsible for fermentative spoilage of dairy products are mostly in either the lactic-acid-producing or coliform group. The genera of lactic acid bacteria involved in spoilage of milk and fermented products include *Lactococcus, Lactobacillus, Leuconostoc, Enterococcus, Pediococcus,* and *Streptococcus.* Coliforms can spoil milk, but this is seldom a problem, since either the lactic acid or psychrotrophic bacteria usually outgrow them. Members of the genera *Enterobacter* and *Klebsiella* are most often associated with coliform spoilage, while *Escherichia* spp. only occasionally exhibit sufficient growth to produce a defect. Lactic-acid-producing bacteria are normal inhabitants of the skin of the cow's teat and the environment. Coliform bacteria are present on udder skin and are also associated with milk residue buildup on inadequately cleaned milking equipment.

Defects of Fluid Milk Products

The most common fermentative defect in fluid milk products is souring. This is caused by the growth of lactic acid bacteria. Lactic acid by itself has a clean, pleasant acid flavor and no odor. The unpleasant sour odor and taste of spoiled milk result from small amounts of acetic and propionic acids. Sour odor can be detected before a noticeable acid flavor develops. Other defects may occur in combination with acid production. A malty flavor results from growth of *Lactococcus lactis* subsp. *lactis* bv. maltigenes. Malty flavor is primarily due to 3-methylbutanal.

Another defect associated with the growth of lactic acid bacteria in milk is "ropy" texture. Most dairy-associated species of lactic acid bacteria produce exocellular polymers that increase the viscosity of milk, causing the ropy defect. The polymer is a polysaccharide containing glucose and galactose with small amounts of mannose, rhamnose, and pentose. Some polymer-producing strains are used to produce high-viscosity fermented products, such as yogurt and Scandinavian ropy milk.

Defects in Cheese

Some strains of lactic acid bacteria produce flavor and appearance defects in cheese. Lactobacilli are a normal part of the dominant microflora of aged Cheddar cheese. If heterofermentative lactobacilli predominate, the cheese is prone to develop an open texture or fissures, a result of gas production during aging. Gassy defects in aged Cheddar cheese are more often associated with the growth of lactobacilli than with the growth of coliforms, yeasts, or sporeformers. *Lactobacillus brevis* and *Lactobacillus casei* subsp. *pseudoplantarum* have been associated with gas production in retail mozzarella cheese. *Lactobacillus casei* subsp. *casei* produces a soft-body defect in mozzarella cheese. Some cheese varieties occasionally exhibit a pink discoloration, which is due to the growth of pigmented strains of propionibacteria and certain strains of *Lactobacillus delbrueckii* subsp. *bulgaricus*. A fruity off-flavor in Cheddar cheese is usually a result of the growth of lactic acid bacteria (usually *Lactococcus* spp.) that produce esterase. The major esters contributing to fruity flavor in cheese are ethyl hexanoate and ethyl butyrate.

Growth of coliform bacteria usually occurs during the cheese-manufacturing process or during the first few days of storage and is therefore referred to as early gas (or early blowing) defect. In hard cheeses, such as Cheddar, this defect occurs when slow lactic acid fermentation fails to rapidly lower the pH or when highly contaminated raw milk is used. Soft, mold-ripened cheeses, such as Camembert, increase in pH during ripening, with a resulting susceptibility to coliform growth. Coliform growth in retail cheese often manifests as swelling of the plastic package. Approximately 10^7 CFU of coliform bacteria/g are needed to produce a gassy defect in the product.

Control of Defects Caused by Lactic Acid and Coliform Bacteria

Defects in fluid milk caused by coliforms and lactic acid bacteria are controlled by good sanitation practices during milking, maintaining raw milk at temperatures below 7°C, pasteurization, and refrigeration of pasteurized products. These microorganisms seldom grow to significant levels in refrigerated pasteurized milk because of their low growth rates compared to those of psychrotrophic bacteria. Control of coliform growth in cheese is achieved by using pasteurized milk, encouraging rapid fermentation of lactose, and good sanitation during manufacture. Controlling defects produced by undesirable lactic acid bacteria in cheese and fermented milks is more difficult, since the growth of lactic acid bacteria must be encouraged during manufacture and the final products often provide suitable growth environments. Undesirable strains of lactic acid bacteria are readily isolated from the manufacturing environment, so their control requires attention to plant cleanliness and protecting the product during manufacture.

Spore-Forming Bacteria

Spoilage by spore-forming bacteria can occur in low-acid fluid milk products that are preserved by substerilization heat treatments and packaged with little chance for recontamination with vegetative cells. Products in this category include aseptically packaged milk and cream and sweetened and unsweetened concentrated canned milks. Nonaseptic packaged refrigerated fluid milk may spoil due to growth of psychrotrophic *Bacillus cereus* and *Bacillus polymyxa* in the absence of more rapidly growing gram-negative psychrotrophs. Hard cheeses, especially those with low interior salt concentrations, are also susceptible to spoilage by spore-forming bacteria. Spore-forming bacteria that spoil dairy products usually come from the raw milk. The spore-forming bacteria in raw milk are predominantly *Bacillus* spp., with *Bacillus licheniformis*, *B. cereus*, *Bacillus subtilis*, and *Bacillus megaterium* most commonly isolated. *Clostridium* spp. are present in raw milk at such low levels that enrichment and most-probable-number techniques must be used for their quantification. Populations of spore-forming bacteria in raw milk vary seasonally; in temperate climates, *Bacillus* and *Clostridium* spp. are at higher levels in raw milk collected in the winter than in the summer.

Defects in Fluid Milk Products

Pasteurized milk can spoil due to the growth of psychrotrophic *B. cereus*. Germination of spores in raw milk occurs soon after pasteurization, indicating that they were heat activated. The defect produced by subsequent growth is described as *sweet curdling*, since it first appears as coagulation without significant acid or off-flavor being formed. Coagulation is caused by a chymosin-like protease. Eventually, the enzyme degrades casein sufficiently to produce a bitter-tasting product. Growth may become visible as "buttons" at the bottom of the carton; these are actually bacterial colonies. Psychrotrophic *B. cereus* also produces phospholipase C (lecithinase), which degrades the fat globule membrane, resulting in the aggregation of the fat in cream. The result is described as *bitty cream defect*. Psychrotrophic *Bacillus* spp. other than *B. cereus*, including *Bacillus circulans* and *Bacillus mycoides*, are also capable of spoiling heat-treated milk. The major heat-resistant species in milk is *Bacillus stearothermophilus*. Other, less heat-resistant *Bacillus* spp. have been isolated from UHT-processed milk, especially *B. subtilis* and *B. megaterium*.

Defects in canned condensed milk. Canned condensed milk may be either sweetened with sucrose and glucose to lower the a_w or left unsweetened. The unsweetened product must be sterilized by heat treatment. *Sweet coagulation* is caused by growth of *Bacillus coagulans*, *B. stearothermophilus*, or *B. cereus*. This defect is similar to the sweet-curdling defect caused by psychrotrophic *B. cereus* in pasteurized milk. Protein destruction, in addition to curdling, can also occur and is usually caused by the growth of *B. subtilis* or *B. licheniformis*. Swelling or bursting of cans can be caused by the growth of *Clostridium sporogenes*. *Flat sour defect* (acidification without gas production) can result from the growth of *B. stearothermophilus*, *B. licheniformis*, *B. coagulans*, *Bacillus macerans*, and *B. subtilis*. Sweetened condensed milk should have sufficiently low a_w to inhibit bacterial spore germination. However, if the a_w is not well controlled, *Bacillus* spp. may produce acid or acid-proteolytic spoilage. Methods for controlling the growth of spore-formers in fluid products mainly involve the use of appropriate heat treatments, including UHT treatments.

Defects in Cheese

The major defect in cheese caused by spore-forming bacteria is gas formation, usually resulting from the growth of *Clostridium tyrobutyricum* and, occasionally, *C. sporogenes* and *Clostridium butyricum*. This defect is often called late blowing or late gas, because it occurs after the cheese has aged for several weeks. Emmental, Swiss, Gouda, and Edam cheeses are most often affected because of their relatively high pH and moisture content, in addition to their low interior salt levels. The defect can also occur in Cheddar and Italian cheeses. Processed cheeses are susceptible to late blowing because spores are not inactivated during heat processing. Late-gas defect results from the fermentation of lactate to butyric acid, acetic acid, carbon dioxide, and hydrogen gas.

Yeasts and Molds

Growth of yeasts and molds is a common cause of spoilage of fermented dairy products, because these microorganisms grow well at low pH. Yeast spoilage is manifested as fruity or yeasty odor and/or gas formation. Hard (or cured) cheeses, when properly made, have very small amounts of lactose, thus limiting the potential for yeast growth. Cultured milks, such as yogurt and buttermilk, and fresh cheeses, such as cottage cheese, normally contain fermentable levels of lactose. They are prone to yeast spoilage. A fermented or yeasty flavor observed in Cheddar cheese spoiled by growth of a *Candida* sp. is associated with elevated levels of ethanol, ethyl acetate, and ethyl butyrate. The affected cheese has a high moisture content (associated with low starter activity and therefore high residual lactose) and a low salt content, which contribute to yeast growth. Yeast spoilage can also occur in dairy foods with low a_w, such as sweetened condensed milk and butter. The most common yeasts present in dairy products are *Kluyveromyces marxianus* and *Debaryomyces hansenii* and their counterparts *Candida famata*, *Candida kefyr*, and other *Candida* spp. Also prevalent are *Rhodotorula mucilaginosa*, *Yarrowia lipolytica*, and *Torulospora* and *Pichia* spp. *Candida* yeasts have been isolated from vacuum-packaged cheese. Mold spores do not survive pasteurization. Fermented dairy products provide a highly specialized ecological niche for yeasts, selecting for those that can utilize lactose or lactic acid and that tolerate high salt concentrations. Yeasts able to produce proteolytic or lipolytic enzymes may also have a selective advantage for growth in dairy products.

The most common molds found on cheese are *Penicillium* spp. Others, such as *Aspergillus*, *Alternaria*, *Mucor*, *Fusarium*, *Cladosporium*, *Geotrichum*, and *Hormodendrum*, are occasionally found. Mold species commonly isolated from processed cheese include *Penicillium roqueforti*, *Penicillium cyclopium*, *Penicillium viridicatum*, and *Penicillium crustosum*. Vacuum-packaged Cheddar cheese supports the growth of *Cladosporium cladosporioides*, *Penicillium commune*, *Cladosporium herbarum*, *Penicillium glabrum*, and *Phoma* spp. Antimycotic chemicals, such as sorbate, propionate, and natamycin (pimaricin), are used to control mold growth. Some *Penicillium* spp. not only are resistant to sorbate but also can degrade it by decarboxylation, producing 1,3-pentadiene. This imparts a kerosene-like odor to the cheese.

SPOILAGE OF PRODUCE AND GRAINS

This section focuses on changes in color, flavor, texture, or aroma brought about by the growth of microorganisms on fruits, vegetables, and grains. For the purpose of this chapter, plant pathology as it relates to the degradation of plant materials is restricted to spoilage problems that arise before harvest, whereas food microbiology deals with spoilage after harvest.

Although according to scientific definitions, there are differences between fruits and vegetables, some confusion does occasionally occur, especially in the minds of consumers. Fruits are defined as the seed-bearing organs of plants and include not only well-known commodities such as apples, citrus fruits, and berries but also items sometimes thought of as vegetables, such as tomatoes, bell peppers, and cucumbers. In contrast, vegetables are defined as all other edible portions of plants, including leaves, roots, and seeds.

Types of Spoilage

In general, three broad types of spoilage exist in plant products. The first type is active spoilage, caused by plant-pathogenic microorganisms actually initiating infection of otherwise healthy and uncompromised products. This reduces sensory quality. A second type of spoilage is passive, or wound-induced, spoilage, in which opportunistic microorganisms gain access to internal tissues via damaged epidermal tissue, i.e., peels or skins. This type of spoilage often occurs soon after the product has been damaged by harvesting, processing equipment, or insects. Similarly, passive spoilage can occur when opportunistic spoilage microorganisms gain entry into internal tissues via lesions caused by plant pathogens or via natural openings, such as stomata.

Spoilage of plant products can be manifested in a variety of ways, depending on the specific product, the environment, and the microorganisms involved. Traditionally, spoilage has been described by the symptoms most often associated with a particular product. Listed in Table 21.4 are types and levels of microorganisms associated with spoilage of fruits and vegetables. Caution should be exercised in using this system, since more than one type of microorganism may produce identical or similar symptoms. For example, the best-known type of spoilage in vegetables is

Table 21.4 Microbiological profiles of selected fresh fruits and vegetables[a]

	Range of microbial load (log CFU/g)			
Commodity	**Psychrotrophic bacteria**	**Yeasts and molds**	**Lactic acid bacteria**	***Enterobacteriaceae***
Freshly cut vegetables	4.3–8.9	2.0–7.8	<1.0–8.5	<1.0–8.0
Arugula	5.7–8.2	5.7–6.1	3.0–5.9	4.3–5.9
Carrot	6.6–8.9	4.7–7.2	4.3–7.6	4.5–7.2
Endive	4.3–7.1	2.0–5.5	<1.0–4.6	3.2–6.2
Lettuce	4.9–7.8	2.9–6.4	1.7–6.3	<1.0–7.1
Spinach	6.1–8.1	4.0–6.0	3.7–6.9	4.7–8.0
Mixed salads	5.2–8.5	3.8–7.8	<1.0–8.5	2.7–7.9
Freshly cut fruits	1.7–7.1	1.7–4.9	1.7–4.8	1.7–4.8
Sprouts	6.3–8.9	2.8–7.6	3.4–7.5	6.3–8.1
Whole vegetables	3.0–7.8	2.2–6.1	<1.0–3.3	<1.0–6.0
Iceberg lettuce	3.2–5.9	2.2–4.5	<1.0–1.2	<1.0–3.3
Lettuce hearts	3.0–5.2	2.6–4.6	<1.0–2.0	1.8–3.6
Romaine lettuce	5.3–6.5	3.7–5.2	<1.0–1.9	<1.0–3.0
Endive	6.2–7.2	4.3–6.1	1.7–2.7	3.9–6.0

[a]Adapted from M. Abadias, J. Usall, M. Anguera, C. Solsona, and I. Viñas, *Int. J. Food Microbiol.* **123:**121–129, 2008.

soft rot, which is usually evidenced by obvious softening of the plant tissue. The disease can be caused by various plant-pathogenic bacteria, most notably *Erwinia carotovora*; however, other bacteria, such as *Bacillus* and *Clostridium* spp., and yeasts and molds are also occasionally implicated in soft-rot spoilage. Types of spoilage are also known by the names of the microorganisms that cause them. For example, *Fusarium* rot and *Rhizopus* soft rot describe not only the symptoms of the spoilage but also the mold that causes them.

Mechanisms of Spoilage

Intact healthy plant cells possess a variety of defense mechanisms to resist microbial invasion. Thus, before microbial spoilage can occur, these defense mechanisms must be overcome. Fruits, vegetables, grains, and legumes have an epidermal layer of cells, i.e., skin, peel, or testa, that provides a protective barrier against the infection of internal tissues. The compositions of the epidermal tissues vary, but the walls of cells in the tissue usually consist of cellulose and pectic materials, and the outermost cells are covered by a layer of waxes (*cutin*). Damage due to relatively uncontrollable factors, such as insect infestation, windblown sand, or rubbing against neighboring surfaces, can occur before harvest. Postharvest damage is often the result of poorly designed or maintained processing equipment. Some microorganisms, especially plant-pathogenic molds, possess mechanisms to penetrate external tissues of plants. Once external barriers are penetrated, not only do these microorganisms quickly invade internal tissues, but also other opportunistic microorganisms often take advantage of the availability of nutrients present in damaged tissue and cause additional spoilage. Once microorganisms penetrate the outer tissues of fruits, vegetables, or grains, they still must gain access to internal areas and individual cells to extract nutrients. Plant tissues are held together by the middle lamella, which is composed primarily of pectic substances. The cell wall consists of the primary layer, composed of cellulose and pectates, and the secondary layer, which consists almost entirely of cellulose.

Degradative enzymes play an important role in the postharvest spoilage of plant products. Five classes of microbial enzymes are primarily responsible for the degradation of plant materials. These are pectinases, cellulases, proteases, phosphatidases, and dehydrogenases. Because pectin and cellulose constitute the main structural components of plant cells, pectinases and cellulases are the most important degradative enzymes involved in spoilage. Pectinases are enzymes that cause depolymerization of the pectin chain. Although pectinases are often discussed as a single enzyme, three main types are recognized for their roles in plant spoilage. Pectinases are produced by several plant pathogens, including *Botrytis cinerea*, *Monilinia fructicola*, *Penicillium citrinum*, and *E. carotovora*. Some pectinases are chain splitting and reduce the overall length of the pectin chain. The ultimate degradation of the pectin chain results in liquefaction of the pectin and complete breakdown of plant tissues. Cellulases are the second major class of degradative enzymes that can lead to spoilage. Cellulases function by degrading cellulose (glucose polymer) to glucose. Like pectinases, several types of cellulases exist, some of which attack native cellulose by cleaving cross-linkages between chains, while others act by breaking the cellulose into shorter chains. Cellulases are of less importance in postharvest spoilage of plant products than pectinases.

Influence of Physiological State

The physiological state of plant products, especially those consisting of fruits or vegetables, can have a dramatic effect on susceptibility to microbiological spoilage. Fruits, vegetables, and grains usually possess some sort of defense mechanism to resist infection by microorganisms. Usually these mechanisms are most effective when the plant is at peak physiological health. Once plant tissues begin to age or are in a suboptimal physiological state, resistance to infection diminishes. Fruits and vegetables differ in the ways in which they change physiologically after detachment from the plant. Nonclimacteric (i.e., no longer exhibiting a burst in respiration rate) fruits and vegetables, such as strawberries, beans, and lettuce, cease to ripen once they have been harvested. In contrast, climacteric fruits and vegetables, such as bananas and tomatoes, continue to mature and ripen after harvest. Ripening can continue to the point where normal cell integrity begins to diminish and tissues deteriorate. This process, known as *senescence*, arises from the accumulation of degradative enzymes produced by the fruit or vegetable and is unrelated to microbial decay. However, the loss of cellular integrity brought about by senescence makes fruits and vegetables even more susceptible to microbial infection and spoilage. Consequently, climacteric fruits and vegetables are usually among the most perishable of plant products.

Microbiological Spoilage of Vegetables

Although vegetables, fruits, grains, and legumes are all plant-derived products, they possess inherent differences that influence both the natural microflora and the type of spoilage encountered. Important intrinsic factors which influence the microfloras that develop on plant products include the pH and a_w. In general, the natural microfloras of vegetables include bacteria, yeasts, and molds representing many genera. Examples of the wide variety of microorganisms associated with fruit and vegetable products are shown in Table 21.4. However, microfloras can vary considerably, depending on the type of vegetable, environmental considerations, seasonality, and whether the vegetables were grown in close proximity to the soil. Both gram-positive and gram-negative bacteria are normally present on vegetables at the time of harvest.

Microfloras of Organically Grown Vegetables

The annual sales of organic foods have increased dramatically in the past decade in the United States. The microbial quality of organically grown vegetables has come under increased scrutiny in recent years, since such products have been linked to a number of outbreaks of foodborne illness. Aside from the human safety factor, levels of microbes associated with a product can decrease shelf life. The methods used in the production of organic vegetables are thought to have a negative effect on the microbial load of the commodity. Although a wide range of microbes are found to be associated with organically grown vegetables, the numbers are similar to those associated with conventionally grown vegetables (Table 21.5). Therefore, the types of microbial spoilage and shelf life are similar for organically and conventionally grown vegetables.

Spoilage Microfloras

A variety of microorganisms can cause spoilage of vegetables. Yeasts, molds, and bacteria cause spoilage of vegetables; however, bacteria are more frequently isolated from initial spoilage defects. The reason for this is that bacteria are

Table 21.5 Levels of mesophilic and psychrotrophic bacteria, yeasts, molds, lactic acid bacteria, and coliforms on vegetables from conventional and organic growers[a]

Study and type of farm	Coliforms	*Escherichia coli*	Yeasts	Molds	Psychrotrophic bacteria	Mesophilic bacteria	Heterofermentative	Homofermentative
Study 1[b]								
Conventional	2.75	ND[e]	5.18	3.54	5.84	5.76	5.10	4.80
Organic	2.97	ND	5.19	3.54	5.85	5.78	5.22	4.79
Study 2[c]								
Conventional	1.5	2.0	ND	ND	ND	ND	ND	ND
Organic, certified	2.3	2.3	ND	ND	ND	ND	ND	ND
Organic, non-certified	2.3	2.4	ND	ND	ND	ND	ND	ND
Study 3[d]								
Conventional	2.9							
Organic, certified	2.9							
Organic, non-certified	2.9							

[a]Values are in log CFU/gram.
[b]C. A. Phillips and M. A. Harrison, *J. Food Prot.* **68:**1143–1146, 2005.
[c]A. Mukherjee, D. Speh, A. T. Jones, K. M. Buesing, and F. Diez-Gonzalez, *J. Food Prot.* **69:**1928–1936, 2006.
[d]A. Mukherjee, D. Speh, E. Dyck, and F. Diez-Gonzalez, *J. Food Prot.* **67:**894–900, 2004.
[e]ND, not determined.

able to grow faster than yeasts or molds in most vegetables and therefore have a competitive advantage, particularly at refrigeration temperatures.

Spoilage can be influenced by the history of the land on which vegetables are grown. For example, repeated planting of one type of vegetable on the same land over several seasons can lead to the accumulation of plant pathogens in the soil and increased potential for spoilage. Soil can also be contaminated by floodwater or poor-quality irrigation water.

Virtually all vegetables receive at least some processing or handling before they are consumed. It is important to realize that many common processing steps increase the likelihood of spoilage. In most cases, processors of vegetables destined to be sold as fresh produce add chlorine to wash water to achieve a concentration of 5 to 250 mg/liter. Chlorine is an effective antimicrobial agent. However, washing with chlorinated water and other disinfectants has only a limited antimicrobial effect on the microflora of the produce. Many processors consider washing fresh produce with chlorinated water a disinfection step, but this is not the real purpose of the inclusion of chlorine. Rather, chlorine is more effective in killing microorganisms in the water and minimizing contamination of the vegetables by the rinse water.

Cutting, slicing, chopping, and mixing are other important processing steps for fresh vegetables. These operations are becoming even more important as the demand for ready-to-eat products increases. The processes can result in increases in populations of microorganisms on fresh vegetables through the transfer of microorganisms from the equipment to the product.

Storage, packaging, MAP, and transportation are other important processing steps that can influence the development of microbiological spoilage of vegetables. Most modified-atmosphere techniques involve reducing the concentration of oxygen while increasing the concentration of carbon dioxide. MAP functions by reducing the respiration and senescence processes of fresh fruits and vegetables, thereby delaying undesirable changes in sensory quality. The reduction of senescence in produce usually requires carbon dioxide concentrations of at least 5%; however, concentrations in excess of 20% can be detrimental to food quality. In general, aerobic gram-negative bacteria are most sensitive to carbon dioxide. Obligate and facultative anaerobic microorganisms are more resistant. Likewise, molds are more sensitive to carbon dioxide than are fermentative yeasts.

More-extensive processing techniques, such as thermal processing (canning) and freezing, have been used for years to preserve vegetables. Improper storage temperatures primarily cause microbiological spoilage of frozen vegetables. The microfloras in frozen vegetables are essentially the same as in raw products. Therefore, the spoilage patterns can be expected to be similar to those of raw products if the products are not properly maintained in a frozen state.

Preservation of vegetables by canning is accomplished by placing the vegetables in hermetically sealed containers and then heating them sufficiently to destroy microorganisms. For most vegetables, processing temperatures exceed 120°C. This eliminates all but the most heat-resistant bacterial spores. Consequently, spoilage of canned vegetables is usually caused by thermophilic spore-forming bacteria, unless the container integrity has been compromised in some way. Acidification without gas production is one of the most common types of spoilage observed in canned vegetables. This defect, called flat sour, is caused by *B. stearothermophilus* or *B. coagulans*. The defect ordinarily occurs when cans receive a marginally adequate thermal treatment or if they are stored at temperatures above 40°C. Another type of spoilage seen with canned vegetables is gas production and consequent swelling of cans. Swelling is caused by thermophilic spore-forming anaerobes, such as *Clostridium thermosaccharolyticum*, which grow and produce large amounts of hydrogen and carbon dioxide. The production of hydrogen sulfide in some canned vegetables can lead to a spoilage problem known as "sulfide stinker." This type of spoilage usually occurs in the absence of can swelling.

Microbiological Spoilage of Fruits

Fresh fruits are similar to vegetables in that they usually have a high enough a_w to support the growth of all but the most xerophilic (growing best at low a_w) or osmophilic (growing best at high a_w) fungi. However, most fruits differ from vegetables in that they have a more acidic pH (<4.4), the exception being melons. Fruits also have a higher sugar content. In addition, fruits usually possess more effective defense mechanisms, such as thicker epidermal tissues and higher concentrations of antimicrobial organic acids.

Normal and Spoilage Microfloras

As with vegetables, the normal microfloras of fruits are varied and include both bacteria and fungi. The sources of microorganisms include all those mentioned for vegetables, such as air, soil, and insects. Unlike that of vegetables, however, spoilage of fruits is most often due to yeasts or molds, although bacteria are also important. Examples of bacterial spoilage are *Erwinia* rots of pears or the production of buttermilk-like flavors in orange juice by lactic acid bacteria. Many of the same handling and processing techniques for fresh vegetables also apply to fresh fruits. Fruits are usually washed or rinsed immediately after harvest and then are usually packed into shipping cartons. The washing step can play an important role in reducing microbial contamination by as much as 3 log CFU/g.

Freshly cut or minimally processed fruit products have become increasingly popular in recent years and will likely become more popular due to their convenience. Both molds and yeasts have the ability to cause spoilage; the latter are more often associated with spoilage of cut fruits due to their ability to grow faster than molds. However, bacteria also play a role in the spoilage of fruits.

Several types of fruit are dried to yield intermediate-moisture products, which normally rely on low a_w as the main mechanism for preservation. Raisins and prunes, and usually dates and figs, are consumed as dried fruit products. Dried apples, apricots, and peaches are also popular with consumers. Depending on the means by which fruits are dehydrated, moisture levels can range from <5 to 35%. In general, most dried fruits have sufficiently low a_w to inhibit bacterial growth. Spoilage is usually limited to osmophilic yeasts or xerotolerant molds. Populations of yeasts and molds on dried fruit usually average <10^3 CFU/g. Yeasts normally associated with spoilage of dried fruits include *Zygosaccharomyces rouxii* and *Hanseniaspora, Candida, Debaryomyces,* and *Pichia* species. Molds capable of growth below an a_w of 0.85 include several *Penicillium* and *Aspergillus* species (especially the *Aspergillus restrictus* series), *Eurotium* spp. (the *Aspergillus glaucus* series), and *Wallemia sebi*.

Fruit concentrates, jellies, jams, preserves, and syrups are resistant to spoilage due to their low a_w. Unlike that of dried fruits, however, the reduced a_w of these products is achieved by adding sufficient sugar to achieve a_w values of 0.82 to 0.94. In addition to reduced a_w, these products are also usually heated to temperatures of 60 to 82°C, which kills most xerotolerant fungi. Consequently, spoilage of these products usually occurs when containers have been improperly sealed or after consumers open them.

Various heat treatments are used to preserve fruit products. Fruits generally require less severe treatment than vegetables due to the enhanced lethality brought about by their acidic pH. Many canned fruits, whether halved, sliced, or diced, are processed by heating the products to a can center temperature of 85 to 90°C. Some processed fruit juices and nectars are rapidly heated to 93 to 110°C and then aseptically placed into containers. In either case, the processes are sufficient to kill most vegetative bacteria, yeasts, and molds. However, several genera of molds produce heat-resistant ascospores or sclerotia. Molds typically associated with spoilage of thermally processed fruit products include *Byssochlamys fulva, Byssochlamys nivea, Neosartorya fischeri,* and *Talaromyces flavus*. Signs of spoilage for heat-processed fruit products include visible mold growth, off-odors, breakdown of fruit texture, or solubilization of starch or pectin in the suspending medium.

In recent years, food microbiologists have become aware that a thermotolerant spore-forming bacterium, *Alicyclobacillus acidoterrestris*, can be a problem in pasteurized juices. This bacterium is sufficiently acid and heat tolerant to survive the pasteurization conditions that kill most other bacteria. *A. acidoterrestris* produces 2-methoxyphenol (guaiacol), which imparts a medicine-like or phenolic off-flavor to several types of juices, most notably apple and orange juices.

Microbiological Spoilage of Grains and Grain Products

Although grains are similar to fruits and vegetables in that they are of plant origin, they differ in many ways. Unlike fruits and vegetables, grains are typically thought of as primarily agronomic or field crops rather than horticultural crops. Grains destined for human food are ground into flour or meal for bakery or pasta products or further processed into snacks or breakfast cereals. The final products, except doughs, often have a_w values of <0.65, below which most microorganisms cannot grow.

Natural Microfloras

Grains and grain products normally contain several genera of bacteria, molds, and yeasts. The specific species depend on the conditions encountered during production, harvesting, storage, and processing. Although the low a_w of grains and grain products might lead one to believe that fungi, especially molds, are the predominant members of the natural microflora, this is not always the case. Indeed, depending on the storage conditions, type of grain, and moisture content, bacterial levels may far exceed mold populations. Of the many different types of microorganisms present on grains, only a few invade the kernel itself. Molds such as *Alternaria*, *Fusarium*, *Helminthosporium*, and *Cladosporium* are primarily responsible for invading wheat in the field. However, infection by these molds is minimal and does little damage unless the grains are allowed to become too moist.

Effects of Processing

Grains intended for human consumption are rarely used in their native state but undergo various processing treatments. Each step in the pretreatment and milling operations reduces populations of microorganisms so that the flours usually contain smaller populations than the grains from which they are made. Aside from fewer microorganisms being present in flour, the profile of the microflora in milled grains is similar to that of whole grains. The final processing step to which most grain products, specifically flours, are subjected is the addition of liquids (e.g., water or milk) and other ingredients to produce doughs.

Types of Spoilage

Properly dried and stored grains and grain products are inherently resistant to spoilage due to their low a_w. However, despite attempts to protect grains from the uptake of water during storage, the a_w can increase to a level that enables xerotolerant molds to grow. Moreover, it should be kept in mind that reduced a_w often only slows the growth of fungi and that spoilage does ultimately occur given enough time. A subtle change in temperature in large bulk quantities of grains can cause condensation to develop in some areas, increasing moisture levels sufficiently to allow mold growth. In such cases, molds such as *Aspergillus* and *Eurotium* spp. and *W. sebi* may grow and cause spoilage. Molds are the primary spoilage microorganisms in grains due to their ability to grow at reduced a_w. The primary concern with mold spoilage is the potential production of mycotoxins. Molds can also cause adverse changes in the color, flavor, and aroma of grain products due to the production of aromatic volatile compounds. The production of volatile metabolites depends more on the fungal species than on the grain type. Another important adverse effect of mold growth on grains is an increase in the free fatty acid value (FAV). Increases in the FAV can result from fungal lipase activity. Because some of these lipases are heat stable, the FAV is sometimes used as an indication of fungal spoilage of grains.

Most grain-based foods receive some type of heat treatment before they are eaten. These treatments are usually sufficient to inactivate all but the most heat-resistant microorganisms. However, development of a spoilage condition known as *ropiness*, caused by the germination of spores of *B. subtilis* or *B. licheniformis*, and a defect referred to as *bloody bread* resulting from the growth of the red-pigmented bacterium *Serratia marcescens* may occur, albeit rarely, in commercially produced breads.

Summary

- *Pseudomonas* spp. are involved in spoilage of dairy, meat, poultry, and produce.
- Attachment of bacteria to muscle tissue is a two-step process.
- Processing equipment can be a major source of spoilage microorganisms.
- Good manufacturing practices are essential in limiting populations of spoilage microorganisms associated with raw materials and the finished product, thereby extending shelf life.
- Control of temperature is critical to prevent spoilage and extend shelf life.
- Proteolytic spoilage occurs prior to lipolytic spoilage.
- Food composition (percent adipose, carbohydrate, and protein) influences microbial growth and ultimately spoilage.
- Growth of one type of microorganism may influence the growth of other microorganisms and result in spoilage.
- The shelf lives of products vary widely depending on composition and processing (Table 21.1).

Suggested reading

Carlin, F. 2007. Fruits and vegetables, p. 157–170. *In* M. P. Doyle and L. R. Beuchat (ed.), *Food Microbiology: Fundamentals and Frontiers*, 3rd ed. ASM Press, Washington, DC.

Frank, J. F. 2007. Milk and dairy products, p. 141–156. *In* M. P. Doyle and L. R. Beuchat (ed.), *Food Microbiology: Fundamentals and Frontiers*, 3rd ed. ASM Press, Washington, DC.

Nychas, G.-J. E., D. L. Marshall, and J. N. Sofos. 2007. Meat, poultry, and seafood, p. 105–140. *In* M. P. Doyle and L. R. Beuchat (ed.), *Food Microbiology: Fundamentals and Frontiers*, 3rd ed. ASM Press, Washington, DC.

Tournas, V. H. 2005. Spoilage of vegetable crops by bacteria and fungi and related health hazards. *Crit. Rev. Microbiol.* **31:**33–44.

Whitfield, F. B. 1998. Microbiology of food taints. *Int. J. Food Sci. Technol.* **33:**31–51.

Questions for critical thought

1. The shelf life of a product is dictated by a number of factors. Explain why hamburger has a shelf life of only 2 days, whereas bacon has a shelf life of 1 week.
2. The microbial floras of fish depend on the temperature of the waters that the fish are harvested from. What types of microorganisms might be expected to be associated with fish harvested from temperate waters? From tropical waters? What other factors influence the spoilage of fish?

3. Raw milk is protected from spoilage by at least three intrinsic systems. Name the three systems and the bases on which they inhibit spoilage.
4. Chlorine is often added to water used to rinse vegetables. What impact does this treatment have on the microbial load of the treated product?
5. Pasteurized milk still requires refrigeration to delay spoilage. What types of microbes may survive pasteurization and potentially result in spoilage of milk? Based on your knowledge of food science, what would happen to milk if the time and temperature of pasteurization were increased dramatically?
6. The a_ws of many products are such that the products are resistant to spoilage. Explain how a_w affects spoilage.
7. Why is it likely that a product will undergo proteolytic spoilage before lipolytic spoilage?
8. What properties of fruits and vegetables support spoilage by yeasts and molds rather than by bacteria?
9. What is the cause of the rancid defect in butter? How does the spoilage of ripened-cream butter differ from that of sweet-cream butter?
10. Comment on the use of irradiation to enhance the microbiological safety of meats. Can this technique be used for fresh produce (fruits and vegetables)? What are its disadvantages?
11. What types of microorganisms are involved in the contamination and spoilage of canned dairy products? Investigate methods that can be used to control sporeformers in canned foods.
12. Bacterial growth on a product can result in organoleptic changes. Changes include the development of off-odor and slime. Which would you expect to occur first and why?
13. How does spoilage differ between climacteric and nonclimacteric fruits? What is senescence, and how can it be controlled?
14. The microbial load of raw vegetables is influenced by growth conditions. What factors would lead to the conclusion that organically grown vegetables have a greater microbial load than conventionally grown vegetables? Would methods designed to reduce the microbial load on organically grown vegetables be effective in reducing the microbial load of conventionally grown vegetables? Explain.
15. Bonus question: Is cheese that has mold growing on its surface considered spoiled? Is it safe to eat?

22 Molds

LEARNING OBJECTIVES

The information in this chapter will enable the student to:

- gain knowledge about common food spoilage molds and those associated with disease
- understand methods used for differentiating and identifying molds
- associate specific mycotoxins with molds that produce toxins of concern
- identify foods that are commonly associated with mold problems
- know conditions that contribute to the growth of molds and toxin production in foods
- discuss methods used for screening of raw and processed materials for the presence of mycotoxins

INTRODUCTION

We have known for centuries that eating certain mushrooms can cause illness or even be fatal. That some common food spoilage molds cause disease and death is a newer realization. Indeed, in 1940, a *Penicillium* species was identified as the source of toxicity in "yellow" rice in Japan. Mycotoxins came to the attention of scientists in the Western world in the early 1960s with the outbreak of turkey "X" disease in England. This disease killed about 100,000 turkeys and other farm animals. The cause of the disease was traced to peanut meal in the feed, which was heavily contaminated with *Aspergillus flavus*. Analysis of the feed revealed that a group of fluorescent compounds, later named aflatoxins, were responsible for the outbreak. Aflatoxins also killed large numbers of ducklings in Kenya and caused hepatoma in hatchery-reared trout in California. International regulations and guidance have been established for mycotoxins in foods and feeds to limit or prevent human and livestock illness (Table 22.1). This chapter focuses on molds that are commonly associated with the spoilage of foods.

ISOLATION, ENUMERATION, AND IDENTIFICATION

Bacteria grow much faster than molds and are often found in contaminated samples. Therefore, media for the enumeration of fungi contain antibacterial compounds and compounds to inhibit mold colony spreading, such as dichloran-rose bengal-chloramphenicol agar or dichloran–18% glycerol agar. Czapek agar, a defined medium based on mineral salts, or a derivative such as Czapek yeast extract agar and malt extract agar are used for the identification of *Aspergillus* species from foods. Growth on Czapek yeast extract–20%

doi:10.1128/9781555817206.ch22

Table 22.1 International regulations and guidance on mycotoxins in food and feed[a]

Country, region, or entity	Mycotoxin(s)	Food or feed	Action level
United States	Patulin	Apple juice, apple juice concentrate, apple juice products	50 ppb
	Fumonisins B_1, B_2, and B_3	Popcorn	4 ppm[b]
		Dry-milled corn bran	4 ppm
		Poultry being processed for slaughter	100 ppb
	Aflatoxins B_1, B_2, G_1, and G_2	Foods	20 ppb
		Peanuts, peanut products	20 ppb
		Brazil nuts, pistachio nuts	20 ppb
	Aflatoxin M_1	Milk	0.5 ppb
	Aflatoxins B_1, B_2, G_1, and G_2	Corn and peanut products intended for:	
		Finishing beef cattle	300 ppb
		Finishing swine ≥100 lb	200 ppb
European Union	Patulin	Apple juice and other foods derived from apples	50 ppb
	Aflatoxin B_1	Spices (nutmeg, ginger, and turmeric)	5 ppb
	Aflatoxins B_1, B_2, G_1, and G_2	Spices (nutmeg, ginger, and turmeric)	10 ppb
	Aflatoxin B_1	Groundnuts, nuts, dried fruit intended for direct consumption	2 ppb
	Aflatoxin M_1	Milk	0.05 ppb
	Ochratoxin A	Raw cereal grains	5 ppb
Codex Alimentarius Commission	Patulin	Apple juice and apple juice ingredients in other beverages	50 ppb
	Aflatoxins B_1, B_2, G_1, and G_2	Peanuts intended for further processing	15 ppb
	Aflatoxin M_1	Milk	0.5 ppb

[a]Based on P. A. Murphy, S. Hendrich, C. Landgren, and C. M. Bryant, *J. Food Sci.* **71:**R51–R65, 2006.
[b]ppm, parts per million.

sucrose agar can also be a useful aid in identifying species of *Aspergillus*. The most effective medium for rapid detection of aflatoxigenic molds is *A. flavus* and *Aspergillus parasiticus* agar. Under incubation at 30°C for 42 to 48 h, *A. flavus* and *A. parasiticus* produce a bright orange-yellow colony reverse (the surface of the colony has a distinct color code, while the reverse [inside] has a different color code), which is diagnostic and readily recognized. The structures of *A. flavus* and *A. parasiticus* are shown in Fig. 22.1.

Unlike *Penicillium* species, *Aspergillus* species are conveniently "color coded," and the color of the *conidia* (nonmotile spores) serves as a very useful starting point in identification. The microscopic morphology is also important in identification. *Phialides* (cells producing conidia) may be produced directly from the swollen apex (*vesicle*) of long stalks (*stipes*), or there may be an intermediate row of supporting cells (*metulae*). Correct identification of *Aspergillus* species is an essential prerequisite to assessing the potential for mycotoxin contamination in a commodity, food, or feedstuff.

General enumeration procedures suitable for foodborne molds are effective for enumerating all common *Penicillium* species. Many antibacterial media give satisfactory results. However, some *Penicillium* species grow rather weakly on very dilute media, such as potato dextrose agar. Therefore, dichloran-rose bengal-chloramphenicol and dichloran–18% glycerol agars are recommended.

Identification of *Penicillium* isolates to the species level is not easy and is preferably carried out under carefully standardized conditions of

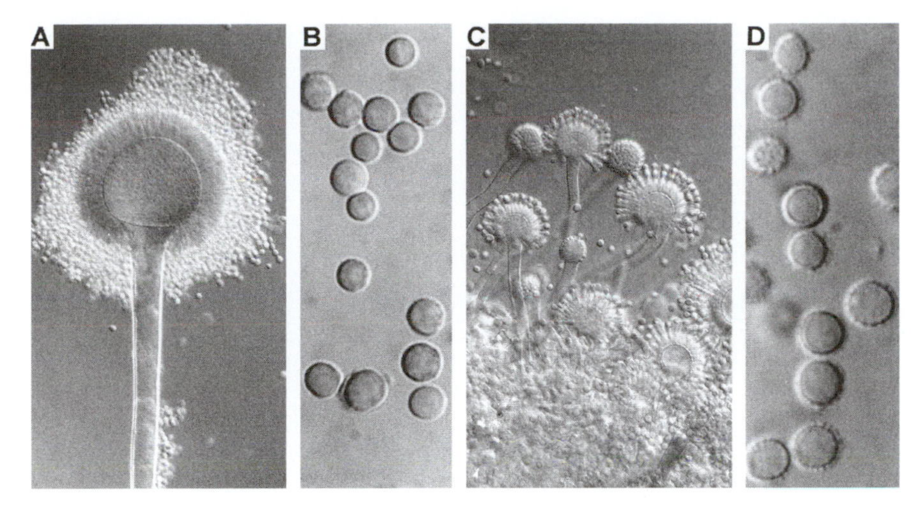

Figure 22.1 Aflatoxigenic fungi. **(A)** *A. flavus* head (magnification, ×195); **(B)** *A. flavus* conidia (magnification, ×1,215); **(C)** young *A. parasiticus* heads (magnification, ×195); **(D)** *A. parasiticus* conidia (magnification, ×1,215). Reprinted from A. D. Hocking, p. 540–550, *in* M. P. Doyle and L. R. Beuchat (ed.), *Food Microbiology: Fundamentals and Frontiers*, 3rd ed. (ASM Press, Washington, DC, 2007). doi:10.1128/9781555817206.ch22.f22.01

medium, incubation time, and temperature. In addition to the microscopic morphology, gross physiological features, including colony diameters and the colors of conidia and colony pigments, are used to distinguish species. A selective medium, dichloran-rose bengal-yeast extract-sucrose agar, is used for the enumeration of *Penicillium verrucosum* and *Penicillium viridicatum*. *P. verrucosum* produces a violet-brown reverse coloration on this agar. Figure 22.2 shows penicillia of four *Penicillium* subgenera.

Fusarium species are most often associated with cereal grains; seeds; milled cereal products, such as flour and cornmeal; barley malt; animal feeds; and dead plant tissue. To isolate *Fusarium* species from these products, it is necessary to use selective media. The basic techniques for the detection and isolation of *Fusarium* employ plating techniques, either plate counts of serial dilutions of products or the placement of seeds or kernels of grain directly on the surface of agar medium in petri dishes, i.e., *direct plating*. Since many *Fusarium* species are plant pathogens and are found in fields where crops are grown, these molds respond to light. Growth, pigmentation, and spore production are most typical when cultures are grown in alternating light and dark cycles of 12 h each. Fluorescent light or diffuse sunlight from a north window is best. Fluctuating temperatures, such as 25°C (day) and 20°C (night), also enhance growth and sporulation.

Several culture media have been used to detect and isolate *Fusarium* species. These include Nash-Snyder medium, modified Czapek Dox agar, Czapek iprodione-dichloran (CZID) agar, potato dextrose-iprodione-dichloran agar, and dichloran-chloramphenicol-peptone agar. Nash-Snyder medium and modified Czapek Dox agar contain pentachloronitrobenzene, a known carcinogen, and are not favored for routine use in food microbiology laboratories. However, these media can be useful for evaluating samples that are heavily contaminated with bacteria and other fungi. The most commonly used medium for isolating *Fusarium* from foods is CZID agar, but rapid identification of *Fusarium* isolates to the species level is

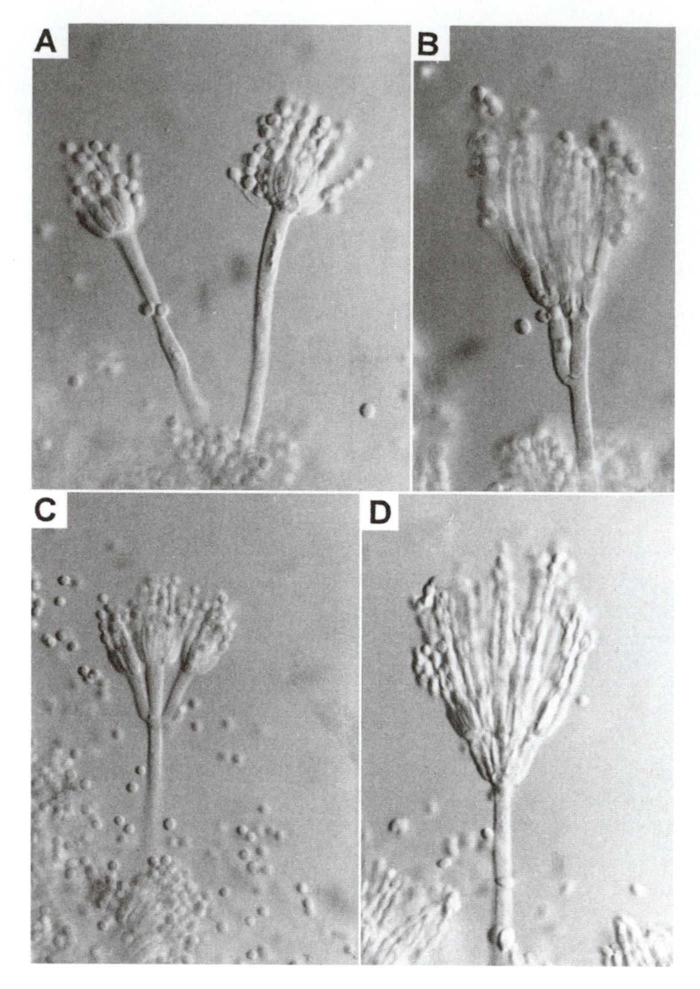

Figure 22.2 Penicillia (magnification, ×550) representative of the four *Penicillium* subgenera. **(A)** *Penicillium* subgenus *Aspergilloides* (*P. glabrum*); **(B)** *Penicillium* subgenus *Penicillium* (*P. expansum*); **(C)** *Penicillium* subgenus *Furcatum* (*P. citrinum*); **(D)** *Penicillium* subgenus *Biverticillium* (*P. variabile*). Reprinted from J. I. Pitt, p. 551–562, *in* M. P. Doyle and L. R. Beuchat (ed.), *Food Microbiology: Fundamentals and Frontiers*, 3rd ed. (ASM Press, Washington, DC, 2007). doi:10.1128/9781555817206.ch22.f22.02

difficult, if not impossible, on this medium. Isolates must be subcultured on other media, such as carnation leaf agar (CLA), for identification. However, CZID agar is a good selective medium for *Fusarium*. Most molds are inhibited on CZID agar, and *Fusarium* species can be readily distinguished.

Identification of *Fusarium* species is based largely on the production and morphology of macroconidia and microconidia (Fig. 22.3). *Fusarium* species do not readily form conidia on all culture media, and conidia formed on high-carbohydrate media, such as potato dextrose agar, are often more variable and less typical. CLA is a medium that supports abundant and consistent spore production. Carnation leaves from actively growing, disbudded young carnation plants free of pesticide residues are cut into small pieces (5 mm^2), dried in an oven at 45 to 55°C for 2 h, and sterilized by irradiation. CLA is prepared by placing a few pieces of carnation leaf on the surface of 2.0% water agar. *Fusarium* isolates are then inoculated onto the agar and leaf interface, where they form abundant and typical conidia

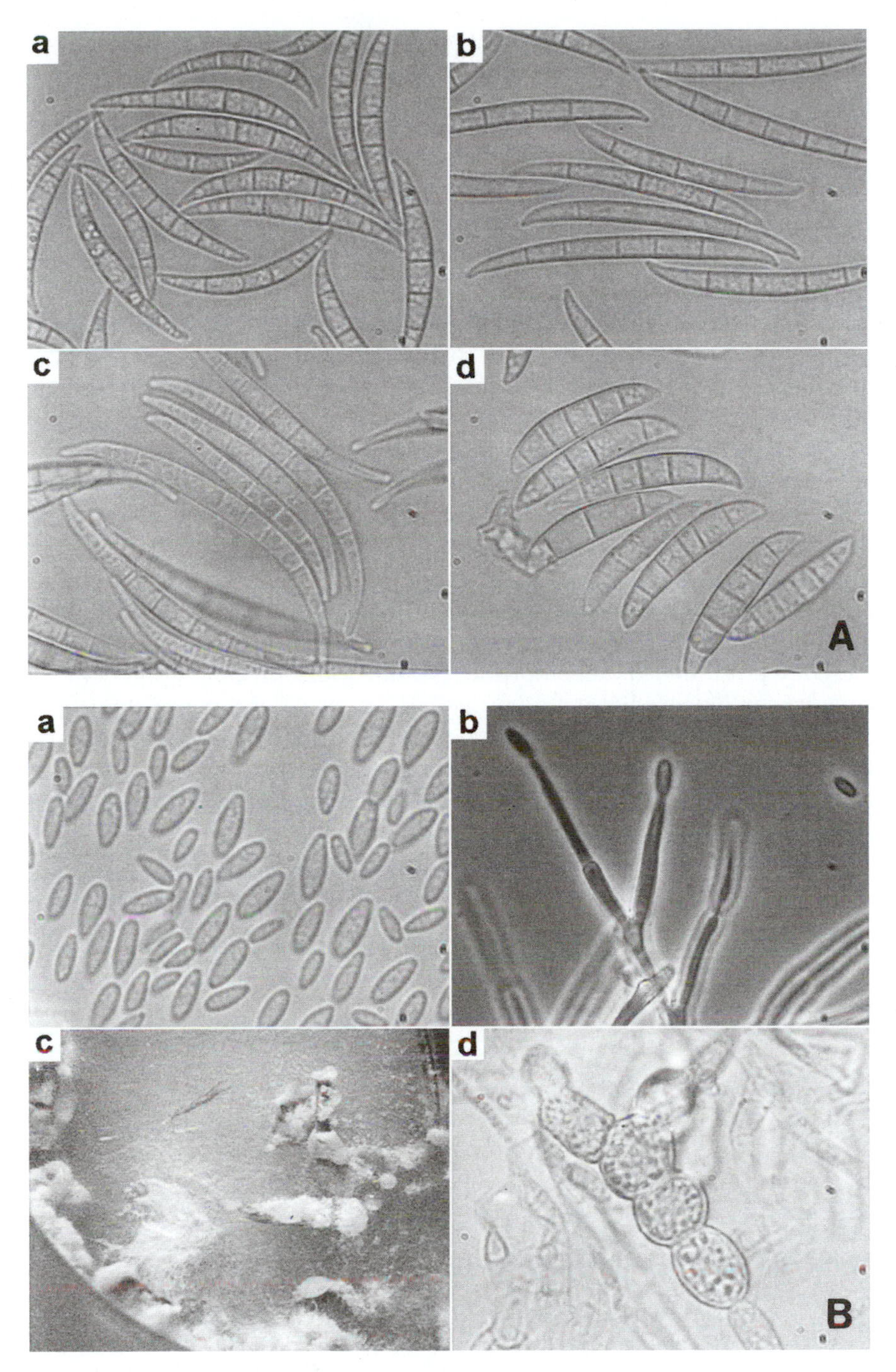

Figure 22.3 **(A)** Examples of macroconidia of *Fusarium* species. **(a)** *F. graminearum;* **(b)** *F. moniliforme;* **(c)** *F. equiseti;* **(d)** *F. culmorum* (magnification, ×1,000). **(B)** Examples of micro- and macroscopic structures of *Fusarium* species. **(a)** Microconidia; **(b)** monophialides; **(c)** sporodochia; **(d)** chlamydospores (magnifications, ×1,000 [a, b, and d] and ×10 [c]). Reprinted from L. B. Bullerman, p. 563–578, *in* M. P. Doyle and L. R. Beuchat (ed.), *Food Microbiology: Fundamentals and Frontiers*, 3rd ed. (ASM Press, Washington, DC, 2007). doi:10.1128/9781555817206.ch22.f22.03

and conidiophores in sporodochia (Fig. 22.4). CLA is low in carbohydrates and rich in other complex, naturally occurring substances that apparently stimulate spore production.

ASPERGILLUS SPECIES

Aspergillus was first described nearly 300 years ago and is an important genus in foods. Although a few species have been used to make food (e.g., *Aspergillus oryzae* in soy sauce manufacture), most *Aspergillus* species occur in foods as spoilage or biodeterioration fungi. They are extremely common in stored commodities, such as grains, nuts, and spices, and occur more frequently in tropical and subtropical than in temperate climates. The genus *Aspergillus* contains several species capable of producing mycotoxins, although the mycotoxin literature of the past 35 years has been dominated by papers on aflatoxins. Since foods contain many toxin-producing *Aspergillus* species, it is important to identify them and determine if they can produce toxin in a particular food.

The genus *Aspergillus* is large, containing >100 recognized species, most of which grow well in laboratory culture. Within the genus *Aspergillus*, species are grouped into six subgenera that are subdivided into sections. There are a number of teleomorphic (sexual-spore-producing) genera that have *Aspergillus* conidial states (*anamorphs*, i.e., asexual states or stages), but the only two of real importance in foods are the xerophilic (growing at low water activity [a_w]) *Eurotium* and *Neosartorya* species. They produce heat-resistant ascospores and cause spoilage in heat-processed foods, mainly fruit products. The most widely used taxonomy for *Aspergillus* is based

Figure 22.4 **(A)** Microscopic structures of *F. sporotrichioides*. **(a)** Macroconidia; **(b)** chlamydospores; **(c)** phialides; **(d)** microconidia (magnifications, ×1,000 [a, b, and d] and ×550 [c]).

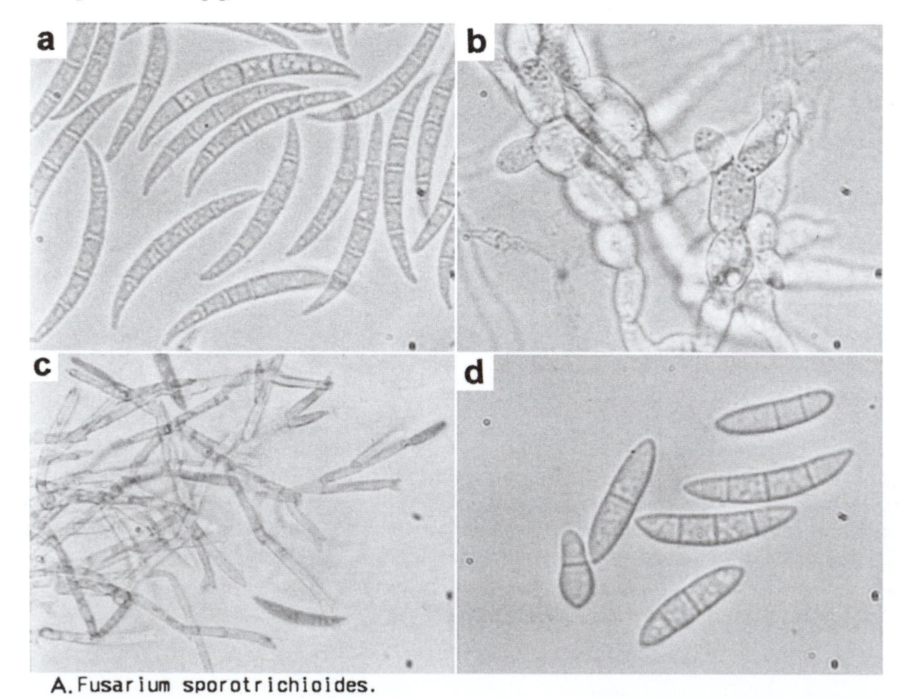

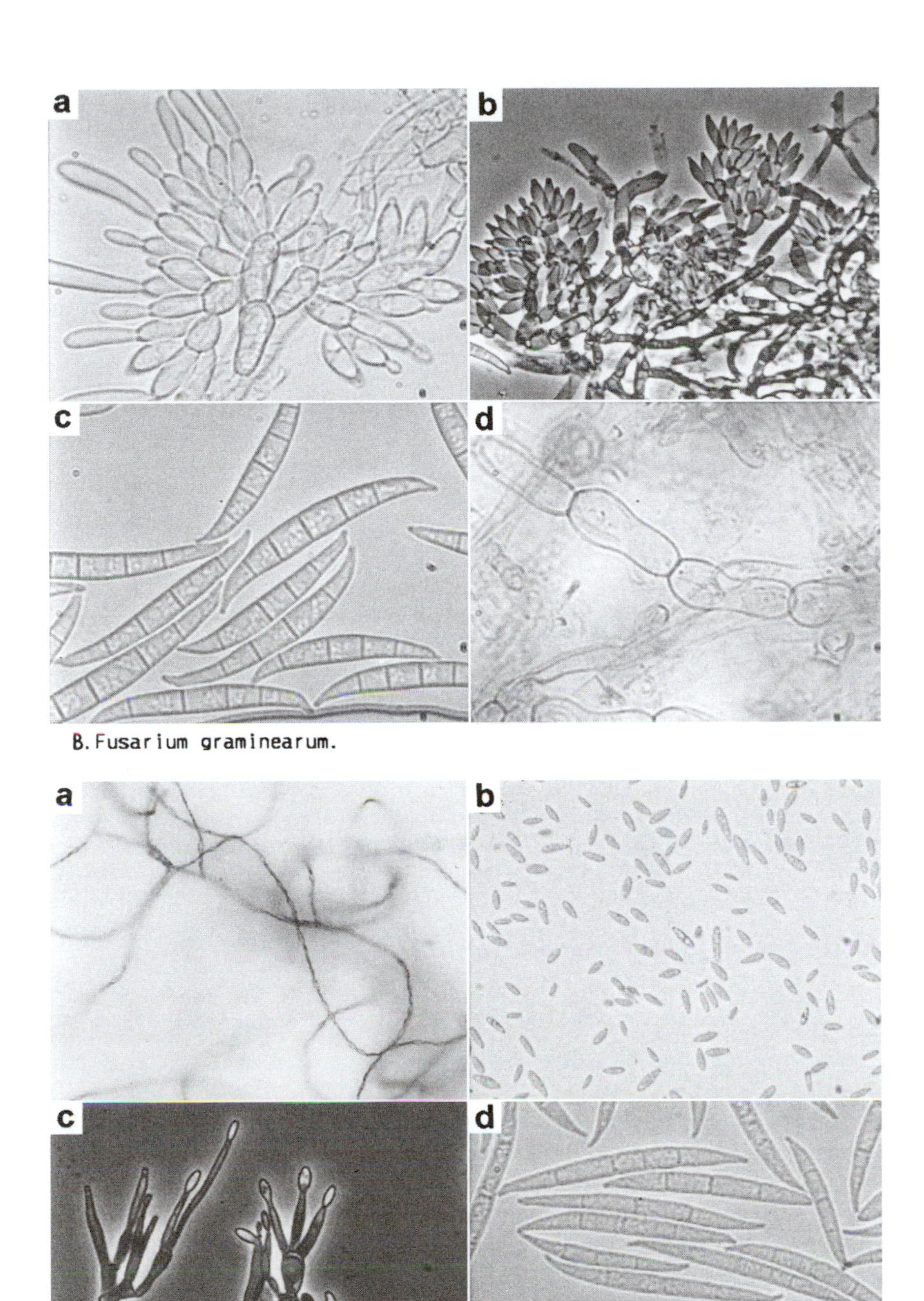

Figure 22.4 *(Continued)* **(B)** Microscopic structures of *F. graminearum*. **(a)** Conidiophores (monophialides); **(b)** monophialides in sporodochia; **(c)** macroconidia; **(d)** chlamydospores (magnifications, ×1,000 [a, c, and d] and ×550 [b]). **(C)** Microscopic structures of *F. moniliforme*. **(a)** Microconidia in chains; **(b)** microconidia; **(c)** monophialides producing microconidia; **(d)** macroconidia (magnifications, ×165 [a], ×550 [b], and ×1,000 [c and d]). *(Figure continues)*

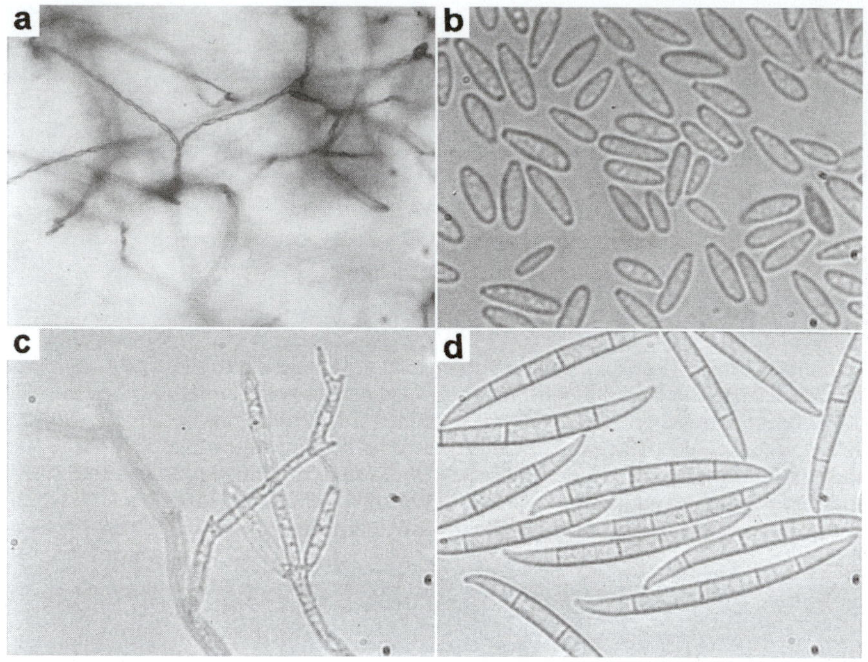

Figure 22.4 *(Continued)* **(D)** Microscopic structures of *F. proliferatum.* **(a)** Microconidia in chains; **(b)** microconidia; **(c)** polyphialides; **(d)** macroconidia (magnifications, ×165 [a] and ×1,000 [b to d]). Reprinted from L. B. Bullerman, p. 563–578, *in* M. P. Doyle and L. R. Beuchat (ed.), *Food Microbiology: Fundamentals and Frontiers*, 3rd ed. (ASM Press, Washington, DC, 2007). doi:10.1128/9781555817206.ch22.f22.04

on traditional morphological taxonomic techniques. However, chemical techniques, such as isoenzyme patterns, secondary metabolites, and ubiquinone systems, and molecular techniques have been used to clarify relationships within the genus.

Almost 50 *Aspergillus* species can produce toxic metabolites, but the *Aspergillus* mycotoxins (Table 22.1) of greatest significance in foods and feeds are aflatoxins (produced by *A. flavus, A. parasiticus,* and *Aspergillus nomius*); ochratoxin A from *Aspergillus ochraceus* and related species and from *Aspergillus carbonarius* and occasionally *Aspergillus niger;* sterigmatocystin, produced primarily by *Aspergillus versicolor* but also by *Emericella* species; and cyclopiazonic acid (*A. flavus* is the primary source, but it is also reported to be produced by *Aspergillus tamarii* [Fig. 22.5]). *Tremorgenic* (i.e., producing involuntary trembling of the body) toxins are produced by *Aspergillus terreus* (territrems), *Aspergillus fumigatus* (fumitremorgins), and *Aspergillus clavatus* (tryptoquivaline). Citrinin, patulin, and penicillic acid may also be produced by certain *Aspergillus* species. Appropriate temperature and a_w must exist for production of aflatoxins (Table 22.2).

Aspergillus species produce toxins that exhibit a wide range of toxicities and cause significant long-term effects. Aflatoxin B_1 is perhaps the most potent liver carcinogen known for a wide range of animal species, including humans. Both ochratoxin A and citrinin affect kidney function. Cyclopiazonic acid has a wide range of effects. Tremorgenic toxins, such as territrems, affect the central nervous system. Table 22.3 lists the most significant toxins produced by *Aspergillus* species and their toxic effects.

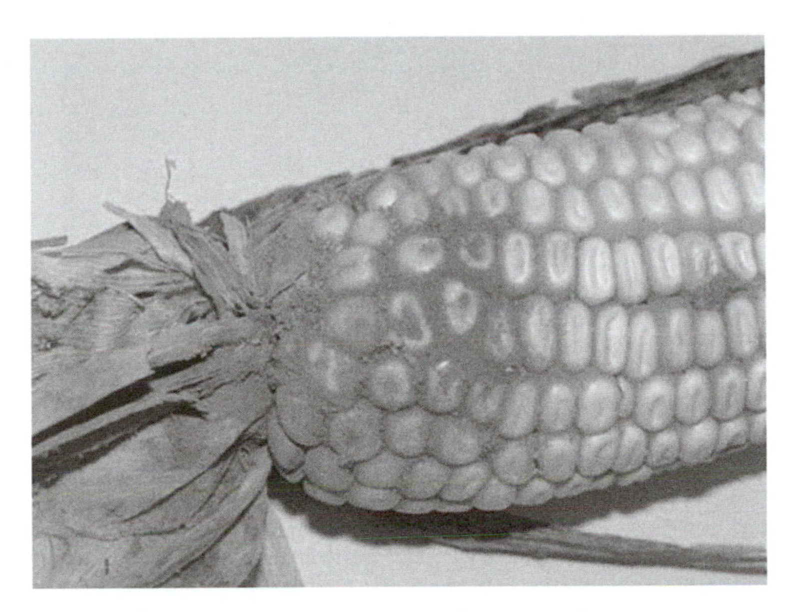

Figure 22.5 *Aspergillus* ear rot in corn before harvest. If the corn is consumed by livestock, it may adversely affect the animals' health. doi:10.1128/9781555817206.ch22.f22.05

A. flavus and *A. parasiticus*

Undoubtedly, the most important group of toxigenic aspergilli are the aflatoxigenic molds, *A. flavus*, *A. parasiticus*, and the recently described but much less common species *A. nomius*, all of which are classified in *Aspergillus* section *Flavi*. Although these three species are closely related and show many similarities, many characteristics may be used in their differentiation (Table 22.4). The toxins produced by these three species are species specific. *A. flavus* can produce aflatoxins B_1 and B_2 and cyclopiazonic acid, but only some isolates are toxigenic. *A. parasiticus* produces aflatoxins B_1, B_2, G_1, and G_2 but not cyclopiazonic acid, and almost all isolates are toxigenic. *A. nomius* is morphologically similar to *A. flavus*, but like *A. parasiticus*, it produces B and G aflatoxins without cyclopiazonic acid. Because this species appears to be uncommon, it has been little studied. The potential toxigenicities of isolates are not known, and the practical importance of the species is hard to assess. *A. oryzae* and *Aspergillus sojae* are related to *A. flavus* and *A. parasiticus* and are used to make fermented foods, but they do not produce toxins. *A. tamarii* produces aflatoxin B and cyclopiazonic

Table 22.2 Optimal conditions for mycotoxin production

Mycotoxin	Species	Temp (°C)	a_w
Aflatoxin	*Aspergillus flavus, A. parasiticus*	33	0.99
Ochratoxin	*Aspergillus ochraceus*	30	0.98
	Aspergillus carbonarius	15–20	0.85–0.90
	Penicillium verrucosum	25	0.90–0.98
Fumonisin	*Fusarium verticillioides, F. proliferatum*	10–30	0.93
Patulin	*Penicillium expansum*	0–25	0.95–0.99

Table 22.3 Significant mycotoxins produced by *Aspergillus* species and their toxic effects[a]

Mycotoxin(s)	Toxicity	Species
Aflatoxins B_1 and B_2	Cause acute liver damage and cirrhosis; carcinogenic (liver), teratogenic, immunosuppressive	*A. flavus*, *A. parasiticus*, *A. nomius*
Aflatoxins G_1 and G_2	Effects similar to B aflatoxins; G_1 toxicity is less than that of B_1 but greater than that of B_2	*A. parasiticus*, *A. nomius*
Cyclopiazonic acid	Degeneration and necrosis of various organs, tremorgenic, low oral toxicity	*A. flavus*, *A. tamarii*
Ochratoxin A	Causes kidney necrosis (especially in pigs); teratogenic, immunosuppressive, probably carcinogenic	*A. ochraceus* and related species, *A. carbonarius*, *A. niger* (occasional)
Sterigmatocystin	Causes acute liver and kidney damage; carcinogenic (liver)	*A. versicolor*, *Emericella* spp.
Fumitremorgins	Tremorgenic (rats and mice)	*A. fumigatus*
Territrems	Tremorgenic (rats and mice)	*A. terreus*
Tryptoquivalines	Tremorgenic	*A. clavatus*
Cytochalasins	Cytotoxic	*A. clavatus*
Echinulins	Feed refusal (pigs)	*Eurotium chevalieri*, *Eurotium amstelodami*

[a]Reprinted from A. D. Hocking, p. 537–550, *in* M. P. Doyle and L. R. Beuchat (ed.), *Food Microbiology: Fundamentals and Frontiers*, 3rd ed. (ASM Press, Washington, DC, 2007).

acid. Obviously, accurate differentiation of related species within the section *Flavi* is important in order to determine the potential for toxin production and the types of toxins likely to be present.

Detection and Identification

A. flavus and *A. parasiticus* (Fig. 22.1) are easily distinguished from other *Aspergillus* species by using appropriate media and methods. The texture of the conidial walls is a reliable differentiating feature: conidia of *A. flavus* (Fig. 22.1B) are usually smooth to finely roughened, while those of *A. parasiticus* (Fig. 22.1D) are clearly rough when observed under an oil immersion lens. The combinations of characteristics most useful in differentiating among the three aflatoxigenic species are summarized in Table 22.4.

Aflatoxins

Aflatoxins are difuranocoumarin derivatives. Aflatoxins B_1, B_2, G_1, and G_2 are produced in nature by the molds discussed above. The letters B and G refer to the fluorescent colors (blue and green, respectively) observed under long-wave ultraviolet (UV) light, and the subscripts 1 and 2 refer to their separation patterns on thin-layer chromatography plates. Aflatoxins M_1 and M_2 are produced from their respective B aflatoxins by hydroxyl-

Table 22.4 Distinguishing features of *A. flavus*, *A. parasiticus*, and *A. nomius*[a]

Species	Conidia	Sclerotia	Toxins
A. flavus	Smooth to moderately roughened, variable in size	Large, globose	Aflatoxins B_1 and B_2, cyclopiazonic acid
A. parasiticus	Conspicuously roughened, little variation in size	Large, globose	Aflatoxins B and G
A. nomius	Similar to *A. flavus* (bullet shaped)	Small, elongated	Aflatoxins B and G

[a]Reprinted from A. D. Hocking, p. 537–550, *in* M. P. Doyle and L. R. Beuchat (ed.), *Food Microbiology: Fundamentals and Frontiers*, 3rd ed. (ASM Press, Washington, DC, 2007).

ation in lactating animals and are excreted in milk at a rate of ~1.5% of ingested B aflatoxins.

Aflatoxins are synthesized through the polyketide pathway, beginning with condensation of an acetyl unit with two malonyl units and the loss of carbon dioxide. Intermediate steps result in the formation of norsolorinic acid. Norsolorinic acid undergoes several metabolic conversions to form aflatoxin B_1. G group aflatoxins are formed from the same substrate as B group aflatoxins, namely, *O*-methylsterigmatocystin, but by an independent pathway. Most of the genes involved in the biosynthesis of aflatoxins are contained within a single gene cluster in the genomes of *A. flavus* and *A. parasiticus*, and their regulation and expression are now relatively well understood.

Toxicity. Aflatoxins are both acutely and chronically toxic in animals and humans, producing acute liver damage, liver cirrhosis, and tumor induction. Even more important to human health are the immunosuppressive effects of aflatoxins, either alone or in combination with other mycotoxins. Immunosuppression can increase susceptibility to infectious diseases, particularly in populations where aflatoxin ingestion is *chronic* (continuous), and can interfere with the production of antibodies in response to immunization in animals, and perhaps also in children.

Acute aflatoxicosis in humans is rare; however, several outbreaks have been reported. In 1967, 26 people in two farming communities in Taiwan became ill with apparent food poisoning. Nineteen were children, three of whom died. Although postmortems were not performed, rice from the affected households contained ~200 μg of aflatoxin B_1/kg, which was probably responsible for the outbreak. An outbreak of hepatitis in India in 1974 that affected 400 people, 100 of whom died, was almost certainly caused by aflatoxins. The outbreak was traced to corn heavily contaminated with *A. flavus* and containing up to 15 mg of aflatoxins per kg. It was calculated that the affected adults may have consumed 2 to 6 mg on a single day, implying that the acute lethal dose for adult humans is on the order of 10 mg. The deaths of 13 Chinese children in the northwestern Malaysian state of Perak were reportedly due to ingestion of contaminated noodles. Aflatoxins were confirmed in postmortem tissue samples. A case of systemic aspergillosis caused by an aflatoxin-producing strain of *A. flavus*, in which aflatoxins B_1, B_2, and M_1 were detected in lung lesions and were considered to have played a role in damaging the immune system of the patient, has been reported. A severe outbreak of aflatoxicoses occurred in Kenya in 2002. More than half of the maize samples (maize flour and maize grains) had aflatoxin B_1 levels of >20 parts per billion (ppb). Nearly 12% had aflatoxin levels of >1,000 ppb (range, 20 to 8,000 ppb). A total of 317 cases were reported, with 125 deaths.

The greatest direct impact of aflatoxins on human health is their potential to induce liver cancer. Human liver cancer has a high incidence in central Africa and parts of Southeast Asia. Studies in several African countries and Thailand have shown a correlation between aflatoxin intake and the occurrence of primary liver cancer. No such correlation could be demonstrated for populations in rural areas of the United States, despite the occurrence of aflatoxins in corn. However, evidence supports the hypothesis that high aflatoxin intakes are causally related to high incidences of cancer, even in the absence of hepatitis B. In animals, aflatoxins cause various syndromes, including cancer of the liver, colon, and kidneys. Regular low-level intake of aflatoxins can lead to poor feed conversion, low weight gain, and poor milk yields in cattle.

Mechanism of action. Aflatoxin B_1 is metabolized in the liver, leading to the formation of highly reactive intermediates, one of which is 2,3-epoxy-aflatoxin B_1. Binding of these reactive intermediates to DNA results in disruption of transcription and abnormal cell proliferation, leading to mutagenesis or carcinogenesis. Aflatoxins also inhibit oxygen uptake in the tissues by acting on the electron transport chain and inhibiting various enzymes, resulting in decreased production of adenosine triphosphate (ATP).

Toxin detection. Mycotoxins can be detected by chemical or biological methods. Chemical determination of aflatoxins has become fairly standardized. Samples are extracted with organic solvents, such as chloroform or methanol, in combination with small amounts of water. The representative sample size(s) depends on the particle size; therefore, a 500-g sample is acceptable for oils or milk powder, but 3 kg of flour or peanut butter is necessary. Extracts can be further cleaned up by passage through a silica gel column, concentrated, and then separated by thin-layer chromatography or high-performance liquid chromatography. Following thin-layer chromatography, aflatoxins are visualized under UV light and quantified by visual comparison with known concentrations of standards. High-performance liquid chromatography with detection by UV light or absorption spectrometry provides a more readily quantifiable (although not necessarily more accurate or sensitive) technique.

Immunoassay techniques, including enzyme-linked immunosorbent assays and dipstick tests, for aflatoxin detection have been developed. These kits are commercially available. Immunoaffinity and fluorimetric detection have been combined in developing biosensors able to detect aflatoxin down to 50 $\mu g/kg$ in as little as 2 min. The polymerase chain reaction (PCR) has been applied to detect the presence of aflatoxigenic fungi in foods, targeting aflatoxin biosynthetic genes.

Occurrence of aflatoxigenic molds and aflatoxins. While *A. flavus* is widely distributed in nature, *A. parasiticus* is less widespread. The actual extent of *A. parasiticus* occurrence in nature is likely greater than reported, since both species are reported indiscriminately as *A. flavus*. *A. flavus* and *A. parasiticus* have a strong affinity for nuts and oilseeds. Corn, peanuts, and cottonseed are the most important crops invaded by these molds, and in many instances, invasion takes place before harvest, not during storage as was once believed (Fig. 22.5). Peanuts are invaded while still in the ground if the crop suffers drought stress or related factors. In corn, insect damage to developing kernels allows the entry of aflatoxigenic molds, but invasion can also occur through the silks of developing ears. Cottonseeds are invaded through the nectaries.

Significant amounts of aflatoxins can occur in peanuts, corn, and other nuts and oilseeds, particularly in some tropical countries, where crops may be grown under marginal conditions and where drying and storage facilities are limited. *A. flavus* grows well on cereals and spices, but aflatoxin production in these commodities can be prevented with proper drying, handling, and storage.

Control and inactivation. Control of aflatoxins in commodities generally relies on screening techniques that separate the affected nuts, grains, or seeds. In corn, cottonseed, and figs, screening for aflatoxins can be done by examination under UV light: those particles which fluoresce may be contaminated. All peanuts fluoresce when exposed to UV light, so this method is not useful for detecting kernels that are contaminated with aflatoxins. Peanuts containing aflatoxins are removed by electronic color-sorting machines, which detect discolored kernels.

Aflatoxins can be partially destroyed by chemical treatments. Oxidizing agents, such as ozone and hydrogen peroxide, remove aflatoxins from contaminated peanut meals. Although ozone is effective in destroying aflatoxins B_1 and G_1 after exposure at 100°C for 2 h, there is no effect on aflatoxin B_2. The treatment also decreases the lysine content of the meal. Treatment with hydrogen peroxide destroys 97% of aflatoxin in defatted peanut meal. The most practical chemical method of aflatoxin destruction is the use of anhydrous ammonia gas at elevated temperatures and pressures, with a 95 to 98% reduction in total aflatoxin in peanut meal. This technique is used commercially for detoxification of animal feeds in Senegal, France, and the United States.

The ultimate method for the control of aflatoxins in commodities, particularly peanuts, is to prevent the plants from becoming infected with the molds. Progress toward this goal is being made through a biological control strategy, which is the early infection of plants with nontoxigenic strains of *A. flavus* to prevent the subsequent entry of toxigenic strains.

Aflatoxins are among the few mycotoxins covered by legislation. Statutory limits are imposed by some countries on the amounts of aflatoxin that can be present in particular foods. The limit imposed by most Western countries is 5 to 20 μg of aflatoxin B_1/kg (parts per billion) in several human foods, including peanuts and peanut products. The amount allowed in animal feeds varies, but up to 300 μg/kg is allowed in feedstuffs for beef cattle and sheep in the United States. There are no statutory limits for aflatoxins in foods or feeds in Southeast Asia.

Other Toxigenic Aspergilli

Although aflatoxins are of greatest concern, there are other toxins produced by other *Aspergillus* species. *A. ochraceus* (Fig. 22.6A) is a widely distributed mold that is particularly common in dried foods, such as nuts (e.g., peanuts and pecans), beans, dried fruit, biltong (a South African dried meat), and dried fish. *A. ochraceus* produces ochratoxin. Ochratoxin A is the major toxin, with minor components of lower toxicity designated ochratoxins B and C. Other reported toxic metabolites are xanthomegnin and viomellein. Other *Aspergillus* species closely related to *A. ochraceus* can also produce ochratoxin A. *Aspergillus sclerotiorum*, *Aspergillus alliaceus*, *Aspergillus melleus*, and *Aspergillus sulphureus* have all been reported to produce ochratoxins, but in *A. ochraceus* and related species, only a portion of isolates are toxigenic. Production of ochratoxin A by *Aspergillus niger* var. *niger* was reported in 1994. *A. carbonarius* also produces ochratoxin A.

Authors' note

Aflatoxin B_1 is considered the most toxic member of the aflatoxin family. The toxin is a potent carcinogen, linked to various types of cancer that are prevalent in China, sub-Saharan Africa, and Southeast Asia.

A. versicolor (Fig. 22.6B) is the most important food spoilage and toxigenic species in *Aspergillus* section *Versicolores*. *A. versicolor* is the major producer of sterigmatocystin, a carcinogenic compound that is a precursor of the aflatoxins. Surprisingly, *A. versicolor* does not produce aflatoxins, but it produces ochratoxin A. Sterigmatocystin has been reported to

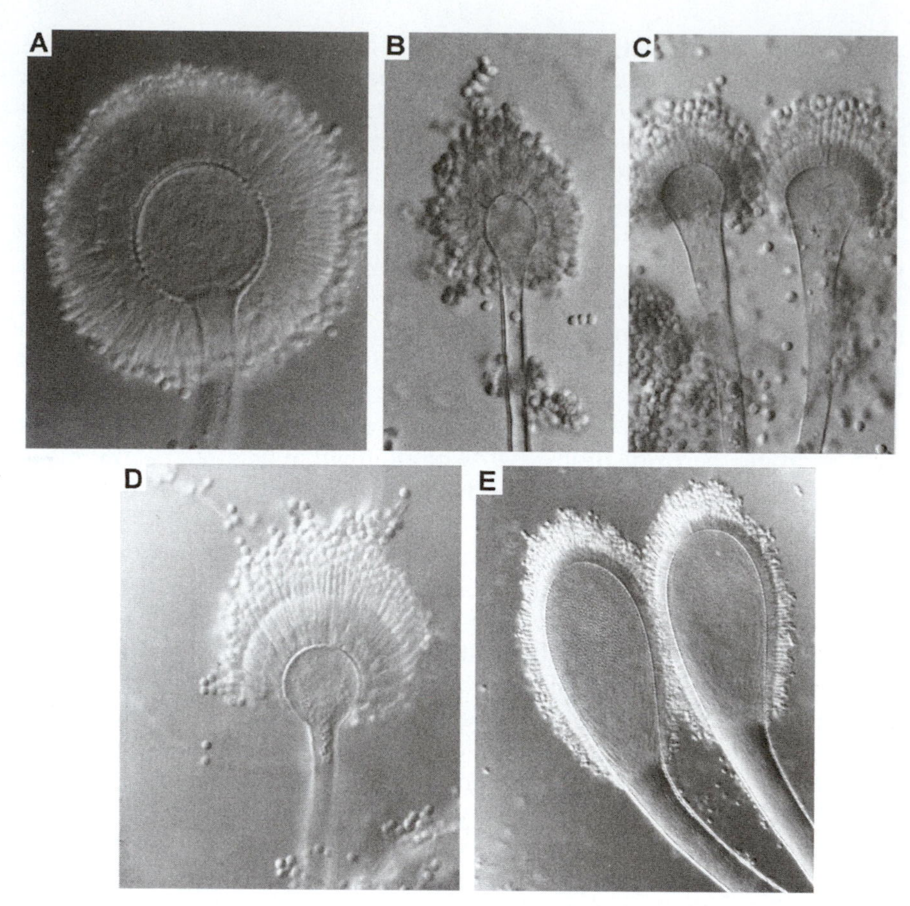

Figure 22.6 Some common mycotoxigenic *Aspergillus* species. **(A)** *A. ochraceus* (magnification, ×485); **(B)** *A. versicolor* (magnification, ×485); **(C)** *A. fumigatus* (magnification, ×485); **(D)** *A. terreus* (magnification, ×485); **(E)** *A. clavatus* (magnification, ×195). Reprinted from A. D. Hocking, p. 537–550, *in* M. P. Doyle and L. R. Beuchat (ed.), *Food Microbiology: Fundamentals and Frontiers*, 3rd ed. (ASM Press, Washington, DC, 2007). doi:10.1128/9781555817206.ch22.f22.06

occur naturally in rice in Japan, wheat and barley in Canada, cereal-based products in the United Kingdom, and Ras cheese. Sterigmatocystin has low acute oral toxicity because it is relatively insoluble in water and gastric juices. As a liver carcinogen, sterigmatocystin is about 1/150 as potent as aflatoxin B_1 but is still much more potent than most other known liver carcinogens.

A. fumigatus, classified in the section *Fumigati* (Fig. 22.6C), is best recognized as a human pathogen that causes aspergillosis of the lung. Its prime habitat is decaying vegetation. *A. fumigatus* is isolated frequently from foods, particularly stored commodities, but is not regarded as a serious spoilage mold.

A. fumigatus is capable of producing several toxins that affect the central nervous system, causing tremors. Fumitremorgins A, B, and C are toxic cyclic dipeptides that are produced by *A. fumigatus* and *Aspergillus caespitosus*. Verruculogen, also produced by these species, has a structure similar to that of the fumitremorgins but with a different side chain. Verruculogen is tremorgenic and appears to cause inhibition of the alpha motor cells of the anterior horn.

A. terreus (Fig. 22.6D) occurs commonly in soil and in foods, particularly stored cereals and cereal products, beans, and nuts, but is not regarded as an important spoilage mold. *A. terreus* can produce a group of tremorgenic toxins known as territrems. *A. clavatus* (Fig. 22.6E) is found in soil and decomposing plant materials. Although *A. clavatus* has been reported to be present in various stored grains, it is especially common in malting barley, an environment particularly suited to its growth and sporulation. *A. clavatus* produces patulin, cytochalasins, and the tremorgenic mycotoxins tryptoquivaline, tryptoquivalone, and related compounds.

PENICILLIUM SPECIES

The discovery of penicillin in 1929 provided impetus to search for other *Penicillium* metabolites with antibiotic properties and ultimately led to the recognition of citrinin, patulin, and griseofulvin as "toxic antibiotics," later called mycotoxins. The literature on toxigenic penicillia is now quite vast. In a comprehensive review of the literature on fungal metabolites, ~120 common mold species were reported as demonstrably toxic to higher animals. Forty-two toxic metabolites are produced by one or more *Penicillium* species. At least 85 *Penicillium* species are listed as toxigenic. Classification within the genus *Penicillium* is based primarily on microscopic morphology (Fig. 22.2). The genus is divided into subgenera based on the number and arrangement of phialides and metulae and *rami* (elements supporting phialides) on the stipes.

Significant *Penicillium* Mycotoxins

As already noted, >80 *Penicillium* species are reported to be toxin producers. The range of mycotoxin classes produced by *Penicillium* species is broader than for any other fungal genus. Toxicity due to *Penicillium* species is very diverse. However, most toxins can be placed in two broad groups, those that affect liver and kidney function and those that are neurotoxins. Generally, the *Penicillium* toxins that affect liver or kidney function are asymptomatic or can cause generalized debility in humans or animals. In contrast, the toxicity of the neurotoxins in animals is often characterized by sustained trembling. However, individual toxins show wide variations from these generalizations.

In fact, the most important toxin produced by a *Penicillium* species, ochratoxin A, is derived from isocoumarin linked to the amino acid phenylalanine. The target organ of toxicity in all mammalian species is the kidney. The lesions can be produced by both acute and chronic exposure. Ochratoxin A is produced by *P. verrucosum* (Fig. 22.7A), which grows in barley and wheat crops in cold climates, especially Scandinavia, central Europe, and western Canada. *P. verrucosum* grows most strongly at relatively low temperatures, down to 0°C, with a maximum near 31°C. The major source of ochratoxin A in foods is bread made from barley or wheat in which *P. verrucosum* has grown. Because ochratoxin A is fat soluble and not readily excreted, it also accumulates in the depot fat of animals that eat feeds containing the toxin, and from there it is ingested by humans eating, for example, pig meat. It has been suggested that ochratoxin A is a causal agent of Balkan endemic nephropathy, a kidney disease with a high mortality rate in certain areas of Bulgaria, the former Yugoslavia, and Romania.

Citrinin is a significant renal toxin affecting domestic animals, such as pigs and dogs. It is also an important toxin in domestic birds, in which it

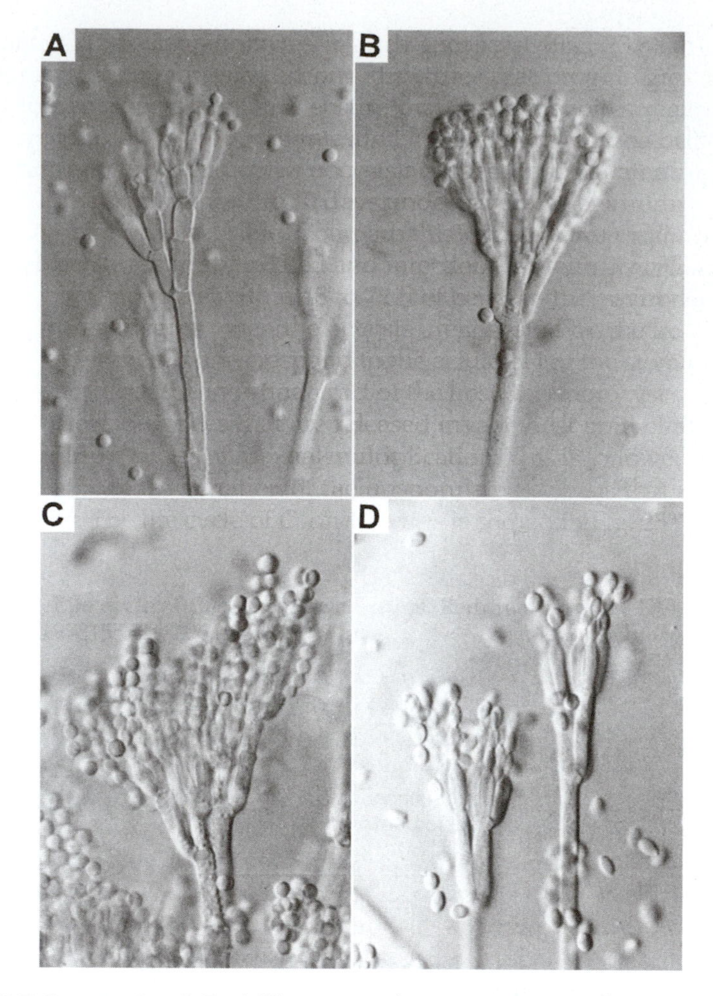

Figure 22.7 Some toxigenic *Penicillium* species (magnification, ×550). **(A)** *P. verrucosum;* **(B)** *P. crustosum;* **(C)** *P. roqueforti;* **(D)** *P. oxalicum.* Reprinted from J. I. Pitt, p. 551–562, *in* M. P. Doyle and L. R. Beuchat (ed.), *Food Microbiology: Fundamentals and Frontiers,* 3rd ed. (ASM Press, Washington, DC, 2007). doi:10.1128/9781555817206.ch22.f22.07

produces watery diarrhea, an increase in water consumption, and reduced weight gain due to kidney degeneration. The importance of citrinin in human health is difficult to assess. Primarily recognized as a metabolite of *Penicillium citrinum,* citrinin is produced by more than 20 other fungal species. *P. citrinum* is a ubiquitous (i.e., found in many environments) mold and has been isolated from nearly every kind of food surveyed for fungi. The most common sources are cereals, especially rice, wheat, and corn; milled grains; and flour. Citrinin is produced over most of the growth temperature range, but the effect of a_w is unknown.

Patulin is a lactone, and it produces teratogenic effects (that is, it causes fetal malformations [birth defects]) in rodents, as well as neurological and gastrointestinal effects. The most important *Penicillium* species producing patulin is *Penicillium expansum,* best known as a fruit pathogen but also widespread in other fresh and processed foods. *P. expansum* produces patulin as it rots apples and pears. Patulin appears to lack chronic toxic effects in humans, and therefore, low levels in juices are perhaps of little concern. However, because apple juice is widely consumed by children as well as adults, some countries have set an upper limit of 50 μg/liter for patulin in apple juice and other apple products. It is also important as an indicator of the use of poor-quality raw materials in juice manufacture. Patulin is produced by several other *Penicillium* species besides *P. expansum* (e.g., *Penicillium griseofulvum* and *Penicillium roqueforti*), but the potential for production of unacceptable levels in foods appears to be much lower. Patulin is quite stable in apple juice during storage; pasteurization at 90°C for 10 s caused a <20% reduction in levels of patulin.

Cyclopiazonic acid, a highly toxic compound, causes fatty degeneration and hepatic cell necrosis in the liver and kidneys of domestic animals. Cyclopiazonic acid is produced by *A. flavus* and at least six *Penicillium* species (e.g., *Penicillium camemberti* Thom, *P. commune*, and *P. griseofulvum*). Along with *A. flavus*, *P. commune*, a frequent cause of cheese spoilage, is probably the most common source of cyclopiazonic acid in foods. Other species producing this toxin have a wide range of habitats.

Acute cardiac beriberi was a common disease in Japan in the second half of the 19th century and has been linked to citreoviridin. The symptoms are heart distress, labored breathing, nausea, and vomiting, followed by anguish, pain, restlessness, and sometimes maniacal behavior. In extreme cases, progressive paralysis leading to respiratory failure occurred. The major source of citreoviridin is *Penicillium citreonigrum* (synonyms, *Penicillium citreoviride* and *Penicillium toxicarium*), a species which usually occurs in rice, less commonly in other cereals, and rarely in other foods. Citreoviridin is produced at temperatures from 10 to 37°C, with a maximum near 20°C.

Several tremorgenic mycotoxins are produced by *Penicillium* species. The most important is the highly toxic penitrem A. Verruculogen, which is equally toxic, is not produced by species common in foods. Less toxic compounds include fumitremorgin B, paxilline, verrucosidin, and janthitrems. At doses of ~1 mg/kg of body weight, penitrem A causes brain damage and death in rats, but the mechanism remains unclear. The only common foodborne species producing penitrem A is *P. crustosum* (Fig. 22.7B). Nearly all isolates of *P. crustosum* produce penitrem A at high levels, so the presence of this species in food or feed is a warning signal. Penitrem A appears to be produced only at high a_w. *P. crustosum* is a ubiquitous spoilage mold, occurring almost universally in cereal and animal feed samples. *P. crustosum* causes spoilage of corn, processed meats, nuts, cheese, and fruit juices, as well as being a weak pathogen on pomaceous fruits and cucurbits.

P. roqueforti (Fig. 22.7C), a species which is used in cheese manufacture but which can also be a spoilage mold, produces two toxins of interest. Roquefortine has a relatively high 50% lethal dose (340 mg/kg intraperitoneally in mice), but it has been reported to be the cause of the deaths of dogs in Canada, with symptoms similar to strychnine poisoning. PR toxin (produced by *P. roqueforti*) is apparently much more toxic (50% lethal dose, 6 mg/kg intraperitoneally in mice) than roquefortine, but it has not been implicated in animal or human disease. Due to their potential occurrence in staple foods, PR toxin and roquefortine are of considerable public health significance.

Secalonic acids are dimeric xanthones produced by a range of taxonomically distantly related molds. Secalonic acid D, the only secalonic acid produced by *Penicillium* species, has significant animal toxicity. Secalonic acid D is produced as a major metabolite of *Penicillium oxalicum* (Fig. 22.7D). A major habitat for *P. oxalicum* is corn at harvest. However, the role of secalonic acid D in human or animal disease remains unclear.

FUSARIA AND TOXIGENIC MOLDS OTHER THAN ASPERGILLI AND PENICILLIA

The most important group of mycotoxigenic molds other than *Aspergillus* and *Penicillium* species are species of the genus *Fusarium*. *Fusarium* species are most often found as contaminants of plant-derived foods, especially cereal grains. Many *Fusarium* species are plant pathogens, while others are *saprophytic*; most can be found in the soil. Consequently, these molds and their toxic metabolites (mycotoxins) find their way into animal feeds and human foods (Table 22.5). In terms of human foods, *Fusarium* species are most often encountered as contaminants of cereal grains, oil seeds, and beans. Another mold genus, other than *Aspergillus* and *Penicillium*, whose members can produce mycotoxins is *Alternaria*. *Alternaria* species are widely distributed in the environment and can be found in soil, decaying plant materials, and dust.

Authors' note

Saprophytic organisms live on dead plant tissue. They do not affect the health of a plant.

Toxigenic *Fusarium* Species

The most common toxic species, discussed below, belong to the sections *Sporotrichiella* (*Fusarium sporotrichioides* and *Fusarium poae*), *Gibbosum* (*Fusarium equiseti*), *Discolor* (*Fusarium graminearum* and *Fusarium culmorum*), and *Liseola (Fusarium verticillioides, Fusarium moniliforme, Fusarium proliferatum*, and *Fusarium subglutinans*). The major mycotoxins produced by these species are summarized in Table 22.5. The genus *Fusarium* is characterized by production of septate *hyphae* that generally range in color from white to pink, red, purple, or brown due to pigment production. The most common characteristic of the genus is the production of large septate, crescent-shaped, fusiform, or sickle-shaped spores known as *macroconidia*. The macroconidia exhibit a foot-shaped basal cell and a beak-shaped or snout-like apical cell (Fig. 22.3A). Some species also produce smaller one- or two-celled conidia known as *microconidia* (Fig. 22.3B), and some produce swollen, thick-walled chlamydospores in the hyphae or in the macroconidia (Fig. 22.3B).

Authors' note

Hyphae are branched or unbranched filaments. A mass of hyphae is referred to as a mycelium.

F. poae and *F. sporotrichioides* are in the section *Sporotrichiella*. *F. poae* is widespread in soils of temperate regions and is found on grains, such as wheat, corn, and barley. It exists primarily as a saprophyte but may be weakly parasitic. It is most commonly found in temperate regions of

Table 22.5 Major mycotoxins that may be produced by *Fusarium* species of importance in cereal grains and grain-based foods[a]

Section	Species	Potential mycotoxin(s)
Sporotrichiella	*F. poae*	Type A trichothecenes, T-2 toxin, diacetoxyscirpenol
	F. sporotrichioides	T-2 toxin
Gibbosum	*F. equiseti*	Unknown
Discolor	*F. graminearum*	Deoxynivalenol (vomitoxin), 3-acetyldeoxynivalenol, 15-acetyldeoxynivalenol, zearalenone, possibly others
	F. culmorum	Nivalenol, zearalenone
Liseola	*F. verticillioides*	Fumonisins and others
	F. proliferatum	Fumonisins, moniliformin, and others
	F. subglutinans	Moniliformin and others

[a]Reprinted from L. B. Bullerman, p. 563–578, *in* M. P. Doyle and L. R. Beuchat (ed.), *Food Microbiology: Fundamentals and Frontiers*, 3rd ed. (ASM Press, Washington, DC, 2007).

Russia, Europe, Canada, and the northern United States. *F. poae* has also been found in warmer regions, such as Australia, India, Iraq, and South Africa. *F. sporotrichioides* is found in soil and a wide variety of plant materials. The mold is found in the temperate to colder regions of the world, including Russia, northern Europe, Canada, the northern United States, and Japan. It can grow at low to very low temperatures, e.g., at −2°C on grain overwintering in the field.

Of the species in the section *Gibbosum*, only *F. equiseti* is discussed here. *F. equiseti* has been found in soils from Alaska to tropical regions and has been isolated from cereal grains and overwintered cereals in Europe, Russia, and North America. However, *F. equiseti* is particularly common in tropical and subtropical areas. For the most part, *F. equiseti* is saprophytic, but it may be pathogenic to plants such as bananas, avocados, and cucurbits. *F. equiseti* may contribute to leukemia in humans by affecting the immune system.

Even refrigerated food can become moldy.

Reprinted with permission from CartoonStock.

"Today's St. Patrick's Day. It'll be a good time to eat up this stuff with the green mold on it."

F. graminearum, classified in the section *Discolor*, is a plant pathogen found worldwide in the soil and is the most widely distributed toxigenic *Fusarium* species. It causes various diseases of cereal grains, including gibberella ear rots in corn and fusarium head blight or scab in wheat and other small grains. These two diseases are important to food microbiology and food safety because the mold and its major toxic metabolites, deoxynivalenol and zearalenone, may contaminate grain and food products made from the grain.

F. culmorum is also widely distributed in soil and causes diseases of cereal grains, among which ear rot in corn is important in food microbiology and food safety, since the mold and its toxins may contaminate corn-based foods. *F. culmorum* produces the mycotoxins deoxynivalenol, zearalenone, and acetyldeoxynivalenol.

The section *Liseola* contains four *Fusarium* species of interest: *F. verticillioides*, *F. moniliforme*, *F. proliferatum*, and *F. subglutinans*. *F. verticillioides* is a soilborne plant pathogen that is found in corn growing in all regions of the world. It is the most prevalent mold associated with corn. It often produces symptomless infections of corn plants but may infect the grain as well and has been found worldwide on food-grade and feed-grade corn. The presence of *F. verticillioides* in corn is a major concern because of the possible widespread contamination of corn and corn-based foods with its toxic metabolites, especially the fumonisins. The main human disease associated with *F. verticillioides* is esophageal cancer. Mycotoxins produced by *F. verticillioides* include fumonisins, fusaric acid, fusarins, and fusariocins.

F. proliferatum is closely related to *F. verticillioides*, yet less is known about this species, possibly because of its frequent misidentification as *F. moniliforme*. It is also frequently isolated from corn, in which it probably occurs in much the same way as *F. verticillioides*. *F. proliferatum* is capable of producing fumonisins but has not been associated with animal or human diseases.

F. subglutinans is very similar to *F. verticillioides* and *F. proliferatum*. It is widely distributed on corn and other grains. Little information is available about this mold, probably because of its misidentification as *F. moniliforme*. *F. subglutinans* has not been specifically associated with any reported animal or human diseases, but it has been found in corn from regions with high incidences of human esophageal cancer.

Detection and Quantitation of *Fusarium* Toxins

Fusarium species produce several toxic or biologically active metabolites. The toxins are trichothecenes, a group of closely related compounds. The trichothecenes are divided into three groups: type A (diacetoxyscirpenol, T-2 toxin, HT-2 toxin, and neosolaniol), type B (deoxynivalenol, 3-acetyldeoxynivalenol, 15-acetyldeoxynivalenol, nivalenol, and fusarenon-X), and type C (satratoxins). Of these, the toxin most commonly found in cereal grains and most often associated with human illness is deoxynivalenol. Other *Fusarium* toxins associated with diseases are zearalenone and the fumonisins. The most common chromatographic separation techniques used to identify these toxins are thin-layer chromatography and high-performance liquid chromatography. Gas chromatography also has some applications, particularly when coupled with mass spectrometry. Immunoassays have been developed for *Fusarium* toxins. Qualitative kits for screening, as well as kits for quantitative analysis, are available for deoxynivalenol, zearalenone, T-2 toxin, and fumonisins.

Other Toxic Molds

Other potentially toxic molds that may contaminate foods include species of the genera *Acremonium, Alternaria, Chaetomium, Cladosporium, Claviceps, Myrothecium, Phomopsis, Rhizoctonia,* and *Rhizopus,* as well as the species *Diplodia maydis, Phoma herbarum, Pithomyces chartarum, Stachybotrys chartarum,* and *Trichothecium roseum.* However, most of these molds are more likely to be present in animal feeds, and their significance for food safety may be minimal. Some have been shown to produce toxic secondary metabolites in vitro which have yet to be found to occur naturally.

The ergot mold *Claviceps purpurea* is the cause of the earliest recognized human mycotoxicosis, known as ergotism. Convulsive ergotism may have been the cause of the behavior that led to the witchcraft trials of 1692 in Salem, Massachusetts. *Alternaria* species infect plants in the field and may contaminate wheat, sorghum, and barley. *Alternaria* species also infect various fruits and vegetables, including apples, pears, citrus fruits, peppers, tomatoes, and potatoes.

Summary

- *Aspergillus, Penicillium,* and *Fusarium* are the mold genera most commonly associated with foods.
- Toxigenic molds are most often found as contaminants of plant-derived foods.
- The characteristics used to differentiate and identify molds are microscopic morphology and gross physiological features, including colony diameters and the colors of conidia and colony pigments.
- Specific growth media that inhibit the growth of bacteria are required for molds.
- Hydrogen peroxide is used commercially for detoxification of grains and other commodities.
- Molds can grow at low to very low temperatures, e.g., at −2°C on grain overwintering in the field.

Suggested reading

Bullerman, L. B. 1996. Occurrence of *Fusarium* and fumonisins on food grains and in foods, p. 27–38. *In* L. Jackson, J. DeVries, and L. Bullerman (ed.), *Fumonisins in Foods.* Plenum Publishing Corp., New York, NY.

Bullerman, L. B. 2007. Fusaria and toxigenic molds other than aspergilli and penicillia, p. 563–578. *In* M. P. Doyle and L. R. Beuchat (ed.), *Food Microbiology: Fundamentals and Frontiers,* 3rd ed. ASM Press, Washington, DC.

Ellis, W. O., J. P. Smith, B. K. Simpson, and J. H. Oldham. 1991. Aflatoxins in food: occurrence, biosynthesis, effects on organisms, detection, and methods of control. *Crit. Rev. Food Sci. Nutr.* **30:**403–439.

Flannigan, B., and A. R. Pearce. 1994. *Aspergillus* spoilage: spoilage of cereals and cereal products by the hazardous species *A. clavatus,* p. 115–127. *In* K. A. Powell, A. Renwick, and J. F. Peberdy (ed.), *The Genus* Aspergillus. *From Taxonomy and Genetics to Industrial Application.* Plenum Press, New York, NY.

Hocking, A. D. 2007. Toxigenic *Aspergillus* species, p. 537–550. *In* M. P. Doyle and L. R. Beuchat (ed.), *Food Microbiology: Fundamentals and Frontiers,* 3rd ed. ASM Press, Washington, DC.

Pitt, J. I. 2000. *A Laboratory Guide to Common* Penicillium *Species*, 3rd ed. CSIRO Food Science Australia, North Ryde, New South Wales, Australia.

Pitt, J. I. 2007. Toxigenic *Penicillium* species, p. 551–562. *In* M. P. Doyle and L. R. Beuchat (ed.), *Food Microbiology: Fundamentals and Frontiers*, 3rd ed. ASM Press, Washington, DC.

Samson, R. A., E. S. Hoekstra, J. C. Frisvad, and O. Filtenborg (ed.). 1995. *Introduction to Food-Borne Fungi.* Centraalbureau voor Schimmelcultures, Baarn, The Netherlands.

Questions for critical thought

1. Temperature and a_w influence mold growth and mycotoxin production. Select a temperature and an a_w that would control the growth of most molds and the mycotoxins they produce. Justify your answer and whether it would be feasible to implement in developing countries.
2. The presence of patulin in apple juice is a human health risk. Why? Detection of patulin may indicate quality control problems. Explain.
3. Which aflatoxins are covered by legislation and why?
4. How would you identify and differentiate *Aspergillus* and *Penicillium*?
5. Based on your knowledge gained from this chapter and other courses you have completed, speculate whether aflatoxin B_1 would be a health risk (toxicity or cancer) if it did not undergo biotransformation in the body.
6. Food microbiologists are concerned with the presence of *F. verticillioides* in corn because of the possible widespread contamination of corn and corn-based foods with its toxic metabolites. What toxic metabolites does this microorganism produce, and what is the main disease of concern for humans?
7. How might molecular identification methods for molds enhance food safety and patient health?
8. In identifying members of the genus *Fusarium*, what common (distinct) characteristics of the microorganism are used?
9. What measures could be initiated to prevent the growth of molds during storage of grains?
10. Compare the characteristics and properties of sterigmatocystin and aflatoxin.
11. List methods that can be used to detoxify grains. Focus on methods that could be implemented in developing countries.
12. Investigate a recent outbreak of foodborne illness related to the growth of toxigenic mold in the product. Describe the characteristics of the product, including processing, handling, and storage practices. Formulate a strategy that could have been implemented to prevent the outbreak.

23 Parasites

LEARNING OBJECTIVES

The information in this chapter will enable the student to:

- identify important food- and waterborne parasites
- understand the differences between protozoa, helminths, and other infectious agents
- appreciate the complicated life cycles of these organisms
- understand transmission of parasites by foods
- identify environmental sources of the organisms

Outbreak

Reports of school and workplace absenteeism increased in Milwaukee County, Wisconsin. Hospital and laboratory tests on stool specimens increased by the hundreds, but surprisingly, they all came back negative for bacterial enteric pathogens. More than 200 tests were performed in only one lab in just a couple of days, exhausting the testing reagent supplies. The stores were advertising Imodium and other antidiarrhea medicines in full-page ads in the newspapers and on billboards (Fig. 23.1), and over this Easter holiday no one was looking forward to the Easter bunny. What happened? The Milwaukee Health Department lab director requested water quality and treatment data from the Milwaukee Water Works. Results indicated spikes in turbidity of treated water at one of the two water treatment facilities. Researchers started thinking "outside the box" a bit and requested that stool specimens be tested for *Cryptosporidium*. Tests came back positive for *Cryptosporidium*. People were suffering with between 5 and 12 bouts of diarrhea each day accompanied by abdominal cramps, vomiting, fever, and muscle aches. The water treatment plant closed, and that evening Milwaukee's mayor issued a boil-water advisory. This is the story that unfolded rather quickly in Milwaukee in early April 1993. When the story was over, 14 laboratories were testing stool and water samples, 403,000 individuals were sick, 4,400 were hospitalized, and more than 100 died. What led to these events? It was the water treatment plant where the contamination originated, but it was not associated with the animal-processing plant nearby, as many had tried to suggest. Analysis of the water treatment process in the Milwaukee Water Works identified several characteristics that really led to the "perfect storm" of events for the largest waterborne outbreak of cryptosporidiosis. After this outbreak, water quality standards and laboratory testing changed, reinforcing the most important aspects of outbreak investigations, control and prevention during and after the outbreak.

doi:10.1128/9781555817206.ch23

Figure 23.1 In 1993 the largest waterborne outbreak of cryptosporidiosis was evident throughout Milwaukee, as can be seen from this billboard advertisement. The sign suggests the grand scale of this outbreak, in which 403,000 people had diarrhea in one city. The picture was published on the front page of the *Milwaukee Sentinel* newspaper on the morning of Thursday, 8 April 1993. The headline noted the boil-water advisory invoked by Mayor John Norquist the previous evening. The photo is courtesy of Jeffrey Davis and is reproduced with permission of the *Milwaukee Journal Sentinel* / JSOnline.com.

INTRODUCTION

The term *parasites* includes organisms that live the parasitic way of life, by living off of their hosts in some way. In fact, more organisms in this world are parasitic than are hosts, a fact that is most underappreciated by the general population. The parasitic way of life is quite successful, and there are parasites that can live with every phylum of animals and plants. Humans can be infected with a myriad of different kinds of parasites, from ciliates and amoebae to worms. While viruses and the rickettsial bacteria are parasitic, we do not include them in our discussion of parasites. At the basic biology level, parasites are eukaryotes, while the bacteria you have been learning about are prokaryotes. The big difference here is the presence of membrane-bound organelles, including an enclosed nucleus, in the eukaryotic parasites.

The science of parasitology grew up alongside bacteriology, and as with bacteria, there are many different types of parasites. The taxonomy changes for parasites on a regular basis as parasitologists learn more about the organisms through DNA analysis and epidemiological characteristics. The parasites that are discussed here are the ones that are most important in transmission via food and water. Parasites are not a problem only within developing countries, which is a popular misconception. In fact, as more individuals live with compromised immune systems, due in part to medical advances, in the developed world, we see an increase in diseases caused by protozoa, including *Cryptosporidium* and *Toxoplasma*. The average person living in the United States is probably under the illusion that worms (or helminths, as we call them here) are not found in locally produced foods or water. Epidemiologists estimate that 55 million children may be infected with helminths in the United States, and this is likely an underestimation. These helminths may not be acquired as readily through food and water in the United States as they are in other countries. There are several regulations in place within the United States that attempt to reduce the transmission of parasites through our food (Fig. 23.2).

Figure 23.2 While the use of night soil applied to agricultural lands may be a theoretically logical practice and is practiced in many places around the world, the application of night soil or raw manure is not recommended in the United States under good agricultural practices. This practice serves as a significant means of distribution of helminth eggs and protozoan cysts around the globe. Here two people spread night soil as fertilizer. The photo was taken in 2005 in a small village in China and is used with permission from photographer Felix Andrews. doi:10.1128/9781555817206.ch23.f23.02

Parasites have unique and complicated life cycles. In fact, some parasites have two different types of hosts, definitive and intermediate hosts. A definitive host is one in which the parasite reaches sexual maturity. An intermediate host is one that is required for parasite development, but the parasite does not reach sexual maturity in this host. Intermediate hosts are not necessary for all parasites. Parasitic life cycles truly are complex; you will notice that some life cycle information is discussed or terms are mentioned, while several points are omitted from this introductory discussion. This is by no means done for any reason other than our focus on the role of parasites that are more commonly transmitted by foods within the United States. Parasites are truly exciting and interesting organisms, and if you think so too, we encourage you to read more in a parasitology textbook.

Some parasites exhibit host specificity and may infect only humans, for example, as appears to be the case for *Cyclospora cayetanensis*, while others may be zoonotic or infect a wide range of animal hosts. An animal that harbors the organism and may transmit it to humans is known as a reservoir. In some cases, these are elusive to epidemiologists who try to determine causes of foodborne outbreaks. While some parasites may live in water and be transmitted by vectors and be extremely important, they are not discussed here. This chapter is limited to the protozoa and helminths that are the biggest players in terms of risk in food- and waterborne illness identified by epidemiological data.

Authors' note

Epidemiologists study the association of sociological factors, climate, local traditions, and global economics as well as pharmacology, pathology, biochemistry, and clinical medicine to understand the transmission of pathogens, including parasites, and ways to better control them.

PROTOZOA

Protozoa are small but generally larger than bacteria. They were first detected by Antonie van Leeuwenhoek (Fig. 23.3) between 1674 and 1716 when he recorded observations of oocysts of a parasite in rabbit livers, later known to be *Eimeria stiedai*. This protozoan is a close relative of some of

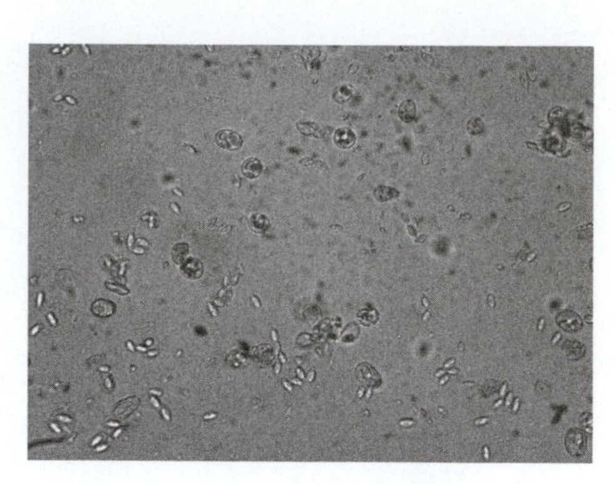

Figure 23.3 Could this have been what the view looked like through the microscope that Antonie van Leeuwenhoek pondered over in 1686? This field of view contains intact *Eimeria* oocysts and sporocysts of *Eimeria* without their oocyst membrane. doi:10.1128/9781555817206.ch23.f23.03

Authors' note

Leeuwenhoek was an excellent parasitologist, and as any good parasitologist knows, you never let an opportunity pass to find new parasites. He identified a second protozoan, Giardia lamblia, *in his own diarrhetic stools. Today some parasitologists give "roadkill" a renewed purpose and make epidemiological findings and identify organisms from a variety of animals that they find on their travels.*

today's most important food- and waterborne protozoa. Over 45,000 species of protozoa have been identified. The organisms within this group cause more disease and death around the world than any other group of organisms. Again, our discussion here is limited to a specific few. Collectively we refer to them as protozoa, but taxonomically they belong to the kingdom Protista and then are further divided into different phyla and orders based on form and function.

Protozoa are single-celled organisms. Their differences in structure, life cycles, and appearance are vast and are what in part make them such fascinating creatures. While each contains different specific structures, there are a few of great importance. The bodies of protozoa are covered with a strong plasma membrane. In fact, most protozoa have more than one membrane as part of the exterior pellicle. These membranes are important for their ability to resist environmental pressures and allow protozoal cysts to survive in the environment—in soil, manure, and water. Cyst formation is particularly common among protozoa that survive in the environment. Protozoa may also have additional coats, such as a thick glycocalyx or glycoprotein surface coat, that may have immunological importance.

An electron microscope reveals the complex layers within the cyst wall, which may contain cellulose or chitin. In some classes of protozoa, the cyst is actually called an oocyst, which is formed after sexual reproduction of different life stages. These serve as the environmentally resistant stage that is transmitted to new hosts. The oocyst or cyst contains the material that will develop into the infective stages during a process called sporulation, a series of fission events (sporogony) with cytokinesis to produce sporozoites (Fig. 23.4). Oocysts (or cysts) are ingested and undergo excystation, which is the release of the infective life stages contained within the oocyst, in the gut of the new host. Excystation is triggered by a series of factors, including a change in pH and the presence of host digestive enzymes that result in swelling of the cyst, secretion of lytic enzymes, and ultimately the release of infective life stages.

Authors' note

Recall your taxonomy from basic biology class as the way in which scientists group organisms on levels of form and function. Starting from the largest grouping down to the smallest, we have kingdom, phylum, class, order, family, genus, and species. You probably know that you are in the kingdom Animalia, phylum Chordata, class Mammalia, order Primates, family Hominidae, genus Homo, *and species* sapiens. *Or at least you know now!*

Protozoa have a variety of other important structures, including those for structural support and locomotion, like flagella. Of course, there are many changes of structure along the life cycle of each organism. As alluded to in this brief discussion, the processes of cyst formation, sporulation, excystation, and infection are complex. It cannot be overemphasized that the numbers and types of protozoa are immense, and only a few types are described here. Just a brief walk through a parasitology text will introduce you to the dozens of phyla and orders within the protozoan phyla.

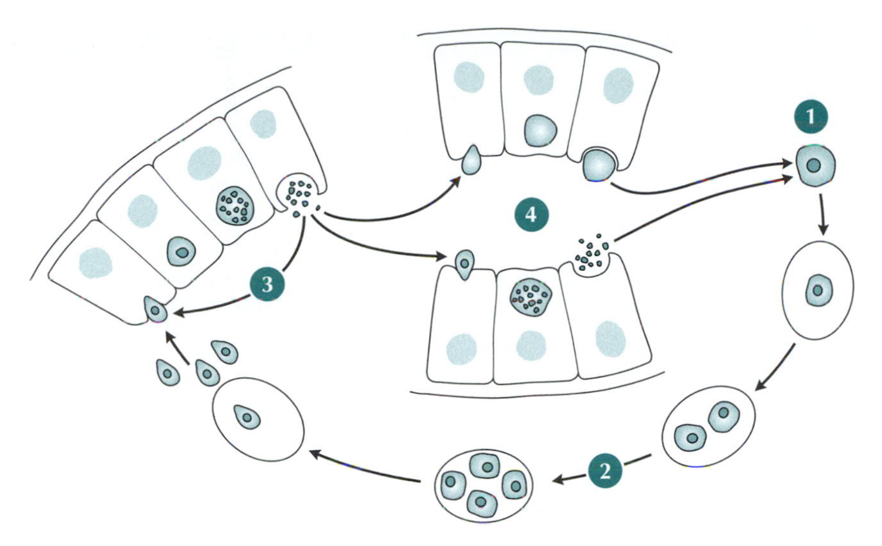

Figure 23.4 The protozoa described here all share similar basic aspects of their life cycles. These are shown here with step 1 represented by the zygote (like an egg, called an oocyst or cyst), step 2 showing infectious life forms called sporozoites found inside a sporulated or infectious (oo)cyst, step 3 showing the merozoites which are formed from asexual reproduction, and, lastly, step 4 showing the gametes forming from sexual reproduction from the latter merozoites. There may be several types of merozoites within a given life cycle, and these may have different names depending on the organism. This figure is intended to give an overview of the basic steps in the life cycle. doi:10.1128/9781555817206.ch23.f23.04

Cryptosporidium spp.

As you read in the outbreak description at the beginning of the chapter, *Cryptosporidium* was the cause of the largest waterborne outbreak in the United States, in 1993 in Milwaukee. Improvements to drinking water reduced this incidence; however, we still have several outbreaks associated with water each year in the United States. Cryptosporidiosis is a gastrointestinal illness that is distributed worldwide; approximately 51% of reported global protozoan waterborne diarrheal outbreaks are associated with *Cryptosporidium*. Now these outbreaks are more likely associated with recreational water, and a seasonal prevalence is certainly present during the summer months. In fact, the incidence of disease transmitted by recreational water has increased steadily over the years since the mid-1990s. Cryptosporidiosis is now the leading cause of outbreaks due to recreational water (Box 23.1). In 2011, the Centers for Disease Control and Prevention (CDC) considered cryptosporidiosis an endemic disease in the United States and estimated that there are approximately 748,123 cases each year in the United States. Some cases may be foodborne as well, associated with ready-to-eat foods and fresh produce. It is estimated that cryptosporidiosis costs the United States $44 million each year from 2,725 hospitalizations with an average stay of 5.8 to 9.3 days and an average cost of $16,203 per hospitalization.

Cryptosporidiosis is transmitted in humans by ingesting oocysts of *Cryptosporidium parvum* or *C. hominis* (Fig. 23.5). There are 19 other species that are species specific and infect specific mammals, birds, reptiles, amphibians, and fish. *Cryptosporidium parvum* is zoonotic, and oocysts may be shed by infected cattle, deer, and other ruminants. *Cryptosporidium*

BOX 23.1

Recreational water cryptosporidiosis

Cryptosporidium appears to be more prevalent now in recreational water than it used to be. Illness is certainly seasonal, and each summer pools close due to contamination and illness of cryptosporidiosis associated with wave pools, fountains, and swimming. Perhaps it is not as easy to keep one's mouth closed when swimming as one might think. The figure below is an epidemiological curve that shows the number of cases of laboratory-confirmed cryptosporidiosis by date of illness onset in a swimming pool in Utah from May through November 2007. This epidemiological curve also provides details of when control measures were adopted. The most important parts of an outbreak are the control and prevention steps to reduce illness and stop it from recurring. Within this one outbreak there were more than 2,000 cases of illness. Why? We can think about this by putting some numbers with the situation. On a hot summer day, hundreds of people may swim in a good-size pool. The median duration of illness is 9 or 10 days, with a range from 3 to 28 days. While the infectious dose is not known, it may be as few as 10 to 30 oocysts, but generally <100. Animal studies indicate that infected calves may excrete up to 10^{10} oocysts daily for up to 14 days during active infection. Now imagine that one toddler in that swimming pool is sick or an ill person using the pool did not follow the best hygienic practices. There could be many hundreds of oocysts in that pool water. But what about the chlorine and the pool filtration systems? It is important that these two prevention measures be in good standing; however, the normal chlorine levels for a swimming pool are 1.0 to 3.0 parts per million, which does not easily inactivate *Cryptosporidium* oocysts. And with a large pool, it takes many hours to filter all the water in the pool. As oocysts are about 5 μm in size, they can be removed by diatomaceous earth filtration systems. Here is a staggering thought: in a pool with 1.0 parts per million of chlorine at pH 7.5 and water at 77°F, inactivation of *Cryptosporidium* would take about 9,600 min (which is 6.7 days). To inactivate *E. coli* O157:H7 under the same conditions would take less than 1 min, and to inactivate hepatitis A virus would take about 16 min. Even the protozoan parasite *Giardia* could be inactivated in less than 1 h under these conditions. Public health services currently recommend that individuals who have diarrhea or who have had diarrhea within the last 14 days do not use a pool.

Epidemiological curve of cryptosporidiosis associated with a contaminated swimming pool in Utah. doi:10.1128/9781555817206.ch23.fBox23.01

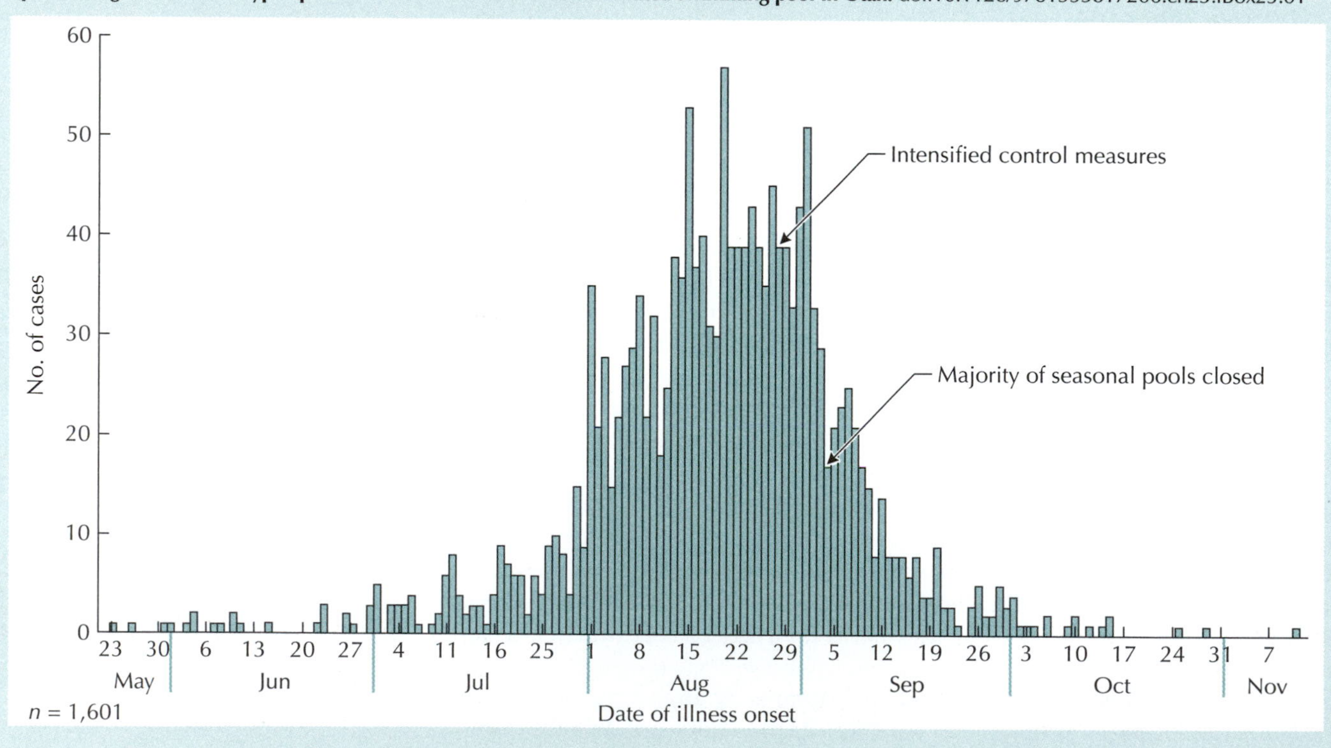

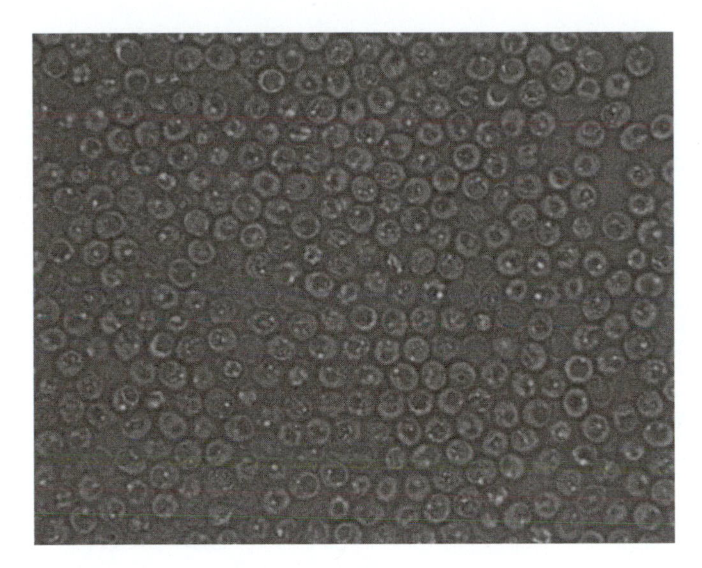

Figure 23.5 Many *Cryptosporidium parvum* oocysts, visualized by differential interference contrast microscopy. doi:10.1128/9781555817206.ch23.f23.05

hominis is infectious only to humans and is a fairly newly named species. Until recently, there were believed to be two different genotypes of *C. parvum*, one of which is now known as *C. hominis*. *Cryptosporidium* oocysts are shed already sporulated and infective. There is great genetic diversity in *Cryptosporidium* species, which is now being tracked by the CDC through a new molecular surveillance program in the United States and in the United Kingdom. Through this program similarities will be noted between animal and human isolates, including environmental and clinical samples. This system has already been successful in source tracking by linking cases with suspected swimming pool outbreaks within the United States.

The spherical oocysts of *Cryptosporidium* are quite small, only about 4 to 5 µm wide. When ingested, the sporozoite life stages excyst in the intestine and invade the epithelial cells of the respiratory system or the intestine from the ileum to the colon. The current life stage undergoes several stages of asexual reproduction safely within the confines of what is termed a parasitophorous vacuole. Here the organisms are safe and cause a blunting of the villi which results in diarrhea in the host, but they do not invade the cells (Fig. 23.6). The asexual life stages lead to the production of microgametes, which then, through sexual reproduction, produce oocysts, which are shed in the feces as early as 5 days postinfection. This series of steps is depicted in the life cycle in Fig. 23.7. Clinical symptoms include a profuse watery diarrhea ranging from 6 to 25 bowel movements each day, with 1 to 17 liters per day in total. Thinking back to the outbreak at the beginning of the chapter, now you can understand how the improperly working water treatment system was overwhelmed by contaminated stools containing *Cryptosporidium* oocysts. Additionally, *Cryptosporidium* oocysts are resistant to normal levels of chlorine. Effective water treatment strategies include physical removal by flocculation and filtration and perhaps treatment with ultraviolet (UV) light or ozone.

As you can see, the life cycle described above is quite complex. Let's add two additional points which increase the infectivity of this pathogen. The first deals with the shedding of oocysts. Approximately 20% of the oocysts that are formed are thin-walled oocysts. These oocysts never exit

Figure 23.6 The center of this image shows a parasitophorous vacuole holding several asexual life stages of *Cryptosporidium parvum* (meront, indicated by the star). The intestinal cells are disrupted and blunted villi are visible (indicated by the arrow). You can see how these villi would not be able to have regular fluid absorption, so the host would have tremendous diarrhea. doi:10.1128/9781555817206.ch23.f23.06

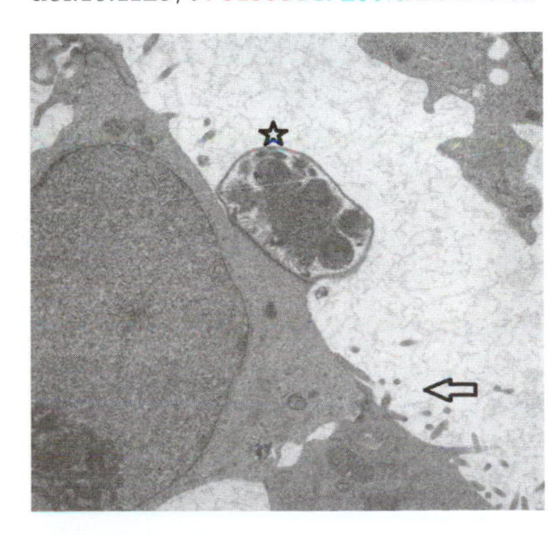

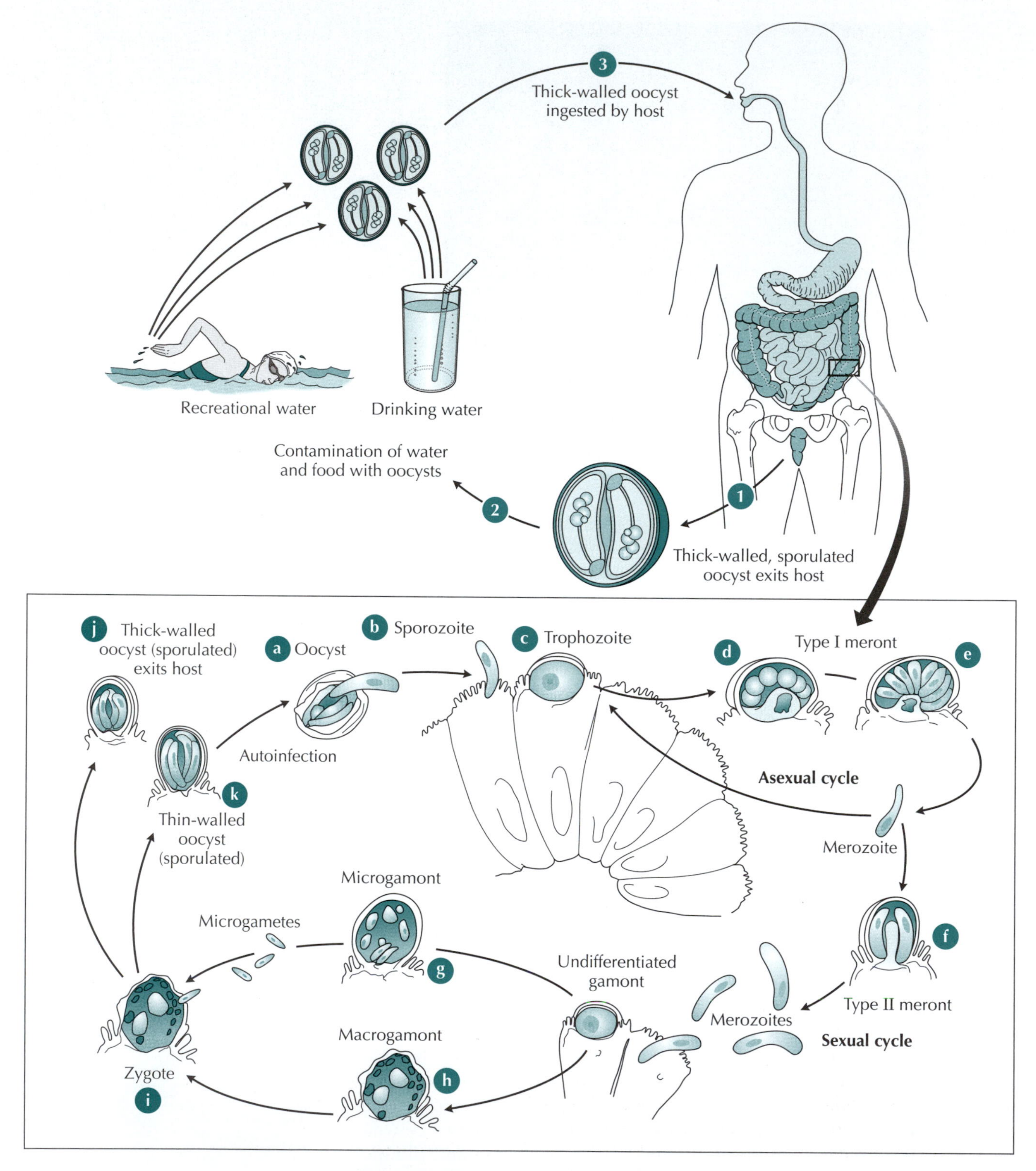

Figure 23.7 Life cycle of *Cryptosporidium*, the causative agent of cryptosporidiosis. Redrawn from a CDC illustration. doi:10.1128/9781555817206.ch23.f23.07

the host and result in a continuous reinfection. The second point is that only recently were drugs discovered that could reduce infection by *Cryptosporidium*. These two facts, coupled with the fact that immunity is largely of the cell-mediated type, allow *Cryptosporidium* to wreak havoc especially on individuals with compromised immune systems. Immunocompetent persons would normally be able to pass the disease after mounting a strong immune system; however, the immunocompromised may suffer from cryptosporidiosis for long periods.

In terms of transmission via foods, cryptosporidiosis has been associated with several food products. Surveys of fresh produce throughout the world have identified *Cryptosporidium* oocysts at different levels. Several ready-to-eat foods and salads have been contaminated by infected food handlers, including prepared salmon, noodle salad, fruit salad, and sandwiches.

During the 1990s and early 2000s, several outbreaks of cryptosporidiosis were associated with consumption of unpasteurized apple cider. *Cryptosporidium* was an important organism of consideration in the adaptation of the juice Hazard Analysis and Critical Control Point (HACCP) rule, which was implemented in 2002. *Cryptosporidium* can be successfully inactivated when apple cider is treated using UV light and pasteurization, when the processes are validated.

Authors' note

It is essential that a food process be tested for the parameters used by the specific producer, including the mixture of ingredients. Process validation should be performed each time a parameter is changed.

Cryptosporidiosis is a disease of substantial economic and health burden within the United States. While its incidence varies seasonally and geographically, the majority of cases are related to contaminated recreational water. Changes in pool filtration and disinfection practices are needed. Food and water containing *Cryptosporidium* have contaminated ready-to-eat foods, molluscan shellfish, and fresh produce.

Cyclospora cayetanensis

Cyclospora cayetanensis became infamous in the United States after being associated with raw raspberries and blackberries imported from Guatemala that were linked to several large outbreaks of gastrointestinal illness in the late 1990s and early 2000s. In 2011, the CDC estimated that there are approximately 19,808 cases each year in the United States. *Cyclospora cayetanensis* is a food- and waterborne protozoan parasite that infects the upper small intestine of humans and can cause severe diarrhea, stomach cramps, and nausea, which may be accompanied by fever. Cyclosporiasis is treatable with the sulfonamide antibiotic trimethoprim-sulfamethoxazole. *Cyclospora* oocysts were first observed in stool samples in Papua New Guinea in the 1970s and were referred to as cyanobacterium-like bodies. Interestingly, *Cyclospora* is still often referred to as an emerging pathogen due to the many unknowns regarding its transmission. Cyclosporiasis is not thought to be associated primarily with immunocompromised individuals like other human protozoan pathogens. *C. cayetanensis* was identified as a new coccidian species in 1993 at Cayetano Heredia University in Lima, Peru, by Ynes Ortega and colleagues when they successfully induced oocyst sporulation and excystation of the sporozoites in vitro.

Cyclospora cayetanensis oocysts are quite large, 7.5 to 10 μm in diameter. The species was once referred to as large *Cryptosporidium* and may be confused with *Cryptosporidium* when oocysts from food samples are viewed microscopically. These oocysts have a strong outer membrane composed of complex carbohydrates and lipids which make these oocysts acid-fast. The oocyst membrane protects two oblong sporocysts that surround the infective

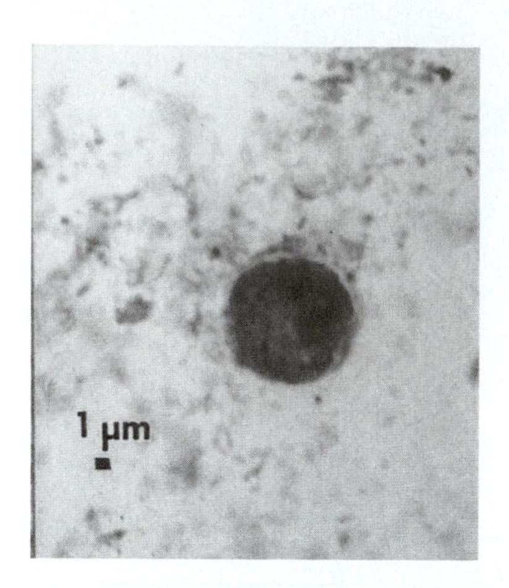

Figure 23.8 A *Cyclospora cayetanensis* oocyst. Reprinted from D. S. Lindsay, S. J. Upton, and L. M. Weiss, p. 2113–2121, *in* P. R. Murray et al. (ed.), *Manual of Clinical Microbiology*, 9th ed. (ASM Press, Washington, DC, 2007). doi:10.1128/9781555817206.ch23.f23.08

life stages, with four sporozoites in each sporocyst (Fig. 23.8). The oocyst and sporocyst membranes are strong structures that provide great stability to environmental pressures and ensure that the sporozoites remain viable along their journey to the small intestine. Like many protozoa but different from *Cryptosporidium*, *Cyclospora* oocysts are shed unsporulated and sporulate outside the host within 7 to 10 days under favorable environmental conditions. Sporulation is variable. Compare this with *Cryptosporidium* oocysts, which are shed already sporulated and infectious, or with *Toxoplasma gondii* oocysts, which sporulate within 48 to 72 h of being in the environment. The infection process begins when the oocysts are ingested by the host. Coccidian oocyst outer membranes respond to the acidic pH of the stomach. When the sporocysts reach the intestinal tract of the host, the sporocyst wall breaks down and the sporozoites that are released invade host epithelial cells and undergo multiple cycles of asexual multiplication. This is followed by sexual development for the formation of the unsporulated oocysts that are shed in the host feces. The life cycle of *C. cayetanensis* is shown in Fig. 23.9.

Figure 23.9 Life cycle of *Cyclospora cayetanensis*. Redrawn from a CDC illustration. doi:10.1128/9781555817206.ch23.f23.09

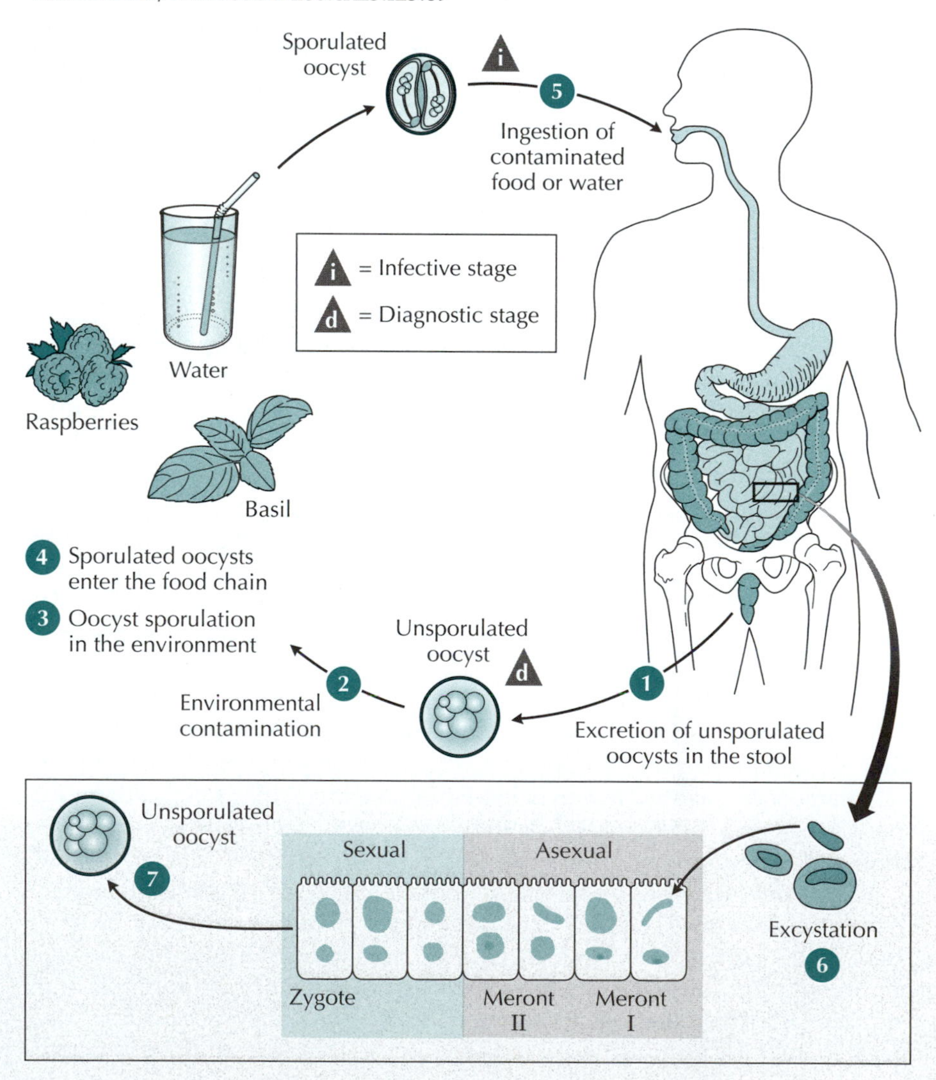

The first reported outbreak of cyclosporiasis in the United States involved contaminated water in Chicago, Illinois, in 1990. Cyclosporiasis has been associated with fresh fruits, vegetables, and herbs likely contaminated by water, soil, or produce handlers (Fig. 23.10). In particular, raspberries, basil, parsley, snow peas, and leafy greens have been implicated as probable transmission vehicles in 19 outbreaks of cyclosporiasis in the United States. As with the hepatitis A virus, *Cyclospora cayetanensis* oocysts are shed by humans and contamination can occur at both preharvest points (soil, feces, irrigation water, dust, insects, or animals) and postharvest points (human handling, equipment, or transport containers).

In total, the 10 events that have involved *C. cayetanensis* and contaminated raspberries accounted for 2,864 illnesses and subsequently eight traceback investigations, including five farm investigations, four of these in Guatemala and one in Chile. The first reported outbreaks of cyclosporiasis associated with raspberries were in New York and Florida in 1995. These outbreaks did not involve traceback investigations of any kind, and approximately 71 individuals were involved. In New York, drinking water from portable coolers at a country club and raspberries that were served during the outbreak period were suspected. In Florida, raspberries were suspected but were a component of a fruit cup and desserts served at various social events. During 1996, more than 1,660 individuals in the United States became ill from raspberries contaminated with *Cyclospora*. There were traceback and farm investigations associated with these three large outbreaks. Together, these events involved 20 states and the District of Columbia. Raspberries were traced back to Guatemala for many of the events, and berries that were implicated had been harvested from between 3 and 30 farms. In the large multistate outbreak in 1996, a majority of the raspberries were traced to one exporter; however, nothing concrete came of the farm investigations, as exporters had included raspberries from different farms in a single shipment. Another large multistate outbreak occurred the following year, again associated with Guatemalan raspberries. Several farms were identified during the investigation by the Food and Drug

Figure 23.10 Even though *Cyclospora* is about 10 times the size of a bacterium, you still cannot tell with the naked eye whether these foods are contaminated.
doi:10.1128/9781555817206.ch23.f23.10

Administration (FDA). In 1998, an outbreak of cyclosporiasis occurred in Massachusetts and no farm investigation was pursued, as it was not possible to determine if the raspberries originated from Chile or from Guatemala. In 2000, raspberries associated with a cake were involved in an outbreak and the berries were traced to three sources (one Guatemalan farm, one Chilean farm, and one unknown U.S. farm). The farm in Guatemala was later implicated in the Pennsylvania outbreak mentioned previously and was also found to be associated with an outbreak in the state of Georgia the same year. Raspberries from Chile were also suspected in the latter outbreak in 2000. In 2002, raspberries from Chile were again suspected in an outbreak that involved 22 individuals.

The role of water has been questioned in the transmission of oocysts to berries and other foods, including mesclun lettuce (1997), basil (1997, 1999, 2004, and 2005), and snow peas (2004), which were associated with outbreaks. The water used to mix pesticides was previously identified as a possible source of contamination in the outbreaks of cyclosporiasis associated with contaminated raspberries. Water and soil are of concern in many parts of the world where cyclosporiasis is endemic and individuals who shed oocysts may be asymptomatic. Contact with soil among health care and farm workers in Guatemala was a risk factor for cyclosporiasis infection. Understanding the risk associated with infection may help us to understand where contamination may be more likely. In several documented outbreaks raspberries were the likely vehicle of contamination, but they were associated with other foods, including wedding cake and lemon tart. While studies attempting to determine the infectious dose have not been successful, epidemiological investigations have suggested that the infective dose is low or that in the berry-associated outbreaks the number of oocysts per berry was high. Compared to other coccidians, *Cyclospora* oocysts require a great amount of time to sporulate in the environment. *Cyclospora cayetanensis* is difficult to study in the laboratory, as humans are its only known host, making access to oocysts and methods to evaluate viability difficult and limited. It is important to note that sporulation can be inactivated by exposing oocysts to extreme temperatures that would be used at home in food preparation or in the food industry.

Toxoplasma gondii

In 2011, the CDC estimated that there are approximately 173,995 cases of *T. gondii* infection each year in the United States. Exactly how much of this incidence is directly foodborne is not well understood at this time; however, toxoplasmosis is the fourth leading cause of hospitalizations due to foodborne illness and the second leading cause of mortality associated with foodborne illness. The last two statistics certainly suggest that toxoplasmosis is a serious disease.

Figure 23.11 Unsporulated **(A)** and sporulated **(B)** *Toxoplasma gondii* oocysts. Courtesy of the CDC. doi:10.1128/9781555817206.ch23.f23.11

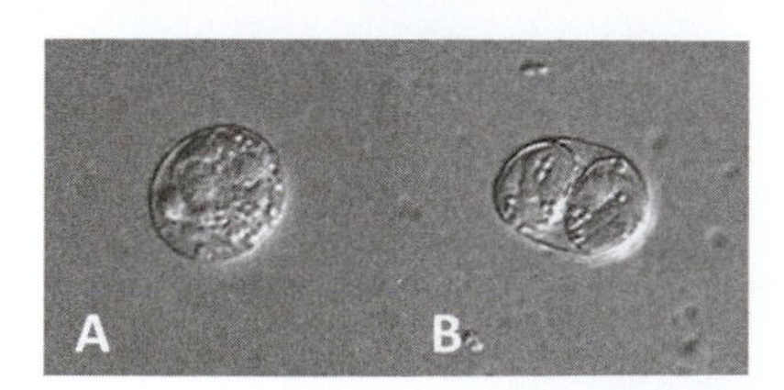

As you may have guessed, toxoplasmosis is caused by the protozoan parasite *Toxoplasma gondii*, which was first observed in 1908 in a desert rodent called the gondi found in Tunisia. This protozoan is different from the others discussed here so far in that it causes an extraintestinal disease, has a more complicated life cycle, and can be transmitted to humans in three different ways. Oocysts (10 to 13 μm by 9 to 11 μm) are excreted unsporulated into the environment by cats, which are the definitive hosts, and then sporulate and become infective in about 24 h (Fig. 23.11). These oocysts are ingested by wild animals, domestic animals, or humans, who serve as

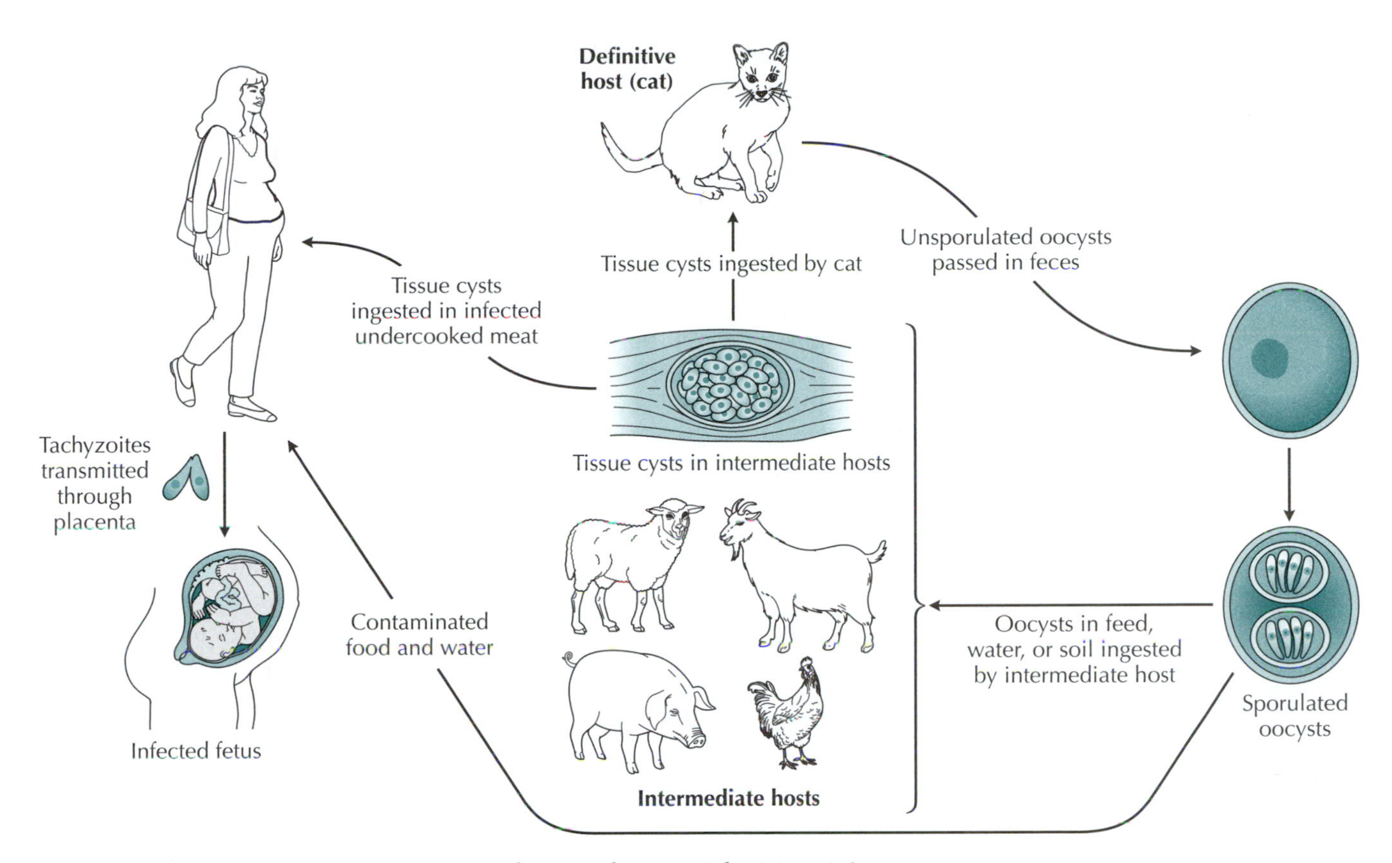

Figure 23.12 Life cycle of *Toxoplasma gondii* showing the role of the feline definitive host that sheds the oocysts and the role of the various potential intermediate hosts. Redrawn from an illustration supplied by J. P. Dubey (Agricultural Research Service, USDA). doi:10.1128/9781555817206.ch23.f23.12

the intermediate hosts in which the organism undergoes a series of asexual reproduction stages that become encysted within muscle tissue (Fig. 23.12). These tissue cysts can contain large numbers of life stages called bradyzoites (Fig. 23.13). As infection becomes chronic, tissue cysts develop in the brain, heart, and skeletal muscles. Cysts may persist for months or even years after the initial infection. A healthy immune system is able to keep an infection in check, so to speak. If immunity lapses, released bradyzoites can boost immunity to its prior level, protecting the body against another infection, and this is called premunition. Immunity for *Toxoplasma*, like other protozoan parasites, involves both antibody- and cell-mediated immunity types. This is one reason why there has been an increase in protozoan parasitic diseases as more people live longer, healthier lives with suppressed immune systems. These latent infections may cause severe health problems in persons with compromised immune systems if the tissue cysts become reactivated. This may occur if tissue cysts break down and the immune system cannot keep the bradyzoites at bay. How are cysts involved in transmission? Tissue cysts can form in domestic animals after they ingest oocysts in their feed or on the pasture. Tissue cysts can form in swine and cattle. If a human ingests contaminated meat that is undercooked, that person ingests tissue cysts which contain viable life stages. From the above information, the first two ways in which *Toxoplasma* can be transmitted is through ingestion of oocysts or tissue cysts.

Figure 23.13 Bradyzoites are present within this tissue cyst from a mouse that was fed an oyster harvested from lab-inoculated seawater. Courtesy of David Lindsay, Virginia-Maryland Regional College of Veterinary Medicine. doi:10.1128/9781555817206.ch23.f23.13

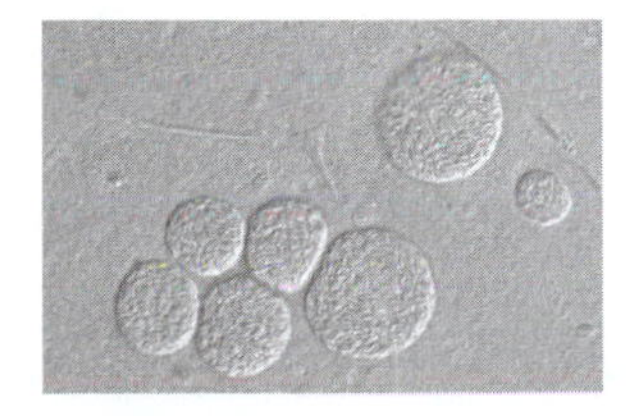

Another form of this disease is quite tragic in nature. Congenital toxoplasmosis occurs when a pregnant woman contracts acute toxoplasmosis at the time of conception or during pregnancy. The organisms can infect the developing fetus by crossing the placental barrier. Neonatal infections may be asymptomatic, but some are not. A significant number of these infections cause death or disability to newborns. Stillbirths or spontaneous abortions may occur. The rate of transmission to the fetus during a maternal infection is approximately 45%. Of these cases, the majority are subclinical; however, around 10% of fetuses may die and 30% may suffer severe damage, including hydrocephalus, intracerebral calcification, retinocarditis, and developmental delay. These issues occur not only in humans but also in sheep, which appear to be particularly susceptible to congenital toxoplasmosis. Early diagnosis and treatment are important. This is why it is essential that women of childbearing age be aware of their toxoplasma serology. If a woman is seropositive before pregnancy, she is protected from congenital infection. Pyrimethamine and sulfonamides can be given together against *Toxoplasma*.

Authors' note

Perhaps this is a good time to remind you that it is important to cook meat properly and thoroughly, be sure to wash hands well after handling meat, and avoid cross-contamination.

Much work has been done to reduce the risk of transmission of oocysts and tissue cysts to animals via contaminated feed and water. It is difficult to reduce this risk to zero; however, the majority of cases of toxoplasmosis now in the United States do not originate from ingesting conventionally raised meat. The origin of the majority of cases of toxoplasmosis today is unknown. Some have originated from the ingestion of improperly cooked meats, including wild boar and other wild game meats (including cougars, black bears, and polar bears—oh, my!). Prevalence studies assessing *T. gondii* antibodies in swine populations in the United States have been conducted (Table 23.1). It is difficult to come to conclusions based on these data and those from the CDC, which place toxoplasmosis as the second leading cause of foodborne mortality and the fourth leading cause of hospitalizations due to foodborne illness. Perhaps wild game animals play a larger role than once thought. Or perhaps oocysts are more prevalent in the environment and are consumed with water or fresh produce. However, the U.S. Department of Agriculture (USDA) estimates that one-half of *T. gondii* infections are caused by the ingestion of raw or undercooked infected meat. According to the CDC in 2000, the presence of immunoglobulin G antibodies to *T. gondii* is about 15 to 20% in persons 12 to 49 years of age. In other places around the world seroprevalance is much higher, for example, at about 80 to 90% among people in France and Austria. Most of the people that are seropositive are asymptomatic. As you read earlier, a healthy immune system is able to keep this protozoan parasite in check. Might poultry play a role in toxoplasmosis? Free-range poultry were tested for the presence of antibodies to *T. gondii* and found to have seroprevalence rates of 86% in Nicaragua, 64% in Ghana, 24% in Indonesia, 12.5% in Italy, 30% in Poland, and 24% in Vietnam. Chickens were also found to be seropositive within the United States, but rates varied by farm, from 16 to 100% positive for free-range birds and 30 to 50% in organically raised chickens. These data were obtained by the USDA.

Giardia intestinalis

Giardia is the most common parasitic cause of diarrhea in the United States. The highly contagious diarrheal disease giardiasis constitutes about 40% of reported waterborne diarrheal outbreaks, and to complicate matters, *Giardia* may be present in asymptomatic carriers. The disease appears to

Table 23.1 Prevalence of *T. gondii* antibodies in sera of pigs from the United States[a]

Yr(s) sampled	Type	Source of sera	No. tested	% Positive[b]
1983–1984	Market hogs	Nationwide	11,229	23
	Sows		623	42
1989–1992	Sows	Iowa	1,000	22.2
1989–1990	All ages	Hawaii (31 farms)	509	48.5
1990	Sows	Nationwide	3,479	20
1991–1992	Sows	Tennessee (343 herds)	3,841	36
1992	Market hogs	Illinois (179 herds)	1,885	3.1
	Sows		5,080	20.8
1992–1993	Market hogs	Illinois (47 herds)	4,252	2.3
	Sows		2,617	15.1
1994–1995	Market hogs	North Carolina (14 herds)	2,238	0.5
1994–1995	Market hogs	Nationwide	4,712	3.2
	Sows		3,236	15
1998	Various ages	New England states (Connecticut, Massachusetts, New Hampshire, Rhode Island)	1,897	47.5
2000	Market hogs	Nationwide	8,086	0.9
	Sows		5,720	6
2002	Market hogs	Massachusetts (1 herd)	55	87.2
2006	Market hogs	Maryland (1 herd)	48	68.7

[a]Adapted from J. P. Dubey and J. L. Jones, *Int. J. Parasitol.* **38:**1257–1278, 2008.
[b]A positive test was indicated by obtaining a 1:20 dilution or higher of antibodies using the modified agglutination test. This type of test uses diluted serum samples taken from the pigs and mixes with a known amount of *T. gondii* antigen and visually identifies a positive test through a binding of the antibody-antigen complex. You might remember this from your basic biology course.

be more prevalent in children than in adults. It is commonly spread in day care centers and, like other gastrointestinal diseases, can be spread easily in institutional environments. While overall prevalence is not too high, infection in the United States is estimated at 2% of the population.

In 2011, the CDC estimated that there are approximately 1.3 million cases each year in the United States. Giardiasis in mammals (including humans) is transmitted by a species known by the names *Giardia lamblia*, *G. intestinalis*, and *G. duodenalis*. While more than 40 species have been described, five are considered valid, and the one above, along with *G. muris*, infects mammals. The other three are specific to birds (*G. ardeae* and *G. psittaci*) and amphibians (*G. agilis*). Why does the infective agent of giardiasis in humans have three names? That is a good question, one with which parasitologists and taxonomists are still wrestling. It appears that the confusion occurred some time ago and continues today; however, now the organism is more routinely referred to as *G. intestinalis*. As mentioned above, *G. lamblia* has been known since 1681, when it was discovered by Leeuwenhoek in his own diarrhetic stools.

Giardia is unique in that it is a flagellated protozoan parasite that exists in two forms: immobile cysts when outside the host and mobile trophozoites when inside the intestine (Fig. 23.14). The cysts are extremely hardy and provide protection from various degrees of heat, cold, and desiccation. Cysts are resistant to conventional water treatment methods such as normal levels of chlorination. Water treatment facilities rely on flocculation and physical removal, which is effective because of the large size of

Figure 23.14 Diagram of a *Giardia* cyst **(left)** and trophozoite **(right)**. Redrawn from a CDC illustration. doi:10.1128/9781555817206.ch23.f23.14

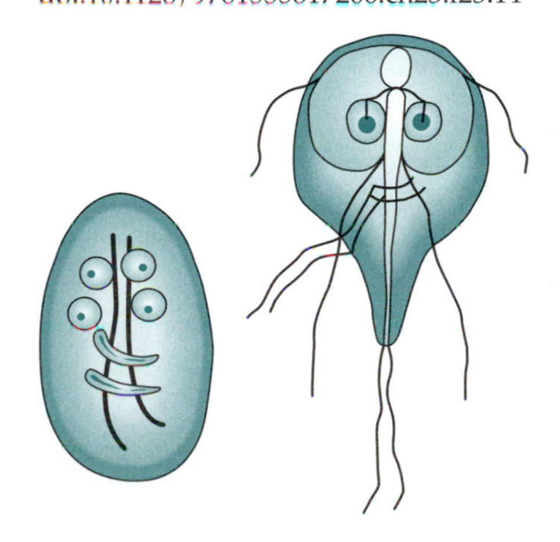

the cysts (8 to 12 μm in length and 7 to 10 μm in diameter). Like *Cryptosporidium* oocysts, *Giardia* cysts are immediately infectious when shed. Once the cyst is ingested with contaminated water or food, trophozoites are released into the small intestine of the host. The trophozoites are unique in shape and structure, which eases their identification in diarrhetic stools. Trophozoites divide by binary fission, similar to bacteria. This occurs so rapidly that one diarrhetic stool can contain up to 14 billion parasites, with the average at about 300 million. As you can imagine, one infected individual can spread disease to many others. After multiple divisions within the host, the active life stages encyst themselves within the colon, dehydrate, and are passed in the feces of the infected individual. The life cycle continues when cysts are swallowed by a new host, safely pass through the stomach, excyst in the duodenum, complete multiple cycles of division, grow flagella, and multiply. The infectious dose is estimated at between 10 and 25 cysts. The incubation period before onset of clinical symptoms is between 3 and 25 days, with a median of 7 to 10 days, and gastrointestinal illness may last for weeks without treatment. Positive identification may be made by microscopy analysis of trophozoites in the stool or by direct fluorescent antibody detection of antigens in the stool. Metronidazole is most commonly prescribed for treatment.

Humans are a major reservoir for *Giardia*, but wild and domestic animals, including cattle, are also documented as possible reservoirs. Giardiasis was once referred to as beaver fever due to its association with campers and drinking water from contaminated streams. This may have catalyzed the development of the many water filtration units available for campers today. The disease has been most frequently associated with the consumption of contaminated water. Foodborne outbreaks have been traced to fresh produce and also to ready-to-eat foods (cold salads and sandwiches) contaminated by infected food handlers.

Giardiasis certainly has a great impact worldwide. *Giardia* is responsible for approximately 40% of reported waterborne diarrheal outbreaks worldwide. As the numbers reported above suggest, it is the most common parasite diagnosed in the United States. Estimated costs within the United States are at $33.8 million per year. This number includes 3,581 hospitalizations per year, with an average stay of 3.9 to 4.4 days at a cost of $9,445 per hospitalization, as determined by the CDC. Interestingly, if we look at the numbers of cases of giardiasis each year since 1992, the numbers are stable. There is a seasonal increase in the summer months each year, but the average number of case reports has remained consistent over the past 10 years. The number of large drinking water-associated outbreaks has decreased due to the regulatory changes that occurred in the late 1980s. Scientists are attempting to better understand the transmission of *Giardia* by assessing the zoonotic potential through molecular genotyping of the strains in the seven recognized assemblages (A to G). In this genetic ranking, assemblages C to G show a marked host specificity (cats and dogs, hoofed livestock, cats, and rodents) and A and B are a bit broader and include humans.

While giardiasis is no longer a major drinking water issue, it is an endemic illness within the United States and places a substantial economic and health burden, particularly for young children. The greatest risk factors are using untreated drinking water, contaminated recreational water, and person-to-person transmission. Animal contact may also be a risk factor. Food can become contaminated through infected food handlers and nonpotable water used for irrigating or washing fresh produce.

Other Protozoa of Interest

Of course, the few protozoa mentioned above are not the only protozoa that may be transmitted by water and food. There are several protozoa that cause human illness and may cause the traveler's diarrhea that you have likely heard about and which elicits warnings not to drink the water or eat fresh produce when traveling. Several types of amoebae may cause gastroenteritis. This disease is also called amoebic dysentery. Free-living amoebae are unicellular protozoa common to most soil and aquatic environments. You may remember watching them in sixth or seventh grade under a light microscope. The organism *Entamoeba histolytica* predominantly infects humans and other primates and is estimated to infect about 50 million people worldwide; however, it is much less prevalent in the United States and affects only <5% of the population. Amoebic cysts are ingested with contaminated food or water and infect the intestinal cells, similar to *Giardia*. The cysts mature and undergo a series of asexual reproductive cycles (Fig. 23.15). Life stages, known as trophozoites, vary with invasive (invading the intestinal cells) or noninvasive infections. Amoebic dysentery can be successfully treated with metronidazole, but it is better to avoid ingesting amoebae if possible. Food may be contaminated by infected food handlers, irrigation water, insects, animals, manure, or night soil fertilizers.

Another group of protozoa that should be mentioned here are the microsporidia. This group includes approximately 150 genera and 1,200 species. These are zoonotic protozoa and can infect a wide array of invertebrate and vertebrate hosts. The main genus that is of concern for foodborne contamination is *Encephalitozoon*. Microsporidia in general are quite different from the protozoa described above. The tiny spores (1 to 4 μm) are highly resistant to environmental pressures. The illness is transmitted by spores covered by a wall of protein and chitin, so they are similar to fungal spores. The spores infect host cells by shooting out a polar tube which carries the genetic material. Microsporidial infections occur more often in immunocompromised persons, and clinical symptoms include diarrhea, fever, malaise, and nausea. The basic life cycle of two species of microsporidia is shown in Fig. 23.16. In 2009 there was an outbreak of gastrointestinal illness associated with cucumbers contaminated by *Enterocytozoon bieneusi*. Contaminated cucumber slices were served in cheese sandwiches and in a salad that were determined to be the probable vehicles of transmission based on epidemiological data. In this outbreak the clinical samples were

Figure 23.15 Trophozoites of *Entamoeba histolytica* contain ingested red blood cells. Reprinted from A. L. Leber and S. M. Novak, p. 1990–2007, *in* P. R. Murray et al. (ed.), *Manual of Clinical Microbiology*, 8th ed. (ASM Press, Washington, DC, 2003). doi:10.1128/9781555817206.ch23.f23.15

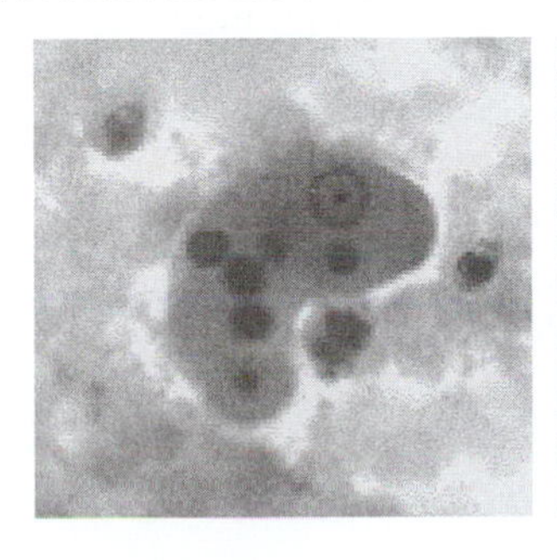
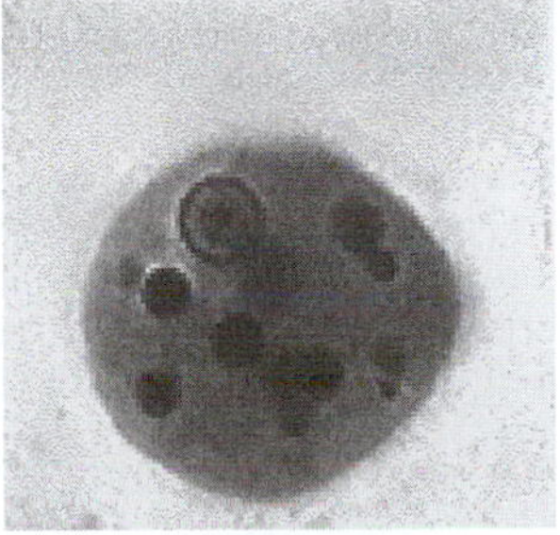

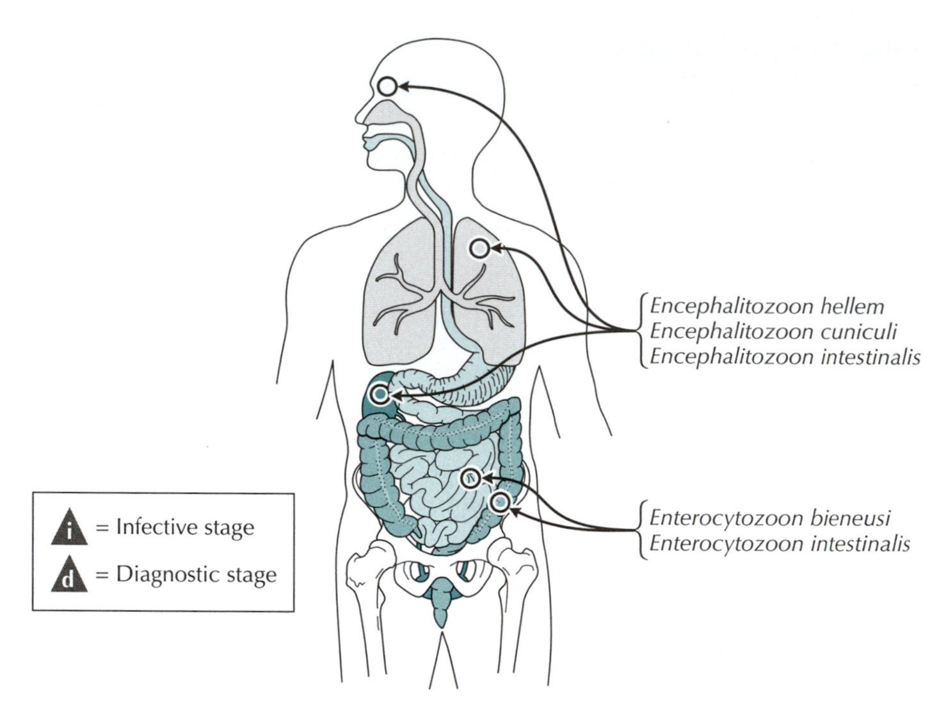

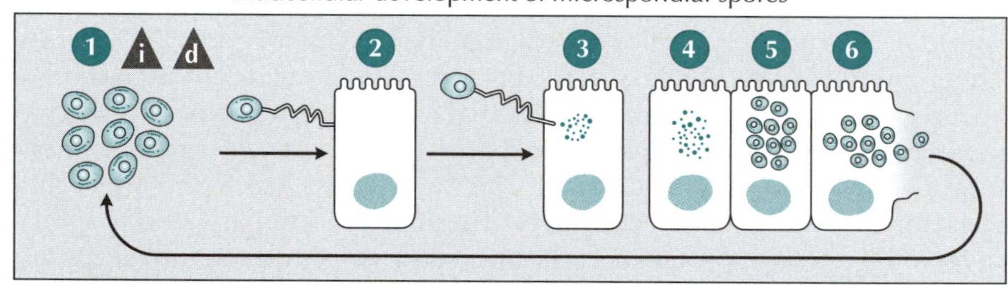

Enterocytozoon bieneusi

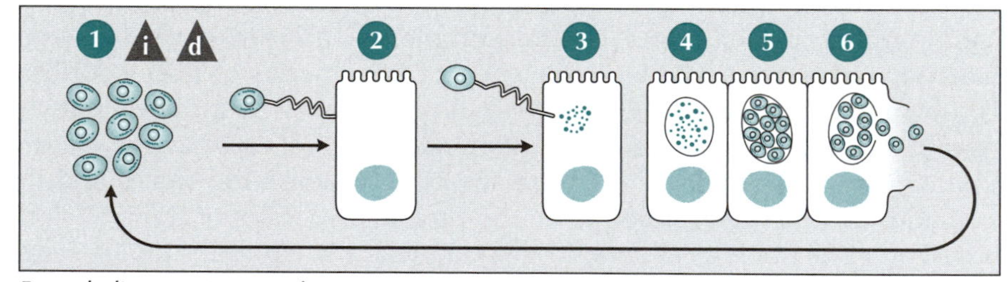

Encephalitozoon intestinalis

Figure 23.16 Life cycle of two common agents of microsporidiosis. The steps are as follows. **(1)** The infective spore can survive for a long time in the environment. **(2)** The spore extrudes its polar tubule and infects the host cell. **(3)** The spore injects the infective sporoplasm into the host cell through the polar tube. **(4)** Inside the cell, the sporoplasm undergoes extensive multiplication through fission and asexual reproduction stages. **(5)** Further development into mature spores can occur either in direct contact with the host cell cytoplasm (*Enterocytozoon bieneusi*) or inside a parasitophorous vacuole (*Encephalitozoon intestinalis*). **(6)** A thick wall is formed around the spore, which provides resistance to adverse environmental conditions; as spores increase in number and completely fill the host cell cytoplasm, the cell membrane is disrupted and releases the spores into the surroundings. **(7)** The free mature spores can infect new cells, thus continuing the cycle. Redrawn from a CDC illustration. doi:10.1128/9781555817206.ch23.f23.16

collected 7 months after the outbreak occurred. In this case it is not known if contamination occurred preharvest or by an infected food handler. This is the first documented produce-linked outbreak due to microsporidia. It is likely that the contamination levels were quite high, so that washing was unsuccessful at removing all the spores from the cucumbers. Another point that is possible and has been shown with other protozoa is that cysts and oocysts adhere quite strongly to fresh produce. While a zoonotic link cannot be completely eliminated, it is likely that since all cases were traced genetically to *Enterocytozoon bieneusi* genotype C, the initial contamination source was human. It is important to reiterate that washing of fresh produce remains an important point in the prevention of foodborne illness.

HELMINTHS

The world of parasites is vast. Perhaps there are more helminths than there are protozoa. Helminths include the roundworms, tapeworms, and flukes. While there are a myriad of helminth genera that cause illness in humans, just a handful are discussed here. These are ones that have been of importance in animal production within the United States. For the most part the removal of these from conventional animal production within the United States is a public health achievement. There are still risks for some of these organisms as we consider the global landscape for food production, and animals raised in outdoor systems inherently confront higher risks of exposure to foodborne parasites. Other helminths discussed here may be risk considerations for production of fruits, vegetables, or seafood. Each type of helminth has its own characteristics. The life cycles of foodborne helminths are still complex but differ from those of the protozoa in the numbers of asexual reproductive cycles. Typically an egg develops into one or more larval stages which grow into adult worms. Sexual reproduction within the definitive host leads to production of eggs that are shed in the feces of the infected host.

Roundworms (Nematodes)

The roundworms, or nematodes, are the most abundant animals on the planet. In general, nematodes are elongated, cylindrical worms that are tapered at both ends and have a thick cuticle to help them survive environmental pressures. Roundworms generally feed on blood, cellular tissues, and fluids. Only a few cause diseases in humans and domestic animals; three of these that can be transmitted by foods are discussed below.

Trichinella spiralis

The most important roundworm from a food microbiology perspective is *Trichinella spiralis*. To start off, this organism is of historical importance because it is the main reason why pork was believed to be "unclean." In stories throughout history, many people became ill after eating undercooked contaminated pork. This is likely part of the scientific basis for why some religions still prohibit eating pork. In the United States, there has been a continual decrease in the incidence of *T. spiralis*. In the 1940s there were more than 400 cases each year. In the 1980s this decreased to 30 to 40 each year, but in the 1990s there were >100 cases each year. The fairly recent increase was likely a result of increased consumption of wild game, foods from other cultures, and organically produced or free-range pork. *Trichinella* is zoonotic and can have a variety of hosts, including pigs, sheep, wild boars, bears, cougars, horses, and dogs. Nearly all of these types of meats have been epidemiologically linked to

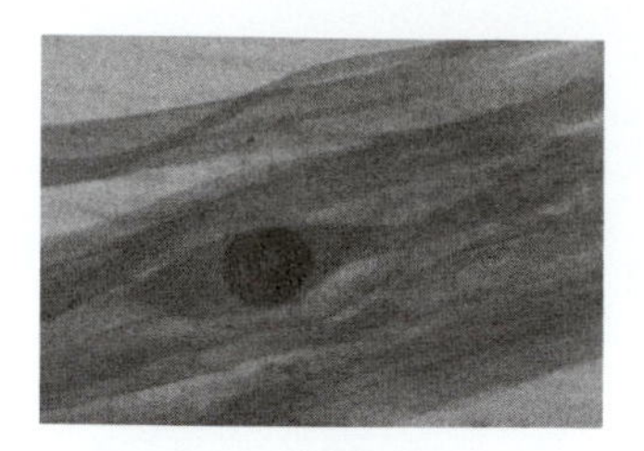

Figure 23.17 Coiled larva of *Trichinella spiralis* in muscle. doi:10.1128/9781555817206.ch23.f23.17

cases of trichinosis within the United States. In 2011, the CDC estimated that there are approximately 162 cases each year in the United States.

To become ill, the host must ingest the encysted larvae that are wound up tightly in contaminated and undercooked meat (Fig. 23.17). The larvae are released from the nurse cells in the stomach in response to the host's digestive enzymes and pH and develop into adult worms in the small intestine of the definitive host. Over a period of 4 to 16 weeks the female adult worm (1.4 to 1.6 mm long) may release many larvae that then migrate, enter the lymph or blood system, and penetrate the tissues. Eventually the spent male and female worms die and pass out of the body. Newborn larvae are carried throughout the bloodstream and may enter skeletal muscle. Most juveniles are carried away by the hepatoportal system through the liver and then to the heart, lungs, and arterial system, which distributes them throughout the body. During this migration, the larvae may be distributed to any muscle within the body. The larva matures in the muscle and forms what are called nurse cell complexes and become encysted. The dedifferentiation of the muscle cell and formation of the nurse cell, also referred to as a cyst, constitute a complex cellular process that involves a rerouting of a network of tiny blood vessels to the cyst. This process is not completely understood.

Trichinellosis has somewhat vague clinical symptoms, which is one reason for the underreporting of this disease. Newborn larvae cause an inflammatory response beginning about 48 h postinfection. Pain may accompany the migration of the larvae to the skeletal tissues, diaphragm, or eyes. The encysted larvae become calcified within 9 to 12 months. Clinical symptoms include edema, muscle atrophy, and potential respiratory problems, depending on the migration of the larvae.

The reduction in trichinae is a success story for swine production and pork safety within the United States (Fig. 23.18). This is such a great story

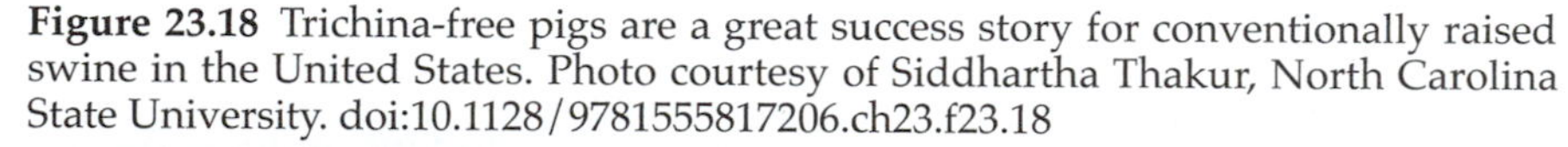

Figure 23.18 Trichina-free pigs are a great success story for conventionally raised swine in the United States. Photo courtesy of Siddhartha Thakur, North Carolina State University. doi:10.1128/9781555817206.ch23.f23.18

that most recently the USDA reduced the recommended cooking temperature to 145°F, since the risk of parasite contamination (*Trichinella* or *Toxoplasma*) is almost zero. The trichina-free certification program has been in development since the mid-1990s and was part of the 2008 farm bill. Voluntary pilot programs were launched in 1995, 1997, and 2000. Certification includes four elements described below.

1. Accredited veterinarians (trained in good production practices for trichinae) work with producers to minimize infection rates on their farms.
2. On-farm audits are performed to document trichina-free status, with periodic audits to prove that good production practices for trichinae are being observed.
3. A statistical sample of the herd is tested regularly at slaughter using enzyme-linked immunosorbent assay or diaphragm digestion test to verify the absence of trichinae.
4. USDA Animal and Plant Health Inspection Service veterinarians conduct random spot audits to ensure completeness and credibility of certifications with trading partners.

It makes sense that the trichina-free certification system is based on certifying management practices rather than testing every pig.

Ascaris lumbricoides

The opposite of the trichinae, which are the smallest nematodes, the ascarids are the largest intestinal roundworms (Fig. 23.19). The adults grow up to 18 inches long and wreak havoc on their human hosts as they block bile and pancreatic ducts. An adult *A. lumbricoides* worm can produce up to 200,000 eggs daily and up to 2.7 million eggs in a lifetime. As the eggs hatch, juveniles penetrate the intestinal wall and enter the circulatory system. If the worms enter the lungs they can cause the disease known as ascaris pneumonia.

From a historical perspective, *Ascaris* was first described by the ancient Greeks, with details of the disease from *A. lumbricoides* and *A. suum* written on papyrus. Infection in North America is estimated at 3 million humans, and infection worldwide is estimated at 600 million. It is believed that *A. lumbricoides* was originally a parasite of pigs and became infectious to humans when pigs and humans lived in close contact with one another.

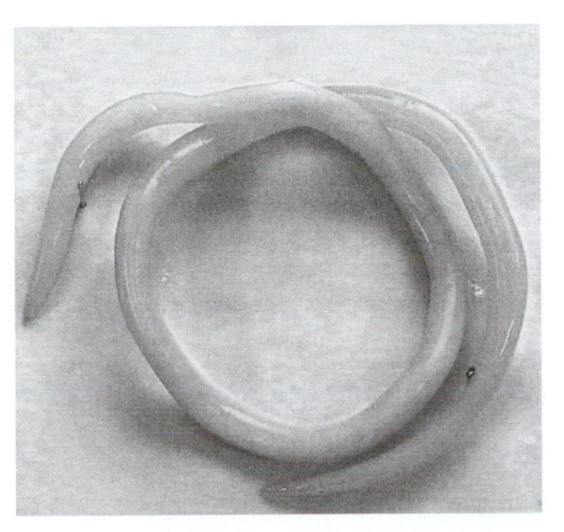

Figure 23.19 An adult roundworm (*Ascaris lumbricoides*). Reprinted from H. Sheorey, B.-A. Biggs, and P. Traynor, p. 2144–2155, *in* P. R. Murray et al. (ed.), *Manual of Clinical Microbiology*, 9th ed. (ASM Press, Washington, DC, 2007). doi:10.1128/9781555817206.ch23.f23.19

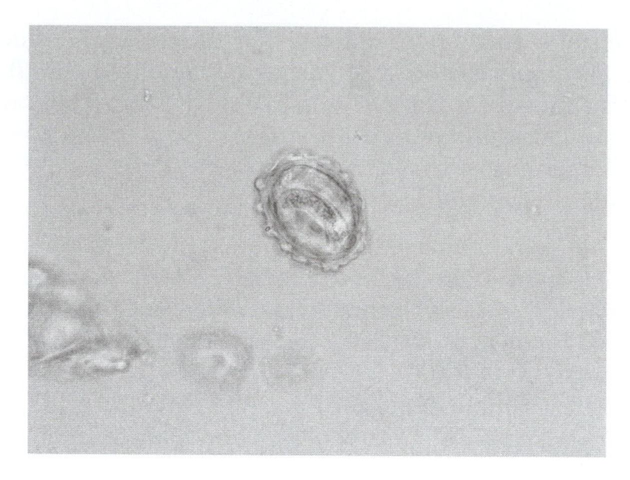

Figure 23.20 An egg of *Ascaris suum*. Imagine one million of these! Courtesy of David Lindsay, Virginia-Maryland Regional College of Veterinary Medicine. doi:10.1128/9781555817206.ch23.f23.20

This is a good example of evolution, as today there are two separate species, *A. suum* in pigs and *A. lumbricoides* in humans. You can imagine that with many eggs being shed, contamination events can be huge. It is difficult to determine the specific number of food- and waterborne cases for several reasons, including the facts that the eggs are voided with feces into the soil, the eggs can remain infectious in the soil for months, and soil may be used in the production of fresh produce (Fig. 23.20). The longevity of the eggs and their extraordinary chemical resistance contribute to the success of the parasite. A higher prevalence was noted in Asia, and it is likely that contamination occurs more often with the use of night soil in agriculture. Due to their ability to survive, when the environment has been so heavily seeded it is difficult to reduce incidence even when proper sanitation habits are initiated at a later date. Children are more likely to become infected, especially if they have conditions like pica (a condition of low levels of minerals that propels one to eat dirt). Blowflies and cockroaches have also been shown to carry *Ascaris* and other eggs.

Anisakis

The codfish worm, a member of the genus *Anisakis*, is a small roundworm. This worm may be small but is still visible with the naked eye and has surprised many seafood connoisseurs. *Anisakis* (Fig. 23.21) is also known as a sushi parasite in Japan, The Netherlands, and the United States; however,

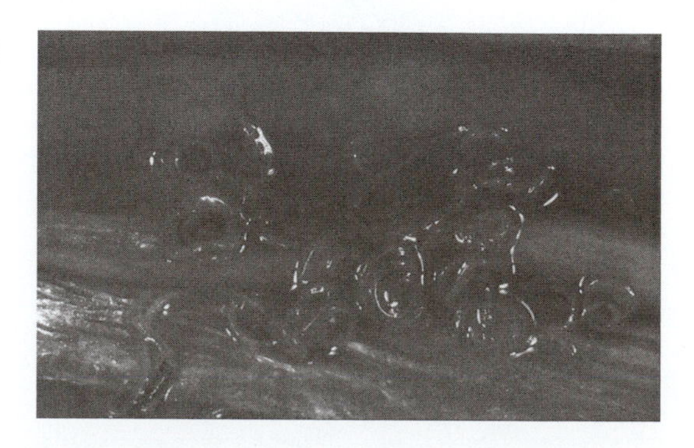

Figure 23.21 Parasitic anisakids on a herring. doi:10.1128/9781555817206.ch23.f23.21

the roundworm is prevalent worldwide. There are fewer than 10 cases of infection in the United States each year, but once you have met *Anisakis*, it is difficult to forget the experience. There are several hundred to 1,000 cases each year reported from Japan. Some researchers speculate that cases may increase in the United States as the sushi trend increases as well. Cooking kills the juveniles, but the increasing popularity of raw fish dishes, such as sushi, sashimi, and ceviche, ensures a continued risk for human infection. The good news is that freezing in a commercial blast-freezer kills the organisms and ensures the safety of the raw fish product.

Clinically, there are two different types of disease caused by *Anisakis*. The first is perhaps the most memorable. This noninvasive form does not display many specific symptoms, and the host is typically asymptomatic—that is, until the worm makes its move and migrates up the esophagus into the pharynx of the host. This is where the name "tingling throat disease" originates. An infected host may spit up the worm due to the tingling in the throat. This is why it is a most memorable disease. The second form of this disease is the invasive form. In this case the worm *Anisakis* penetrates the stomach or the small intestine and causes nausea, pain, vomiting, and diarrhea in the host.

Tapeworms (Cestodes)

The majority of tapeworms infect wild and domestic animals, but a few infect humans or domestic animals. Even still, many tapeworms, also called cestodes, cause no medical or economic problems; however, tapeworms are quite interesting, and this section serves as an introduction to them. In some cases, the life cycles of cestodes are not completely understood, and there is still much room for research on cestodes, including those described below.

Tapeworms have a distinct body profile (Figure 23.22). The head, called a scolex, has suction cups for adhering to the host. The body is composed of many segments called proglottids. Each proglottid has a complete set of systems, including reproductive organs, and can grow into its own organism in time. Like the roundworms, tapeworms feed on blood and cellular tissue.

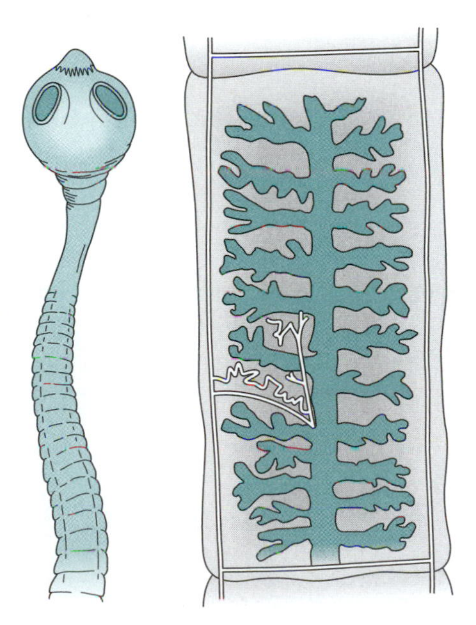

Figure 23.22 Drawing of a typical tapeworm. The scolex has suction cups for adhering to the host, and the body is composed of many segments called proglottids; each proglottid has a complete set of systems, including reproductive organs, and can grow into its own organism in time. doi:10.1128/9781555817206.ch23.f23.22

Taenia spp.

The genus *Taenia* contains two species of zoonotic importance. For *Taenia solium* the intermediate hosts are pigs, and for *T. saginata* the intermediate hosts are cattle. The larval stage in beef cattle is referred to as *Cysticercus bovis* or beef measles. The disease produced in cattle is known as cysticercosis bovis, and the flesh that is riddled with the juvenile life stages is called measly beef. A person who eats infected beef that is not cooked sufficiently to kill the juveniles becomes infected. The folded scolex and neck of the cysticercus excyst (in a way) in response to bile salts in the host. As the adult worm grows (Fig. 23.23), the scolex absorbs to the intestinal cells, and within 2 to 12 weeks the worm begins to shed gravid proglottids. USDA inspectors check each cattle head in federally inspected abattoirs for the visible presence of beef measles in the inner cheek meat of the animal. Humans are the definitive host for these two species, so the human host is the one who sheds eggs in the feces. The disease begins with the ingestion of raw or undercooked contaminated meat that contains the cysts or encysted larvae. The encysted larvae are released by the digestion process of the human host, and adult worms grow and attach to the intestinal cells. Humans shed the eggs or proglottids that break off and release with the feces. Some proglottids develop into larvae, grow, and migrate to tissues, causing disease there. Cattle and pigs are the intermediate hosts and become infected by ingesting vegetation contaminated by eggs or gravid proglottids.

In terms of clinical disease, mild infections may be asymptomatic. Human hosts may feel dizziness and abdominal pain and have diarrhea. As the adult worms grow, they may cause intestinal obstruction and pain.

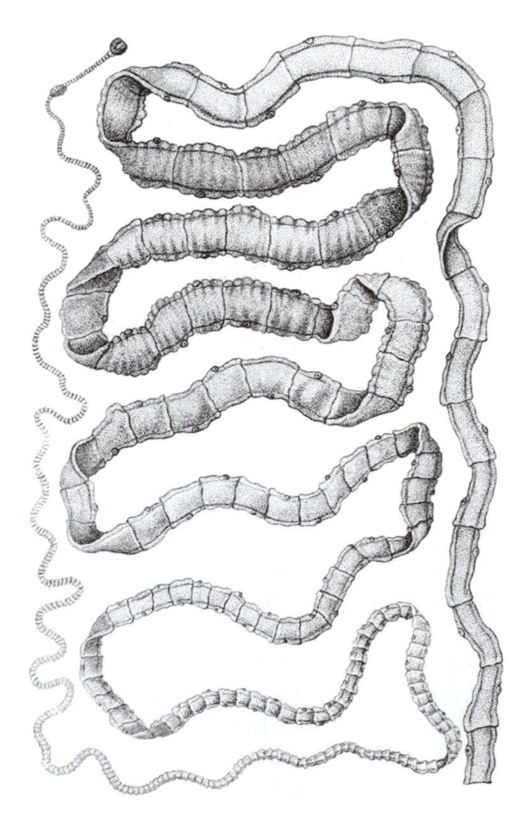

Figure 23.23 An adult *Taenia solium* tapeworm. doi:10.1128/9781555817206.ch23.f23.23

Some individuals react more strongly to the antigens on the worms and may display what looks like an allergic reaction. It is believed that *T. solium* is the most dangerous adult tapeworm of humans because of the possibilities of self-infection with the juvenile cysticerci. Unlike those of *T. saginata*, the cysticerci of *T. solium* can develop readily in humans. Infection occurs when shelled larvae pass through the stomach and hatch in the intestine. Virtually every organ and tissue in the body may harbor cysticerci, but especially connective tissue, eyes, brain, heart, liver, and lungs. As mentioned above, some individuals have an allergic reaction, which may be worsened with the death of a cysticercus. This elicits a strong inflammatory response, resulting in the eventual calcification of the parasite. If this occurs in the eye, there is little chance for corrective surgery.

The rate of human infection is highest in areas of the world where beef is a major food and sanitation is deficient. In several developing nations in South America and Africa, it is more common for people to ingest meat that is charred on the outside but raw inside. It is always important to cook meat thoroughly and at temperatures recommended by the USDA. Prevention of cysticercosis with *T. solium* depends on early detection and personal hygiene. The risk of fecal contamination of food and water must be reduced, including contamination by infected food handlers. The epidemiology of neurocysticercosis indicates that cases are decreasing. The majority of cases within the United States are associated with travel or immigrant workers.

Diphyllobothrium latum

The largest human tapeworm, also known as the broad fish tapeworm, is *Diphyllobothrium latum*. This worm grows up to 10 m in length! Like *Anasakis*, this worm is found in raw or undercooked fish and is certainly memorable when found in a human host. It is unique in particular for its size. The larvae first develop in a crustacean and then may be found in whitefish, trout, pike, or salmon (Fig. 23.24). Its prevalence has been documented across Europe (including the Baltic area), within the Great Lakes in the United States, and in Canada. This cestode is most common in fish-eating carnivores, particularly in northern Europe. The worm may have hitched a ride to the other areas where it has been isolated, perhaps in a fish or in a dolphin. It is not surprising that clinical symptoms include nausea, pain, and diarrhea. Interestingly, some infected individuals are left with a vitamin B_{12} deficiency following infection. This cestode apparently can absorb up to 44% of the vitamin B_{12} available to a human host. This failure of absorption of vitamin B_{12} in the intestine leads to pernicious anemia.

Again, the human is a definitive host through which the adult worms develop and eggs are passed in the feces. An adult worm can shed up to a million eggs a day. The eggs develop into larvae, which live in crustaceans that are ingested by fish, which are then undercooked or pickled and ingested by humans. This worm is not picky in its definitive hosts; bears and other mammals can also fill this spot in the life cycle. Epidemiologically speaking, cases occur more often in countries where raw fish is consumed on a regular basis. *Diphyllobothrium latum* builds up in local fish in communities that dispose of sewage by draining it into lakes or rivers. These worms have been shown to survive preparation of gefilte fish and may be transmitted in sushi.

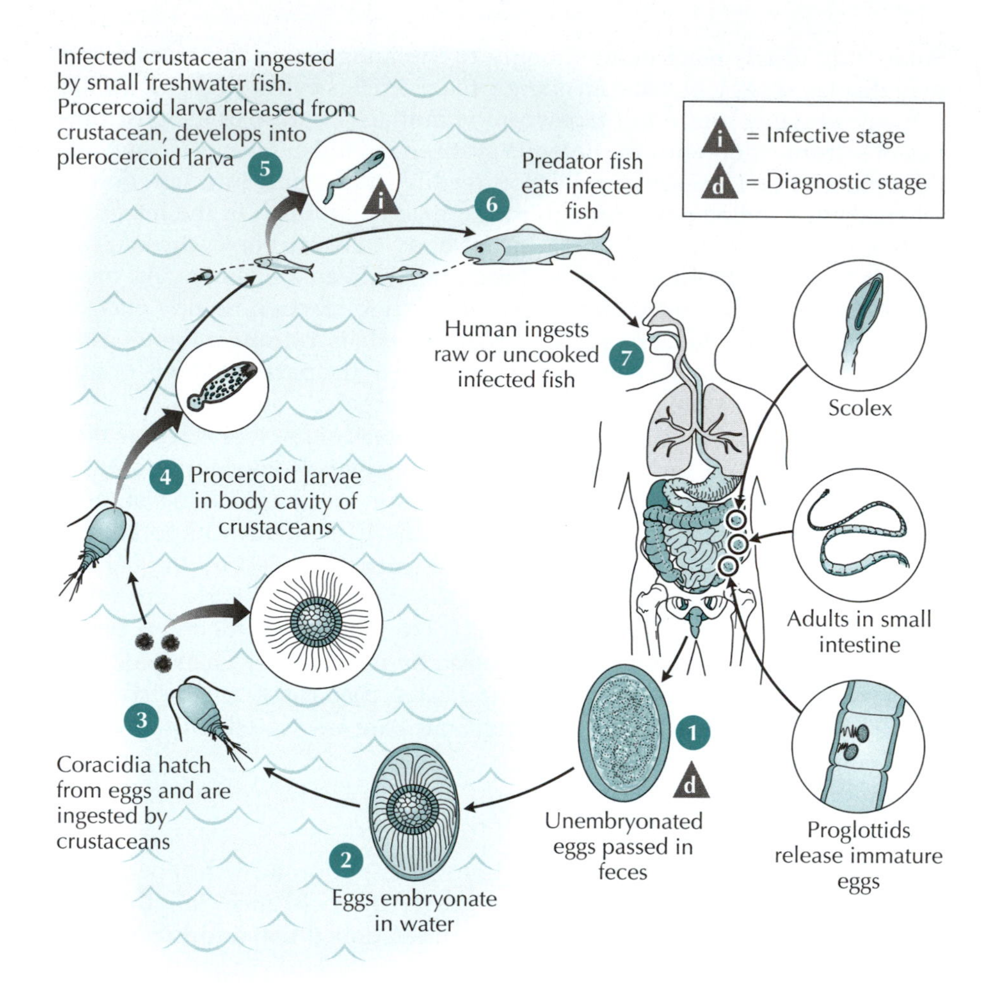

Figure 23.24 Life cycle of *Diphyllobothrium latum*. Redrawn from a CDC illustration. doi:10.1128/9781555817206.ch23.f23.24

Flukes (Trematodes)

Trematodes are flukes and are among the most abundant parasites (second only to nematodes, but that's another story). There is a fluke to every animal species on the planet. Flukes have complicated life cycles, with at least two hosts, one of which tends to be a mollusc. The typical life cycle (Fig. 23.25) involves a ciliated free-swimming larva that hatches from its shell and penetrates the first intermediate host, which is the snail. Within the snail the miracidium undergoes cycles of asexual reproduction and erupts from the snail as cercariae, which after more asexual reproduction stages become metacercariae and can infect the definitive host. Recall that sexual reproduction can only occur within the definitive hosts and that they are the ones to shed the eggs in their feces.

Fasciola hepatica

Fasciola hepatica is rare in humans in many countries, but it is an important parasite of sheep and cattle and has been for hundreds of years. The first recorded account of *F. hepatica* was in 1379 by Jean de Brie in his description of the sheep disease called liver rot. The species name *hepatica* comes from the fact that this organism infects the liver of its host. These flukes are large

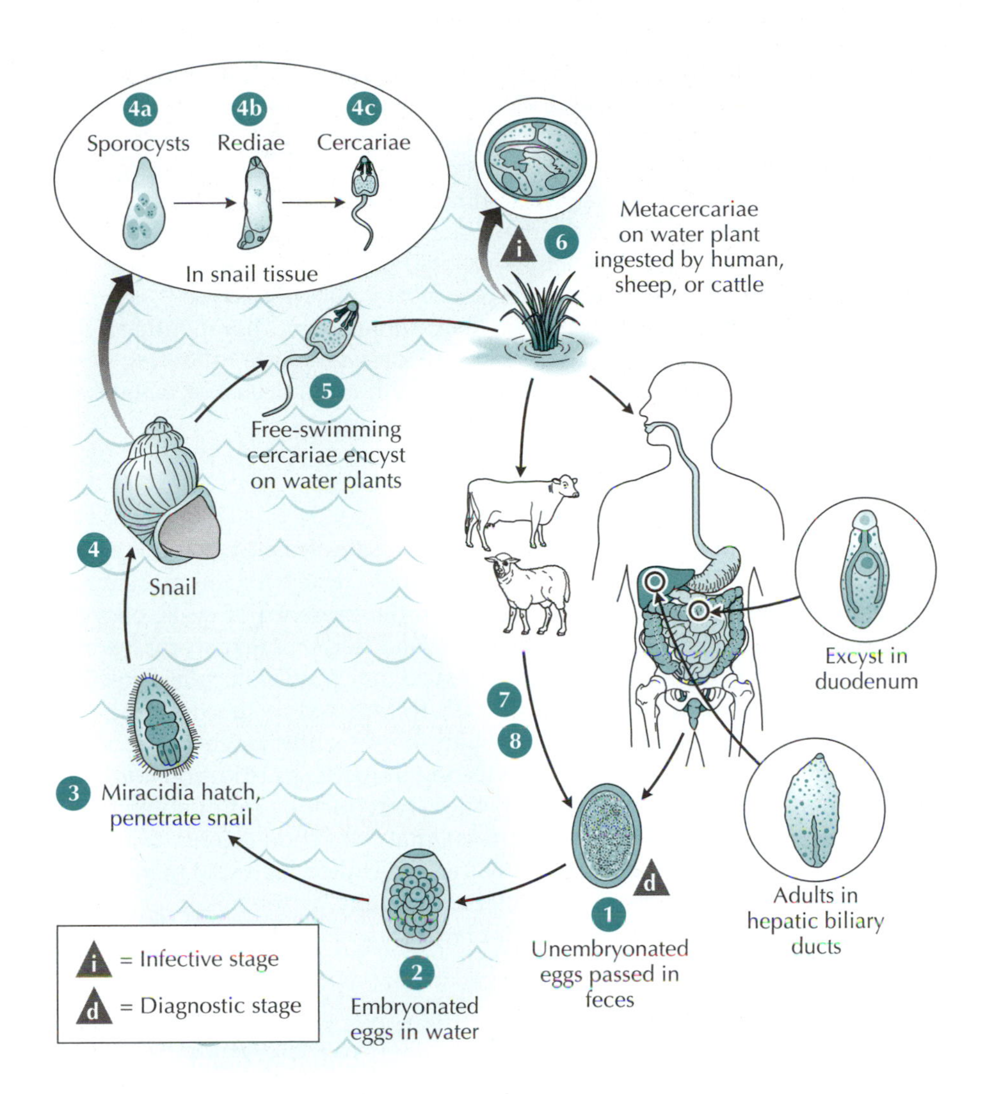

Figure 23.25 Life cycle of *Fasciola hepatica*. Life stages invade the snail intermediate host and undergo several developmental stages. The cercariae are released from the snail and encyst as metacercariae on aquatic vegetation or other surfaces. Mammals acquire the infection by eating vegetation containing metacercariae. In humans, maturation from metacercariae into adult flukes takes approximately 3 to 4 months. The adult flukes reside in the large biliary ducts of the mammalian host. Redrawn from a CDC illustration. doi:10.1128/9781555817206.ch23.f23.25

and leaf shaped and infect mainly herbivores, as they require vegetation for encysting. *Fasciola hepatica* is one of the largest flukes in the world, reaching 30 mm in length and 13 mm in width (Fig. 23.26). The adult flukes live in the bile passages of the liver of many kinds of mammals, causing damage with the small sucker located on their anterior end. Their eggs are passed out of the liver with the bile and into the intestine and mixed with feces. In the United States, it seems that two specific snail species (*Fossaria modicella* and *Stagnicola bulimoides*) are involved in this life cycle. These snails prefer warm waters, such as those found in the southeastern United States.

While human infections are rare, they do occur and more often are associated with consumption of watercress. Cases are more common in Europe, the Middle East, and Asia and may be associated with other types of vegetation. While sheep infections tend to be more common, beef cattle within

Figure 23.26 An adult liver fluke, *Fasciola hepatica*. Courtesy of Melvin Kramer. doi:10.1128/9781555817206.ch23.f23.26

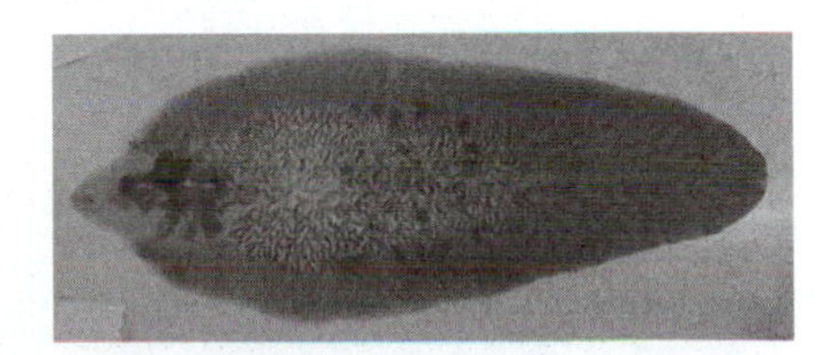

the United States may show variable infection rates. In California 58% of beef cattle were positive, in Florida 68% were positive, and in Montana 17% were positive as determined by liver biopsy at slaughter. Acute-phase symptoms include abdominal pain, hepatomegaly, fever, vomiting, and diarrhea. Infections can persist for months. The chronic phase occurs when the adult flukes block the bile ducts in the liver. In epidemiological studies, diagnosis has often been slow and correlated only with a history of watercress consumption. Blood serum tests confirm infection with *F. hepatica*. Several drugs have been shown to have effective chemotherapeutic effects on *Fasciola*. Strategies to reduce the risk of *Fasciola* include following good irrigation and sanitation practices in the growth and harvest of watercress and similar leafy greens where snails may be common.

DETECTION

Detection is traditionally performed on clinical samples and in foods when an outbreak is suspected. Foods are not routinely tested for the presence of parasites. Testing procedures rely heavily on the presence of cysts, oocysts, eggs, or life stages in the clinical isolates as visualized by microscopy. Several of the oocysts stain acid-fast and can be viewed that way. Molecular detection assays using polymerase chain reaction (PCR) targeted to specific genes are also important and may be combined with fluorescently tagged antibodies, depending on the sample type. When testing in foods, purification and concentration methods are typically used. These may include immunomagnetic separation, which is used by the Environmental Protection Agency for water testing of *Cryptosporidium* and *Giardia* within the United States. In many cases the larvae of the helminths can be detected visually by the naked eye or by practices like candling in fish to look for larvae.

PREVENTATIVE MEASURES

Prevention of infection by parasites is generally similar to that for bacteria, including following good manufacturing practices (GMPs) and good agricultural practices (GAPs). Good sanitation and personal hygiene are always important to practice if you are a line chef or a home cook. Water quality is certainly an issue, and the use of potable or filtered water is recommended. While these issues sound like common sense, the connections between them are quite complex (Fig. 23.27). In terms of parasites in meats, freezing generally kills most helminths and proper cooking of all foods inactivates any parasites present. Irradiation also inactivates parasites. Keeping feral animals away from pasture lands will reduce the risk of transmission of several types of parasites. Several regulations have reduced the incidence of parasites in our food supply, including laws against garbage feeding to animals and trichina-free certification programs for swine. Of course it is important to cook all meats properly, following USDA guidelines, and to wash fresh produce. Other basic preventative measures are listed below.

- Wash hands thoroughly with soap and water after using the toilet and before handling or eating food.
- Do not swim in public waters if you have diarrhea.
- Avoid water that might be contaminated. Do not swallow recreational water. Do not drink untreated water from shallow wells, lakes, rivers, springs, ponds, or streams.

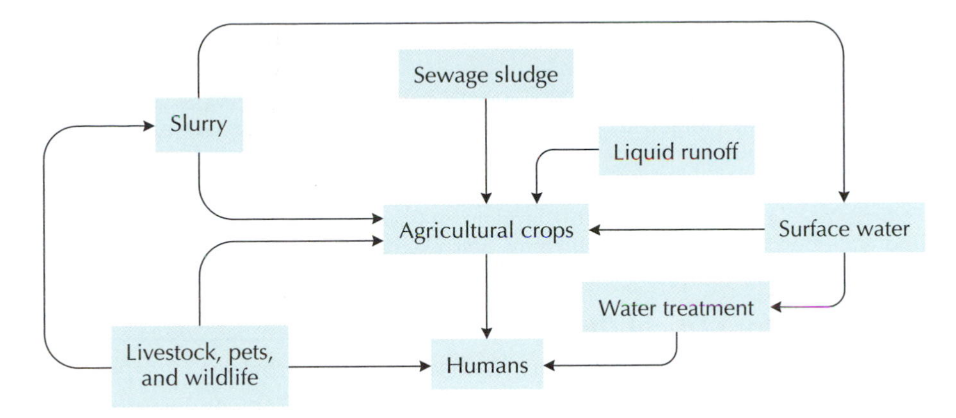

Figure 23.27 Parasite transmission in the environment is highly complex, as shown in this image adapted from D. Dawson, *Int. J. Food Microbiol.* **103:**207–227, 2005. doi:10.1128/9781555817206.ch23.f23.27

- Water can be made safe by heating it to a rolling boil for at least 1 min or by using a filter that has an absolute pore size of 1 µm or less or one that has been NSF rated for "cyst removal."

Summary

- Protozoa have a relatively low infectious dose and have caused outbreaks associated with water, fresh produce, and ready-to-eat foods.
- Both protozoa and helminths have complicated life cycles. Transmission via food or water may include (oo)cysts, encysted larvae, free-swimming larvae, or eggs.
- *Cryptosporidium* and *Giardia* (oo)cysts are shed already infectious and sporulated, while those of *Cyclospora* and *Toxoplasma* require time in the environment to sporulate.
- *Toxoplasma gondii* causes an extraintestinal disease and can be transmitted in three ways: ingestion of oocysts, ingestion of tissue cysts in undercooked infected meat, and congenitally.
- Parasitic diseases such as those caused by *Fasciola* and *Trichinella* are difficult to diagnose, especially when the clinical symptoms are vague and adult helminths are not obvious in clinical samples.
- Parasites can be transferred in a preharvest environment from livestock, soil amendments, or water.
- Helminths can be inactivated by heat and freezing. Protozoa can be inactivated by heat. The risk of parasite transmission can be reduced by using USDA cooking temperatures and good sanitation and hygiene practices.

Suggested reading

Davies, P. R. 2011. Intensive swine production and pork safety. *Foodborne Pathog. Dis.* **8:**189–201.

Dawson, D. 2005. Foodborne protozoan parasites. *Int. J. Food Microbiol.* **103:**207–227.

Dubey, J. P. 2010. *Toxoplasmosis of Animals and Humans*, 2nd ed. CRC Press, Boca Raton, FL.

Fayer, R., and L. Xiao (ed.). 2007. Cryptosporidium *and Cryptosporidiosis*, 2nd ed. CRC Press, Boca Raton, FL.

Newell, D. G., M. Koopmans, L.Verhoef, E. Duizer, A. Aidara-Kane, H. Sprong, M. Opsteegh, M. Langelaar, J. Threfall, F. Scheutz, J. van der Giessen, and H. Kruse. 2010. Food-borne diseases—the challenges of 20 years ago still persist while new ones continue to emerge. *Int. J. Food Microbiol.* **139**(Suppl. 1)**:**S3–S15.

Ortega, Y. R., and R. Sanchez. 2010. Update on *Cyclospora cayetanensis*, a food-borne and waterborne parasite. *Clin. Microbiol. Rev.* **23:**218–234.

Questions for critical thought

1. What role does taxonomy play in characterizing protozoa?
2. How are oocysts and helminth eggs transmitted in the environment?
3. Describe the differences in the transmission of *Cryptosporidium*, *Cyclospora*, and *Toxoplasma* oocysts.
4. Describe the three ways in which *Toxoplasma* can be transmitted.
5. A case of giardiasis can last for several weeks if untreated. In some cases, when the infection passes the individual is left lactose intolerant (unable to completely process the lactose sugar in dairy foods, which leads to feelings of bloat and gas). Why do you think this is?
6. Describe the differences between a roundworm, a tapeworm, and a fluke.
7. Immunocompromised individuals are more prone to infection with protozoan parasites. In particular, several parasitic diseases have increased greatly in acquired immunodeficiency syndrome (AIDS) patients, who generally have limited cell-mediated immunity. Explain why.
8. Speculate on the detection frequencies of bacteria, protozoa, roundworms, tapeworms, flukes, and viruses in the food supply of the United States.
9. Explain why trichina-free certification is based on certifying pig management practices rather than on testing pigs alone.
10. Describe how current preventative measures are working to reduce transmission of parasites and incidence of disease.
11. Explain the role that *Cryptosporidium* and *Escherichia coli* O157:H7 had in contaminated apple cider. To understand this better, look up the Hazard Analysis Critical Control Point (HACCP) rule for juice.

24 Viruses and Prions

LEARNING OBJECTIVES

The information in this chapter will enable the student to:

- differentiate among viruses, bacteria, protozoa, and prions
- recognize the characteristics of viruses and their life cycles
- understand the positive role of viruses in pathogen control, their detrimental effect in dairy fermentations, and their role in foodborne illness
- compare and contrast control strategies for viruses, bacteria, protozoa, and prions in foods
- explain what a prion is and how it is transmitted
- appreciate that science is constantly changing and is not always clear

INTRODUCTION

Most of food microbiology, and the bulk of this book, is devoted to bacteria. However, viruses cause approximately 60% of the foodborne illness cases in the United States, compared to the 39% caused by bacteria. One of the most common foodborne viruses is norovirus, which causes 11% of all deaths linked to foodborne illness. There are several prominent foodborne viruses that can be infectious to humans; however, bacteria may be infected by viruses too. Bacteriophages are viruses that infect bacteria, and these may be useful or harmful; for example, some may be used to reduce contamination, while others may cause problems in the dairy industry.

Another type of infectious agent is discussed in this chapter. At one time scientists questioned if prions were similar to viruses, and some scientists still ponder this question. Prions (pronounced pree-ons) are infectious protein particles. We have learned more about prions as the causative agents of mad cow disease, a new variant of Creutzfeldt-Jakob disease (nvCJD), and CJD. nvCJD is epidemiologically linked to mad cow disease, more properly named bovine spongiform encephalopathy (BSE), and nvCJD was responsible for 171 deaths and another 172 illnesses in those still surviving with the disease in the United Kingdom (as of June 2011). Although they have not caused illness in the United States, they are covered here because they are biological and can get into the food supply (although the U.S. Department of Agriculture [USDA] is trying to keep them out and has several active firewalls in place).

Bacteria, protozoa, viruses, and prions are fundamentally different (Table 24.1). While bacteria and protozoa are complex, foodborne viruses consist of RNA and a fairly simple protein coat composed of a few proteins

doi:10.1128/9781555817206.ch24

Table 24.1 Comparison of bacteria, protozoa, viruses, and prions

Trait	Disease-causing agents			
	Bacteria	**Protozoa**	**Viruses**	**Prions**
Composition	Complex structures of lipid, carbohydrates, proteins, and genetic material	Complex structures of lipid, carbohydrates, proteins, and genetic material	Single strand of RNA in a relatively simple structure	A specific type of protein
Types important in food	More than a dozen gram-positive and gram-negative organisms	Less than a dozen of true significance to the United States	Two most important are noroviruses and hepatitis A virus	Only one that we know of
Control in food	Control growth by changing intrinsic or extrinsic factors. Kill by applying energy.	These do not grow in food. Prevent entry by good hygiene and sanitation.	These do not grow in food. Inactivate by heat. Prevent entry by good hygiene and sanitation.	These do not grow in food. Not inactivated by heat. Public health measures prevent consumption of meat from "mad cows."
Disease	Infection or intoxication	Infection	Infection	Protein conformational conversion
Transmission	Eating food on which bacteria have grown	Fecal-oral	Fecal-oral	Eating meat that contains prions
Detection methods	Relatively easy to culture in the microbiology lab	Microscopy, molecular biology, mammalian cell culture, or animal models	Electron microscopy, molecular biology, or mammalian cell culture	Examination of tissue at autopsy
No. of cases/yr in United States	>4 million	>2 million	>30 million	None yet from origin within the United States[a]

[a]Case reports in the United States are believed to be associated with individuals who were infected while living outside of the United States.

repeated in specific patterns. A prion is even simpler, consisting of a single protein molecule that undergoes a structural change and becomes infectious to its host. Viruses and prions and even protozoa do not grow in foods, for the most part cannot be cultured in the laboratory, and are difficult to detect. All three agents are different from bacteria in these ways, but like bacteria, they may be transferred from feces to food through poor hygiene and sanitation practices. For viruses, the food is a vehicle to get to the intestine, where they infect the person. Prions are proteins that are part of the food, and they are not inactivated by heat. Bacteria and viruses cause millions of cases of foodborne illness per year. No cases of prion-related nvCJD in the United States due to consumption of food derived from cattle in the United States have been reported to date.

VIRUSES

Elementary Virology

There are just a few basics to understanding viruses.

1. Viruses are obligate intracellular parasites. This means that they can live *only inside* a host, *to its detriment*. Viruses have no life outside a living cell. There is even some debate as to whether they are alive at all! Thus, viruses are frequently called particles. For reasons outlined below, viruses are very "picky" about what living cells (or hosts) they infect. Viruses have limited host specificity. A virus whose host is a bacterium cannot infect a human. Such viruses are collectively referred to as bacteriophages. Nor can a virus like the

one that causes foot-and-mouth disease infect a human. However, sometimes a mutation allows a virus to jump the species barrier. Such a jump is the presumed origin of human immunodeficiency virus (HIV).

2. Viruses are very simple, consisting only of DNA or RNA (but never both), repeated protein in a pattern as the protein coat called a viral capsid, and occasionally a small amount of glycoprotein in what is referred to as a lipid bilayer or envelope. The majority of viruses that are of concern for causing foodborne diseases do not have an envelope and may be referred to as naked viruses.
3. The basic principle of viral infection is that their attachment proteins (also called ligands) bind to receptor proteins on the host cell (Fig. 24.1). The ligand-receptor protein interaction is specific, much like an enzyme-substrate interaction. Bacteriophages bind only to bacteria. The viruses that cause human foodborne diseases (e.g., the noroviruses that cause gastrointestinal upset) can bind only to specific carbohydrates expressed on cells. It is the ligand-receptor interaction that creates viral specificity.
4. The viral "life" cycle consists of four simple steps. First, the virus attaches to the host cell through the ligand-receptor interaction. Then, the genetic information in the virus (i.e., RNA or DNA) enters the host cell. Once in the host cell, the viral genome commandeers the host's metabolic machinery to replicate viral genes and synthesize viral protein. Finally, the viral parts spontaneously assemble into viruses that rupture the host and are released. Unlike bacteria, viruses do not replicate by doubling; they replicate in a "lytic burst."

Authors' note

When an organism has the ability to infect humans and nonhuman animals, it is called zoonotic and the disease it causes may be called a zoonosis.

This simple explanation chronicles the life of a *lytic* phage. The lytic phage enters the host cell, does its dirty work, and bursts out. There is another type of phage, the *temperate* phage. When these tricksters enter a host, their genes enter (or integrate) into the host genes. In this "hiding place," they are replicated as part of the host genome. These integrated phages are in the *lysogenic state*. The lysogenic phages can be dormant for a long time

Figure 24.1 A T-even phage infecting a cell. Reprinted from http://www.armageddononline.org with permission. doi:10.1128/9781555817206.ch24.f24.01

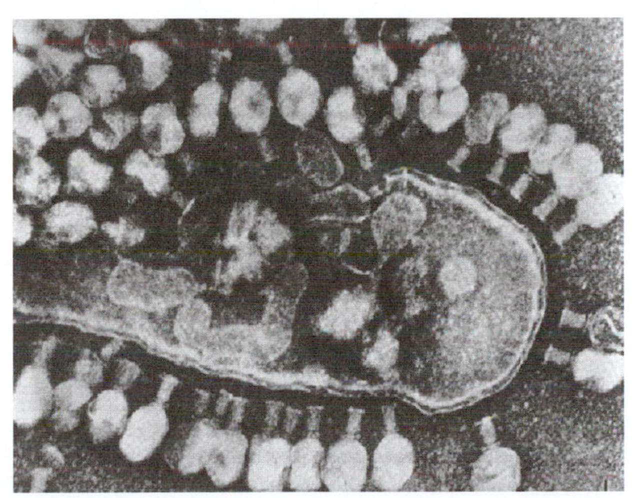

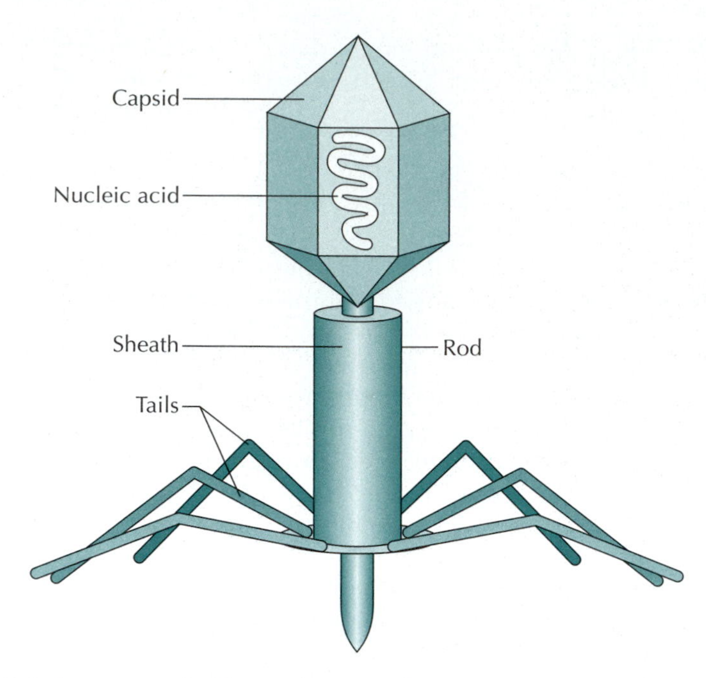

Figure 24.2 Structure of a T-even bacteriophage. Redrawn from http:// www.armageddononline.org with permission. doi:10.1128/9781555817206.ch24.f24.02

and then reactivate, as in the case of shingles or cold sores. At some point, the lysogenic phages "go lytic." When they leave the host genome, they can take some host DNA with them and transfer these genes into the next host they infect. The process is called *transduction*. Genetic engineering sometimes uses transducing phages as a mechanism of gene transfer.

Viruses that attack bacteria are called *bacteriophages* and have a relatively complex structure. Bacteriophages look like a lunar landing module. The head contains the viral genome, which is pushed through the tail when the receptor protein ligands interact with the host-specific receptors (Fig. 24.2). Human viruses, such as the norovirus (Fig. 24.3), are structurally simpler, but at the same time the protein capsid is more complex. DNA enters the host by a more complicated mechanism, but the cycle still follows the one explained above. The complete way in which norovirus binds to host cells, invades, and replicates is still not well understood. The study of bacteriophages and viral surrogates is a popular method used to learn more about human-pathogenic viruses.

Figure 24.3 A norovirus. Courtesy of B. V. V. Prasad, Baylor College of Medicine. doi:10.1128/9781555817206.ch24.f24.03

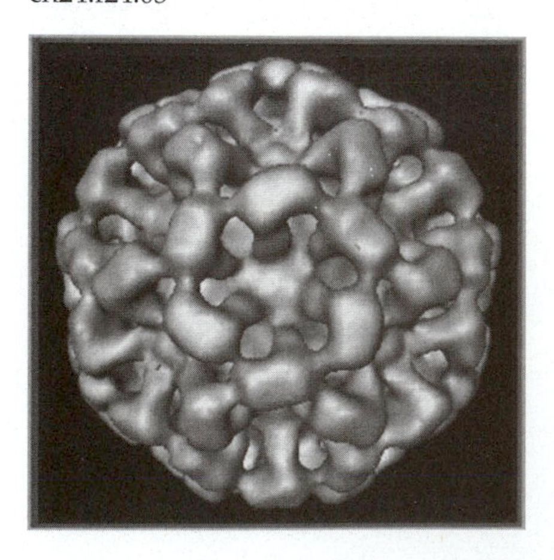

Bacteriophages are counted in a fashion similar to the bacterial plate count. The "plaquing" of viruses exploits their lytic nature. The host bacteria are spread in a confluent layer on an agar plate. (The idea is to create a lawn of bacteria on the plate rather than producing a countable number of bacteria.) The viruses are serially diluted (in the same fashion as bacteria for plate counts) and spread over the surface of bacteria. The plates are incubated. Each virus infects and spreads through all the bacteria around it, lysing them and creating a clearing zone, or plaque (Fig. 24.4). The number of plaques is directly related to the initial number of virus particles and is expressed as plaque-forming units (PFU).

In a confluent monolayer or lawn of human cells grown in the laboratory, some viruses can make plaques in the cells, which look similar to those made by bacteriophages. Some viruses do not make complete plaques by lysing their host cells in this neat pattern but still kill the cells, and this can be visualized microscopically and is called cytopathic effect. Cytopathic cells look very different from healthy cells under the microscope (Fig. 24.5).

Other than plaques on lawns of bacteria or mammalian cells, it is difficult to visualize these microscopic organisms. Viruses may be 100 times smaller than a bacterium. Viruses can be visualized using electron microscopy (Fig. 24.6), and in the many years of science prior to molecular biology, electron microscopy was the preferred method of detection from feces or vomitus samples. Since the introduction of reverse transcriptase polymerase chain reaction (RT-PCR), viral genomes of RNA can be detected in environmental, food, and clinical samples. Molecular biology tools like RT-PCR have increased the ability to work with viruses and learn how they survive in and are transmitted by foods.

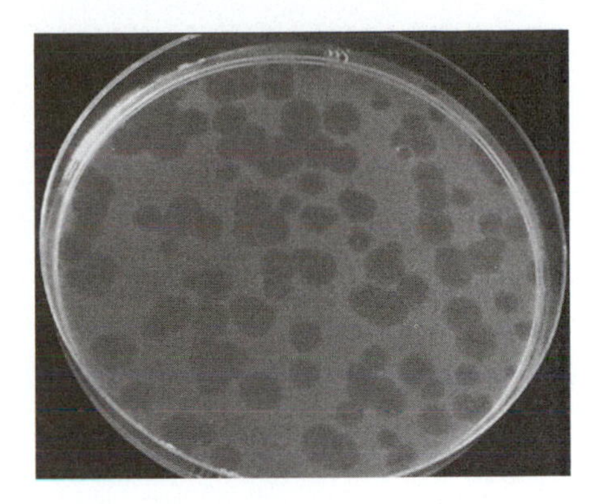

Figure 24.4 Clearing zones on the lawn of sensitive bacteria indicate plaques caused by viruses. From the ASM Microbe Library, courtesy of D. Sue Katz, Rogers State University, and Marie Panec, Moorpark College. Photograph taken by Grisel Quiroz. doi:10.1128/9781555817206.ch24.f24.04

Viruses as Agents of Foodborne Illness

If you have ever suffered with a gastrointestinal illness for approximately 24 h and then felt better (called acute gastroenteritis), you may have been infected with norovirus, the most common cause of foodborne illness. Viruses are spread throughout many environments, including in kitchens, in irrigation water, on crops, and on human hands. The majority of viruses are transmitted directly through contamination with human fecal matter, demonstrating once again the importance of good personal hygiene and proper sanitation. While humans are the host for the most important human gastrointestinal viruses, scientists are finding more information on how noroviruses survive and spread through swine and bovine manure. An important way to use manure and biosolids is to treat them and apply them as soil amendments. Since viruses can survive well under various conditions, it is essential that these treatment practices be validated for the ability to inactivate viruses and other pathogenic microorganisms. It is generally believed that the majority of norovirus and other enteric virus infections are spread by humans through food handling or person-to-person transmission; however, this is likely not how all cases are spread. Raw shellfish from water polluted with feces are a classic cause of viral food poisoning. Through their filtering action, bivalves concentrate viruses from the water by a factor of 100 to 1,000. The first reported outbreak (630 cases) of

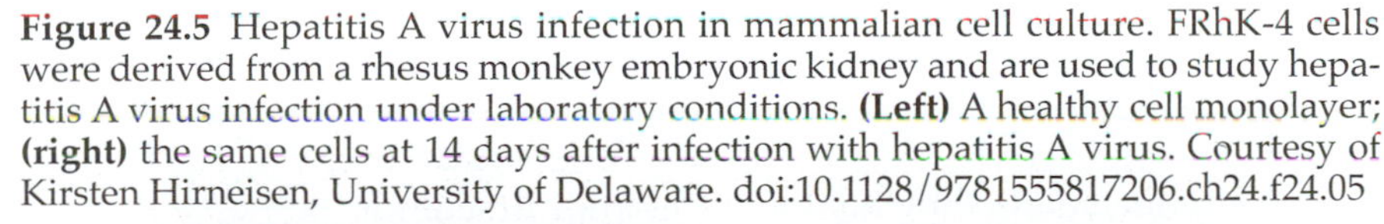

Figure 24.5 Hepatitis A virus infection in mammalian cell culture. FRhK-4 cells were derived from a rhesus monkey embryonic kidney and are used to study hepatitis A virus infection under laboratory conditions. **(Left)** A healthy cell monolayer; **(right)** the same cells at 14 days after infection with hepatitis A virus. Courtesy of Kirsten Hirneisen, University of Delaware. doi:10.1128/9781555817206.ch24.f24.05

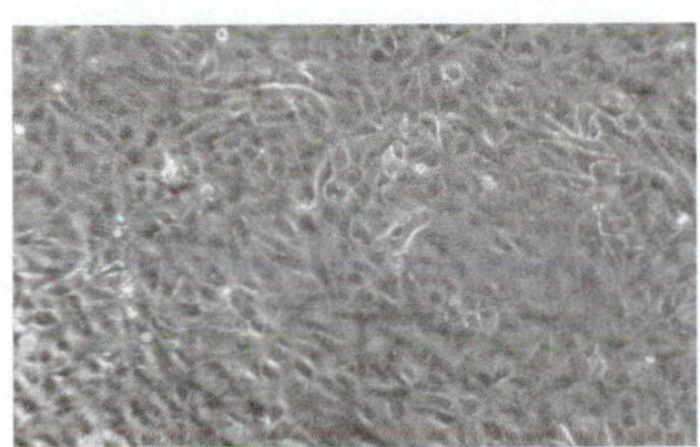

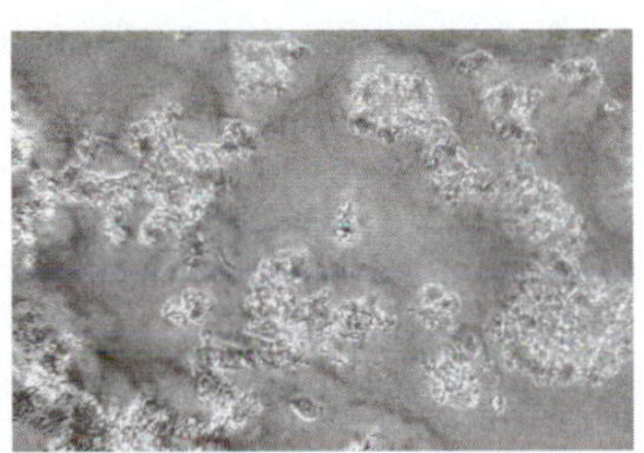

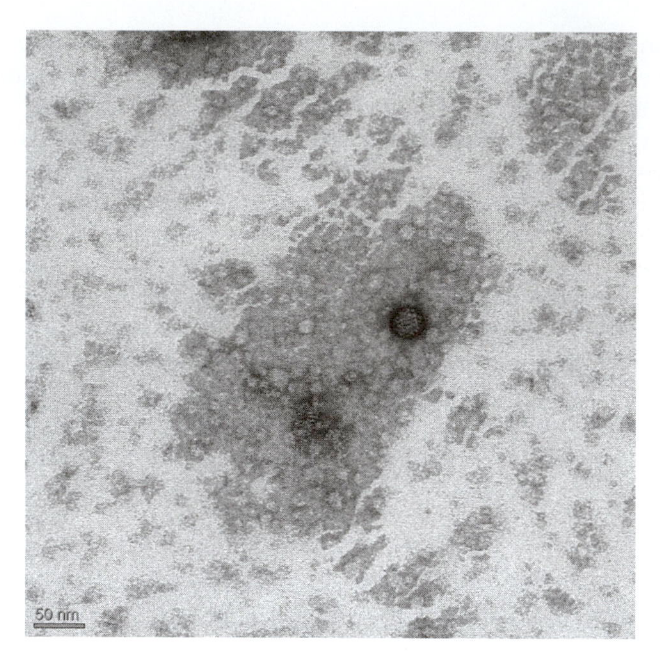

Figure 24.6 Scanning electron micrograph of mouse norovirus. Courtesy of Jie Wei, University of Delaware. doi:10.1128/9781555817206.ch24.f24.06

Authors' note

Before the crops were planted, the field in Fig. 24.7 was amended with manure and compost to enrich the soil and provide nutrients for the growing plants. Can you imagine how many organisms could be there if the manure was not treated properly? And not only viruses, right?

hepatitis A from oysters was caused when feces from an outhouse drained into an oyster bed. In 1988, clams harvested from polluted waters in China were eaten raw and caused 300,000 (!) cases of hepatitis A.

What is an infectious dose? Through studying the epidemiology of foodborne outbreaks we know that viruses can be spread very easily. In particular, some viruses, like norovirus, are spread easily by aerosols. The infectious dose of norovirus may be as low as 18 virus particles. Can you imagine 18 virus particles, each being a size of only 30 nm? (A nanometer is 10 to 12 times smaller than a meter.) Viruses infect humans, grow in intestinal cells, and release large numbers of progeny into the feces. A food handler's feces can easily contain 10^8 viruses per gram. Vomit can also carry viruses, with 10^7 particles in a single droplet. A simple exercise illustrates the power of the fecal-oral route. Imagine a food handler who is negligent in washing his hands after defecating. (Sorry, guys, but studies show that many more men than women fail to wash their hands after toileting.) A 0.01-g piece of feces left under his fingernails would contain 10^6 virus particles. If the infectious dose were 1,000 particles, the feces could get mixed into 1,000 servings of salad dressing, doughnut glaze, or whipped cream, causing illness in 1,000 people (1,000 particles/serving × 1,000 servings = the million virus particles originally under the fingernail). This example is one you do not want to experience firsthand.

Similar incidents of fecal contamination occurred in the following places:

- St. Paul, Minnesota, where 3,000 people got sick after eating buttercream frosting prepared by an infected food handler
- cruise ships, where hundreds of people per ship became sick from eating contaminated fruits and ice
- a central school lunch kitchen in Japan, where 3,200 children and 120 teachers were stricken

Figure 24.7 What is the story behind this field? doi:10.1128/9781555817206.ch24.f24.07

- Michigan schools, where frozen strawberries distributed by the school lunch program sickened 230 people

There are several types of enteric viruses, all of which are nonenveloped, which allows them to survive well under many environmental pressures. The most common cause of acute gastroenteritis is norovirus. Outbreak cases are discovered by their symptoms, since to this date the tests available for virus detection are not routinely used by industry; however, new Food and Drug Administration (FDA)-approved tests and other laboratory methods are used in clinical labs for the detection of viruses in human samples. The clinical criteria for noroviruses and for another important virus, hepatitis A virus, are listed in Table 24.2. Other viruses may cause acute gastroenteritis as well and are found in both developing and developed nations. Rotaviruses are a leading cause of gastroenteritis in children and are usually transmitted by contaminated water, causing 130 million cases of diarrhea and 6 million deaths per year on a worldwide basis. In fact, rotaviruses are the causative agents for approximately one-third of all diarrhea-associated hospitalizations in children under the age of 5 years. Astroviruses and enteric adenoviruses also cause gastroenteritis, survive well in the environment, and may be transmitted through water, food, and person-to-person transmission. Adenovirus is interesting in that

Table 24.2 Characteristics of the two major foodborne viruses

Characteristic	Norovirus	Hepatitis A virus
Disease symptoms	Nausea, projectile vomiting, diarrhea due to action on intestinal cells; lasts for 12–60 h	Jaundice, anorexia, vomiting, prolonged malaise; lasts for weeks to months
Incubation period	Usually 24–48 h, but can be as quick as 10 h	15–50 days (median, 29 days)
Shedding of viral particles in feces (or vomit)	Simultaneous with symptoms; continues a few weeks after illness	Shed for 10–14 days before symptoms appear
Immunity	Exposure does not generate immunity	Exposure confers durable immunity[a]
Type of virus	Calicivirus, spherical, 32-nm diam, with depressions on surface	Picornavirus, spherical, 28-nm diam
Estimated no. of foodborne cases/yr	>8 million	>3,500

[a]However, there can be a relapse years later. It is noteworthy that children get less severe symptoms and can have the disease with no symptoms at all. These children have lifelong immunity. Thus, in developing countries where most people are exposed as children, adult cases are rarer. In industrialized countries, adults are at greater risk.

it contains DNA as its genetic material, while the other major viruses of food and water origin all contain RNA. Its shape and unique spikes on the capsid help with ease of identification through electron microscopy.

Enteric viruses have simple structures. A coding strand of RNA is surrounded by a few structural proteins in a pattern that makes up the viral capsid. The RNA makes more RNA inside the host cell by using an RNA-dependent RNA polymerase. The viral genome codes for the polymerase, the capsid proteins, and perhaps other virus-specific enzymes. There is no DNA in the replicative cycle of foodborne viruses.

Intervention

Viral particles are very persistent in the environment. They survive for hours on hands, for weeks on dry surfaces or vegetables, and for months in water and soil. Nonetheless, in theory, viral foodborne diseases can be prevented through a few simple measures. The risks of viruses being in food can be reduced by good food handler hygiene and by not using shellfish from polluted water. Using potable (i.e., drinking) water in all areas of food processing also prevents viral contamination. Heating food or sanitizing food contact surfaces kills viruses. Nonetheless, approximately 30 million cases of foodborne viral diseases per year suggest that these simple measures are not being followed or that viruses are able to bypass these strategies.

Other antibacterial food-processing methods may or may not kill viruses. Viruses are relatively acid resistant and can survive stomach acid. More radiation is needed to kill viruses than bacteria. Traditional heating methods kill viruses as easily as bacteria; however, in terms of food production many companies are moving away from thermal processing to retain fresh-like sensory attributes of food. The antiviral effectiveness of new, nonthermal methods such as high pressure, ultraviolet light, and ozone is still not well understood, but some discussion of this is provided in chapter 28.

While some processes are not well understood, depuration is a process that does not work well for the removal of viruses from shellfish but is still used to various degrees by the shellfish industry. Depuration is a process that attempts to cleanse shellfish of viruses by moving them from the polluted water to clean water. Over time, the shellfish should wash the viruses out with the clean water. This process is monitored using fecal coliforms as indicators. However, the viruses are frequently retained after the indicator coliforms have been washed out. This makes it hard to determine when depuration is complete. What is the big deal about shellfish contamination anyway? Recall that shellfish are filter feeders and they are highly efficient at filtering water and as a result concentrate the solutes from the water. Researchers have shown experimentally that oysters can concentrate hepatitis A virus 100-fold when grown in experimentally contaminated water samples. On top of that, many of the fresh oysters harvested and sold in the United States are consumed in their raw state, without any thermal inactivation step for the viruses.

Noroviruses

Noroviruses belong to the family *Caliciviridae* and were originally called Norwalk or Norwalk-like viruses from their first outbreak at an elementary school in Norwalk, Ohio. Noroviruses cause acute diarrheal disease with an incubation period of 1 to 2 days and, as stated earlier, are a major cause of foodborne disease. Noroviruses occupy a perennial position in the "top 10" agents of foodborne illness, occasionally winning the number 1 spot. They, like all viruses, are difficult to study because they cannot be cultured like bacteria or grown in the laboratory. In fact, human noroviruses do not grow in cell culture or other animals and so are particularly difficult to work with in the lab. Scientists rely on molecular biology to be able to study noroviruses. Like other viruses, noroviruses are detected by electron microscopy, antibodies, and genetic probes.

Noroviruses are small, round caliciviruses that contain a single strand of RNA. They are transmitted by the fecal-oral route. The viral particles are hardy, surviving storage, refrigeration, freezing, and passage through stomach acid. Upon reaching the small intestine, the virus infects mucosal cells, multiplies inside, kills the cells, and contaminates more feces. Viruses continue to be spread by fecal shedding for more than 1 week after the person recovers. These viruses generally have a low infectious dose, as few as 10 particles, and are highly infectious. It has been determined that an infected individual may shed as much as 9.5×10^9 virus particles/g of feces! Norovirus infection may affect any age group, and outbreaks are difficult to prevent and control, as demonstrated by the regular occurrence of outbreaks on cruise ships and in large institutions, despite rigorous preventive actions. The low infectious dose combined with the high attack rate allows norovirus to be spread easily through communities, including college campuses. Add to that the fact that there is no long-lasting immunity for norovirus and you have a significant infectious virus, causing an estimated 20 million cases each year in the United States.

Hepatitis A

Only a small percentage (~5%) of all hepatitis cases are food related, making them insignificant in the overall control of hepatitis. Hepatitis A virus uses food as its vehicle and is often transmitted by infected food handlers. One part of this problem is that virus shedding occurs before the onset of clinical symptoms. Again, this reinforces the need for constant good hygiene and sanitation practices.

Hepatitis A has an incubation period of 15 to 50 days. This is troublesome because an infected person sheds the virus in their feces for 10 to 14 days *before* becoming sick. The shedding can also continue for weeks after the person recovers. This makes it easy to unwittingly transmit the virus. Hepatitis is not a traditional gastroenteritis. It causes jaundice (yellowing of the skin), anorexia, vomiting, and a profound melancholy. These symptoms can last several weeks, and since there is no way to culture the virus, the diagnosis can be confirmed only by the detection of hepatitis-specific antibodies in the person's blood. Hepatitis can cause permanent or reoccurring kidney problems. When the virus infects hepatocytes (liver cells), the body's immune system responds with T cells that are toxic to liver cells. This results in liver damage. In countries where fecal contamination is common, most children get a mild case of hepatitis (and acquire immunity!) by age 5. In more developed countries, hepatitis A is more common among adults, who get more serious cases. Hepatitis A virus is very different from norovirus in that there seems to be only one serotype of hepatitis A virus and lifelong immunity occurs after infection.

Hepatitis A virus is a member of the family *Picornaviridae*. (Interestingly, poliovirus is in the same family.) The spherical coat, consisting of four different proteins, surrounds and protects the coding strand of RNA. The RNA codes for the RNA-dependent RNA polymerase mentioned previously, a protease, and a large protein.

Noroviruses and hepatitis A virus are compared in Table 24.2.

Food Relatedness of Other Viruses

There are a myriad of different viruses that can be transmitted through water (and food) worldwide (Box 24.1), and a few of the main ones that cause illness in the United States are mentioned here.

Authors' note

Don't forget that water and food often serve as vehicles for the same pathogenic organisms. In fact, ice is often called the forgotten food!

Astroviruses cause an estimated 3 million illnesses each year in the United States. They are round, with star patterns on their coat, and carry a single coding strand of RNA. There is an incubation period of 3 to 4 days. Diarrhea is the primary symptom, usually lasting 2 to 3 days; sometimes it can last as long as 2 weeks. As mentioned above, rotaviruses contain double-stranded RNA and are a common cause of infant death in developing countries, but they also cause an estimated 3 million illnesses each year in the United States.

It is important to note that many viruses are not transmitted by foods. These include many clinically important viruses like rabies, hantavirus, rhinoviruses (which cause the common cold), and HIV. These and many other viruses are not shed in the feces or transmitted by the fecal-oral route. HIV-positive people pose no risk as food handlers, and even those symptomatic with acquired immunodeficiency syndrome (AIDS) are not a threat if they do not have diarrhea. (Although HIV *is not* transmitted by the fecal-oral route, diarrheal fluid can contain *Salmonella*, *Escherichia coli* O157:H7, and noroviruses, which *are* transmitted this way.)

Are Specific Food Commodities More at Risk?

People often ask which foods are most connected with viral contamination. The answer to this question is not easy for several reasons. The first is that foods are not often tested for viral pathogens, as it is not easy to detect them. Scientists are developing new methods to extract viral nucleic acid from

BOX 24.1

Hepatitis E virus

Hepatitis E virus (HEV) (see figure) is a member of the family *Hepeviridae*, which is composed of nonenveloped, single-stranded, positive-sense RNA viruses. Phylogenetic analyses of HEV isolates from various parts of the world indicate that at least four different genotypes of HEV exist. Genotypes 1 and 2 are associated with HEV epidemics via human-to-human transmission and are found predominantly in Asia and the Indian subcontinent. Genotypes 3 and 4 are common in U.S. and European swine and other animals and are associated with sporadic and cluster cases of zoonotic transmission to humans. What does all this mean? Well, there may be zoonotic risk of HEV, and this virus appears to be an emerging pathogen in the United States. Sporadic clusters of acute hepatitis E associated with consumption of raw or undercooked swine livers have been reported in the United States. A total of 11% of grocery-sold swine livers were contaminated with infectious HEV. The presence of HEV in swine livers and potential existence in other pork products, such as sausage, scrapple, and ground pork, raise concern for consumer safety. Though there have been no documented HEV infections associated with shellfish consumption in the United States, HEV has been associated with shellfish in other regions of the world, suggesting that HEV-contaminated shellfish may be a risk in the United States. An outbreak of virulent human-to-human transmissible genotype 2 HEV in Mexico suggests international trade as a possible means of introduction into the United States. Furthermore, in southern and western regions of the United States, feral swine populations are rapidly increasing, suggesting increased transmission potential of zoonotic HEV to humans. Infection with virulent HEV strains can be medically serious, with pregnant women having up to a 20% fatality rate in some geographic areas. Approximately 1 in 5 people in the United States are seropositive for HEV. Like hepatitis A virus and human norovirus, HEV is an environmentally stable and persistent virus and may be an emerging threat to food safety in the United States.

Hepatitis E virus visualized by a scanning electron microscope. Obtained with permission from the CDC Public Health Image Library.
doi:10.1128/9781555817206.ch24.fBox24.01

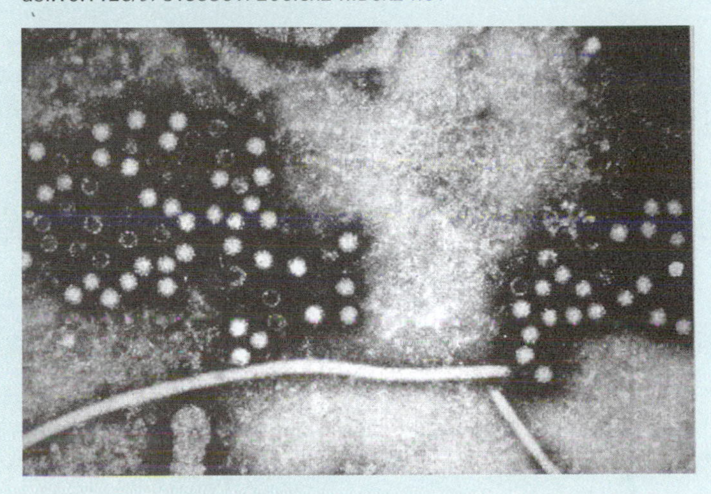

foods and detect just a few virus particles, but this is not routinely performed today. Since there is no model for assessing infectivity of human noroviruses, we cannot tell if these particles are infectious and could have caused disease. The second reason is that it is often difficult to retrieve food that has been involved in an outbreak. It is important to find food that has not previously been opened and detect the pathogen. Those points aside, ready-to-eat foods, molluscan shellfish consumed raw, and fresh produce consumed raw appear to be more often connected with viral transmission. What about berries? The following epidemiological studies discuss the role of berries in the transmission of viruses. Like with the protozoa, viruses seem to adhere to berries well and have caused disease associated with the consumption of fresh berries. Both raspberries and strawberries (raw and frozen) have been associated with outbreaks of hepatitis A virus and norovirus.

Hepatitis A outbreaks with berries. Hepatitis A virus, which is spread by human feces, is thought to have contaminated the berries by contact with infected harvesters or contaminated irrigation water. Frozen and fresh raspberries have also been associated with illness due to norovirus, also spread through human feces and food handlers. Processing berries, including freezing and mild cooking, may be an important issue in the case with virally contaminated berries. These processing steps do not necessarily clear berries from viral contamination. The stems of strawberries destined for freezing are removed in the field, using either a metal device or a thumbnail. The berries are then transported at ambient temperature to a processing facility, where they are washed with water, sliced if applicable, and often mixed with up to 30% sucrose before being frozen. The extra human handling during harvesting and comingling in the processing facility is believed to place these berries at greater risk for contamination.

One of the earlier recorded berry outbreaks associated with viral contamination was an outbreak of hepatitis A in Scotland linked to consumption of raspberry mousse prepared from frozen raspberries. The mousse was prepared from two 3-pound containers of frozen raspberries, gelatin, sugar, and pasteurized cream. Some of the leftover mousse was sent home with the staff or was served on the "sweet trolley" in the dining room the next day. Twenty-four individuals were diagnosed with jaundice, deranged liver functions, fever, malaise, nausea, and "flu-like" symptoms approximately 24 to 28 days after consuming the mousse. The raspberries were blast-frozen at the distribution center. The raspberries had been obtained from several farms, including small holdings and large private gardens. Contamination of raspberries apparently occurred at the time of picking or packing, probably by a food handler who was unknowingly shedding hepatitis A virus. In fact, it was later noted that one of the pickers had a hepatitis A virus infection at the time of picking, as stated by a local physician. This restates the importance and impact of good personal hygiene and sanitation practices, along with the need for good education of food handlers at each stage from the farm to the consumer.

A multistate outbreak of hepatitis A was traced to frozen strawberries processed in a single plant in California in 1990. Nine hundred students, teachers, and staff in Georgia and Montana developed hepatitis A virus infection from eating strawberry shortcake and other desserts. Epidemiological data indicated that contamination did not occur from an infected worker within the processing plant but most likely from an infected picker, perhaps when the stems were being removed by hand rather than with a metal tool. Strawberries are still often stemmed prior to being brought into the processing facility.

In the months of February and March of 1997 in Michigan and in Maine there was a similar outbreak at first linked to frozen raspberries and strawberries. More conclusive epidemiological evidence from case-control studies determined that illness was associated only with frozen strawberries; the cases involved schoolchildren and employees. In Michigan, as in the outbreak in 1990, the frozen strawberries were consumed in strawberry shortcake desserts served in the school cafeterias. A total of 287 cases of hepatitis A were reported from 23 schools in Michigan and 13 schools in Maine. Traceback analysis implicated strawberries grown in Mexico and processed and distributed through a California processing facility. There was no indication of specific lots that were contaminated, as these records were not maintained at the schools at this time. A thorough investigation of the California processing facility did not identify any problems and showed good sanitation and manufacturing practices and limited employee hand contact with the berries, and there was no record of employees with illnesses at the time the strawberries were processed. The FDA also conducted an investigation in the fields in Mexico. The fields were drip irrigated rather than spray irrigated, which eliminated the likelihood that berries were contaminated by contaminated water. This investigation revealed several potential problems, including limited slit latrines for the workers and limited access to hand-washing facilities that were on trucks circulating through the fields. While no records of worker illnesses were maintained, the workers did not wear gloves and removed the stems from the strawberries with their fingernails in the fields. The direct hand contact with the berries combined with poor hygiene practices is a possible source of contamination. Other strawberries from the same distributor were placed on hold, and among the 13 other states that received frozen strawberries, two cases of hepatitis A were reported in Tennessee, nine cases in Arizona, five cases in Wisconsin, and four cases in Louisiana. All of these cases with the exception of those from Louisiana were associated with state school lunch programs. The Louisiana cases were traced back to consumption of a commercially prepared smoothie drink. For these clusters of cases no epidemiological studies were conducted. The viruses isolated from the majority of cases of hepatitis A described in this multistate outbreak showed high genetic similarity. It is likely that contamination was not uniform and perhaps at low levels, due to the relatively low number of cases compared to the large quantity of frozen strawberries that were consumed.

Norovirus-associated outbreaks with raspberries. While detection is difficult, epidemiological investigations implicated consumption of raspberries contaminated with norovirus as the cause of several outbreaks. Perhaps as virus detection methods improve, more outbreaks associated with these viruses will be detected. The specific outbreaks discussed here occurred in Europe. In November 2001, an outbreak of norovirus in 30 individuals involved baked raspberry cakes. At first there was an apparent association with both pear and raspberry cakes, but epidemiological evidence indicated that raspberry cakes had a stronger association with illness. The pink cakes were made with a cream topping containing whole frozen raspberries. Multiple norovirus strains were detected in the raspberries after a complex series of extraction coupled with PCR and genetic sequencing methods was implemented.

In France in March 2005, 75 students and teachers reported symptoms of nausea, vomiting, and diarrhea lasting for 1 to 2 days. Epidemiology showed that illness was strongly associated with the consumption of raspberries blended with *fromage blanc*, a fresh cheese similar to cottage cheese that was served with lunch in the school cafeteria. Stool samples were positive for norovirus, Musgrove strain. Virus was not successfully isolated from raspberry samples. As in the cases described above, the raspberries in this outbreak were deep frozen and were blended with the cheese while frozen. The blended desserts were topped with individual frozen berries placed by hand; however, the workers were not ill before or at the time of the outbreak.

In May 2005, nearly 200 patients and employees at two hospitals in Denmark fell ill with symptoms of norovirus. Again, epidemiology linked illness to consumption of a *fromage blanc* cheese dessert made with raspberries. Again, fecal samples were found to be positive for norovirus. When these illnesses occurred, the Regional Food Inspectorate called for withdrawal of the frozen raspberries. Unfortunately, the recall did not happen quickly enough, as just shortly after this, in early June, nearly 300 cases of norovirus were associated with the same dessert served to approximately 1,100 people in a "meals on wheels" system. As with the case described above, many of the fecal samples were found to be positive for norovirus. The three outbreaks described above were not believed to be linked to each other, as the raspberries came from a different producer in the outbreak in France than in Denmark. The exact cause of the outbreak was not determined, but it is clear that contamination was spotty due to the relatively small number of illnesses compared to the numbers that consumed the frozen raspberries.

During the summer months of 2006, 43 individuals became ill with norovirus associated with the consumption of contaminated raspberries in four outbreaks. A homemade cake containing raspberries and cream was the cause of one outbreak. Another was associated with cheesecake and raspberries. Norovirus was detected in fecal samples from these patients, and the raspberries were of the same brand and imported from China. In a third outbreak at a school, drinks made from the same brand of imported raspberries caused 30 illnesses. In the fourth outbreak, a homemade raspberry parfait, also made from the same brand of imported raspberries, was served to nine participants of a meeting, who all became ill with norovirus.

It is likely that the majority of norovirus infections go undetected and the majority are associated with contamination via food handlers. Have you ever gotten sick from eating berries?

Bacteriophages in the Dairy Industry

We have established that viruses can lyse their hosts. Sometimes that host is bacterial. Now imagine a bacteriophage attack on a 5,000-gallon fermentor full of lactic acid bacteria and milk. The bacteriophages attack and infect the lactic acid bacteria. The lactic acid bacteria lyse, and with no bacteria to convert lactose into lactic acid, the fermentation becomes "stuck." In plants that process 1 million pounds per day, bacteriophage attacks become unthinkable.

There are several ways to combat bacteriophage attacks (Table 24.3). The least sophisticated method is asepsis. Fogging the plant with 20 parts per million of chlorine kills environmental bacteriophages. The control of airflow from the processed to the unprocessed side of the plant also helps prevent contamination. Asepsis (keeping organisms out) works well for the production of starter cultures but becomes impractical for large manufacturing facilities.

Culture rotation is the next level of bacteriophage control (Fig. 24.8). The processor starts with a collection of different starter culture strains (A to K). Instead of using a single strain of bacterium for the fermentation, a pool of three strains (A, B, and C) is used. Samples from the fermentor are plaqued against each strain on a regular basis to detect whatever phages are present. If one strain, C, is attacked by bacteriophages (as evidenced by positive plaques), that strain is removed from the starter culture pool and replaced with phage-insensitive strain D. If strain A is subsequently attacked by phages, it is replaced with strain E, etc. The strains used in the fermentation are rotated to keep them ahead of the bacteriophage attack. To keep the pool of bacteriophage-resistant strains from drying up, the phage-sensitive strains are mutagenized to create more phage-insensitive strains that can be added back into the pool.

Table 24.3 Defenses of lactic acid bacteria against bacteriophages

Defense	Mechanism
Biological	
Adsorption interference	Lactic acid bacteria can prevent bacteriophage adsorption if they lack specific proteins or mask the receptors that the bacteriophages recognize.
Restriction/modification	Resistant lactic acid bacteria have enzymes that modify their own DNA to distinguish it from phage DNA, which they destroy using restriction enzymes.
Abortive infection	These systems inhibit infection after the bacteriophages have adsorbed, penetrated, and started the lytic process.
Commercial	
Phage-inhibitory media	Starter cultures are grown in media that lack Ca, which is required for phage adsorption.
Frozen concentrated starter cultures	Starters are grown in phage-inhibitory media, concentrated, and added to the fermentation at such high levels that no further growth is required.
Strain rotation	Lactic acid bacterial strains are constantly rotated so that bacteriophage-sensitive strains are identified and replaced with bacteriophage-insensitive strains.
Good manufacturing practices	Minimize contamination with viruses

Keeping the starter culture bacteriophage free is another way to protect the fermentation. Here, phage-inhibitory media are used to grow the starter cultures. Bacteriophages require Ca^{2+} for absorption onto lactic acid bacteria. When phosphate is added to the media to chelate divalent cations, the bacteriophages cannot bind and the starter culture is protected.

The final method of protecting the fermentation is to use starter cultures at such high concentrations that they can carry out the fermentation without needing to multiply. If the bacteria are not growing, the bacteriophages cannot lyse them. This application uses lactic acid bacteria that cannot metabolize proteins and therefore cannot grow. When the bacteria are added at a final concentration of 10^7 to 10^8 colony-forming units (CFU)/ml, there is enough metabolic activity to conduct the fermentation in the absence of bacterial growth.

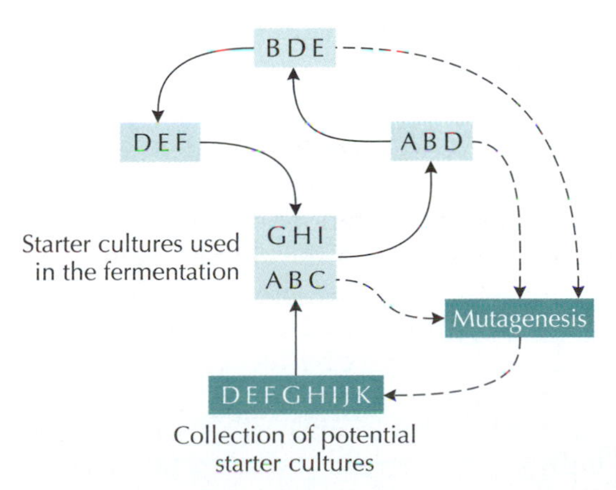

Figure 24.8 Diagrammatic explanation of culture rotation. A pool of bacteriophage-resistant lactic acid bacterial strains is used to construct a three-strain fermentation culture. As one strain becomes attacked by bacteriophages, it is replaced by another member of the pool. The phage-sensitive strain is mutagenized to phage resistance and returned to the pool for later use. (See the text for additional information.) doi:10.1128/9781555817206.ch24.f24.08

Beneficial Uses of Viruses

Bacterial Detection

Bacteria can also be used in positive food safety applications. Because bacteriophages can multiply more rapidly than their bacterial hosts, they can be used in rapid detection systems for specific pathogens. When phages are treated with a fluorescent dye and are exposed to a phage-sensitive host, they become concentrated (and visible) on the target bacteria. Phages can also carry reporter genes such as the β-galactosidase gene. When a temperate phage integrates into the target pathogen's chromosome, the phage reporter genes are expressed and the presence of the pathogen is reported by the production of a color. The luminescence (*lux*) gene also makes a good reporter gene. Such detection systems have been developed for *E. coli* O157:H7, *Listeria monocytogenes*, and *Salmonella* species.

Bacterial Control

Not only can bacteriophages (see box on next page) be used to detect pathogens, but also they can be used to control them! The FDA recently approved the use of a bacteriophage preparation, LMP-102, as a food additive to control *L. monocytogenes* in ready-to-eat meat and poultry. Intralytix, Inc., petitioned for the approval after demonstrating that no toxic material was carried over from the *L. monocytogenes* cells used to produce the phage. The development of phage resistance is minimized by using a mixture of six phages specific to different *L. monocytogenes* strains. The FDA had no toxicity or environmental issues with this antimicrobial since the phage is specific to *L. monocytogenes*. Intralytix, Inc., further assured the FDA of the additive's safety by providing the complete sequence of this double-stranded DNA virus. Since the bacteriophage preparation contains lytic rather than lysogenic phages, there was no concern that it might transfer undesirable genes into new hosts. The Center for Science in the Public Interest (a consumer watchdog group) commended the safety of the virus and its utility in *Listeria* control. A similar phage product, ECP-100, is under development for control of *E. coli* O157:H7.

The use of bacteriophages as sanitizers is another intriguing application. The application takes an ecological approach. With traditional sanitizers, all the bacteria on a manufacturing surface are killed, leaving a vacant niche. Pathogens can rapidly fill this niche. By using pathogen-specific bacteriophages as a sanitizer, only the pathogen is killed and the remaining bacteria in the niche can prevent the pathogen's reintroduction. The Environmental Protection Agency (EPA) has issued an experimental permit so that this application can be tested in a manufacturing setting.

PRIONS

Switching gears, now it is time to discuss a nonviral infectious protein particle called a prion. Prions do not replicate like bacteria or commandeer the host's metabolic machinery as do viruses. However, several important lessons about the nature of science can be learned from the prion story.

- Science-based public policy can be based only on what is "known."
- The fact that something has not been discovered does not mean that it does not exist.
- Established scientific "fact" is often theory, subject to modification.
- New findings can overthrow the most established dogma.

Bacteriophages

MANAN SHARMA, Agricultural Research Center, USDA, Beltsville, Maryland

Bacteriophages (see figure) are viruses which infect bacterial cells. They were discovered by two scientists working separately: Frederick Twort in 1915 and Felix d'Herelle in 1917. Bacteriophages (meaning "bacteria eaters," from the Greek *phagein* [to eat]) were frequently used as agents against a variety of bacterial infections from the 1920s through the 1940s. The rise of antibiotic-resistant pathogenic bacteria and interest in natural or "green" antimicrobials have refocused attention on the use of bacteriophages as antimicrobials. Phages are found in the same varied environments as their bacterial hosts; it is estimated that there are 10^{32} bacteriophages on Earth. Bacteriophages are present in various foods; they have been isolated from sausages, ground beef, freshwater and saltwater fish, raw skim milk, cheese, various deli meats, mushrooms, lettuce, refrigerated biscuit dough, and frozen chicken pot pies. Phages are specific for groups of species of bacteria, which allows them to inactivate pathogens while not affecting the remaining overall microbial ecology and balance in a specific food (e.g., a phage can display lytic specificity against *E. coli* O157:H7 but not *E. coli* O104:H4). Several U.S. regulatory agencies have approved bacteriophages to be used on various foods and food contact surfaces, leading to their potential use as antimicrobials in foods. Bacteriophages adsorb to the surface of cells, inject their DNA, and direct the cell's machinery to replicate phage particles before lysing the cell from within and releasing more phage particles. Lysins and holins are important enzymes in this process, which allows the release of the phage particles from the cell. Bacteriophages can also kill bacterial cells using a mechanism termed lysis from without (LO). LO is the lysis of bacterial cells through adsorption of a high number of bacteriophages to the cell, which lyses the cell without proceeding through the full cycle of infection. At 4°C, most pathogenic bacteria will not be metabolically active and the cycle of phage infection cannot be completed. However, LO can occur at refrigeration temperatures. The multiplicity of infection, the ratio of bacteriophages to bacterial cells, also influences the efficacy of phage-mediated inactivation. Using multiple phages specific for the same pathogen also reduces the potential for a pathogen to become resistant to infection with a single bacteriophage. Mixtures of bacteriophages specific for respective pathogens have been shown to be effective against *E. coli* O157:H7, *Salmonella* spp., and *L. monocytogenes* on a variety of produce and meat products.

Bacteriophage P2 imaged with a transmission electron microscope. Magnification, ×45,700. (Imaged by Mostafa Fatehi and reproduced with permission.) Imagine these bacteriophages lysing pathogenic bacteria on lettuce and cantaloupes, making your salads safer to eat. For more information, see the research paper by Sharma et al. (researchers at the the USDA's Agricultural Research Service) listed in "Suggested reading." doi:10.1128/9781555817206.ch24.fGuestBox

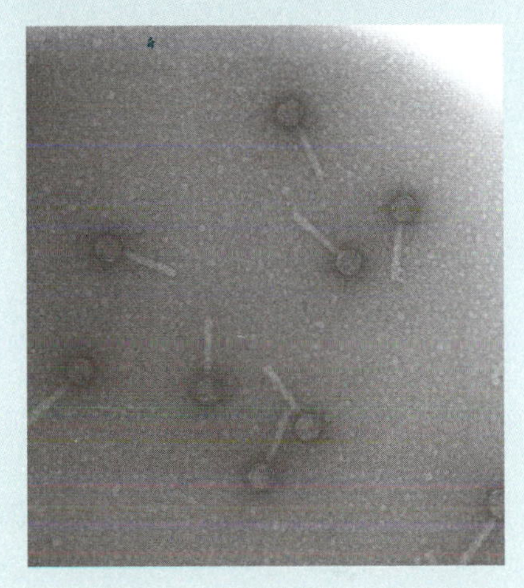

A Short History of the Prion

The first animal with symptoms of what we now know as transmissible spongiform encephalopathy (TSE) was reported in 1732, but the disease was not recognized as such at the time. Today, most consumers have heard of mad cow disease. This disease, more formally known as BSE, brought the British meat industry to its knees when 182,000 suspect cattle were destroyed. Regulators fear its appearance in the United States.

BSE did not appear (or perhaps exist?) in cattle until 1986, but the prion story started much earlier. In 1972 a young doctor named Stanley Prusiner (Fig. 24.9) became fascinated with a patient whose brain was being eaten away. However, there was no immune response and the rest of her body was untouched. Prusiner's studies of her disease were considered ridiculous and almost cost him his job. Animals that die from spongiform encephalopathy (SE) have decayed, spongy brains. SE was recognized as a symptom of scrapie in sheep, kuru in tribal natives in Papua New Guinea who practice rituals of cannibalism, chronic wasting disease in cattle and elk, and CJD in humans. However, until Prusiner, no one had linked the causes of

Figure 24.9 The Nobel laureate Stanley B. Prusiner, American neurologist and biochemist. doi:10.1128/9781555817206.ch24.f24.09

these diseases, which have no treatment. They were thought to be caused by mutations, genetic predispositions, or "slow" viruses (since the incubation period was 3 to 5 years). Prusiner proposed the prion (coined by combining "proteinaceous" and "infectious") concept to explain scrapie in 1982.

Authors' note

How or why "proin" became "prion" we do not know, but it did.

According to Prusiner, a prion is an infectious protein particle devoid of nucleic acid. PrP^c is the prion protein normally found in healthy brain and backbone membranes. When it is contacted by the infectious modified prion, PrP^{sc}, PrP^c refolds into the PrP^{sc} configuration.

The prion concept contradicted the dogma that only DNA and RNA transmit biological information. It was met with derision. Furthermore, by postulating that prion proteins can exist in two forms, Prusiner broke the central tenet of molecular biology: DNA sequence is transcribed to RNA sequence, which is translated into a protein with a specific amino acid sequence that dictates the protein's (only possible) structure. Protein structure ultimately determines protein function.

Authors' note

Rendering is a process that converts waste animal tissues into value-added materials, including meat and bonemeal, tallow or soap, and other by-products. Rendering became a continuous process in the 1980s, and meat and bonemeals became important additives to animal and plant feed.

This would have remained an obscure piece of scientific history except for one thing: in 1986 BSE broke out in cattle in the United Kingdom. Since that time, it has been detected in over 200,000 cattle in the United Kingdom. Although the disease was attributed to slow viruses at the time, prions had jumped the species barrier from sheep to cows. PrP^{sc} is transmitted from animal to animal when they are fed offal or the rendered remains of infected animals, since the prions are not inactivated even with the extreme heat of the rendering process. (The United States banned the use of offal in animal feed in 1989, and the ruminant feed ban was expanded in 1997. Thus, there should be no prions left in the U.S. meat system.)

Over the next 10 years, the "statistical correlation" of beef consumption with a new form of CJD became clear in Britain. There were 155 probable human cases resulting in 150 deaths in the United Kingdom through February 2005. (Note the high fatality rate.) The link between CJD and beef caused those in the British beef industry to change their feeding regimens and kill all animals in herds afflicted with BSE. Through the end of 2010 184,500 cases of BSE had been confirmed in the United Kingdom alone in more than 35,000 herds.

There were strong epidemiological and laboratory links for the association of a new human prion disease called nvCJD that was first reported in the United Kingdom in 1996. The peak of the outbreak in the United Kingdom was in 1993, with almost 1,000 new cases each week. In terms of human disease, the interval between the most likely period for the initial extended exposure of the population to the potentially BSE-contaminated food was between 1984 and 1986 and the onset of the initial cases of nvCJD was between 1994 and 1996, consistent with the incubation periods of these diseases. The United States confronted this issue, and around the world countries tried to put new firewalls in place to protect their food from potential BSE contamination. Well over 100,000 cattle were sacrificed as a preventive measure. On the human side, the initially ridiculed Stanley Prusiner was awarded the Nobel Prize in medicine in 1997 for his discovery of the prion.

The U.S. experience with BSE has been different from that of the United Kingdom due to the United States' early banning of offal in animal feed and the strategic testing of high-risk animals. In 2003, 1 cow of 394,000 tested in the United States was identified as having BSE. American authorities quickly determined that the cow came from Canada and had never entered the food chain. Canadian authorities destroyed 2,700 head of cattle to prevent the possible spread of BSE-infected animals for export or for human consumption.

Due to this quick action, countries that had banned Canadian beef in May 2003 eased the ban a few months later. Unfortunately, two more cases of BSE were detected in Canada in 2005. In March 2006, the USDA announced confirmation of a BSE-positive cow in Alabama in the United States. This animal never entered the food supply and was thought to have contracted BSE in an atypical form associated with a rare genetic abnormality. USDA scientists are still studying the gene mutation associated with this case. As of March 2011, 19 BSE cases had been identified in Canadian-born cattle, of which the majority were born after the feed bans put in place in 1997. Based on what is known about BSE and nvCJD, it appears that the feed bans put in place in North America have protected to the food supply from BSE contamination.

Prion Biology

The prion concept remains controversial, especially the "devoid of nucleic acid" part of the definition. It is difficult to prove that something does not exist. Maybe it exists but cannot be found. That is the argument against prions: there "must" be a small amount of nucleic acid, but it cannot be detected. However, treatments that destroy nucleic acids do not destroy prions, and nucleic acids have not been detected in purified prions. The concept of protein conformational conversion explains all of the *infectiously transmissible* SEs. However, protein conformational conversion is only one of three mechanisms that cause SEs. A hereditary defect in PrP^c, such as the one that causes Libyan Jews to get CJD 30 times more often than other Jews, causes a familial form of CJD. A mutation in PrP^c causes sporadic CJD in the elderly. Finally, there is the conversion of PrP^c into PrP^{sc} by PrP^{sc} consumed in the diet.

One must understand a bit of protein chemistry to understand how prions work. The *primary sequence* of amino acids in a protein determines its *secondary structure*. Common secondary structures are α helices and β-pleated sheets. A protein molecule may contain several helices, pleated sheets, and unordered sections. These arrange themselves to form the protein's overall shape, or conformation. This is called its tertiary structure. PrP^c has a conformation with several α helices. When it contacts PrP^{sc}, it refolds to match that conformation, which has large runs of pleated sheets (Fig. 24.10). In this conformation, the prion protein destroys brain tissue.

Authors' note

Think about a chicken egg, which is composed of many proteins. When you crack open an egg on a skillet, the heat begins to denature the proteins, causing them to coagulate and cook. The scrambled egg you eat for breakfast has a different structure from the initial cracked egg. Proteins are also denatured and changed by physical forces, like whisking. When you whisk egg whites to make a delicious sugary meringue, you break the initial bonds and rearrange the structure, again changing the original conformation of the egg proteins. With proteins, structure confers function in many examples, including prions.

Figure 24.10 Structures of the normal prion protein, PrP^c **(left)**, and the infectious form, PrP^{sc} **(right)**. Note the changes in the secondary structures (coils and arrows) as well as the change in the overall shape (tertiary structure). The two forms have the same amino acid sequence. Reprinted with permission from Sarit Helman. doi:10.1128/9781555817206.ch24.f24.10

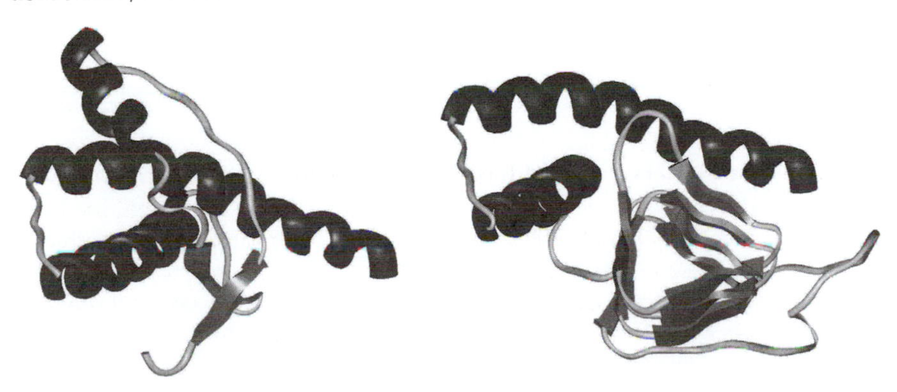

Prions are not inactivated by heat, irradiation, or food preservatives. Once they are in the food, they are there for good. The only way to prevent TSEs is to prevent the converted form from touching the normal form. This is done by making sure that animals do not eat animals. Most industrialized countries now ban the use of animal by-products as animal feed. The United States has banned importation of beef from countries where BSE is present. However, wild animals such as mink and deer also carry TSEs. Theoretically, these prions might jump the species barrier into domestic cattle.

Summary

- Viruses have a simple life cycle of adsorption, entry, replication, assembly, and release in a lytic burst.
- Viruses are host specific due to unique ligand-receptor binding.
- Noroviruses cannot be grown in culture, but bacteriophages can be detected by plaques on plates of sensitive bacteria and some viruses can be grown in the laboratory on mammalian cells in culture.
- Viruses can cause as much foodborne disease as bacteria (if not more) but cannot grow in foods.
- Fecal-oral transmission is the main route of human infection with enteric viruses.
- Proper heating and good hygiene should minimize the risk of contamination from viruses.
- Noroviruses and hepatitis A virus are the major foodborne viruses.
- Viruses have useful applications such as pathogen detection and control.
- There are several methods to prevent bacteriophage attacks on fermenting dairy cultures.
- Prions are infectious proteins devoid of RNA and DNA and were discovered by Stanley Prusiner in the 1980s.
- Prions cause TSEs such as scrapie, kuru, mad cow disease, and CJD.
- TSEs are caused when normal PrP^{c} is converted into infectious PrP^{sc} by mutation or by contact with an ingested PrP^{sc} molecule.
- Effective management strategies and firewalls have reduced the risk of BSE-contaminated meat entering the food supply in the United States.

Suggested reading

Cliver, D. O. 1997. Virus transmission via food (an IFT Scientific Status Summary). *Food Technol.* **51:**71–78.

D'Souza, D. H., C. L. Moe, and L.-A. Jaykus. 2007. Foodborne viral pathogens, p. 581–610. *In* M. P. Doyle and L. R. Beuchat (ed.), *Food Microbiology: Fundamentals and Frontiers*, 3rd ed. ASM Press, Washington, DC.

Goyal, S. M. (ed.). 2006. *Viruses in Food*. Springer Science + Business Media, LLC, New York, NY.

Huston, W., and C. M. Bryant. 2005. Transmissible spongiform encephalopathies—an IFT Scientific Status Summary. *J. Food Sci.* **70:**R77–R87.

Kniel, K., and A. E. H. Shearer. 2009. Berry contamination: outbreaks and contamination issues, p. 271–305. *In* G. M. Sapers, E. B. Solomon, and K. R. Matthews (ed.), *The Produce Contamination Problem*. Academic Press, New York, NY.

Prusiner, S. B. 1998. Prions. *Proc. Natl. Acad. Sci. USA* **95:**13363–13383.

Sharma, M., J. R. Patel, W. S. Conway, S. Ferguson, and A. Sulakvelidze. 2009. Effectiveness of bacteriophages in reducing *Escherichia coli* O157:H7 on fresh-cut cantaloupes and lettuce. *J. Food Prot.* **72:**1481–1485.

Questions for critical thought

1. The origin of BSE is still disputed, as are adequate testing measures within the United States. What are the current thoughts on its origin and the usefulness of current testing programs?
2. How are bacteria, viruses, protozoa, and prions different?
3. What intervention can be used to make food safe from all of them?
4. There are no routine lab tests to detect viruses or prions in foods. How does this make their control different from controlling foodborne pathogens? How would you design an effective test for foods?
5. What is a prion? What are the three ways in which it can be formed?
6. Much of this book deals with how to prevent microbial growth in food. How can viruses be prevented from growing in foods?
7. Think of three terms that could be used to convey to your less sophisticated peers the concept of fecal-oral transmission.
8. Why is dilution *not* the solution for preventing viral contamination by shellfish? Diluting viruses in large volumes of water should bring them below the infectious dose. Consider the case of the outhouse over the oyster bed. Assume 50 people, all infected at 10^8 particles per g of feces, depositing 2 kg of fecal material per day into a body of water that is 1 km by 1 km and 10 m deep. How many viral particles per milliliter of water would there be?
9. How are biological viruses like computer viruses?
10. Compare the ways that computers can be protected from viruses to the ways that humans can be protected.
11. Draw a diagram of the life cycle of a temperate bacteriophage.
12. Write a paragraph (or two) designed to convince consumers that using viruses to kill *L. monocytogenes* is "safe and natural."

SECTION

IV

Control of Microorganisms in Food

25 Chemical Antimicrobials

LEARNING OBJECTIVES

The information is this chapter allows the student to:

- link chemical preservatives to their ability to prevent food spoilage
- distinguish positive and negative aspects of food preservatives
- characterize traits that distinguish "natural" from "chemical" preservatives
- describe factors that affect antimicrobial preservation systems and how these factors interact
- understand how organic acids inhibit microbes and the influence of pK_a and pH on their antimicrobial action
- understand the effect of water activity on microbial growth and microbial ecology
- use the water activity concept to distinguish between "bulk water" and water that is unavailable for microbial growth
- identify enzymes and inorganic food preservatives, how they work, and in what foods they are useful

INTRODUCTION

Preservatives are good. Without them, food spoils. Spoilage can be caused by microbiological, enzymological, chemical, and physical changes. Preservatives can prevent or delay these changes. Most antimicrobials inhibit growth without killing the microbe. Chilling, freezing, water activity (a_w) reduction, acidification, fermentation, energy (e.g., thermal, irradiation, or pressure), and antimicrobial compounds all inhibit microbial growth.

Antimicrobials are sometimes called preservatives. However, when used alone, "preservative" is a broader term than "antimicrobial." Preservatives include antibrowning agents, antioxidants, stabilizers, etc. The major antimicrobial targets are foodborne pathogens and spoilage microorganisms. They cause off-odors, off-flavors, liquefaction, and discoloration, and they produce slime. Spoilage can be a good thing. It warns people not to eat foods that may contain pathogens. However, pathogens can grow without obvious signs of spoilage. (If pathogens always spoiled food, no one would eat the food and there would be no foodborne illness.) Preservation systems that allow pathogen growth without spoilage are inherently dangerous.

Most antimicrobials only *inhibit* growth. They are bacterio*static* rather than bacteri*cidal*. Bacteriostatic compounds stop microbial growth; bactericidal compounds kill the bacteria. Antimicrobials act at specific metabolic

doi:10.1128/9781555817206.ch25

targets. Antimicrobials having different mechanisms can be combined to give a synergistic effect where the sum effect is more than the effect of the individual parts (e.g., 1 + 1 = 5). This is called *hurdle technology*. The cell wall, cell membrane, enzymes, and genes are all targets for antimicrobial action. Because metabolism is an interlocking web, it is often hard to determine an antimicrobial's exact mechanism of action.

Antimicrobial classification is arbitrary. This chapter divides antimicrobials into two classes, traditional "chemical" antimicrobials and naturally occurring ones.

Chemical antimicrobials

- have been used for years
- are approved by many countries
- are made by synthetic or natural means
- are usually considered "bad" by consumers

Chemical antimicrobials are also found in nature. These include acetic acid from vinegar, benzoic acid from cranberries, and sorbic acid from rowanberries.

Naturally occurring food preservatives refer to compounds that are

- extracted from natural sources
- "organic"
- not usually subject to regulatory approval as new food additives
- "label friendly" (i.e., they do not scare consumers when consumers see them on labels)

It should be noted, however, that except for flavors, there is no federal regulatory definition of "natural" in reference to food ingredients.

FACTORS THAT AFFECT ANTIMICROBIAL ACTIVITY

Foods are complex systems. Many variables influence microbial growth and inhibition. For instance, the food's pH, a_w, storage temperature, and gas composition have an impact on antimicrobial effectiveness. These factors must be considered when developing preservation systems. Antimicrobial activity is affected by *microbial, intrinsic, extrinsic,* and *process* factors.

Microbial factors include

- the organism's inherent resistance (e.g., vegetative cells versus spores) to the preservation method
- initial cell number
- growth rate
- growth phase
- interaction with other microorganisms (e.g., antagonism)
- cellular composition (which may be a function of the cells' history)
- physiological status (injury, adaptation, and metabolic state)

There is an explosion of knowledge about how cells respond to stress. Tolerance gained by exposure to one stressor, such as cold, may make cells more tolerant to another, such as acid. This is called adaptation, stress response, or acquired tolerance. Special proteins are produced when cells are stressed. They are responsible for the acquired tolerance. Some stress responses are specific to a given stressor. Others give broad resistance to a range of stressors.

Intrinsic factors are part of the food itself. These include composition, pH, buffering capacity, oxidation-reduction potential, and a_w. These can be specified during product development to formulate "food safety" into the food. When a product is reformulated to reduce salt, acidity, sugar, or other potentially inhibitory substances, the impact on microbial safety should be considered. For example, pathogens have grown in previously stable products when they were reformulated to replace sucrose with an artificial sweetener. This removed the a_w barrier that prevented pathogen growth.

Extrinsic factors also influence antimicrobial activity. These variables are external to the food. They include temperature, storage time, atmosphere, and relative humidity. The time-temperature combination is perhaps the most important consideration for microbial growth in food. It is arguably the most difficult to control. (Consider the shopper who forgets that he left his refrigerated entrée in the trunk of his car and then eats it the next day.) Extrinsic factors also need consideration during product development. A manufacturer of frozen foods expanding into refrigerated entrées faces new microbiological challenges.

The microbiology of a processed food is often different from that of an unprocessed food. Processing can change food composition. For example, drying changes the a_w of a food. The addition of soluble compounds can also decrease a_w. Heating foods can kill vegetative cells but allow spores to thrive. Food processing can also change food microstructure.

ORGANIC ACIDS

The short-chain organic acids (Table 25.1) have similar antimicrobial mechanisms and applications. They are all found in nature but are often manufactured chemically for use in foods. They have "generally recognized as safe" status and are label friendly. Due to their mechanism of action (see below), organic acids are most useful below their pK_a (explained in the next paragraph). They are rarely used in foods with a pH of >5.5. This is a major limitation.

Table 25.1 Short-chain organic acids

Organic acid	Formula	pK_a	Food in which it is found	Organisms against which it is active	Applications	Concn (%)
Acetic	CH_3COOH	4.5	Vinegar	Bacteria and yeasts	Bakery, cheese, olives, gravies, sauces, meat, and poultry	<0.1–0.8
Benzoic	C_6H_5COOH	4.2	Cranberries	Fungi	Beverages, jams, jellies	<0.1
Lactic	$CH_3OHCOOH$	4.8	Fermented food	Bacteria	Used mainly for pH control and flavor; also used for sanitation, spraying on meat	<0.4–2.0
Propionic	CH_3CH_2COOH	4.9	Swiss cheese	Yeasts, molds, rope-forming bacteria	Baked goods and cheeses	0.3–no limit
Sorbic	$CH_3CH{=}CHCH{=}CHCOOH$	4.7	Rowanberries	Fungi and certain bacteria	Baked goods, cheese, wine	0.1–0.3

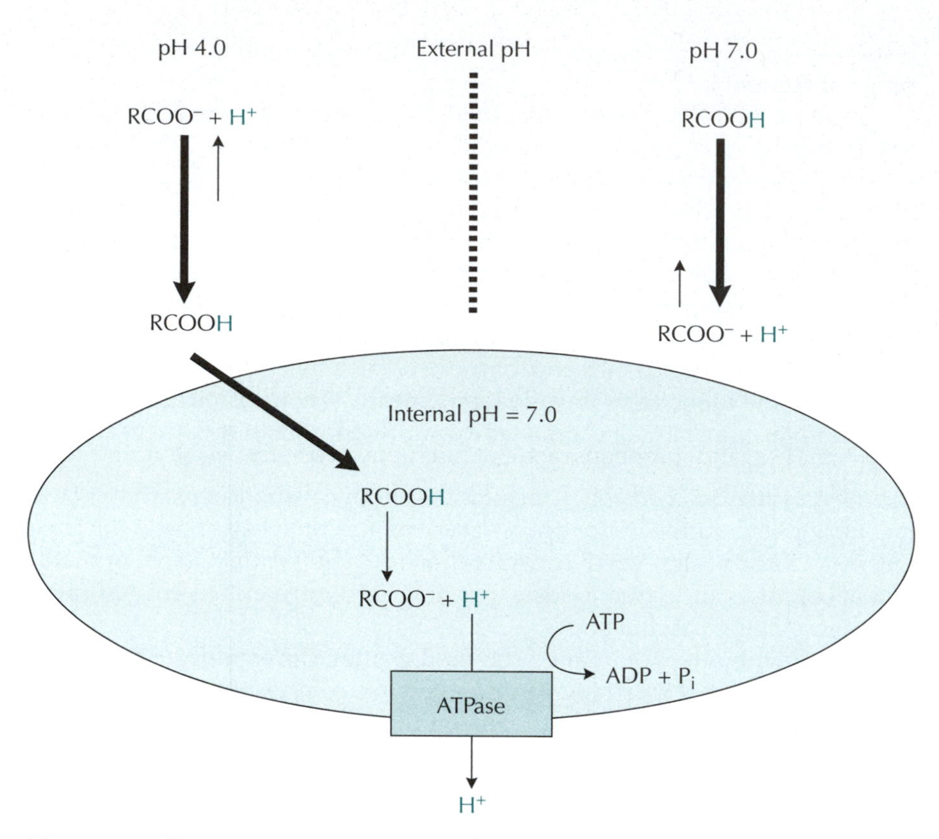

Figure 25.1 Organic acids are effective inhibitors only at low pH, where their protonated form can cross the cell membrane. At the higher pH inside the cell, the acid dissociates and the cell must use energy to pump out the proton to deacidify the cytoplasm. At high external pH, the dissociated acid cannot cross the membrane and get into the cell and therefore is not effective. doi:10.1128/9781555817206.ch25.f25.01

To be effective, organic acids must cross the bacterial cell membrane and get into the cytoplasm (Fig. 25.1). Organic acids exist in charged and uncharged forms. The charged form cannot cross the bacterial membrane and is essentially useless. It is the uncharged (protonated) acid that gets into the bacterium and inhibits growth. The pH at which the concentrations of charged and uncharged acids are equal is the pK_a. Since pH is a log scale, at one pH unit above the pK_a, there is 10-fold less protonated acid. This low concentration of protonated acid would not be inhibitory. This is why organic acids are most useful in acidic foods near or below their pK_a. At low pH, the protonated acid crosses the cell membrane. The neutral cytoplasmic pH then causes the acid to dissociate. The release of the proton acidifies the cytoplasm. The cell uses adenosine 5′-triphosphate (ATP) to pump protons out of the cell. This maintains the neutral intracellular pH needed for viability but uses up the ATP energy needed for growth.

PARABENZOIC ACIDS

Benzoic acid has a pK_a similar to that of the organic acids. However, the carbon chains between the benzene and the acid parts of the molecule increase the pK_a. These parabenzoic acids, commonly called parabens, remain undissociated up to their pK_a of 8.5. This gives them an effective pH range of 3.0 to 8.0. The methyl, propyl, and heptyl parabens are approved antimicrobials in most countries.

The antimicrobial activity of parabens increases with increasing chain length. Unfortunately, solubility in water decreases with increasing chain length. The oppositional influence of chain length and solubility determines which parabenzoate can be used in a given application. Methyl and propyl parabenzoates are often used in a 2:1 to 3:1 ratio to capitalize on their respective solubility and increased activity. The *n*-heptyl ester is used in beers, noncarbonated soft drinks, and fruit-based beverages. Parabens are generally more active against molds and yeasts than against bacteria (Table 25.1). Parabens are used in baked goods, beverages, fruit products, jams and jellies, fermented foods, syrups, salad dressings, wine, and fillings.

Phenolics and parabens act on microbial membranes, but the specific mechanisms are unclear. The membrane's barrier properties may be degraded, allowing intracellular compounds to leak out. Respiration may also be inhibited, perhaps due to changes in membrane fluidity.

NITRITES

Nitrites are used in cured meats such as bacon, hot dogs, ham, and bologna. Curing solutions contain nitrite, salt, sugar, spices, and ascorbate or erythorbate. Nitrite has many uses in cured meats. Nitrosomyoglobin makes cured meat pink. In addition to being antimicrobials, nitrites contribute to cured meat flavor and texture and serve as antioxidants. Nitrite's effectiveness depends on many factors. Nitrites work better at a lower pH. Nitrite is more inhibitory in the absence of air. The reducing agents ascorbate and isoascorbate enhance nitrite's antibotulinal action. Temperature, salt, and microbial load also influence nitrite's effectiveness.

Sodium nitrite inhibits *Clostridium botulinum* in cured meats. Nitrite also inhibits bacteria such as *Achromobacter*, *Enterobacter*, *Escherichia coli*, *Flavobacterium*, *Micrococcus*, *Pseudomonas*, and *Listeria monocytogenes*. Governmental regulations limit its use to 156 parts per million (ppm) for most cured meat products and 100 to 120 ppm in bacon. Sodium erythorbate or isoascorbate is required as a cure accelerator and to inhibit nitrosamine formation.

Nitrosamines are carcinogens formed by reactions of nitrite with amines. The discovery that nitrosamines are produced from nitrite and meat protein at high temperature triggered extensive research on nitrite substitutes. However, no other compound could replace nitrite's antimicrobial, taste, and color functions. Concern about nitrosamines decreased upon learning that green vegetables such as spinach and broccoli were a larger source of dietary nitrites.

PHOSPHATES

Sodium acid pyrophosphate (SAPP), tetrasodium pyrophosphate (TSPP), sodium tripolyphosphate (STPP), sodium tetrapolyphosphate, sodium hexametaphosphate (SHMP), and trisodium phosphate (TSP) are active against microbes in foods.

Gram-positive bacteria are more susceptible to phosphates than gram-negative bacteria. TSPP, STPP, and SHMP at 0.5% inhibit *Bacillus subtilis*,

Clostridium sporogenes, and *Clostridium bifermentans*. SAPP inhibits *C. botulinum* toxin production but not growth. Outgrowth of *Bacillus* spores is prevented by 0.2 to 1.0% SHMP.

Polyphosphates' ability to chelate metal ions is responsible for their antimicrobial action. Polyphosphates probably inhibit gram-positive bacteria and fungi by removing essential cations from the cell wall binding sites. Divalent cations can reverse polyphosphate inhibition. Inhibition is also reduced at lower pH due to protonation of the chelating sites. Orthophosphates have no chelating ability and thus no inhibitory activity.

SODIUM CHLORIDE

Sodium chloride (NaCl), or common salt, is the oldest preservative known. A few foods, such as raw meats and fish, are preserved solely by high salt concentrations, i.e., by "salting." More often, salt is combined with processes such as canning, pasteurization, or drying. Foodborne pathogens are usually inhibited by an a_w of 0.92 or less (equivalent to 13% [wt/vol] NaCl) under otherwise optimum conditions. The exception is *Staphylococcus aureus*, which can grow at an a_w of 0.83 to 0.86 under aerobic conditions. Under anaerobic conditions, the minimum a_w is >0.90. *L. monocytogenes* is also salt tolerant. It survives in saturated salt solutions at low temperatures.

Sodium chloride's primary lethal mechanism is plasmolysis (shrinkage of the cell due to water loss). Salt reduces a_w, creating unfavorable growth conditions. As the a_w decreases, cells undergo osmotic shock and rapidly lose water through plasmolysis. During plasmolysis, a cell dies or remains dormant. To resume growth, the cell must concentrate specific solutes to equilibrate with the external a_w. However, they must use up ATP to concentrate these solutes, and that loss of cellular energy results in growth inhibition or death.

WATER ACTIVITY

Water is a major factor in controlling food spoilage. However, it is the *availability* of the water, not the amount, that affects microbes. Sugar and salt addition is commonly used to lower a_w. These, and other food molecules, chemically bind water molecules. The water is bound by ionic interactions, hydrogen bonding, etc., with chemical constituents of the food. The bound water is not available for microbial growth. How much water is unbound or available determines if microbes can grow.

The measure of available water in foods is water activity (a_w). a_w is defined as the ratio of the vapor pressure of water in a food, P, to the vapor pressure of pure water, P_0, at the same temperature:

$$a_w = \frac{P}{P_0}$$

The movement of water vapor from a food to the air depends on the food's moisture content and composition, the temperature, and the humidity. At a constant temperature, the water in the food equilibrates with water vapor in the air. This is the food's *equilibrium moisture content*. At the equilibrium moisture content, the food neither gains nor loses water to the air. The relative humidity of the surrounding air is thus the equilibrium relative humidity (ERH). ERH is defined as

$$\text{ERH (\%)} = a_w \times 100.$$

The third way to conceptualize water activity is expressed by Raoult's law. Since the addition of solutes depresses vapor pressure, one can calculate the a_w from the equation

$$a_w = n_1/(n_1 + n_2)$$

where n_1 is moles of solute and n_2 is moles of solvent.

The a_ws of various foods are shown in Fig. 25.2. High-moisture foods such as fruits, vegetables, meats, and fish have a_ws of ≥0.98. Intermediate-moisture foods (e.g., jams and sausages) have a_ws of 0.7 to 0.85. Additional preservative factors (e.g., reduced pH, preservatives, and pasteurization) are required for their microbiological stability.

The difference between a_w and moisture content is illustrated in Table 25.2. Foods at the same a_w may have different moisture contents due to their chemical composition and water binding capacity.

Dehydration preserves food by removing available water, i.e., reducing a_w. Hot air removes water by evaporation. Freeze-drying removes water by sublimation (the conversion of ice to vapor without passing through the liquid stage) after freezing. These processes reduce the a_ws of the foods to levels that inhibit growth.

Various microbes have different a_w requirements. Decreasing the a_w increases the lag phase of growth, decreases the growth rate, and decreases the number of cells at stationary phase. Foodborne microbes are grouped by their minimal a_w requirements in Table 2.7. Gram-negative species usually require the highest a_w. Gram-negative bacteria such as *Pseudomonas* spp. and most members of the family *Enterobacteriaceae* usually grow only above a_ws of 0.96 and 0.93, respectively. Gram-positive non-spore-forming bacteria are less sensitive to reduced a_w. Many *Lactobacillaceae* have minimum a_ws near 0.94. Some *Micrococcaceae* grow below a_ws of 0.90. Staphylococci

Figure 25.2 Illustration of the effect of a_w on shelf life and microbial growth.
doi:10.1128/9781555817206.ch25.f25.02

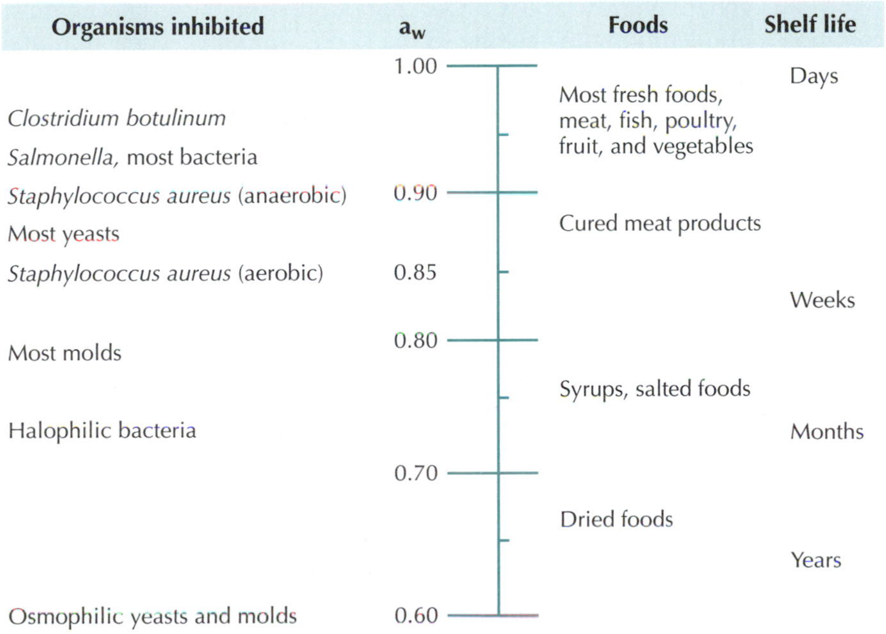

Organisms inhibited	a_w	Foods	Shelf life
	1.00		Days
Clostridium botulinum		Most fresh foods, meat, fish, poultry, fruit, and vegetables	
Salmonella, most bacteria			
Staphylococcus aureus (anaerobic)	0.90		
Most yeasts		Cured meat products	
Staphylococcus aureus (aerobic)	0.85		
			Weeks
Most molds	0.80		
		Syrups, salted foods	
Halophilic bacteria			Months
	0.70		
		Dried foods	
			Years
Osmophilic yeasts and molds	0.60		

Table 25.2 Moisture contents of various dry or dehydrated food products when their a_w is 0.7 at 20°C[a]

Food	Moisture content (%)
Grains	4–9
Milk powder	7–10
Cocoa powder	7–10
Whole-egg powder	10–11
Skim milk powder	10–15
Dried, fat-free meat	10–15
Rice and legume seeds	12–15
Dehydrated vegetables	12–22
Dried soups	13–21
Dried fruits	18–25

[a]Reprinted from J. Farkas, p. 567–591, *in* M. P. Doyle, L. R. Beuchat, and T. J. Montville (ed.), *Food Microbiology: Fundamentals and Frontiers*, 2nd ed. (ASM Press, Washington, DC, 2001).

are unique among foodborne pathogens because they can grow at a minimum a_w of about 0.86 under aerobic conditions. However, they do not make toxins below an a_w of 0.93. Most spore-forming bacteria do not grow below an a_w of 0.93. Spore germination and outgrowth of *Bacillus cereus* are prevented at a_ws of 0.97 to 0.93. The minimum a_w for *Clostridium perfringens* spore germination and growth is between 0.97 and 0.95.

Osmotolerant yeast species grow at a_ws lower than those of bacteria. Halotolerant species (i.e., osmotolerant species specifically resistant to salt) such as *Debaryomyces hansenii*, *Hansenula anomala*, and *Candida pseudotropicalis* grow well on cured meats and pickles at NaCl concentrations of up to 11% (a_w = 0.93). Some "xerotolerant" (tolerant of very dry conditions) species such as *Zygosaccharomyces rouxii*, *Zygosaccharomyces bailii*, and *Zygosaccharomyces bisporus* grow on and spoil foods such as jams, honey, and syrups having a high sugar content (and correspondingly low a_w).

Molds generally grow at a_ws lower than foodborne bacteria. The most common xerotolerant molds belong to the genus *Eurotium*. Their minimal a_w for growth is 0.71 to 0.77, while the optimal a_w is 0.96. True xerophilic molds such as *Monascus* (*Xeromyces*) *bisporus* do not grow at a_ws of >0.97 to 0.99. The relationship of a_w to mold growth and toxin formation is complex. At some a_ws, molds can grow but not make toxins.

The varied a_w growth limits of bacteria and fungi reflect the mechanisms that help them grow at low a_ws. Bacteria protect themselves from osmotic stress by accumulating compatible solutes intracellularly. Compatible solutes equilibrate the cells' intracellular a_w to that of the environment but do not interfere with cellular metabolism. Some bacteria accumulate K^+ ions and amino acids, such as proline. Halotolerant and xerotolerant fungi concentrate polyols such as glycerol, erythritol, and arabitol.

DISINFECTANTS

Disinfectants are not food preservatives in a strict sense, but they do kill or inhibit microorganisms that can spoil food. They can be indirect or "incidental" food additives used on food contact surfaces. These applications are regulated by the Environmental Protection Agency rather than the Food

Preservatives can be used to extend the expiration dates of food but unfortunately not of people.
Reprinted with permission from CartoonStock.

"Now that I'm in midlife, I read all the food labels. I can use all the preservatives I can get."

and Drug Administration (FDA). Disinfectants are becoming increasingly important as the U.S. Department of Agriculture (USDA) begins to require microbial reduction steps in the processing of raw meat and poultry. Most often, these take the form of an antimicrobial wash.

Sulfites

Sulfur dioxide (SO_2) has been used as a disinfectant since ancient times. Antimicrobial salts of sulfur dioxide include potassium sulfite (K_2SO_3), sodium sulfite (Na_2SO_3), potassium bisulfite ($KHSO_3$), sodium bisulfite ($NaHSO_3$), potassium metabisulfite ($K_2S_2O_5$), and sodium metabisulfite (NaS_2O_5). Sulfites are used primarily in fruit and vegetable products to control spoilage yeasts and molds and bacteria. Bacteria are generally more sensitive than yeasts or molds. This allows wine to be treated with sulfites to inhibit the bacteria that cause the malolactic fermentation, without interfering with the yeast's alcoholic fermentation. Sulfites also act as antioxidants that inhibit enzymatic browning.

The most important factor for sulfites' antimicrobial activity is pH. In water, sulfur dioxide and its salts exist in a pH-dependent equilibrium:

$$SO_2 \cdot H_2O \leftrightarrow HSO_3^- + H^+ \leftrightarrow SO_3^{2-} + H^+$$

Once again, dissociation of the acid plays a major role. As the pH decreases, the proportion of $SO_2 \cdot H_2O$ increases and the bisulfite (HSO_3^-) ion

Across

3. Organic acids ________ when they pass into the cell and thereby lower the intracellular pH
10. Type of natural antimicrobial oil
13. Acid associated with the oxidation of ethanol to inhibit bacteria and yeast
14. Preservatives that kill bacteria are ________
15. An enzyme that inhibits the growth of mostly gram-positive bacteria
16. Type of factor in the food environment that can influence the effectiveness of an inhibitor
17. Acid produced by *Propionibacter* spp. that inhibits yeasts and molds

Down

1. A specific type of preservative
2. Compounds that, in addition to inhibiting microbes, give cured meat its color and flavor
4. Antimicrobials that are also antioxidants
5. Technology that uses multiple inhibitors to prevent the growth of bacteria
6. Compounds that prevent microbial growth
7. Natural preservatives can be extracted from ________
8. Type of preservative that is made by synthetic means
9. Organic acid whose effectiveness increases but whose solubility decreases with chain length
11. Organic acids are not very useful at pH values ________ their pK_a
12. Allows bacteria that have been exposed to one stress to become more resistant to a second stress

concentration decreases. The pK_a values for sulfur dioxide are about 1.8 and 7.2. Sulfite is most inhibitory when the acid or $SO_2 \cdot H_2O$ is undissociated. Therefore, their most effective pH range is <4.0. $SO_2 \cdot H_2O$ is 100 to 1,000 times more active than HSO_3^- or SO_3^{2-} against *E. coli*, yeasts, and *Aspergillus niger*. Sulfites, especially the bisulfite ion, are very reactive. They form addition compounds (α-hydroxysulfonates) with aldehydes and ketones.

Low concentrations of sulfur dioxide are fungicidal. The inhibitory concentration of sulfur dioxide is 0.1 to 20.0 ppm against *Saccharomyces*, *Zygosaccharomyces*, *Pichia*, *Hansenula*, and *Candida* species. Sulfur dioxide at 25 to 100 ppm inhibits *Byssochlamys nivea* growth and patulin production in juices. Sulfites may inhibit spoilage bacteria in wines, fruits, and meats.

Sulfur dioxide controls microbial spoilage in fruits, fruit juices, wines, sausages, fresh shrimp, and pickles. It is added at 50 to 100 mg/liter to grape juice used for making wine to inhibit molds, bacteria, and yeasts. If the right concentration is used, sulfur dioxide does not interfere with wine yeasts or with wine flavor. During fermentation, sulfur dioxide also serves as an antioxidant, clarifier, and dissolving agent. Sulfur dioxide (50 to 75 mg/liter) also prevents postfermentation spoilage by *Acetobacter* spp. which oxidize ethanol into acetic acid (i.e., vinegar).

The targets for sulfite inhibition are the cytoplasmic membrane, DNA replication, protein synthesis, and enzymes. In *Saccharomyces cerevisiae*, *Saccharomyces ludwigii*, and *Z. bailii*, sulfites diffuse into the cell. Other fungi have active transport systems for sulfites.

Chlorine

Chlorine in the form of hypochlorous acid is probably the most widely used sanitizer. It is added to wash waters and water used to convey fruits and vegetables. It is used to sanitize food contact surfaces and to reduce the microbial load on the surfaces of meat and poultry.

Sodium hypochlorite (bleach) is the most commonly used form. A strong oxidizer, it is active against bacteria, yeasts, molds, and viruses. Unfortunately, it is very corrosive, even to metals such as stainless steel.

The use of chlorine dioxide (ClO_2) has important advantages over hypochlorite. While it has similar antimicrobial properties, it is odorless and also degrades undesirable compounds like phenols, sulfites, and mercaptans (which taste and smell bad).

Quaternary Ammonium Compounds

Quaternary ammonium compounds, commonly called "quats," are a group of chemicals in which four compounds surround a nitrogen atom (Fig. 25.3). They are very versatile because the R groups can be any number of saturated or unsaturated, cyclic or noncyclic, substituted or unsubstituted alkyl groups. The quaternary ammonium compound is positively charged and is accompanied by a negatively charged compound, often chloride. These compounds are colorless, relatively odorless, and noncorrosive. Quats are active against bacteria, yeasts, fungi, and viruses. As you might guess, they are more expensive than hypochloride-based disinfectants.

Figure 25.3 Structure of a quaternary ammonium compound, where the R ligands can be the same or different compounds. See the text. doi:10.1128/9781555817206.ch25.f25.03

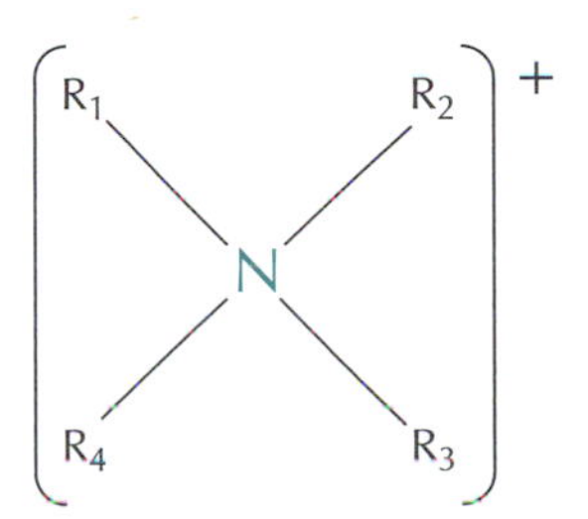

Peroxides

Peroxides are compounds that have covalently bound oxygens (-O-O-) at their core. When represented as R_1-O-O-R_2, the addition groups (R_1 and R_2) can be inorganic (i.e., hydrogen) or organic (acetic acid) compounds.

Hydrogen peroxide and peroxyacetic acid are the two most commonly used peroxides.

The action of both compounds depends on dissociation. In the case of hydrogen peroxide,

$$O_2^- + H_2O_2 \rightarrow OH\cdot + OH^- + O_2$$

The lethal action of both compounds comes from the OH˙ radical. This very strong oxidizer reacts with membrane lipids, DNA, and proteins.

Hydrogen peroxide is used as a sanitizer that kills a wide variety of bacteria, yeasts, molds, and viruses, with the exception of bacteria which have strong catalase activity. Hydrogen peroxide is used medicinally in 3 to 5% solutions to disinfect wounds. In the food industry, similar concentrations are used to sterilize equipment and food contact surfaces. Because it whitens meats, it cannot be used in that application. Hydrogen peroxide is not approved as a food additive per se. Thus, the FDA has limited its use to produce whose endogenous catalase activity can remove hydrogen peroxide residues.

Peroxy acids are much stronger than hydrogen peroxides and can be used at much lower concentrations that do not affect poultry or meat quality. The reductions of microbial loads on poultry, meat, and produce are important applications of peroxy acids. Peroxy acids are also used to sterilize food contact surfaces and containers used in aseptic processing. Peroxidases and catalase do not affect peroxy acids.

Ozone

Ozone is another disinfectant that is a strong oxidant. Its primary target is nucleic acids. Ozone has "generally recognized as safe" status for bottle washes. Ozone has a great deal of potential for use with raw fruits and vegetables, where it has been shown to reduce *E. coli* O157:H7 viability over 1,000-fold. However, as of this writing, these applications have not been approved. Ozone is not effective for meats, because meat reactions reduce the ozone concentration below the inhibitory levels.

NATURALLY OCCURRING ANTIMICROBIALS

Many foods contain natural antimicrobials. They may extend the food's shelf life, but their use as direct food additives presents many challenges. Ideally, they should work when added as food ingredients, without purification. The cost of using the natural antimicrobial should be as low as for chemical preservatives. They should not affect the food's taste, nutrition, or safety. However, most natural antimicrobials are normally present in amounts too low to work when mixed into foods. If higher levels are used, the taste and smell of the food are affected.

Lysozyme

Lysozyme is an antimicrobial enzyme produced in eggs, milk, and tears. Commercial lysozyme is made from dried egg white. The lysozyme in hen's eggs has 129 amino acids. It is stable to heat (100°C) at low pH (<5.3). It is inactivated at lower temperatures at higher pH values. The enzyme breaks the β-1,4 glycosidic bonds between *N*-acetylmuramic acid and *N*-acetylglucosamine of peptidoglycan. This breaks down the bacterial cell wall.

Lysozyme is most active against gram-positive bacteria, probably because their peptidoglycan is more exposed. It inhibits *C. botulinum*,

Clostridium thermosaccharolyticum, Clostridium tyrobutyricum, Bacillus stearothermophilus, B. cereus, and *L. monocytogenes.* Lysozyme is the main antimicrobial in egg albumen. Ovotransferrin, ovomucoid, and alkaline pH enhance its activity in eggs. Lysozyme is less effective against gram-negative bacteria because they have less peptidoglycan and a protective outer membrane.

The chelator ethylenediaminetetraacetate (EDTA) enhances lysozyme activity. EDTA helps give lysozyme access to the peptidoglycan. The susceptibility of gram-negative cells can be increased by chelators (e.g., EDTA) that bind Ca^{2+} or Mg^{2+}. These cations are needed to maintain a functional lipopolysaccharide layer. Gram-negative cells are also sensitized to lysozyme by pH shock, heat shock, osmotic shock, drying, and freeze-thaw cycling.

Lysozyme is one of the few natural antimicrobials subject to regulatory approval for food use. In Europe, it is used to prevent gas formation ("blowing") in cheeses. Lysozyme is used in Japan to preserve seafood, vegetables, pasta, and salads.

Lactoferrin and Other Iron-Binding Proteins

Milk and eggs contain antimicrobial iron-binding proteins. Lactoferrin is the main iron-binding protein in milk. Milk also contains low levels of the iron-binding protein transferrin. There are two iron-binding sites per lactoferrin molecule. One bicarbonate (HCO_3^-) is required for each Fe^{3+} bound by lactoferrin. Citrate, another chelator in milk, inhibits lactoferrin's activity, while bicarbonate reverses the inhibition. To be effective, lactoferrin must be in a low-iron environment where bicarbonate is present. Milk has these conditions.

Microorganisms with a low iron requirement are not inhibited by lactoferrin. Iron stimulates growth in many genera, including *Clostridium, Escherichia, Listeria, Pseudomonas, Salmonella, Staphylococcus, Vibrio,* and *Yersinia.* Lactoferrin's depletion of iron may cause inhibition.

Egg albumen has another iron-binding molecule, ovotransferrin. This compound is sometimes called conalbumin. It makes up 10 to 13% of the total egg white protein. Ovotransferrin has an amino acid sequence that is very similar to that of lactoferrin. Each ovotransferrin molecule has two iron-binding sites. Like lactoferrin, it binds anions, such as bicarbonate, with each bound ferric iron. To be inhibitory, there must be more ovotransferrin molecules than iron molecules and the pH must be alkaline. Ovotransferrin inhibits gram-positive and gram-negative bacteria. Gram-positive bacteria are generally more sensitive. *Bacillus* and *Micrococcus* species are very sensitive. Some yeasts are also sensitive. Ovotransferrin is bacteriostatic. Like for lactoferrin, the primary inhibitory mechanism is probably iron depletion.

Avidin

Avidin is another egg albumen protein. It makes up about 0.05% of the total egg albumen protein. It is stable to heat and a wide pH range. Avidin binds biotin at a ratio of four biotin molecules per avidin molecule. Avidin inhibits bacteria and yeasts that require biotin for growth. However, avidin also binds transport proteins in the *E. coli* outer membrane. This suggests that avidin may inhibit bacteria by interfering with transport.

Table 25.3 Antimicrobial activities of spices

Spice(s)	Antimicrobial compound(s)	Concn	Target of activity
Cloves Cinnamon	Eugenol Cinnamon aldehyde	Spices are 10 to 20% volatile oil. Clove is active at 1/100 dilution.	Clove inhibits *B. subtilis*, *E. coli*, *Salmonella*, *L. monocytogenes*, *S. aureus*
Oregano Thyme	Carvacrol Thymol	MIC,[a] 0.02–0.05%; MBC,[b] 0.03–0.1% essential oil	*E. coli*, *Salmonella*, *L. monocytogenes*, *S. aureus*, *B. cereus*, some yeasts and molds
Sage, rosemary	Terpenes, borneol, camphor, thujone	MIC, 0.3%; MBC, 0.5%	More effective against gram-positive bacteria; *B. cereus*, *S. aureus*, *L. monocytogenes*
Vanillin	4-Hydroxy-3-methoxy-benzaldehyde	500–2,000 μg/ml	Molds and nonlactic gram-positive bacteria

[a]MIC, minimum inhibitory concentration.
[b]MBC, minimum bactericidal concentration.

Spices and Their Essential Oils

Spices are usually added to foods as flavoring agents. However, they also have some antimicrobial activity. The ancient Egyptians used spices for food preservation and embalming around 1550 B.C.E. Cloves, cinnamon, oregano, thyme, and, to a lesser extent, sage and rosemary are the strongest antimicrobial spices. Table 25.3 lists their active chemical components, the concentrations required, and sensitive organisms. Not all spices have antimicrobial activity.

There are not many studies on the inhibitory mechanism(s) of spices. The terpenes in spice oils are the primary antimicrobials. Many of the most active terpenes, such as eugenol, thymol, and carvacrol, are phenolic. Therefore, their modes of action are probably related to those of other phenolic compounds. They interfere with membrane function.

Onions and Garlic

Onions and garlic contain the best-characterized plant antimicrobials. These compounds inhibit *B. subtilis*, *Serratia marcescens*, *Mycobacterium*, *B. cereus*, *C. botulinum* type A, *E. coli*, *Lactobacillus plantarum*, *Leuconostoc mesenteroides*, *Salmonella*, *Shigella*, and *S. aureus*. The fungi *Aspergillus flavus*, *Aspergillus parasiticus*, *Candida albicans*, *Cryptococcus*, *Penicillium*, *Rhodotorula*, *Saccharomyces*, *Torulopsis*, and *Trichosporon* are also inhibited.

The major antimicrobial in garlic is allicin, which is formed by allinase when garlic cells are disrupted. Allicin probably inhibits sulfhydryl-containing enzymes in bacteria. A similar reaction occurs in onions, forming thiopropanal-*S*-oxide. Onions also contain the antimicrobial phenolic compounds protocatechuic acid and catechol.

Isothiocyanates

Isothiocyanates (R-N=C=S) are potent antimicrobials. They are made from glucosinolates in plant cells from the Cruciferae, or mustard, family. This family contains cabbage, kohlrabi, Brussels sprouts, cauliflower, broccoli, kale, horseradish, mustard, turnips, and rutabaga. Fungi, yeasts, and bacteria are inhibited by 0.016 to 0.062 μg/ml in the vapor phase or 10 to 600 μg/ml in liquid. Isothiocyanates may inhibit cells by reacting with disulfide bonds or inactivating sulfhydryl enzymes.

Phenolic Compounds

Phenolic compounds have an aromatic ring with one or more hydroxyl groups. Phenolic compounds are classified as simple phenols and phenolic acids, hydroxycinnamic acid derivatives, and the flavonoids. Many inhibitors already discussed are phenolics.

Simple phenolic compounds include monophenols (e.g., *p*-cresol), diphenols (e.g., hydroquinone), and triphenols (e.g., gallic acid). Gallic acid occurs in plants as quinic acid esters or hydrolyzable tannins (tannic acid). The only preservative use of simple phenols is in wood smoke. Smoking meats, cheeses, fish, and poultry imparts both a desirable flavor and a preservative effect. Phenol and cresol contribute to smoke's flavor and antioxidant and antimicrobial actions. Several commercial smoke preparations at 0.25 and 0.5% reduce *L. monocytogenes* viability in buffer. The phenolic glycoside oleuropein, or its aglycone, inhibits *L. plantarum*, *L. mesenteroides*, *Pseudomonas fluorescens*, *B. subtilis*, *Rhizopus*, and *Geotrichum candidum*.

Summary

- Microbial spoilage is just one type of spoilage.
- Foods that permit pathogens, but not spoilage organisms, to grow are especially dangerous.
- Antimicrobials can be "-cidal" or "-static" but are usually the latter.
- Antimicrobial activity is influenced by microbial, intrinsic, and extrinsic factors.
- Organic acids are effective below their pK_a values because they cross the bacterial membrane, dissociate, and acidify the cytoplasm.
- Parabenzoic acids act similarly to organic acids but have higher pK_a values.
- Lysozyme is an enzyme that kills gram-positive cells by breaking down their cell walls.
- Nitrites prevent *C. botulinum* growth in cured meats and confer color and flavor.
- Sodium chloride lowers a_w and causes cellular plasmolysis.
- Phosphates and lactoferrin chelate metal ions that microbes need to grow.
- Sulfites are used primarily on fruits and vegetables to inhibit yeasts and molds.
- The amount of available water (a_w) determines which organisms can grow in a food and influences lag time, growth rate, and final cell density.
- The lowest a_w that allows growth of pathogens is 0.85. This makes it a cardinal value in food microbiology.

Suggested reading

Ash, M., and I. Ash. 1995. *Handbook of Chemical Preservatives*. Grower, Brookfield, VT.

Branen, A. L., P. M. Davidson, and S. Salminen (ed.). 1990. *Food Additives*. Marcel Dekker, Inc., New York, NY.

Davidson, P. M., and T. M. Taylor. 2007. Chemical preservatives and natural antimicrobial compounds, p. 713–746. *In* M. P. Doyle and L. R. Beuchat (ed.), *Food Microbiology: Fundamentals and Frontiers*, 3rd ed. ASM Press, Washington, DC.

Questions for critical thought

1. If an antimicrobial were bacteriostatic and inhibited microbial growth "forever," would foods last "forever"? Explain.
2. You are director of microbiology at Freda's Family of Fine and Fancy Foods. The marketing department has directed that all those nasty preservatives be removed from all products. How can you maintain the safety and quality of Freda's products?
3. Diagram the process by which organic acids work in high-acid foods but not in low-acid foods.
4. If you were assigned the task of developing the "ideal" antimicrobial, what criteria would it have to meet?
5. Calculate the amount of protonated acetic acid at pH 3.5 and pH 5.5 if you started with a 1 M solution. How much acetic acid would you have to add at pH 5.5 to get as much protonated acetic acid as there is at pH 3.5? You will need this equation: $pH = pK_a + \log [A^-]/[HA]$.
6. You are an extension specialist at Big State University. A local inventor approaches you for advice on marketing a new antimicrobial. Called Superlator, it chelates all cations and prevents the growth of bacteria that require any cations. What advice do you give her?
7. What concepts from Chemistry 101 are important for understanding how preservatives work?
8. Famous Frank's Freshly Frozen Fabulous Pot Pies is losing market share and rapidly becoming less famous as the demand for "fresh, never frozen" food increases. Frank knows that he has to change and figures that going fresh is no big deal. (Just skip the freezing step!) He hires you, the company's first microbiologist, to advise him on what, if any, special challenges might be presented by these fresh products. What advice do you give him?
9. Take the following ingredient list and make it label friendly: "This product contains dihydrogen oxide, sodium chloride, ascorbic acid, acetic acid, and 4-hydroxy-3-methoxy-benzaldehyde."

Authors' note

In the world of industry, "challenges" is often a code word for "problems." No one likes to have problems, but we are all excited by challenges (or so the thinking goes).

26

Biologically Based Preservation and Probiotic Bacteria

LEARNING OBJECTIVES

The information in this chapter will help the student to:

- understand that biological methods can be used to "naturally" enhance food safety without changing the food (as fermentations do)
- appreciate the potential antimicrobial uses of the small proteins called "bacteriocins"
- relate the organization of bacteriocin genes to the way bacteriocins are made
- describe the characteristics of probiotic bacteria, suggest possible health benefits, and critically evaluate commercial claims for probiotic bacteria
- understand that the scarcity of validated data about the health benefits of probiotic bacteria is due to the complexity of the gastrointestinal tract's ecosystem

INTRODUCTION

It is easy to tell if foods are processed or contain preservatives. Just look at the food or information on the labels. However, this information is often misused. Consumers are wary of preservatives and "processed" foods, even though they give us a safe and diverse diet. As a result, consumers increasingly rely on refrigeration to ensure the safety of "fresh" foods. (Note well that the opposite of "fresh" is not "processed" or even "frozen"; it is "spoiled.") There are two reasons that refrigeration should not be used as the sole preservation method. The first is that 20% of home and commercial refrigerators are at >10°C (50°F). (You *do* remember the 40-140 rule that foods should be held below 40 or above 140°F to prevent the growth of most foodborne pathogens.) The second reason is that *Listeria monocytogenes* can grow at refrigeration temperatures of <10°C. Because of this, additional means of keeping refrigerated foods safe are required. Lactic acid bacteria (LAB) can provide this added protection. LAB are accepted by consumers as "natural" and "health promoting." They are, through fermentation, among the oldest forms of food preservation. Biologically based preservation methods that do not ferment the foods are among the newest forms of food preservation.

This chapter provides an overview of biopreservation. Biopreservation is the use of LAB, their metabolic products, or both to improve or ensure the safety and quality of foods that are not fermented.

Some bacteria produce antimicrobial proteins called bacteriocins. Bacteriocins inhibit spoilage and pathogenic bacteria without changing the food (i.e., acidifying it, producing gas, solidification, etc.). The use of bacteriocins is a newer and emerging area of food microbiology.

doi:10.1128/9781555817206.ch26

The food industry is market driven. Advertisers try to turn technical needs that might be seen as marketing negatives (i.e., the need to preserve a food) into something that will help sell the food (i.e., the preservation method improves consumer health). Thus, consumers want foods that in addition to being nutritious are distinctly health promoting. Since LAB are natural and perceived as health promoting, foods preserved by biological techniques may have an edge if marketed as containing "probiotic" bacteria.

BIOPRESERVATION BY CONTROLLED ACIDIFICATION

Organic acids (such as acetic, lactic, and citric acids) can be added to food to inhibit microbial growth. LAB, however, can produce lactic acid *in* the food. Many factors determine the effectiveness of acidification "in place." These include the food's initial pH, its buffering capacity, the pathogen of greatest concern, the nature of the fermentable carbohydrate, and the growth rates of the LAB and of the pathogen at refrigeration and abuse temperatures. Biopreservation can require customization and research for each product application. Bacteriocins, diacetyl, and hydrogen peroxide may enhance acid's inhibition. For example, MicroGard, a cultured milk product that is generally recognized as safe (GRAS), is frequently added to cottage cheese in the United States as a biopreservative. MicroGard is made by fermenting milk with *Propionibacterium shermanii* to produce acetic acid, propionic acid, low-molecular-weight proteins, and a bacteriocin.

The idea of using LAB acid production to prevent the production of botulinal toxin dates back to the 1950s. This technology exploits *Clostridium botulinum*'s inability to grow at a pH of <4.8 to prevent its growth if bacon is left unrefrigerated. LAB and a fermentable carbohydrate (such as glucose or lactose) are added to the food. The LAB grow and produce acid only when the food is temperature abused. Under proper refrigeration, the LAB cannot grow, no acid is formed, and the preservation system is invisible to the consumer.

Authors' note

Green vegetables such as spinach and broccoli were subsequently found to have much higher levels of nitrosamines than that found in cured meat. However, there is a consensus that there is no cancer risk from nitrite, regardless of source.

Nitrites are powerful inhibitors of botulinal spores. When it was found that the nitrites used to cure meats form cancer-causing chemicals called nitrosamines, research was initiated to find nitrite substitutes with antibotulinal activity.

Nobi Tanaka at the University of Wisconsin reduced the amount of nitrite added to bacon by using controlled acidification. When bacon was inoculated with 10^3 botulinal spores/g and incubated at 28°C, toxin was produced in 58% of the bacon samples prepared with the standard 120 parts per million of nitrite. When the nitrite was reduced to 80 or 40 parts per million and supplemented with sucrose and starter cultures, the pH dropped and $\leq 2\%$ of bacon became toxic. (This may not sound very effective, but in the control experiment using conventional bacon, more than half of the samples became toxic. If you like bacon, be sure to refrigerate it!) The U.S. Department of Agriculture (USDA) approved the "Wisconsin process" for bacon manufacture in 1986.

BACTERIOCINS

General Characteristics

The continual generation of new data has gradually changed the definition of *bacteriocins*. They are first and foremost antimicrobial proteins which are made by bacteria. They do not kill the bacteria that produce them and

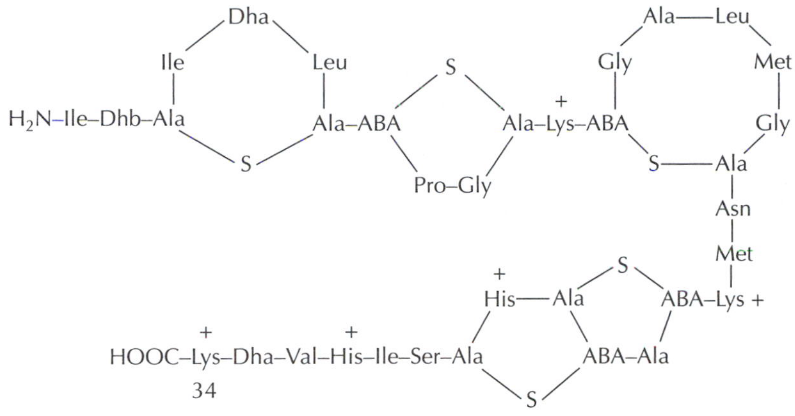

Figure 26.1 Structure of nisin showing positions of unusual amino acids (dehydroalanine [Dha], dehydrobutyrine [Dhb], lanthionine [Ala-S-Ala], and methyl lanthionine [ABA-S-Ala]) as well as regular amino acids. doi:10.1128/9781555817206.ch26.f26.01

are ribosomally synthesized. Their action was once thought to be restricted to gram-positive bacteria, but some bacteriocins also inhibit gram-negative bacteria. Bacteriocins are not enzymes but may have some unknown cellular function.

The lantibiotics and the "pediocin-like" bacteriocins are the two most important bacteriocin groups. If you know the basic biochemistry of amide linkages in proteins, you can see that the lantibiotic nisin has a very strange structure (Fig. 26.1). Nisin contains unusual amino acids other than the "normal" 20. After the protein is made, some amino acids react with cysteine to form thioether (single sulfhydryl) lanthionine rings. Bacteriocins containing lanthionine rings are called lantibiotics. There are many structurally similar lantibiotics. Nisin, the first and best-characterized LAB bacteriocin, is produced in two related forms. Nisin A contains a histidine at position 27 and is the most widely used nisin. Nisin Z has an asparagine at position 27 and is also commercially available. Nisins with other amino acid substitutions are not in common usage. Another lantibiotic, subtilin, is produced by *Bacillus subtilis*. It also contains five lanthionine rings and has a conformation similar to that of nisin.

Pediocin-like bacteriocins are small heat-stable proteins made of the usual 20 amino acids. They all have the same leader amino acid sequence containing a glycine-glycine-↓-any amino acid cleavage site (where the arrow indicates the cleavage site). The upstream amino acids are cut off during the process of excreting the bacteriocin. Pediocin-like bacteriocins act against *L. monocytogenes*. They have a tyrosine-glycine-asparagine-glycine-valine-(any amino acid)-cysteine amino-terminal sequence (Fig. 26.2). Pediocin PA-1, pediocin AcH, sakacins A and P, leucocin A, bavaricin MN, and curvacin A are members of this group.

Authors' note

Proteins are made by linking amino acids' amide and carboxy groups. There is an amine "left over" at the starting end and an unlinked carboxyl group at the ending end.

The antimicrobial peptide polylysine is in a class by itself. As the name suggests, polylysine is a polymer consisting of 25 to 35 lysine monomers (Fig. 26.3). It has no secondary or tertiary structure. Polylysine is produced on an industrial scale in Japan by a fermentation using *Streptomyces albulus*. It is effective at concentrations of less than 100 μg/ml against bacteriophages, fungi, yeasts, and gram-positive and gram-negative bacteria.

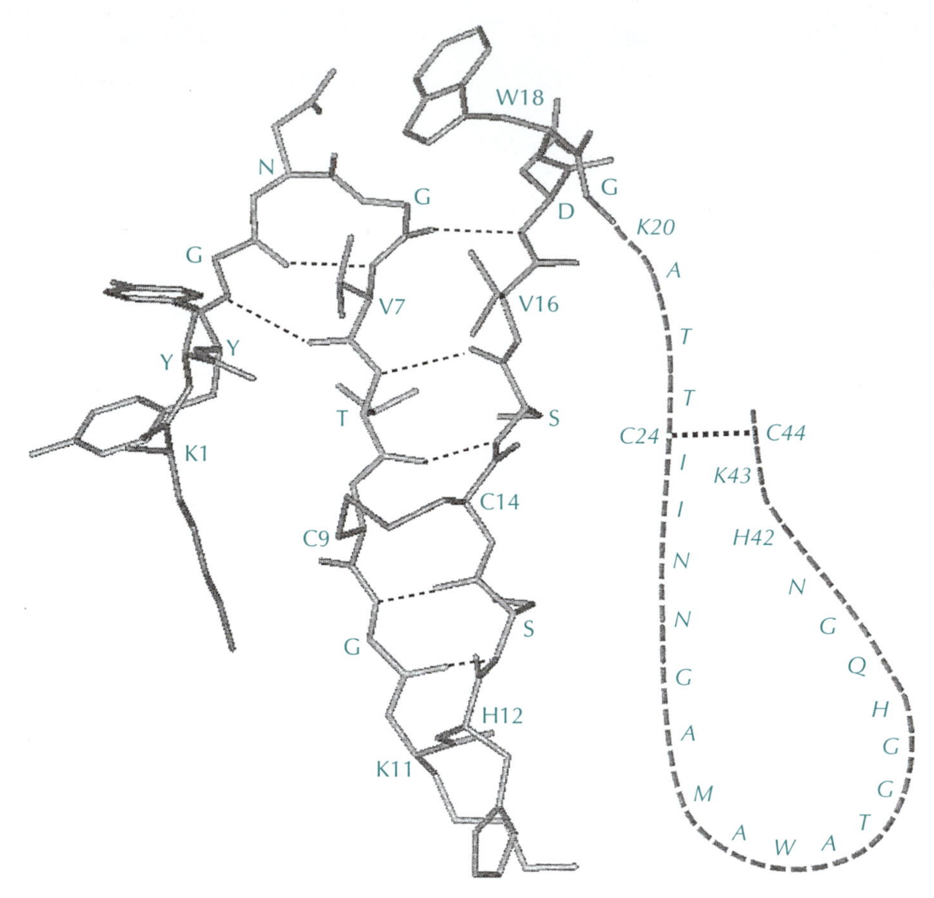

Figure 26.2 Structure of pediocin PA-1 showing the KYYNGV carboxy terminus, disulfide linkages, hairpin loop, and unstructured amino terminus. doi:10.1128/9781555817206.ch26.f26.02

Figure 26.3 Structure of polylysine, where n = 25 to 35 lysine molecules. doi:10.1128/9781555817206.ch26.f26.03

$$H\text{-}[NH\text{-}CH_2\text{-}CH_2\text{-}CH_2\text{-}CH_2\text{-}\underset{}{\overset{NH_2}{\overset{|}{C}}}H\text{-}CO]_n\text{-}OH$$

Bacteriocin Applications in Foods

Bacteriocins or the bacteria that make them can be added to foods to inhibit pathogens. Only nisin is commercially available as an ingredient, but pediocin addition is also effective. Many applications add the LAB to produce the bacteriocin in situ rather than adding the pure bacteriocin. Bacteriocins can also be used to improve the quality of fermented foods by inhibiting unwanted bacteria and promoting growth of the starter culture. There are many benefits to using defined starter cultures to make fermented foods. However, the native bacteria of the food usually outcompete the starter cultures. Using starter cultures that produce bacteriocins to kill the native bacteria solves this problem.

Nisin is added to milk, cheese, sauces, salad dressings, dairy products, many canned foods (although not in the United States), mayonnaise, and baby foods throughout the world. It is GRAS for the inhibition of botulinal spores in many foods. Nisin also sensitizes spores to heat, so thermal treatments can be reduced. However, this application is not approved in the United States because it might mask poor process control.

Nisin is often used with other inhibitors. Nisin may be used with food packaged in a modified atmosphere. Nisin increases the shelf life and delays toxin production by type E botulinal strains in fresh fish packaged in a carbon dioxide atmosphere. The combination of nisin and modified atmosphere to prevent *L. monocytogenes* growth in pork is more effective

Across

2. Type of bacterium that confers a health benefit when eaten
4. Small antimicrobial protein
6. This type of bacteriocin is generally more effective in meat than is nisin
7. Because the food industry is ________ driven, it is very desirable for a preservative to be "natural"
10. A bacteriocin that is generally recognized as safe for use in foods
11. Bacteriocins usually act against this type of bacteria (2 words)
13. Resistance to ________ is a highly desirable trait for probiotics that act in the intestinal tract
14. Method for preserving food through acid production
15. Bacteria that have colonized the intestine are called ________

Down

1. Bacteriocin genes are frequently carried on a ________
3. Use of bacteria or their metabolites to preserve food without acidification
5. Inadequate method of preserving food
8. Probiotics may limit ________ infections
9. Class of bacteriocin that undergoes posttranslational modification
12. Bacteriocin genes are usually organized on a single ________

than either used alone. When nisin (5 mg/liter) is added to liquid whole eggs before pasteurization, their refrigerated shelf life doubles.

Pediocins inhibit *L. monocytogenes* vegetative cells but not clostridial spores. European patents cover the use of pediocin PA-1 as a dried powder or culture liquid to extend the shelf life of salads and salad dressing and as an antilisterial agent in cream, cottage cheese, meats, and salads.

Pediocins are more effective than nisin in meat. Dipping meat in pediocin PA-1 decreases the viability of attached *L. monocytogenes* 100- to 1,000-fold.

Pretreating meat with pediocin reduces subsequent *L. monocytogenes* attachment. Pediocin AcH can cause a 10- to 1,000,000-fold reduction of listeriae in ground beef, sausage, and other products. In most cases, pediocin kills listeriae rapidly and delays growth of the survivors. Emulsifiers such as Tween 80 or entrapment in other lipids makes pediocin work better in fatty foods.

Bacteriocin-Producing Starter Cultures Can Improve the Safety of Fermented Foods

Foods that are usually made by fermentation are easily improved using bacteriocin-producing starter cultures. For example, including a nisin-producing starter culture to make Cheddar cheese increases the shelf life from 14 to 87 days at 22°C.

Pediococci that make pediocins are especially effective in fermented meats. Pediocin production by *Pediococcus acidilactici* PAC 1.0 during the manufacture of fermented dry sausage reduces *L. monocytogenes* viability >10-fold. When *Pediococcus acidilactici* H is used to ferment summer sausage, *L. monocytogenes* viability drops more than 1,000-fold.

Organization of Bacteriocin Genes: a Generic Operon

The DNA or amino acid sequences of many bacteriocins are known. There is a general model for the genetic organization of bacteriocin genes, although the genetics of polylysine are unknown. The organization of a generic *operon* is shown in Fig. 26.4. An operon keeps the genes containing the structural information close to the genes involved in bacteriocin

Figure 26.4 A generic bacteriocin operon. The structural gene (*struct*) codes for a prepeptide which is modified and excreted by the products (P_1 and P_2) of the processing genes (*process*$_1$ and *process*$_2$). The operon also contains the immunity gene (*immun*). The operon may be regulated by a signal transduction pathway coded for by *reg1* and *reg2*, producing a histidine kinase (R_1) and response regulator (R_2) subject to phosphorylation (R_2~P). Regulatory molecules P and P* affect transcription of the operon. ADP, adenosine diphosphate. Adapted from T. J. Montville, K. Winkowski, and M. L. Chikindas, p. 629–647, *in* M. P. Doyle, L. R. Beuchat, and T. J. Montville (ed.), *Food Microbiology: Fundamentals and Frontiers*, 2nd ed. (ASM Press, Washington, DC, 2001). doi:10.1128/9781555817206.ch26.f26.04

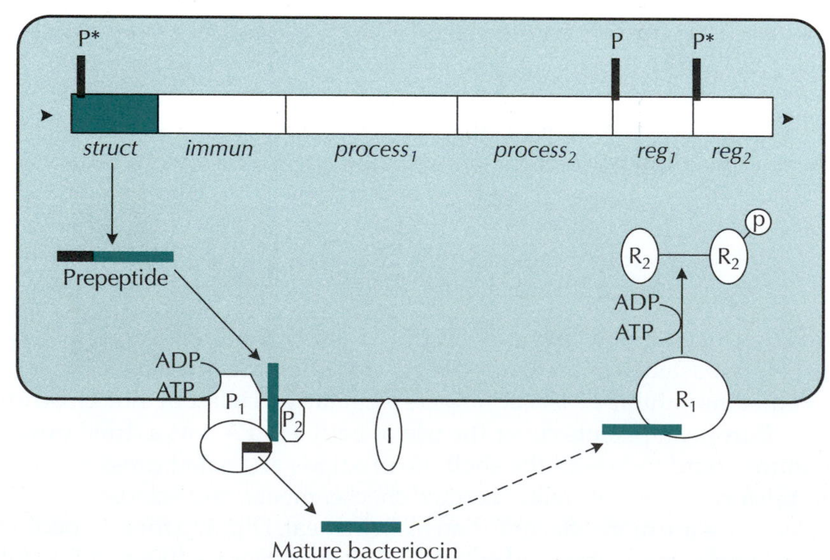

Even uncultured people contain bacterial cultures.
Reprinted with permission from CartoonStock.

immunity, maturation, processing, and export. Not all bacteriocin operons contain all of these genes, which may be located on the chromosome or on a plasmid. The organization of the genes is not identical for every bacteriocin. The generic operon presented in Fig. 26.4 shows many of the genetic similarities.

The *structural gene* usually codes for a prepeptide. The prepeptide is made of the excreted bacteriocin plus an N-terminal "leader sequence" that is cut off during transport. "Pediocin-like" bacteriocin genes code for a prepeptide containing a glycine-glycine cleavage site. The leader sequence may be needed for recognition by export machinery. The *immunity gene* protects cells that make bacteriocins from being killed by them. Immunity is specific to the bacteriocin being made and is coordinated with its production. The *processing and export genes* code for at least two proteins that mature the bacteriocin and export it from the cell. The *regulatory genes* code for proteins that turn bacteriocin synthesis on and off.

Authors' note
In this case, "immunity" means "protection from" and has nothing to do with antigens, antibodies, or the immune response.

Bacteriocin Action

LAB bacteriocins target the cell membrane of sensitive bacteria. They disrupt the membrane by making pores. This increases its permeability to small compounds, causing a rapid efflux of preaccumulated ions, amino acids, and, in some cases, adenosine 5′-triphosphate (ATP) molecules. The equilibration of compounds across the membrane destroys the gradients required for many vital cellular functions. The mechanism of polylysine action is quite different. Because it is so cationic, there is electrostatic absorption of polylysine onto the cell surface, where it strips off the cell membrane.

Legal Status

The legal status of bacteriocins and the bacteria that make them is unclear. LAB are GRAS for the production of fermented foods. GRAS status, which is given by the U.S. Food and Drug Administration (FDA), is especially desirable because it lets a compound be used in a specific application without additional approval. Nisin is the only bacteriocin that has GRAS status. The 1988 GRAS affirmation for the use of nisin in pasteurized processed cheese was supported by toxicological data. This affirmation is the foundation for many additional GRAS affirmations. Polylysine received GRAS affirmation for use in cooked rice and sushi rice in 2003.

In the nonregulatory sense, bacteriocins are widely considered safe. However, bacteriocins produced by GRAS organisms are not automatically GRAS themselves. Bacteriocins that are not GRAS are regulated as food additives and require premarket approval by the FDA. A food fermented by bacteriocin-producing starters can be used as an ingredient in a second food product. Its use *as an ingredient* might coincidentally extend shelf life of the product without its being listed as a preservative on the label. However, if the ingredient was added *for the purpose of extending shelf life,* the FDA would probably consider it an additive and require both premarket clearance and label declaration. Purified bacteriocins used as preservatives definitely require premarket approval by the FDA.

PROBIOTIC BACTERIA

The idea that eating LAB improves health first appeared in print in 1907. In his book *Prolongation of Life,* the Nobel laureate Elie Metchnikoff attributed the longevity of people in the Balkans to the bacteria in the yogurt they ate. He hypothesized that the bacteria in the yogurt suppress "bad" bacteria in the gut. The term probiotic was rediscovered in the 1970s and used to describe the healthful effects of feeding microbial supplements to animals. In 1989, Fuller proposed a narrower definition where viable "good" bacteria are consumed and act in the gastrointestinal (GI) tract to benefit the host organism (i.e., us!). "Probiotics" have been defined in a variety of ways by different authors and organizations. For example, the FAO/WHO (Food and Agriculture Organization and the World Health Organization) define probiotics as "live microorganisms which when administered in adequate amounts confer a health benefit to the host." While we know that fermentations improve digestibility, generate amino acids, and produce vitamins in food, there are fewer peer-reviewed data indicating that probiotic bacteria promote "intestinal well-being" or are otherwise health promoting. There are many reports of the probiotic effect of LAB.

Authors' note

There is no agreement as to what constitutes "intestinal well-being."

Some of the species involved are listed in Table 26.1. Applications are given in Table 26.2 and probiotic products in Table 26.3. Although there

Table 26.1 Examples of human probiotic species and strains with research documentation[a]

Bifidobacterium breve	*Lactobacillus delbrueckii* subsp. *bulgaricus*	*Lactobacillus reuteri*
Bifidobacterium lactis	*Lactobacillus fermentum*	*Lactobacillus rhamnosus*
Bifidobacterium longum	*Lactobacillus johnsonii*	*Lactobacillus salivarius*
Lactobacillus acidophilus	*Lactobacillus paracasei*	*Saccharomyces boulardii*
Lactobacillus casei	*Lactobacillus plantarum*	*Streptococcus thermophilus*

[a]Adapted from Table 39.2 of T. R. Klaenhammer, p. 797–811, *in* M. P. Doyle, L. R. Beuchat, and T. J. Montville (ed.), *Food Microbiology: Fundamentals and Frontiers*, 2nd ed. (ASM Press, Washington, DC, 2001). Information in the source table was compiled from M. E. Sanders and J. Huis in't Veld, *Antonie van Leeuwenhoek* **76:**293–315, 1999.

are relatively few well-controlled studies in humans, the therapeutic use of LAB is gaining acceptance in mainstream medicine.

Cultures used to make fermented products are usually chosen for their technological traits rather than their ability to promote health. For example, the *Lactobacillus bulgaricus* and *Streptococcus thermophilus* traditionally used to make yogurt are hardy and acidify rapidly. But they have poor resistance to acid and bile salts and do not survive stomach passage. To capitalize on their health-promoting potential, many yogurt manufacturers now use *Lactobacillus acidophilus* and *Bifidobacterium* spp. that have been isolated from humans and are resistant to acid and bile salts. Other characteristics important for commercial cultures include the ability to be cultured on a large scale, GRAS status, and demonstrated clinical efficacy.

The Human GI Tract Is a Microbial Ecosystem

The complexity of the human GI tract makes research on probiotic bacteria difficult. The average human contains over 400 species of bacteria. Thirty or forty of these comprise 99% of the bacteria in healthy humans.

Table 26.2 Possible health benefits and mechanisms of probiotic bacteria

Health effect	Possible mechanism(s)
Promotes ability to digest lactose	Probiotic cultures make lactose-using enzymes
Fights foodborne disease	Occupies colonization sites, promotes immunity, decreases severity of diarrhea
Anticancer	Binds cancer-causing compounds, decreases levels of natural enzymes that promote cancer, stimulates immune function
Enhances immune system	Strengthens defenses against infections and tumors, increases antigen-specific immune responses, lowers inflammatory responses
Fights heart disease	Increases bile activity, may reduce cholesterol
Reduces blood pressure	Bacterial action on milk proteins produces a compound that lowers blood pressure in animals
Reduces ulcers	Production of inhibitor against *Helicobacter pylori*[a]
Limits urogenital infections	Adhesion to urinary and vaginal cells, competitive exclusion, production of inhibitors (surfactants, hydrogen peroxide)

[a]It is now widely accepted that ulcers are caused by *H. pylori*, not stress.

Table 26.3 Commercially available probiotic products[a]

Product	Manufacturer	Market(s)	Claims	Microorganism	Uniqueness
Yakult	Yakult	Japan, Australia, India, Indonesia, Europe, South Korea, Brazil, United States, etc.	Immunomodulation, control of some GI tract infectious microorganisms, regulation of the GI tract healthy microbiota	*Lactobacillus casei* Shirota (identified as *L. paracasei* based on DNA homology analysis)	The company emerged in 1935 as a single-product business with the best-in-industry scientifically documented positive effects on human health.
Activia	Dannon	Sold in more than 30 countries worldwide	The product was initially advertised with an emphasis on reducing long intestinal transit time. Currently, it is claimed to support proper functioning of the digestive system.	*Bifidobacterium animalis* DN173010 (but marketed as "*Bifidus regularis*")	Targets individuals with constipation
DanActive	Dannon	Market similar to that of Activia. The company claimed "limited launch" of the product on the U.S. market in 2004.	Claimed to support the immune system, although a lawsuit in 2008 questioned the scientific validity of this statement. The only NIH-approved clinical study on day care children (3–6 years old) was conducted by Georgetown University Department of Family Medicine (2006–2007).	*L. casei* DN-114 001 (taxonomically improperly marketed as "*L. casei* Immunitas")	May help in combating diarrhea and *Clostridium difficile* in elderly patients and as an adjuvant for treatment of *Helicobacter pylori* infections. The published data lack consistency.
GoodBelly	Next Foods	United States	The company claims, "GoodBelly is specifically designed to be enjoyed daily in order to boost your immune system and improve digestive health by replenishing your body's healthy microflora. GoodBelly's patented Lp299v probiotic strain has been clinically proven in over 15 years of research."	*Lactobacillus plantarum* 299v. The strain/product is licensed from Probi AB (Sweden) exclusively for the U.S. market.	According to the manufacturer, this is the first nondairy probiotic beverage on the U.S. market delivered in several attractive flavors.
ProBugs	Lifeway	United States	Claims are very general, such as "health support"	*Lactobacillus lactis, L. rhamnosus, Streptococcus lactis* subsp. *diacetilactis, L. plantarum, L. casei, Saccharomyces florentinus, Leuconostoc cremoris, Bifidobacterium longum, B. breve, L. acidophilus, Bifidobacterium lactis* HN019, *L. acidophilus* NCFM	First U.S.-marketed probiotic product containing, according to its claims, more than 10 probiotic bacteria and yeast. First product specifically marketed for children.
Kefir	Lifeway	United States	Same as the claims for ProBugs	Same microorganisms as in the ProBugs line of products	Variety of flavors and fat contents in products
Flapjack	Reflex Nutrition	United Kingdom	High-protein energy bar enriched with microelements, vitamin E, and a probiotic	*Lactobacillus sporogenes* (*Bacillus coagulans*)	A representative of a new generation of nondairy, probiotic-containing foods and supplements containing a spore-forming, health-promoting bacterium
Attune Probiotic Bar	Attune Foods	United States	Probiotic bar available in several flavors and in white, milk, and dark chocolate. Two probiotics are claimed to benefit healthy GI microfloras. A similar product is marketed by Lifeway.	Marketed by Danisco as HOWARU Dophilus and HOWARU Bifido, they are *L. acidophilus* NCFM and *B. lactis* HN019, respectively.	According to Danisco, *L. acidophilus* NCFM is, perhaps, the most studied probiotic that is likely to help with abdominal pain, especially in patients with irritable bowel syndrome.

[a]Courtesy of Michael Chikindas, Rutgers University.

It is existentially noteworthy that we contain as many bacterial cells as human cells (~10^{13} to 10^{14}). Because it has so many organisms and environments, the GI tract must be studied as an ecosystem. (Various sections of the GI tract have different acidity levels, different surfactants, different resident bacteria, etc.) Humans, like other ecosystems, experience a progression of inhabitants. We are born more or less sterile inside. We are colonized by bifidobacteria when breast-fed or by lactobacilli when fed with cow's milk. Strict anaerobes come with the introduction of solid food.

Authors' note

A baby's feces change as its diet and intestinal microbiota change. I was fascinated to watch my first child's poop change from black meconium to mustardy yellow cream to foul brown solid as his diet progressed.

Once established, the composition and distribution of the human microbiota are very hard to change. In ecological terms, the microbiota of the human GI tract is a climax community.

Authors' note

A climax community is an ecosystem in which the progression of species has stopped, all niches are filled, and a characteristic group of organisms is maintained.

The metabolic activity of the climax community is quite variable and strongly influenced by the host physiology and by everything that affects host physiology. (Think about what happens when you eat too many beans.)

The human GI tract is about 350 cm long from the oral orifice to the anal orifice. The GI tract is divided into three major sections. Each has its own distinct microbiota. The stomach is highly acidic (pH 1 to 3) and is populated by $<10^3$ colony-forming units (CFU) of aerobic gram-positive organisms per g. The intestinal pH (6.8 to 8.6) favors microbial growth. The small intestine is a transitional zone inhabited by 10^3 to 10^4 CFU of *Lactobacillus, Bifidobacterium, Bacteroides,* and *Streptococcus* per g. Microbial growth in the large intestine is luxuriant. There are 10^{11} to 10^{12} CFU/g, with anaerobes outnumbering aerobes 100- to 10,000-fold. A diverse microbial population populates the large intestine. Bacteria from the genera *Bacteroides, Fusobacterium, Lactobacillus, Bifidobacterium,* and *Eubacterium* are present in large numbers. *Lactobacillus acidophilus* is especially important. The *Enterobacteriaceae* are present at relatively lower levels.

The normal human biota contains two different bacterial populations. There are indigenous bacteria that have colonized the host by adhering to the intestine. There are also transient bacteria that are just passing through. Dietary factors such as a carnivorous or vegetarian diet, or even starvation, make surprisingly little difference in the distribution of bacteria in the GI tract.

The complexity of the GI ecosystem makes colonization hard to study. Colonization is influenced by gastric acidity, bile salt concentration, peristalsis, digestive enzymes, and the immune response. Feces are 33 to 50% bacteria on a dry weight basis and are their own ecosystem. The fecal ecosystem is subject to dehydration and has high levels of enzyme activities. From a commercial standpoint, the use of transient populations has an advantage. They have to be consumed (and purchased) on a continuing basis. This ensures a continuing revenue stream for the company.

The future use of probiotic bacteria is more difficult to predict. *Something* happens in the GI tract when LAB are consumed in large numbers. Exactly *what* happens and *why* it happens are unclear. It is very important that these benefits be studied in the context of diet, well-characterized probiotic bacteria, individual human intestinal ecosystems, and the established principles of microbial ecology. Much of this lies outside the scope of pure culture microbiology and may be the work of future food microbiologists.

Summary

- Refrigeration alone is insufficient to ensure microbial food safety.
- LAB may be used to produce acid "in place" when a food is held at temperatures that are too warm. The acidification prevents pathogens from growing and making toxin.
- Bacteriocins are small proteins that kill sensitive bacteria by making pores in them.
- Lantibiotic bacteriocins have unusual amino acids and single sulfur rings.
- Pediocin-like bacteriocins have a common leader sequence and cleavage site and are active against *Listeria monocytogenes*.
- Bacteriocins can be produced by GRAS LAB in the food or added to the food as "natural preservatives" if they are approved as food additives or have GRAS affirmation.
- The genes for bacteriocins, their regulation, and the enzymes that process them are grouped together in an operon.
- Polylysine is another peptide antimicrobial but is different from lantibiotics of pediocin-like bacteriocins.
- Probiotic bacteria are thought to give health benefits by improving the balance of bacteria in the intestines.
- Possible health benefits for probiotics include decreased cholesterol, increased ability to use lactose, increased immune function, decreased symptoms of diarrheal illness, and anticancer action.
- The human GI tract is a complex ecosystem. This makes the study of probiotics difficult.

Authors' note

GRAS status can be conferred in three ways. (i) The compound was in use before 1958. (ii) The FDA affirms that a compound is GRAS in response to an industry request. (iii) A company "self-affirms" that a compound is GRAS. (Welcome to the wild world of food law!)

Suggested reading

Cleveland, J., T. J. Montville, and M. L. Chikindas. 2001. Bacteriocins: safe food preservatives of the future. *Int. J. Food Microbiol.* **71:**1–20.

Klaenhammer, T. R. 2007. Probiotics and prebiotics, p. 891–907. *In* M. P. Doyle and L. R. Beuchat (ed.), *Food Microbiology: Fundamentals and Frontiers*, 3rd ed. ASM Press, Washington, DC.

Kolida, S., D. M. Saulnier, and G. R. Gibson. 2006. Gastrointestinal microflora: probiotics. *Adv. Appl. Microbiol.* **59:**187–219.

Montville, T. J., and M. Chikindas. 2007. Biopreservation of foods, p. 747–766. *In* M. P. Doyle and L. R. Beuchat (ed.), *Food Microbiology: Fundamentals and Frontiers*, 3rd ed. ASM Press, Washington, DC.

Sanders, M. E. 1999. Probiotics—scientific status summary. *Food Technol.* **52:**67–77.

Yoshida, T., and T. Nagasawa. 2003. ε-Poly-L-lysine: microbial production, biodegradation, and application potential. *Appl. Microbiol. Biotechnol.* **62:**21–26.

Questions for critical thought

1. What is the temperature of your refrigerator? (If you are living in a dorm room without a refrigerator, check the temperature of the refrigerator at the store where you buy food.)
2. Why does the pH drop in Wisconsin process bacon?
3. What were the original six characteristics used to determine if an inhibitor was a bacteriocin?

4. Why is it a good thing that a bacteriocin does not act against the organism that makes it?
5. Why does it benefit the bacterium to have all the genes involved in bacteriocin synthesis and regulation clustered in an operon?
6. Choose one health benefit of probiotic bacteria. Find, read, and evaluate three research articles that relate to it.
7. If humans contain 10^{13} to 10^{14} human cells and 10^{13} to 10^{14} bacterial cells, why don't we look more like slimy bacterial colonies?
8. List and discuss factors that influence the conclusions of studies on the health benefits of probiotics.
9. What are the differences between Fuller's definition and the FAO/WHO definition of "probiotic"? Why does it matter how probiotics are defined?
10. Why would it be advantageous to produce a bacteriocin in, e.g., cheese, rather than adding the pure compound?
11. How can a small peptide (~10 to 15 amino acids) be made if it is not ribosomally synthesized?

27

Physical Methods of Food Preservation

LEARNING OBJECTIVES

The information in this chapter will help the student to:

- appreciate the roles of temperature, controlled atmosphere, and freezing in microbial growth
- differentiate among different types of heat processing, their objectives, and their industrial uses
- quantitatively determine, using D and z values, the lethality of a given heat process on an organism

INTRODUCTION

Conditions too stressful for microbial growth can damage cells. Minor stresses inhibit cell growth. Major stresses kill cells. Thus, food environments can be manipulated to inhibit or kill microbes. Physical manipulation of foods can inhibit microbial growth, kill cells, or mechanically remove them from the food. Dehydration, refrigeration, and freezing inhibit microbial growth. Microbes can be killed by heating. This chapter discusses these physical preservation methods.

PHYSICAL DEHYDRATION PROCESSES

Drying

Heat evaporates water during the drying of foods. Both the drying temperature and the decreased water activity (a_w) affect the microbes. Factors such as the size and composition of food pieces and the time-temperature combinations used in drying influence the microbiological effect of drying.

When air is used for drying, initially the temperature is low and the relative humidity is high. The size and composition of the food particle determine the length of this phase. If the particle is large and the material dense, this phase can be long, allowing microbes to grow. During later phases of drying, the temperature is high but the relative humidity is low. This provides no opportunity for growth, but it is not very lethal either. The higher temperatures result in lower moisture content, and dry heat is less lethal than wet heat. If the drying lasts long enough (at least 30 min), there are time-temperature combinations that kill microbial cells. With certain drying methods, the surface temperature may reach 100°C, while the internal temperature remains lower. This kills surface microbes but not those in the food's interior. Lethal combinations of high temperature and

doi:10.1128/9781555817206.ch27

high humidity are rare during drying. In fact, the major cause of microbial death is the high-temperature, low-humidity conditions which last long after the actual drying. Cell populations decrease during storage because injured cells that cannot recover at low a_w gradually die.

The a_w values of dried foods are usually much lower than the minimum a_w required for microbial growth. Therefore, dried products are microbiologically stable. If the relative humidity increases or the temperature drops, condensation on the product surface may permit mold growth.

Freeze-Drying

Freeze-drying (lyophilization) combines two preservation methods. It freezes the food and then dehydrates the frozen food through vacuum sublimation of the ice. That is, water moves from the ice state to the vapor state without becoming liquid. This is a gentle way to remove water. In other drying methods, the solution moves to the food surface, where the water leaves and the solutes concentrate. During freeze-drying, the sublimation front moves into the food and the ice sublimates where it is formed. Thus, solutes remain inside the food at their original location, and the food retains its original structure. As a result, freeze-dried foods rehydrate rapidly to 90 to 95% of their original moisture content. Unfortunately, the large surface area of freeze-dried foods makes them vulnerable to oxidation during storage. Packaging freeze-dried foods in inert atmospheres delays oxidation. Vapor-impermeable packaging prevents their rehydration during storage.

The freeze-drying of foods inhibits microbial growth by decreasing a_w during freezing and by the sublimation of the ice. The extents of cell damage in these two phases may be different. Microbial survival also depends on the composition of freeze-dried food. Carbohydrates, proteins, and colloidal substances are usually protective. Cell viability slowly decreases during storage of freeze-dried foods.

Freeze-drying is optimized for cell survival during the preservation of stock cultures.

Low dehydration temperature, protective additives (such as glycerol or nonfat dried milk solids) in the microbial suspension, and storage of lyophilized cultures under vacuum increase viability. Gram-positive bacteria survive freeze-drying better than gram-negative bacteria.

Authors' note

While beyond the scope of this book, the preservation of bacteria in "culture collections" is a neglected aspect of food microbiology. Many microbiologists are amateur naturalists, maintaining their own private culture collections and trading cultures with their colleagues like baseball cards. However, bacterial traits can "drift" over years of repeated transfer. It is better to get cultures in the freeze-dried form from a recognized culture collection such as the American Type Culture Collection. Secure forms of culture preservation are essential for process security and research reproducibility.

COOL STORAGE

Cool (or chill) storage refers to storage at temperatures from about 16 to −2°C. While pure water freezes at 0°C, most foods remain unfrozen until ≤−2°C. However, many fruits and vegetables suffer from a chilling injury when kept at <4 to 12°C. The length of cool storage may vary from a few days to several weeks (Table 27.1).

Chemical reaction rates decrease as temperatures decrease. This decreases microbial growth rates. Refrigeration temperature is below the minimal growth temperature of most foodborne microbes (see chapter 2). Even psychrotrophic microbes such as *Yersinia enterocolitica*, *Vibrio parahaemolyticus*, *Listeria monocytogenes*, and *Aeromonas hydrophila* grow very slowly at low temperatures. A temperature increase of only a few degrees can dramatically increase microbial growth rates. Refrigerated foods can also spoil rapidly if their initial microbial populations are high. Since the temperature requirements for various microbes differ, refrigeration may change the ecology of the microbial population.

Authors' note

While the importance of refrigeration is well recognized, 20% of refrigerators have temperatures of >50°F! Do not count on them to ensure food safety.

Table 27.1 Shelf life extension of raw foods by cool storage[a]

Food	Avg useful storage life (days) at:	
	0°C (32°F)	22°C (72°F)
Meat	6–10	1
Fish	2–7	1
Poultry	5–18	1
Fruits	2–180	1–20
Leafy vegetables	3–20	1–7
Root crops	90–300	7–50

[a]Adapted from J. Farkas, p. 567–591, *in* M. P. Doyle, L. R. Beuchat, and T. J. Montville (ed.), *Food Microbiology: Fundamentals and Frontiers*, 2nd ed. (ASM Press, Washington, DC, 2001).

The temperature range over which an organism can grow depends on its ability to regulate membrane fluidity. At low temperatures, psychrotrophs have more unsaturated fatty acid residues and more branched-chain fatty acids in their lipids. This leads to a decrease in lipid melting point (i.e., increased fluidity). The fluid state allows membrane proteins to function. The idea that cells alter their fatty acid compositions to maintain membrane fluidity is termed *homeoviscous adaptation*.

Controlled-Atmosphere Storage

The minimal growth temperature is lowest when other factors are optimal. When other factors are unfavorable, the minimal growth temperature increases. The most important factors are pH, a_w, and oxygen concentration. Adding refrigeration to this suboptimal combination of factors extends food stability. Controlled-atmosphere storage is widely used for certain fruits and vegetables. The oxygen content is reduced (2 to 5%) and the carbon dioxide content increased (8 to 10%). These conditions are maintained in airtight chilled storage rooms. Modified-atmosphere storage slows respiration of the fruit or vegetable. This limits quality deterioration and microbial growth.

Carbon dioxide inhibits growth for several reasons. When dissolved in water, carbon dioxide reduces the pH. However, the primary mechanism is direct inhibition of microbial respiration (see chapter 2). The reduced oxygen levels contribute to carbon dioxide's inhibitory effect. Psychrotrophic spoilage bacteria such as *Pseudomonas* and *Acinetobacter* are particularly sensitive to carbon dioxide, while lactic acid bacteria and yeasts are not.

Modified-Atmosphere Packaging

Modified-atmosphere packaging (MAP) is similar to controlled-atmosphere storage. In the case of vacuum-packaged products, the air pressure rather than composition is changed. The residual air pressure of only 0.3 to 0.4 bar (1 bar = 10^5 pascals = 14.5 pounds per square inch [lb/in^2]) reduces the available oxygen. If packaging films with very low gas permeability are used for fruits and vegetables, the carbon dioxide concentration may increase. This is because oxygen is used and carbon dioxide is produced during respiration. This modified atmosphere is very inhibitory to certain microbes and increases the keeping quality of foods. However, *Clostridium botulinum* (and other anaerobes) grows well in the absence of O_2 and may be a hazard. Because of this, some vacuum-packed foods, especially pasteurized foods, should be kept below 3.3°C and may require secondary barriers to microbial growth.

MAP for respiring products like fresh fruit and vegetables is complicated. The gas permeability of the packaging film must equilibrate the atmosphere in the package. The package atmosphere is affected by the rate at which respiration and gas permeability change with temperature. MAP could extend the shelf life of many perishable products, but its adaptation is hindered by concern about growth of pathogens at refrigeration temperatures. Fish and fish products pose a risk for *C. botulinum* growth in MAP products. The U.S. National Academy of Sciences recommends that fish not be packed under modified atmospheres unless its safety under such conditions has been proven experimentally.

FREEZING AND FROZEN STORAGE

Many convenience foods are frozen to maintain their high quality. Freezing is, however, a highly energy-intensive process. Commercially, foods are frozen to <-18°C in cold air, by contact with a cooled surface, by submersion into the cold refrigerant liquid (such as liquid nitrogen), or by spraying the refrigerant onto the food.

Foods freeze over a broad temperature range rather than a single well-defined temperature. Depending on the foods' composition, water starts freezing at −1 to −3°C. This initial freezing increases the solute concentration in the water not yet frozen. This further decreases the freezing point. This continues until the temperature where the solutes reach their solubility limit, and the residual water freezes. This is called the *eutectic* temperature. The totally frozen state is a complex system of ice crystals and crystallized soluble substances. It occurs at −15 to −20°C for fruits and vegetables and below −40°C for meats.

Freezing produces an osmotic shock in the microbes. Intercellular ice crystal formation causes mechanical injury. The concentration of cellular liquids changes the pH and ionic strength. This inactivates enzymes, denatures proteins, and inhibits metabolic processes. Cell membranes suffer the major damage. In addition, thawing reexposes surviving cells to these effects. The injury of microbial cells may be reversible or irreversible. The extents of injury and repair and rates of microbial death and survival vary according to the freezing, frozen storage, and thawing conditions.

The freezing rate has a major impact on microbial viability. During slow freezing, crystallization occurs extracellularly and the cell's cytoplasm becomes supercooled at −5 to −10°C. Cells lose water that then freezes extracellularly. Water crystallization increases the concentration of solutes in the external solution. This crystallization also removes water from the cells. Overall, the microbes are exposed to osmotic effects for a relatively long time. This increases injury. Increased freezing rates decrease the duration of the osmotic effects. This increases microbial survival. When freezing rates are too high, crystal formation also occurs intracellularly. This injures the cells drastically and decreases survival rates. Freezing rates also affect food quality. Faster freezing results in higher quality. However, while food manufacturers can optimize freezing, they have little control over thawing, which has an equally important effect.

The food environment has the most important influence when microbes are frozen. Certain compounds enhance, while others diminish, freezing's lethal effects. Sodium chloride reduces the freezing point of solutions, thereby extending the time during which cells are exposed to high solute concentrations before freezing occurs. Compounds such as glycerol, saccharose, gelatin, and proteins generally protect cells from freeze damage.

Microbes do not grow at temperatures below about −8°C. The fate of microbes surviving freezing may change during storage. Often the death of survivors is fast initially and slows gradually, and finally the survival level stabilizes. Death during frozen storage is probably due to the unfrozen, very concentrated residual solution. The residual solution may change during storage. At fluctuating temperatures, such as those in a home freezer with automatic defrost, the size of the ice crystals may increase. Fluctuating temperatures are more lethal than stable ones.

Although freezing and frozen storage reduce microbial viability, freezing is not considered a lethal process. Under some conditions, survivors can grow during thawing. Their levels may then equal or exceed the level before freezing. Thawing releases a nutrient-rich solution from the food cells. During thawing, microbes can penetrate damaged food tissue more easily, and liquid condenses on the food surface. These conditions favor microbial growth. Refreezing thawed products can be dangerous; they are especially vulnerable to rapid microbial spoilage.

PRESERVATION BY HEAT TREATMENTS

Heat is the most widely used method for killing microbes. For a food microbiologist, a microbe's heat resistance is one of its most important traits.

Pasteurization (named after Louis Pasteur [Box 27.1]) is a relatively mild heat treatment. Pasteurization's purpose is to kill non-spore-forming *pathogenic* bacteria. It also inactivates enzymes and kills many (99 to 99.9%) spoilage organisms. Because of this, pasteurized foods take longer to spoil, especially when refrigerated to delay the growth of surviving organisms.

Sterilization kills *all* microbes. The goal of commercial sterilization is freedom from pathogens and shelf stability rather than absolute sterility. A product may contain a viable spore that cannot grow (due to low pH, low a_w, etc.) and still be "commercially sterile." *Clostridium botulinum* spores, for example, may be present in high-acid foods. Because they cannot germinate and grow, they do not present a hazard. This level of commercial sterility can be obtained in a small pressure cooker or in the giant retorts used in commercial canning (Fig. 27.1).

Authors' note

Note that foods can be "canned" in glass jars, in foil pouches, and in other types of containers besides cans.

Technological Fundamentals

Foods can be heat processed after or before packaging. The most common method is to heat after packaging by canning. Canning is the commercial sterilization of food in hermetically sealed containers.

Nicolas Appert invented the canning process in the early 1800s. His research was in response to Napoleon's offer of a prize for the development of a food preservation method that would extend the range of his troops. Canned foods, the C rations of World War II, or today's MRE (meal, ready to eat) remain the staple of soldiers around the world.

Foods and their packaging material can be sterilized separately. The food is then placed aseptically (in a sterile environment) into the package. This process is called *aseptic packaging*. Postprocess contamination can occur if bacteria are not rigorously excluded from the packaging area. Aseptic technology is widely used for fruit juices, dairy products, creams, sauces, and soups that would suffer quality loss if heated for long periods.

BOX 27.1

Louis Pasteur, the first food microbiologist

Louis Pasteur (Fig. 1) (1822–1895) is one of the mythic giants of science but was also a man of great humility and integrity. He was not a particularly good student as a child. He preferred to go fishing or to paint. But his life bent toward science when, at 26, he "discovered" stereochemistry, i.e., that some acids could exhibit "handedness" by bending light to the left or to the right. He concluded that life was asymmetric and abandoned chemistry to study biology.

Figure 1. Portrait of Louis Pasteur in his laboratory (1885), by the Finnish artist Albert Edelfelt. doi:10.1128/9781555817206.ch27.Box27.1.f1

Pasteur the biologist "fathered" several more disciplines. Until this time, many people thought that life "popped up" spontaneously and that fermentation was a chemical process. Pasteur disproved this using a simple but elegant experiment with curved-neck flasks (Fig. 2). They were full of culture medium that would support bacterial growth, but the curved neck prevented bacteria from entering. These flasks, some of which are still on exhibit at the Institut Pasteur, remained sterile. By showing that "life comes from life," he killed the idea of spontaneous generation. From this, he postulated that fermentation was a biological rather than chemical process. Milk, wine, and beer spoilage caused by "ferments" could be stopped by heating for a few minutes at relatively low temperatures, a process we now know as "pasteurization." Pasteur was the first food microbiologist!

Pasteur is also the father of immunology, having developed animal vaccines for cholera and anthrax. He did not want to test vaccination on humans, fearing the guilt and ignominy that he would receive if someone died. His hand was forced, however, when a woman named Madame Meister appeared at his laboratory with her 9-year-old son, Joseph. Joseph had been mauled by a rabid dog 2 days earlier and would certainly die without a rabies vaccination. With fear in his heart, Pasteur treated the boy, who recovered and worked at the Institut Pasteur for the rest of his life. Joseph Meister's loyalty to Pasteur is legendary. There is a story that when the Germans overran France in 1940, they sought to disinter Pasteur in retribution for his refusal to accept an honorary degree from the University of Bonn. Meister took his own life rather than open the crypt of his savior.

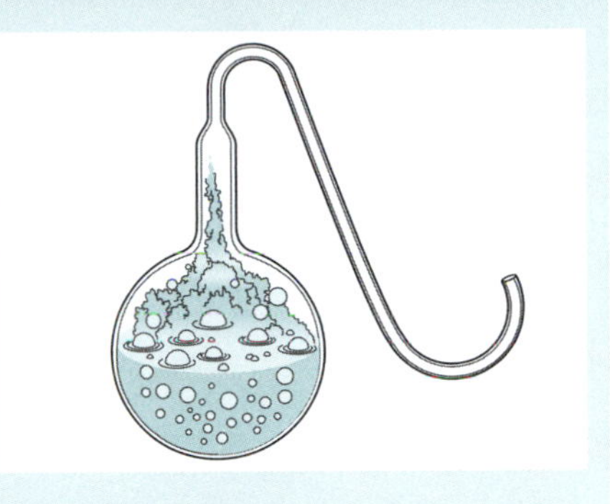

Figure 2. Illustration of a curved-neck flask of the type used by Pasteur. doi:10.1128/9781555817206.ch27.Box27.1.f2

We live in an age that debates the worth of "basic" versus "applied" science. Some feel that basic science is "better" science. Others see no need to pay for research unless it addresses some current problem. We would do well to remember the words of Louis Pasteur: "There does not exist . . . a science one can give the name applied science. There are science and the application of science, bound together as the fruit and the tree which bears it."

These products are most often packaged in boxes (like the juice boxes children take to lunch). Aseptic processing of liquids that contain particulates (individual food pieces) is technically challenging.

High-temperature, short-time (HTST) heating consists of rapid heating to temperatures of about 140°C, holding for several seconds, and then rapid cooling. This produces shelf-stable food. HTST processes improve product quality, process efficiency, and the shelf life of a food. HTST heating preserves nutrients and sensory attributes better than conventional processing which subjects foods to lower temperatures for long periods. Juice boxes, drink pouches, and a variety of condiments are made using HTST technology.

The mild heating of packaged foods combined with well-controlled refrigeration produces a product of very high quality. Such products, also known as REPFEDs (i.e., refrigerated processed foods of extended

Figure 27.1 A pressure cooker that is used in home canning is actually a small retort. doi:10.1128/9781555817206.ch27.f27.01

Authors' note
Should food be thought of as "durable"?

durability), cook-chill products, or *sous vide* meals (foods mildly heated within vacuum packs), are becoming very popular.

However, some types of *Clostridium botulinum* can grow at temperatures as low as 3.0°C and are a hazard in these foods. The Food and Drug Administration (FDA) requires a second hurdle (i.e., a preservative) in addition to refrigeration to ensure the safety of these foods.

Thermobacteriology

"Wet" heat (e.g., steam) kills microbes by denaturing nucleic acids, proteins, and enzymes. DNA damage may be the key event in killing vegetative cells. In spores, the germination system is the most heat-sensitive component. During mild heating, cytoplasmic membranes are a major site of injury. Dry heat is less lethal than wet heat and kills more slowly by dehydration and oxidation. Because of these different mechanisms, dry heat needs higher temperatures and longer times to cause the same lethality as wet heat.

The death of a bacterial population by constant-temperature heating has logarithmic kinetics. That is, an equal time at a given temperature kills an equal percentage of the bacteria, regardless of the number of bacteria present. For example, if 5 min of boiling kills 90% of the bacteria, 10 min will kill 99% and 15 min will kill 99.9%. In theory, the kill can never reach 100%. Another way to look at this is that if the initial bacterial number were 100,000, after 5 min the number would be reduced to 10,000. Similarly, if the initial number were 10,000,000, after 5 min 1,000,000 would still be alive. It should be evident that when many bacteria are present, a 90 or 99%

The information needed to solve this puzzle may be found in chapters 27 and 28.

Across

1. The final cell number after a thermal treatment is ________ to the initial number
5. ________ sealed to exclude air
6. This physical method of food preservation is not considered lethal for bacteria
8. Method of using heat to kill pathogenic bacteria, invented by a Frenchman
9. Freeze-drying is not very ________ to bacteria and can even be used to preserve bacteria in the laboratory
10. Science and its ________ are bound together like a grape and a vine
13. One mechanism by which heat kills bacteria is ________ of proteins
15. Perhaps the oldest form of physical preservation
17. Dried grapes
18. ________ processes preserve foods without heating them
19. ________ manipulation of food can kill or inhibit microbes

Down

2. The use of radiation to kill pathogens in food is conceptually similar to this other preservation method
3. Kinetics by which bacteria grow and die
4. The thermal resistance of ________ is greater than that of vegetative cells from the same species
7. Irradiating food does not make it ________
8. Bacteria that can grow under refrigerated conditions
11. Type of energy that could solve many food safety problems if there were not such consumer resistance to it
12. The low ________ of UV light limits its use in foods
14. Early patron of food processing
15. The *D* value is also known as the ________ reduction value
16. Ionizing radiation kills cells by damaging their ________

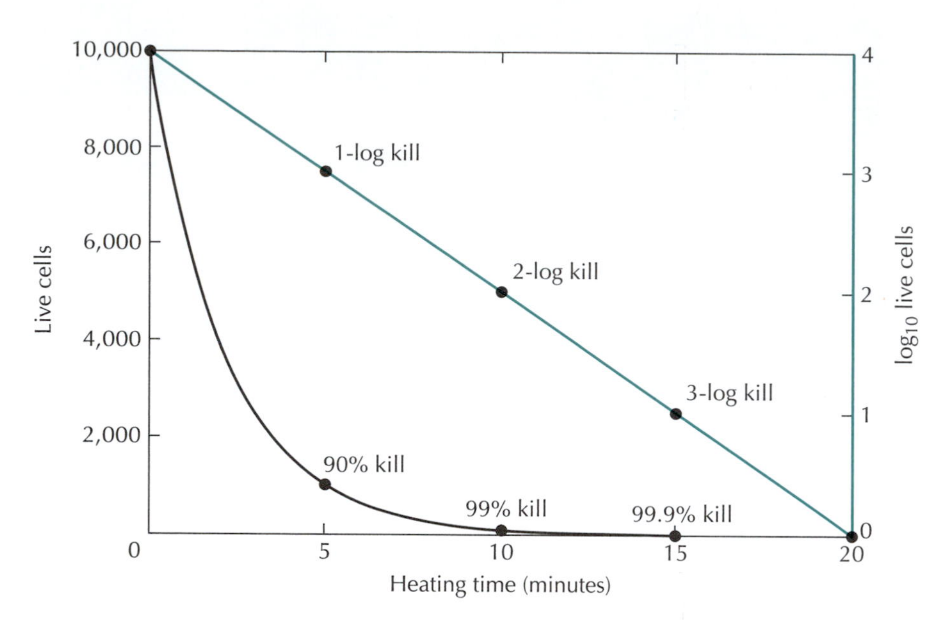

Figure 27.2 Graph illustrating linear and logarithmic plots of the same data for viability of a bacterial population that is heated at a constant temperature. The black plot should be read using the linear (left) *y* axis. The green plot corresponds to $\log_{10}$ values and should be read using the $\log_{10}$ (right) *y* axis. doi:10.1128/9781555817206.ch27.f27.02

reduction may not increase the safety of the product. Figure 27.2 shows the arithmetic and log transform of lethality data.

This negative exponential curve is expressed by the equation

$$N = N_0\, e^{-k \cdot t}$$

In this equation, N is the number of viable cells after any time (t) of heating, N_0 is the initial number of viable cells, and k is the rate constant (time^{-1}) for destruction at temperature T. If N_0, k, and t are known, N can be calculated. Similarly, if N_0 and k are known, the time (t) required to reduce a population to N bacteria can be determined.

The above may seem complicated. Fortunately, when the $\log_{10}$ number of viable bacteria is graphed against the time of heating, the resultant survival curve is linear (Fig. 27.3). In this log-linear curve, the time it takes to reduce the number of viable bacteria 10-fold (i.e., 1 log, or 90%) is called the *D* value, or decimal reduction value. In food microbiology, the *D* value is similar to the concept of the chemical rate constant, *k*.

Authors' note

The D value is inversely related to k by the equation $D_T = 2.3/k_T$.

Survival curves show the rate of destruction of a specific organism in a specific medium or food at a specific temperature. Under given conditions, the death rate at any given temperature is constant and independent of the initial cell number. The logarithms of *D* values plotted against the heating temperatures give the thermal death time curve (Fig. 27.4). The thermal death time is the time necessary to kill a given number of organisms at a specified temperature.

The *D* value gives the microbe's heat resistance at a single temperature. However, in heat processing, microbes are exposed to many temperatures as the product heats up to the processing temperature, holds it, and then cools off. Higher temperatures have greater lethality (smaller *D* values) than lower temperatures. The thermal death time curve shows how heat

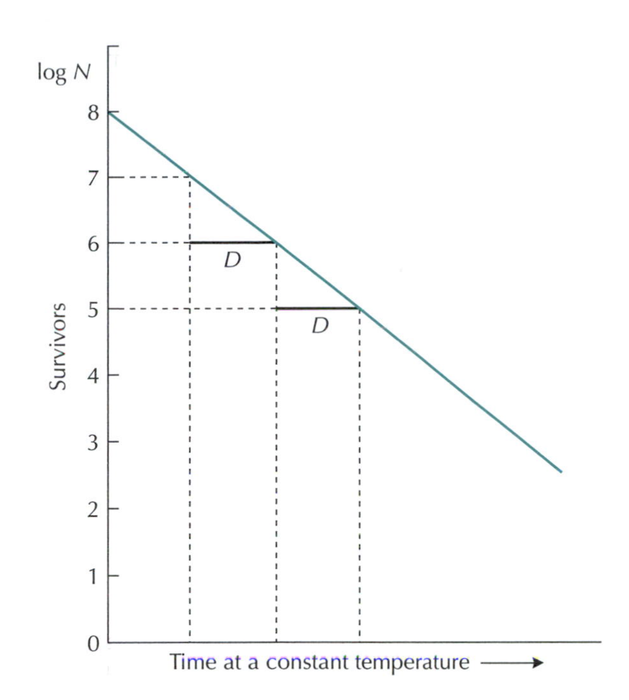

Figure 27.3 The bacterial survival curve shows the logarithmic order of bacterial death from which the *D* value is derived. Redrawn from J. Farkas, p. 567–591, *in* M. P. Doyle, L. R. Beuchat, and T. J. Montville (ed.), *Food Microbiology: Fundamentals and Frontiers*, 2nd ed. (ASM Press, Washington, DC, 2001). doi:10.1128/9781555817206.ch27.f27.03

Figure 27.4 The thermal death time curve illustrates the *z* and *F* values. T_r, reference temperature; τ, death time. Redrawn from J. Farkas, p. 567–591, *in* M. P. Doyle, L. R. Beuchat, and T. J. Montville (ed.), *Food Microbiology: Fundamentals and Frontiers*, 2nd ed. (ASM Press, Washington, DC, 2001). doi:10.1128/9781555817206.ch27.f27.04

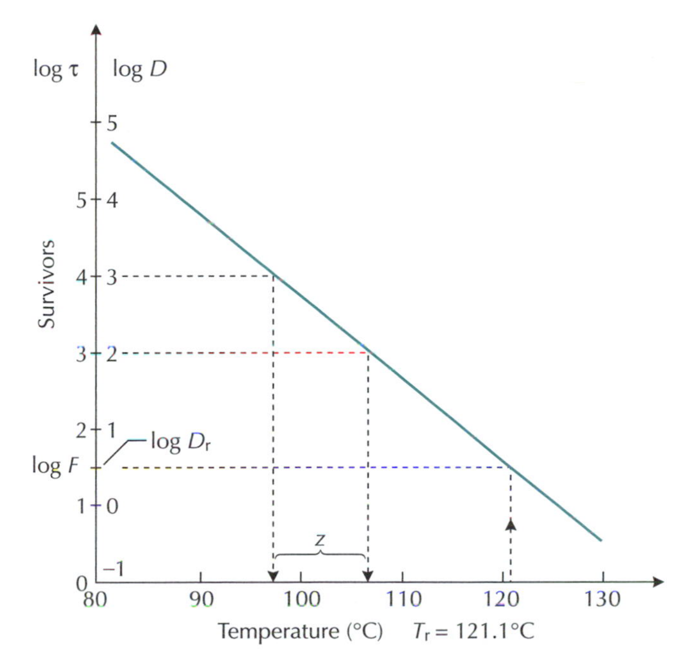

resistance changes at different temperatures. This enables one to calculate the cumulative lethality of each time-temperature combination encountered during the process.

The z value describes how lethality changes with temperature. The z value is the temperature increase (i.e., number of degrees) required to reduce the D value by a factor of 10. Graphically, it is equal to the slope of the thermal death time curve. The z value can also be calculated from the equation

$$t_2 = t_1 \times 10^{(T_1 - T_2)(z) - 1}$$

where T_1 is the higher of the two temperatures at which the D value is t_1, T_2 is the lower temperature at which the D value is t_2, and z is the number of degrees required to change the D value by a factor of 10. For example, if the D value at 250°F were 2 min and the z value were 18°F (typical for many spores), then the D value at 232°F would be 20 min. It takes an 18° change in temperature to cause a 10-fold change in D value. What would be the D value at 268°F?

Know the difference:

The *D* value is the time (in minutes) it takes at a given temperature to decrease viability by a factor of 10.

The *z* value is the number of degrees it takes to change the D value by a factor of 10.

The heat resistances (D values) and how they change with temperature (z value) are influenced by many factors (e.g., pH and a_w) and differ among microbial species. Bacterial spores are more heat resistant than vegetative cells of the same species. Some spores survive for minutes at 120°C or for hours at 100°C. Less heat-resistant spores have $D_{100°C}$ values of $<$1 min. Vegetative cells of bacteria, yeasts, and molds can have D values of $<$0.1 min at 70 to 80°C.

The age of the cells, their growth phase, the growth temperature, growth medium composition, exposure to prior stressors, and the heating environment also affect heat resistance. For example, one would expect the D value of a *Salmonella* sp. to be different in chocolate than it is in milk because chocolate has a higher fat content and lower a_w. Thus, unlike chemical constants such as pK_a or molecular mass, D and z values depend on the situation of use. Since they depend on so many factors, great caution should be used in applying values obtained in one environment to some other food environment.

Generally speaking, the heat resistance of vegetative cells (Table 27.2) is much less than that of spores (Table 27.3). Vegetative cells of spore-forming bacteria are not unusually heat resistant. The high heat resistance of spores is caused by their specific structures, which result in a relatively dehydrated spore core. (Remember that dry heat is much less lethal than wet heat.) The molecular basis of heat resistance in spores is covered in chapter 3.

Vegetative forms of yeasts are killed by 10- to 20-min heat treatment at 55 to 60°C. Increasing the temperature 5 to 10°C gives similar lethality for yeast ascospores. Vegetative propagules (a reproductive bud or offshoot) of most molds and conidiospores are inactivated by 5 to 10 min of wet heat at 60°C. Several mold species forming sclerotia or ascospores are much more heat resistant. They cause serious spoilage problems. Mold spores survive dry-heat treatment for 30 min at 120°C and cause spoilage of baked foods.

Table 27.2 Relative heat resistances of some vegetative bacteria

		***D* value (min) at:**				
Organism	**Heating medium**	**70°C**	**65°C**	**60°C**	**55°C**	***z* value (°C)**
Escherichia coli	Ringer solution, pH 7.0				4	
Lactobacillus plantarum	Tryptic soy broth + sucrose (a_w = 0.95)		4.7–8.1			
	Tomato juice, pH 4.5	11.0				12.5
Pseudomonas aeruginosa	Nutrient agar				1.9	
Pseudomonas fluorescens	Nutrient agar				1–2	
Salmonella enterica serovar Senftenberg (strain 775W)	Skim milk			10.8		6.0
	Phosphate buffer (0.1 M), pH 6.5		0.29			
	Buffer plus sucrose, 30:70 (%, wt/vol)		1.4–43			
	Same as above, plus glucose, 30:70 (%, wt/vol)		2.0–17.0			
	Milk chocolate	440				18.0
Salmonella enterica serovar Senftenberg	Heart infusion broth (a_w = 0.99), pH 7.4			6.1		6.8
	+ NaCl (a_w = 0.90, pH 7.4)			2.7		13
	+ sucrose (a_w = 0.90, pH 7.4)			75.2		8.9
Salmonella enterica serovar Typhimurium	Phosphate buffer (0.1 M), pH 6.5		0.056			
	Milk chocolate	816				19.0
Staphylococcus aureus	Custard or pea soup			7.8		4.5

The death rates of bacteria are also influenced by heating rates and their history of temperature exposure. When cells are exposed to sublethal temperatures above their growth maxima (i.e., to a sublethal heat shock), they become more heat resistant. This is because the production of heat shock proteins is induced by the sublethal temperature. In addition to heat, other environmental stresses, chemicals, or mechanical treatments can trigger the production of "heat shock" proteins. This confers cross-resistance among stressors. For example, cells that form heat shock proteins in response to sublethal temperatures may become more resistant to acid. The heat shock response caused by food processing can protect the microbes.

Calculating Heat Processes for Foods

Foods in containers do not reach processing temperatures instantly. Furthermore, all temperatures (above a minimum) encountered on the way to the process temperature contribute to microbial inactivation. Because of this, many time-temperature combinations can yield the same process lethality. Six tenths of a minute at 268°F can be as lethal as 6 min at 250°F. There must be some "common denominator" or "reference value" used to compare these lethalities. The *F* value is used for this purpose. The *F* value is the time at a specific temperature required to kill a specific number of cells having a specific *z* value.

Since the *F* value is the time to decrease a population with a specific *z* value at a specific temperature, some temperature must be (arbitrarily) set as the universal reference *F* value.

This universal reference is the F_0 value, the temperature is 121°C (250°F), and the *z* value is 10°C (18°F). The F_0 value of a heat treatment is called its

Authors' note

The temperature chosen is actually not so arbitrary, since it is the temperature produced by low-pressure (15-lb/in²) steam in canneries.

Table 27.3 Approximate heat resistances (*D* values) of some bacterial spores

Types of food and typical organisms	*D* value (min) at: 121°C	*D* value (min) at: 100°C	*z* value (°C)
Low-acid foods (pH > 4.6)			
Thermophilic aerobes			
Geobacillus stearothermophilus[a]	4.0–4.5	3,000	7
Thermophilic anaerobes			
Clostridium thermosaccharolyticum	3.0–4.0		12–18
Desulfotomaculum nigrificans	2.0–3.0		
Mesophilic anaerobes			
Clostridium sporogenes	0.1–1.5		9–13
C. botulinum types A and B	0.1–0.2	50	10
Clostridium perfringens		0.3–20	10–30
Mesophilic aerobes			
Bacillus licheniformis		13	6
Bacillus subtilis		11	7
Bacillus cereus		5	10
Bacillus megaterium		1	9
Acid foods (pH ≤ 4.6)			
Thermotolerant aerobes			
Bacillus coagulans	0.01–0.1		
Mesophilic aerobes			
Bacillus polymyxa		0.1–0.5	
Bacillus macerans		0.1–0.5	
Clostridium butyricum (or *Clostridium pasteurianum*)		0.1–0.5	

[a]Formerly classified as *Bacillus stearothermophilus*.

sterilization value. F_0 measures the lethality of a given heat treatment and relates it to the time required for the same lethality at some other temperature, *T*. Thus, if one knows the time it takes to kill a population of cells with a *z* value of 10°C at 121°C (the reference temperature), one can get the same result at other times or temperatures using the following equation:

$$\text{Time}_{\text{new}} = \frac{F_0 \text{ (minutes at 121°C)}}{10^{[\text{new temperature (°C)} - 121\text{°C}]/z\text{ (°C)}}}$$

Various foods have different F_0 requirements. Heat sensitivity and thermal death curves are affected by many factors, so thermal death curves should be made for the actual food being processed. Because the spores of *C. botulinum* types A and B are the most heat-resistant spores of a foodborne pathogen, commercial sterilization of low-acid (pH > 4.6) foods must kill these spores. A "botulinum cook" is a heat process that reduces the population of *C. botulinum* spores by an arbitrarily established factor of 12 decimal ($\log_{10}$) values. This "12*D*" concept provides a large safety margin for low-acid canned foods. If there were one spore in each of 10^{12} cans that received a 12*D* botulinum cook, there would only be one can with a surviving spore. Since botulinal spores have a *D* value of 0.20 min at 121°C, a botulinum cook is equal to 2.4 min (12*D* × 0.20 min/*D*). Any combination of time and temperature that yields an F_0 of 2.4 min is acceptable for canned foods. Spoilage organisms may have spores that are more heat resistant than botulinal spores. The higher heat resistance of the spoilage organisms often determines the commercial process.

Summary

- Physical treatments of food can inhibit, kill, or remove microbes in foods.
- Drying may not kill microbes, but it prevents their growth.
- Refrigeration prevents the growth of many pathogens and slows the growth of all microbes.
- Freezing prevents all microbial growth but cannot be considered lethal.
- Controlled and modified atmospheres can extend the shelf life of fruits and vegetables but may create conditions favorable for specific pathogens.
- Pasteurization is designed to kill all pathogens but not spoilage organisms.
- Commercial sterilization aims to kill all microbes that can grow in a product.
- Heat kills microbes with negative exponential kinetics.
- The D value is the *time* required for a 90% (1-log) reduction in viability of a population.
- The z value is the *number of degrees* required to change the D value 10-fold.

Suggested reading

Farkas, J. 2007. Physical methods of food preservation, p. 685–712. *In* M. P. Doyle and L. R. Beuchat (ed.), *Food Microbiology: Fundamentals and Frontiers*, 3rd ed. ASM Press, Washington, DC.

Questions for critical thought

1. Microbiological kinetic constants have their counterparts in chemical kinetic constants. If the D value is analogous to the chemical rate constant (k), what chemical constant is analogous to the z value?
2. You may remember that low-acid canned foods must receive a thermal process equivalent to an F_0 of 2.4 min. This is based on a 12D reduction of *C. botulinum* spores, which have a $D_{250°F}$ of 0.2 min. Imagine two process deviations. In deviation A, the time is 10% less than it should be (i.e., the product receives 2.16 min at 250°F). In deviation B, the temperature is 10% too low (i.e., the product receives 2.40 min at 225°F). Which process deviation represents a greater threat to public safety? Be quantitative in your answer (i.e., calculate the log reduction in *C. botulinum* spores as the result of each deviation, assuming a z value of 25°F). Show all of your work.
3. Define z value. If an organism has a $D_{80°C}$ of 20 min, a $D_{95°C}$ of 2 min, and a $D_{110°C}$ of 0.2 min, what is its z value?
4. The year is 2027 and you are director of research for Peter's Premium Preservative-Phree Phoods. One of your product lines is a tomato sauce.

Authors' note

Don't be discouraged. Some of these problems took me 45 minutes to solve.

Your procurement department has identified a very inexpensive source of tomatoes from Israel, but your microbiology department reports that they typically contain spores of "*Bacillus citricus*" (a species made up for this question) at about 100 colony-forming units (CFU)/g. This organism can metabolize citric acid and thereby increase the pH of the product. Anticipating your every need, the microbiologists scan the World Wide Web for data on *B. citricus* heat resistance and present you with the information in the table below.

	CFU/g at:		
Time (min)	**80°C**	**85°C**	**90°C**
0	2.2×10^6	2.2×10^6	3.3×10^6
10	7.0×10^5		2.9×10^5
20	4.4×10^5	3.2×10^5	2.5×10^4
30		1.8×10^4	3.5×10^3
40	2.0×10^5	3.2×10^4	1.7×10^3
60	4.1×10^4		

a. You decide to show those microbiologists that you are still technically competent, so you determine the $D_{90°C}$ and the z value (in degrees Celsius) for *B. citricus* from the data. (Each survivor curve has five data points; ignore the blank cells.) What values do you get?

b. Your current process requires that the jars of sauce be held for 7 min at an internal temperature of 100°C. Will this process be adequate for tomatoes contaminated with *B. citricus*?

c. Do you have any other concerns about this product or the data upon which you are making your judgment? Are there other factors you should consider?

Note: You should document your decision in case the FDA wants to inspect your records. Show all your work. If you solve the problem graphically, attach your graph to your answer. Justify your "yes" or "no" answers to parts b and c.

Authors' note

"Potted" is an old-fashioned word for "canned."

5. You have taken a position as director of microbiology in equatorial Africa with Angelica's Premium Potted Products. One of your product lines is canned kudu (an African antelope) meat. The main spoilage problem is a gram-positive spore-forming rod. The person who had the job before you decided that product quality could be vastly improved by using a higher temperature for a shorter period. (The product is currently given a 7*D* process of 791 min at 220°F.) The previous microbiologist had prepared spores and determined that the $D_{250°F}$ was 2.3 min. She had also conducted thermal resistance studies at 230 and 240°F, but then she ate some bad fugu fish and tragically died. You find her data (below). How long would it take to achieve a 7*D* process at 260°F? Would this process to sufficient to meet the 12*D* botulinum cook requirement if you wanted to export the product to the United States?

Note: You should document your decision in case the FDA wants to inspect your records. Show all your work, including the calculation of the *D* value at 230 and 240°F and the *z* value determination. If you solve the problem graphically, attach your graph to your answer. Justify your "yes" or "no" answers for the botulinum cook. This problem is difficult. Ask your instructor if you can work on it with a friend.

Data on heat resistance of spores that cause product spoilage

	CFU/g at:	
Time (min)	**230°F**	**240°F**
0	1.0×10^4	12,500
5		3,400
10	4.7×10^3	650
15		190
20	2.4×10^3	45
35	7.0×10^2	

6. Your company makes a pasteurized juice product with an initial (postpasteurization) microbial load of 10^5 CFU/g. Storage studies show that after 5 days at 35°C, the microbial load is 10^8 CFU/g. Your process engineer finds that by increasing the pasteurization temperature a few degrees, the initial (postpasteurization) microbial load can be lowered to 10^3 CFU/g. If this process were used, what would the final number be after 5 days at 35°C? (Assume that all conditions except the initial number are identical.) Justify your answer. (Hint: You may wish to consider the equation for the exponential growth of microbes.)

28 Nonthermal Processing

LEARNING OBJECTIVES

The information in this chapter will enable the student to:

- learn the difference between thermal and nonthermal processing
- identify strengths and weaknesses of nonthermal processing technologies used in food production
- gain a basic understanding of the functionality of the most popular nonthermal processing methods
- understand what makes a food a good candidate for nonthermal processing

INTRODUCTION

Louis Pasteur first showed the role of microorganisms in food quality and spoilage in the 1860s. Soon after Pasteur's work, A. K. Shriver of Baltimore, Maryland, invented the pressure cooker, and by 1900, Samuel Cate Prescott and Wiliam Lyman Underwood established the relationship between time and temperature in the thermal processing of foods. In 1920 the National Canners' Association established the logarithmic nature of the thermal death time curve and also the importance of the bacterium *Clostridium botulinum* in determining the standards of safety for the canning process. Research continues today regarding thermal processing technologies, which are used for many foods. As you can see, innovation is pivotal to the development of processing methods, for increased food production and for better food production over time. While thermal processing remains important today, researchers continue to make advancements in the technology and instrumentation. Starting in the 1980s food manufacturers, scientists, and consumers began to want more from their foods. From that point onward, food scientists have been discovering and evaluating methods of food preservation that are collectively called nonthermal processing methods. Why did it take so long for the development of nonthermal methodology and even longer for it to be used in the commercial production of foods? As you will see throughout this chapter, the technology and instrumentation used in nonthermal processing are quite intricate and advanced. Today, new technologies that are less disruptive to food quality than traditional thermal processing are increasingly common. And this is a good thing, since consumers are increasingly demanding.

According to leading scientists, nonthermal processing technologies promise to revolutionize the food industry. While thermal processing is effective at killing microorganisms that cause illness or food spoilage, it also

doi:10.1128/9781555817206.ch28

reduces the flavor, color, and texture of the food. These changes in the sensory characteristics and organoleptic profile are important to consumers. In many cases this is the basis for consumer brand loyalty and for repeat purchases of a food product. As you can imagine, this is even more important in a time when we are bombarded with advertisements for hundreds of food products. Nonthermal processing methods are advantageous, as they do not alter the nutrient content, color, or texture of a food. Nonthermal processing does less damage to the food and leaves it with "fresh-like" characteristics. Collectively, these are the great strengths of nonthermal processing methods.

Another benefit of nonthermal processing technologies is that they are typically applied under room temperature (20 to 25°C) or cooler conditions. This can also help retain those fresh-like characteristics of foods. This can also save energy. Some scientists are conducting research on combining heat to increase microbial inactivation and to get closer to treatments like pasteurization. There are several different types of processing methods, and they all exploit different physical hurdles like pressure, light, electromagnetic radiation, and sound to inactivate spoilage microorganisms and pathogens. Enzyme inactivation is also a product of nonthermal processing. The same is true of thermal processes, like blanching in boiling water to inactivate enzymes that might cause browning or decrease shelf life. Nonthermal treatments can maintain the important sensorial characteristics of foods while enhancing shelf life compared to that of a fresh or untreated product. Most nonthermal technologies require short processing times, i.e., seconds or minutes. This reduction in processing time is reflected in energy costs. This is desirable for companies when considering integrating a new process into their current methods and in part balances out the initial investment costs in the new machinery.

ACCEPTANCE

Before we look at the various types of treatments available, it is important to consider their acceptability. If a method is to be acceptable, it must be validated by federal agencies *and* accepted by consumers. The latter may seem more important. If a technology is not acceptable by consumers, it will never make it to the marketplace. This has been seen for irradiation, for example. Processes must be scientifically tested and shown to be as effective as previously accepted processing techniques. Part of this acceptance includes regulatory acceptance, which will allow future commercialization of a nonthermal processing method. Food companies and researchers will validate a treatment before it can be widespread in use or application. One example of this is changes that were made to the definition of pasteurization in legal terms so that companies could include the ability to use a treatment other than heat to "pasteurize" a food. For example, ultraviolet (UV) light and ozone may be used as alternatives to heat pasteurization of apple cider, according to the juice Hazard Analysis and Critical Control Point (HACCP) rules. Researchers set out to identify nonthermal processes that might be equivalent to thermal pasteurization. The definition used by the regulatory agencies needed to be rewritten to allow companies to utilize nonthermal methods that have been validated and scientifically proven to work as effectively as thermal pasteurization. The National Advisory

Authors' note

The NACMCF identified a problem area (described here) that continues to plague researchers. This is the question of test organisms. What is the best way to determine the most resistant microorganism(s) of public health significance? How much of that organism do you use in your testing procedures? Researchers are still seeking answers to these questions. Maybe you can discover an answer.

Committee on Microbiological Criteria for Foods (NACMCF, pronounced "nac-mff") adopted a new definition for pasteurization in 2004: "any process, treatment, or combination thereof, that is applied to food to reduce the most resistant microorganism(s) of public health significance to a level that is not likely to present a public health risk under normal conditions of distribution and storage." The NACMCF went on to include a list of potential alternatives, including high hydrostatic pressure, pulsed electric fields (PEFs), UV light, pulsed light, cold plasma, oscillating magnetic fields (OMFs), and ultrasound.

Consumers do not want food processors disturbing their favorite foods. Consumers have shared uneasy feelings about genetically modified ingredients and irradiation. Both of these are considered safe by scientists; however, they bring about great emotions in many consumers. The acceptance of a technology depends on the consumers' perception of the risks and benefits. This may be difficult, as consumers often do not understand the risks associated with food. Consumer acceptance is influenced by the perceived credibility of the data, regulatory policy, and demonstrated responsibility of the industry. In general, consumers do not think about how foods are processed but expect that the foods that they consume are safe and of high quality.

Consumer attitudes towards foods drive marketing and development of new food products and new processes. For example, in chapter 27 you read about how the process of irradiation can inactivate pathogens and create a safe food product; however, due to consumer perceptions of irradiation as hazardous, this technology is not widely utilized. Consumer attitudes are addressed in focus groups or small meetings where consumers are asked to speak about their opinions on food preservation technologies or other factors. Focus groups are composed of consumers of various demographics in order to obtain the most useful information. The key benefits identified by consumers include good flavor, convenience, and health-enhancing properties. In a national telephone survey, 80% of consumers reported that convenience was an important consideration when choosing food in the grocery store. The media also influences consumer perceptions. This has become more noticeable over the past few years; media stories have increased on several topics, including dietary fiber, beneficial fatty acids, lycopene, vitamin C, and probiotic cultures. Scientists can show how nonthermal processing technologies can give consumers more of the foods they want, but how they communicate that effectively is quite important. Consumers may associate some technologies with risks that may include exposure to workers or the environment, harm to consumers, or changes in the food product. Several studies have shown that when consumers are provided with basic scientific information about a new technology, their level of concern decreases. This information must be delivered using familiar terms. For this reason, nonthermally processed foods may be referred to as "minimally processed." The future research, marketing, and development of nonthermal technologies should include consideration of consumer beliefs and attitudes. This may include premarket consumer analysis on a broad range, as current models suggest a wide discrepancy in the perception of risks. Concerns vary by population demographics and by food process. False impressions can be lessened by good consumer communication, effective utilization of labels, and various forms of media communications.

Authors' note

It is important that scientists are trained to communicate effectively with the public. The best information is that which is presented in a variety of sources. There are many researchers working in this area, and scientific writing and communication constitute a growing field. Two leading organizations in this are the Institute of Food Technologists and the International Food Information Council. The combination of written and oral communication and science is a great field for career development too.

Cold Plasma: Overview of Plasma Technologies and Applications

Brendan A. Niemira, Eastern Regional Research Center, Agricultural Research Service, U.S. Department of Agriculture, Wyndmoor, Pennsylvania

Cold plasma is a new food processing technology which can sanitize fragile and heat-sensitive surfaces, such as fresh and freshly cut fruits and vegetables. This process is in the first stages of development, and the terminology used to describe it is new to many in the food-processing industry.

To answer the question "what is plasma?" it is useful to consider the three states of matter we commonly see in daily life. As materials acquire energy, they change state, from solid (lowest energy) to liquid and then to gas. At each transition, the interactions among the molecules become more energetic. At extremely high energies, matter undergoes a further transformation into plasma, the "fourth state" of matter. In the broadest sense, plasma can be thought of as an ionized gas consisting of neutral molecules, electrons, and positive and negative ions.

There are two processes that occur during the creation of plasmas which give them their unique properties: ionization (atoms absorbing energy and coming apart) and recombination (atoms coming back together by releasing energy into the target). It takes significant energy to create cold plasma from normal gases or liquids. This energy is retained by the plasma for some defined period of time, and the amount of energy a given volume of plasma contains is related to its chemical composition, density, and temperature. For example, helium is easier to ionize than oxygen. Also, gases are easier to ionize when under a vacuum and harder to ionize when at normal atmospheric pressure.

Unlike conventional plasmas, which can cause thermal damage to tissues, cold plasma has high energy density and chemical reactivity but at cooler temperatures that can be applied to food products. Although the technologies used to create and control cold plasmas can be complex, the inputs to the system are simple: energy and a gas to turn into plasma. This gas can be a pure gas (such as helium, nitrogen, or oxygen), a defined mixture of gases, or normal air. The output is plasma that is extremely energetic and reactive but is also "self-quenching"—the active chemical species within the plasma recombine with each other. The end results are UV light and chemical recombination products such as ozone.

For food-processing and food microbiology purposes, the general mechanisms of sanitization are defined by the physical and chemical nature of plasma, rather than by the nature of the substrate. Cold plasma inactivates microbes by three primary mechanisms:

1. chemical interaction of cell membranes with radicals (O, OH, etc.), excited or reactive molecules (O_2^-, O_3, NO, etc.) or charged particles (electrons and atomic or molecular ions) produced in the plasma
2. erosion of cell membranes and cellular constituents by UV radiation
3. destruction of DNA by UV light

Cold plasma treatment which results in a combination of multiple mechanisms will have the greatest sanitizing efficacy. From the standpoint of the mechanisms of interaction with the treated surfaces, cold plasma antimicrobial treatment systems can be divided into three groups, described below.

Remote treatment systems generate plasma which is then moved by the flow of the feed gas. The surface to be treated can be physically separated from the electrodes used to generate the plasma, which simplifies the design and operation of the device. However, this flow of "decaying" plasma (sometimes called an "afterglow") relies on lower-activity, long-living chemical species in concentrations lower than in "active" plasma.

Direct treatment systems supply "active" plasma in higher concentrations. UV radiation on the treated surface is higher, as there is essentially no intervening normal atmosphere between the plasma and the surface. A challenge with these plasma systems is that food materials with high moisture content can conduct electricity at sufficiently high voltages. Also, extended exposure increases the potential for negative sensory impacts. These cold plasma systems are therefore somewhat more challenging to build and operate than those of the first group.

Electrode contact systems place the food surface to be sterilized in very close contact with one of the electrodes. Pathogens are exposed to the broadest combination of active antimicrobial agents, at the highest possible intensity. The positioning of the material to be treated is critical to avoid localized heat buildup and tissue damage. Although apparently the most promising of the types of cold plasma from an antimicrobial efficacy standpoint, these systems are among the most technically challenging.

Cold plasma is a flexible new technology for treating foods and food contact surfaces. Still in technological infancy, it promises to be a fascinating area of continued research.

Niemira, B. A., and A. Gutsol. 2011. Nonthermal plasma as a novel food processing technology, p. 271–288. *In* H. Q. Zhang, G. V. Barbosa-Cánovas, V. M. Balasubramaniam, C. P. Dunne, D. F. Farkas, and J. T. C. Yuan (ed.), *Nonthermal Processing Technologies for Food*. IFT Press, Wiley-Blackwell, Ames, IA.

HIGH-PRESSURE PROCESSING

Many argue that the movement towards the development of high-pressure processing (HPP) for foods was also the beginning of a developmental period of rapid growth in nonthermal processing technologies. HPP is considered one of the most important food-processing innovations in the last 50 years. HPP may also be referred to as "high-hydrostatic-pressure processing" or "ultrahigh-pressure processing." In HPP, foods are placed in sealed pouches, placed within a water bath in a sealed vessel, and then subjected to high pressure (Fig. 28.1). HPP-treated foods can have fewer additives and do not have the thermally induced flavor changes that can occur in thermal processing. HPP preserves the freshness of foods and does not break covalent bonds. Many thermal processes induce structural changes in the food or denature proteins by disrupting covalent bonds.

HPP works under Le Chatelier's principle (named for the French chemist Henry Louis Le Chatelier, who discovered it), which states that the application of pressure can shift the system equilibrium toward the state that occupies the smallest volume. This means that any phenomenon (or molecular reaction) that is accompanied by a decrease in volume is enhanced by pressure; therefore, pressure stimulates reactions that result in a decrease in volume but opposes reactions that involve an increase in volume. It is believed that pressure is exerted at the macroscopic level in a quasi-instantaneous manner equally throughout the sample volume. In other words, pressure is exerted equally and uniformly throughout a product. This is based on yet another principle that you may have learned in high school physics, the Pascal principle. These are the basic chemical and physical theories behind HPP, but how does it work?

Figure 28.1 Conventional HPP equipment may be used for a range of products. Image courtesy of Avure Technologies (http://www.avure.com). doi:10.1128/9781555817206.ch28.f28.01

Historical Aspects of HPP

Dallas G. Hoover, University of Delaware, Newark, Delaware

Application of high pressure in preservative food processing was first attempted by Bert Hite at the Agricultural Experimental Station of West Virginia University. In the mid-1880s, publications by France's P. Regnard and colleagues reported the effects of oceanic pressures on living systems. In these articles, the inactivation of marine bacteria by hydrostatic pressures was also described. Apparently, from these reports Hite was able to recognize the potential of hydrostatic pressure to inactivate food microorganisms and consequently preserve foods. Hite constructed his prototype pressure unit in Morgantown from parts purchased from cannon manufacturers in Harpers Ferry, West Virginia. Using toothpaste tube-like containers, he was able to make fruit-based low-pH foods shelf stable. His 1899 publication noted the early equipment failures, some nearly catastrophic. One machine exploded due to equipment failure and scattered shrapnel around the laboratory, leaving the machine in fragments. Of course at Hite's time, laboratory safety was not considered the way it is today. Hite noted that during microbial experimentation, student workers scrubbed their skin up to their elbows with mercuric chloride to improve their aseptic technique.

HPP is studied in academia, industry, and government. Here, Dallas Hoover gets ready to conduct research on the inactivation of bacterial spores by HPP at the Food Microbiology Laboratory at the University of Delaware. doi:10.1128/9781555817206.ch28.fGuestBox02

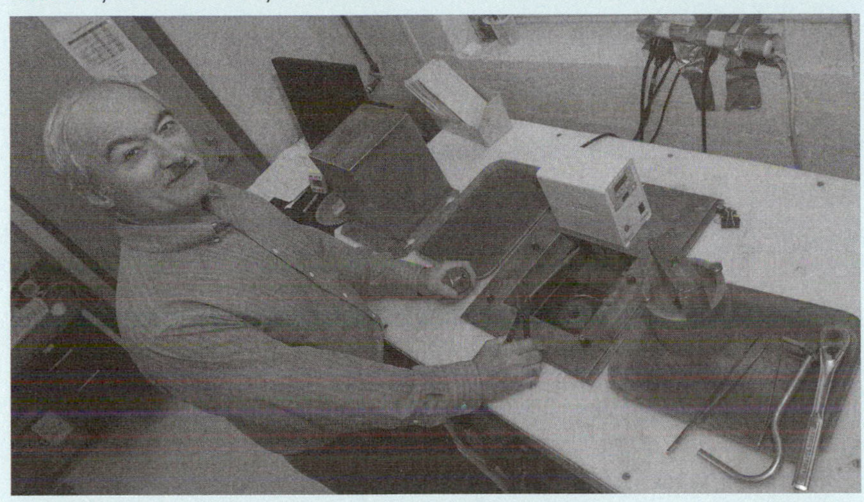

Following the work of Hite, occasional articles appeared over the next 50 or so years that were related to the development of commercial HPP, but none generated enough momentum to move HPP from research novelty to marketplace success. That began to change in the early 1980s, when concurrently and independently, Dan Farkas at the University of Delaware and R. Hayashi in Japan began efforts to explore the possibilities of HPP (see figure). In 1982, Farkas borrowed an old Autoclave Engineers pressure unit originally designed to fabricate ceramic parts for the conductor industry. He was motivated to do this by a 1974 Institute of Food Technologists meeting abstract by D. C. Wilson noting the significant reduction in *D* values (see chapter 27) of spores treated with a combination of elevated temperature and pressure. Farkas' development of HPP with this pressure unit was the beginning of the rise of HPP as a commercially viable technique.

Each HPP machine is a bit different, but in general, food products are sealed in pouches and pressure is conducted through water. For the greatest effectiveness, as much air as possible should be removed from the pouch. All parts of the food receive the pressure; however, each food component may react differently to the pressure and microorganisms may be inactivated at different rates within these different components. Foods are chemically complex, composed of carbohydrates, proteins, fats, and water. Each of these responds differently under the physical compression of HPP. Try to imagine how this works on a molecular level of inactivating microorganisms and enzymes. Imagine pressure causing shifts in the protein coat of a virus and then that virus is no longer able to bind to the receptor on the host cell and infect it. With even greater pressure, the coat protein may be disrupted and even lyse. Certain foods may also show macromolecular changes. For example, pressure treating a food which contains a lot of air, such as a marshmallow, causes changes in compressibility after the air escapes from the product after treatment. It is no longer light and pillowy. A food with a high moisture content may not experience much distortion visually. A food that has a high level of moisture is a good candidate for HPP, but as you can imagine, a marshmallow is not. This change can be viewed in a positive manner for some foods; for example, HPP treatment can help shuck oysters by popping open the shell (Fig. 28.2).

Commercial-scale equipment can be used for batch or semicontinuous processing. In all systems, a large piston moves back and forth to generate the pressure. The commercial batch vessels may hold up to 600 liters of water. Food containers are placed into baskets that are then loaded into the vessel, which fills with water prior to being pressurized. After the pressure is held for a desired length of time, the vessel is decompressed and the water released. Semicontinuous systems have two or more vessels in line with free-floating pistons to compress the liquid as needed. A pump is used to move the free piston toward the discharge port. Picture a system with three vessels; the first vessel is complete and is discharged, the second one is being compressed, and the third one is being loaded. The placements change and the system is maintained in close to a continuous fashion. HPP is likely the most integrated nonthermal technology used in commercial processing today.

Imagine how much pressure needs to be applied to a food product in order to disrupt the enzymes that cause browning (phenol oxidase) or to disrupt bacterial membranes by altering hydrogen bonds. In HPP, applied pressure is more than four or five times that at the bottom of the Marianas Trench (100 megapascals [MPa]) at the bottom of the ocean. Commercial HPP is at pressures of 300 to 550 MPa. Pressure is held for seconds up to a few minutes. Scientists have identified a means to pasteurize food or sterilize food using HPP. Pressure-assisted thermal sterilization subjects food to a combination of elevated pressures and moderate heat for 1 to 5 min. Would you be surprised to know that you have likely enjoyed eating HPP-treated

Figure 28.2 A 100-liter HPP unit is ready to treat oysters. HPP-treated oysters have reduced pathogen contamination with the added benefit of being shucked mechanically by the HPP treatment. doi:10.1128/9781555817206.ch28.f28.02

food? Recall that one of the best examples of the latter is pressurized avocado sold in sealed plastic pouches for the preparation of guacamole or other delicious treats. Examples of the former include ready-to-eat deli meats. HPP protects the deli meats from postprocessing contamination, for example, from *Listeria monocytogenes* or norovirus that could be present on the hands of food service workers or in the environment. Inactivation of vegetative bacterial cells and viruses requires pressure in the range of 300 to 700 MPa, which is the range for conventional processing. Bacterial inactivation varies with the species and the food, reaffirming the need for process validation. Increasing pressure increases the death rate of microorganisms, but increasing the duration of pressure treatment does not correlate with enhanced inactivation. The unique aspect of pressure-assisted thermal sterilization is that the increase in time is combined with an increase in temperature and can inactivate spores and all vegetative cells. But the addition of heat keeps this from being a truly nonthermal processing method.

OZONE

Ozone is also a technology that has gained increasing popularity due to its effectiveness and ability to be termed a "natural" disinfectant. The use of ozone is specifically effective for water treatment and has also penetrated the food industry. Ozone was given the status of a GRAS (for "generally recognized as safe") substance by the U.S. Food and Drug Administration in 1982 for bottled water and in 1997 for foods. Ozone is used across Europe for water treatment and is becoming a better alternative for water treatment within the United States. In 2005, the Metropolitan Water District of Southern California replaced its chlorination water treatment system with ozone. The benefits are substantial, including elimination of the creation of chlorine disinfection by-products, which are potential carcinogens and may also be environmentally hazardous. Ozone may also be replacing older methods for cleaning applications. Ozone can be used to disinfect surfaces from bacteria, viruses, and parasites; however, its effectiveness may be limited by the presence of organic compounds. This can limit its usefulness in foods as well. Ozone was first found to prevent the growth of yeasts and molds during the storage of fruits in 1939. Its use by processors of freshly cut produce is growing. Ozone is touted by some scientists to be one of the most powerful disinfectants known.

When you hear the word ozone, you likely think about the "ozone layer" in Earth's atmosphere. Within this layer UV light is dampened and shorter wavelengths of sunlight are filtered out for our safety on the earth below. Ozone is a naturally occurring triatomic molecule, O_3 (Fig. 28.3), and a very reactive form of oxygen. It is generated in the air by electric discharges such as lightning and other types of radiation. It is a colorless gas at room temperature and has the aroma of fresh rain. As you may recall from your high school chemistry class, the oxygen you breathe is O_2, which is a stable gas, while the three oxygen molecules bound together are unstable. Because it is highly reactive, ozone commercially must be produced onsite continuously or as needed. A process called corona discharge is used to generate ozone in small or large tanks. A high-voltage electrical spark is fired across a gap in the presence of oxygen or dry air to turn oxygen into ozone. The high-voltage alternating current excites oxygen electrons and induces splitting of oxygen molecules into atoms which combine with

Figure 28.3 A cartoon representation of the triatomic ozone molecule.
doi:10.1128/9781555817206.ch28.f28.03

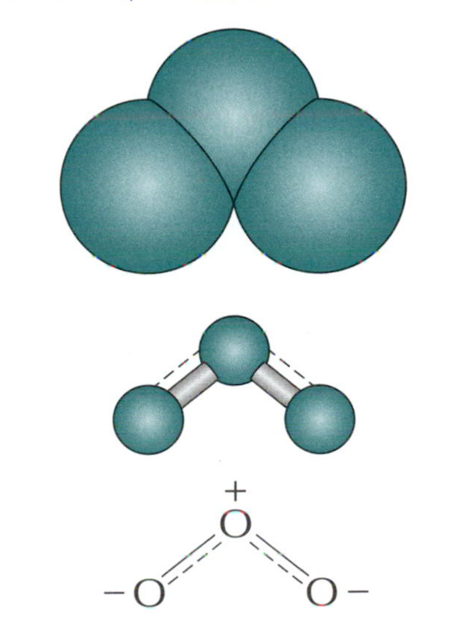

other oxygen molecules to produce ozone. Various generators can produce ozone at different rates and measured as parts per million.

Gaseous ozone is quite soluble in water and is more soluble at lower temperatures. This is good news for the majority of food-processing production sites, which are cool (~4 to 12°C) environments. Solubility of ozone gas in water used for food processing can be improved by the following seven parameters:

1. decreasing water temperature
2. increasing water purity
3. increasing the ozone concentration delivered
4. increasing the pressure of the gas above the water
5. decreasing the size of gas bubbles in the injection system
6. improving gas mixing and distribution
7. extending ozone residence time

In clean distilled water at room temperature (20 to 25°C), ozone is stable for only about 20 min. It reacts rapidly in the presence of organic material, including food, microorganisms, and even dissolved substances in water. This is quite different from pure ozone gas in the environment, which at the same temperature (25°C) has a calculated half-life of 19.3 years. The stability of aqueous ozone can be influenced by the presence of organics in the water. These organics are ozone-demanding materials, including metal ions and oxygen radical scavengers or antioxidants. Just as the water pH influences the stability of chlorine, water pH has an effect on the stability of ozone. The pH of wash water or flume water is constantly monitored; the disinfectant is also monitored and chlorine or ozone added as needed. At pH 5.0, ozone is more stable, with a loss of stability at increasing pH. In food-processing environments the water source is important. Water may contain easily oxidized dissolved organic and inorganic materials. By knowing the water quality in advance, a production supervisor can adjust the level of ozone delivered into the water. This would be part of the validation study for using a specific water treatment system. Ozone output and residual need to be monitored in any ozone treatment system. For water applications, the efficacy of treatment is commonly expressed as ozone concentration delivered in parts per million or milligrams per liter and contact time in minutes. The product of multiplication of these two parameters is the *Ct* value and indicates the concentration and time that are needed to inactivate a given microbial population. Recall the definition used by the NACMCF committee, which indicates that the ozone treatment process should be validated for the most pertinent pathogen of importance depending on the food.

The safety of ozone has been questioned in the past; however, modern generators have reduced these fears. Proper use of ozone generators does not put workers or the environment at risk. Interestingly, ozone has several applications in chronic infectious diseases, dentistry, and blood vessel disorders. At very high concentrations, ozone can be hazardous. It is important that the ozone be measured and delivered properly. Any treatment process should be validated to determine the efficacy of the process for a given food, place, and time. Some scientific studies have shown that ozone is 3,000 times faster than chlorine at killing bacteria and other microorganisms. Cost comparisons between ozone and chlorine often include the transport, cleanup, and storage of large amounts of liquid chlorine or chlorine gas, which all come with some risk of spillage or containment.

Ozone is an attractive disinfectant for many food industries, including fresh produce. The benefits include fast decomposition into simple oxygen because of the high reactivity of ozone. Thus, there are no safety concerns about residual ozone in the treated food products. The decomposition also prevents the accumulation of waste products in the environment. Ozone is an effective disinfectant against bacteria, spores (Box 28.1), fungi, protozoa, and viruses. Cellular damage has been observed by electron microscopy. The mechanisms of microbial inactivation by ozone are attributed to its oxidation reactions with cellular components, including lipid, protein, and nucleic acid. Reactive ozone yields many reactive oxygen species, including hydroxyl (·OH), superoxide anions ($\cdot O_2^-$), and hydroperoxyl ($\cdot HO_2^-$) radicals. An important benefit

BOX 28.1

Spores and ozone

Spore-forming bacteria, including *Bacillus* and *Clostridium* species, are especially problematic in the food industry due to their extreme resistances to heat and environmental pressures. When the environment surrounding the vegetative bacteria becomes less ideal or conditions become harsh, these bacterial species are able to convert into a dormant state in which the cell dehydrates and all metabolic processes cease. Through this survival mechanism, spore-forming bacteria are able to survive food-processing technologies, including cooking, freezing, drying, pressure, extreme pH, chemical disinfectants, infectious agents, and many others. After food is processed and conditions become less harsh, spores are able to germinate and convert back into growing cells which have the ability to grow in foods and ultimately cause foodborne illness.

Much research has been focused on the use of ozone and other oxidizing agents in food-processing techniques in order to eliminate the presence of bacterial spores in foods. Oxidizing agents are able to eliminate microbial pathogens, including viruses, bacteria and their spores, fungi and their spores, and protozoa via the activity of free radicals released by the oxidizing agent. Aqueous ozone, or ozone that is bubbled into water, has been observed to have a higher potential than most oxidizing agents to inactivate spores. Studies involving the inactivation of bacterial spores by oxidizing agents suggest that inactivation is a result of oxidative damage to a spore's inner membrane.

Damage to the spore's inner membrane via oxidization can have several effects, including inability to germinate, spore death after germination, and cell lysis. More interestingly, researchers have found that spore survivors of ozone treatment exhibit increased sensitivity to inactivation by a normally minimal heat treatment. This could mean that spores present in foods that have been pretreated with ozone may be more easily inactivated during the cooking process. Spores treated with ozone were also observed to be more sensitive to the presence of NaCl (salt) in laboratory media than nontreated spores. Therefore, spores may be more responsive to the presence of salt in foods when pretreated with ozone.

A graduate researcher prepares spores for ozone treatment.
doi:10.1128/9781555817206.ch28.fBox28.01

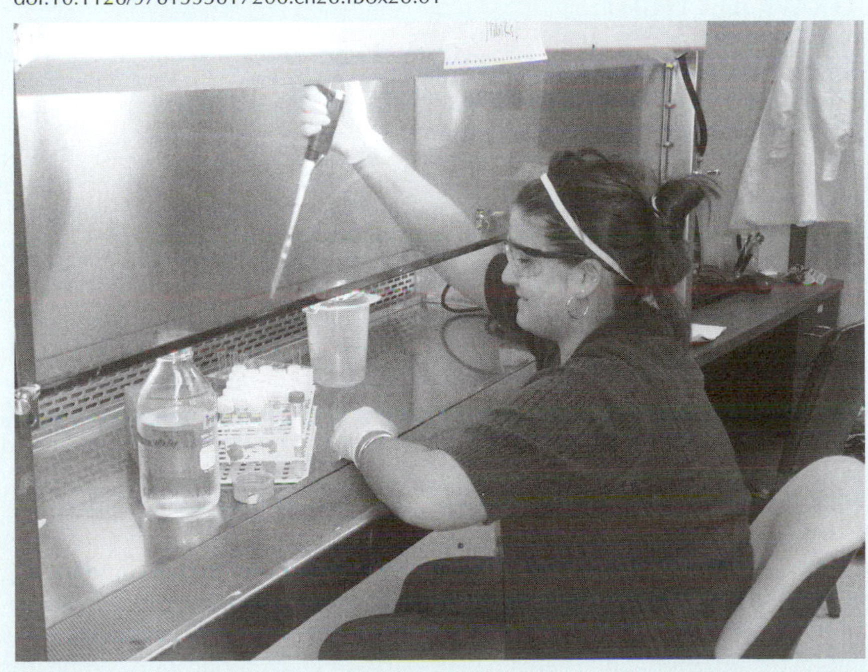

Markland, S. M., K. E. Kniel, P. Setlow, and D. G. Hoover. 2011. Characterization of superdormant spores of *Bacillus cereus* and *Bacillus weihenstephanensis*. M.S. thesis. University of Delaware, Newark.

of ozone is the ability to inactivate the protozoan parasites *Giardia*, *Cryptosporidium*, and *Cyclospora*. These three protozoa have caused many waterborne outbreaks and are resistant to normal levels of chlorine. Many water treatment plants across Europe and within the United States have started using ozone in order to reduce the risk of transmission of these protozoa. The mechanism of inactivation is extensive protein degradation of the (oo)cyst wall, resulting in great structural modifications that end the life cycle.

Ozone has been evaluated for use on several foods. For fruits and vegetables, ozone has been tested in aqueous solutions for use in wash or flume water. Aqueous ozone can be delivered by injection into water pipes, like chlorine is, or gaseous ozone can be pumped into a closed facility. In these systems, ozone reduced *Escherichia coli* O157:H7 on apples and the natural microflora of lettuce. Gaseous ozone was tested on fruits and vegetables during cold dry storage. This treatment suppressed fungal growth on blackberries, grapes, onions, potatoes, and sugar beets. For meat and poultry, ozone has been tested as an alternative disinfectant to chlorine for carcass washes, in poultry-processing chiller water, and on hatchery equipment. One drawback to using ozone to decontaminate beef or poultry is that the carcasses have an ozone demand due to the large amounts of proteins and fats. These rapidly react with the ozone before it can inactivate microorganisms that might be present. Gaseous ozone has been used to successfully decontaminate the outside of shell eggs, thereby reducing the risk of illness from handling eggs or by cross-contamination. Aqueous ozone successfully reduced the natural microbiota of fish fillets. Also, aqueous ozone was found to reduce levels of bacteria on live catfish entering a processing facility. Researchers have also reported use of gaseous ozone on spices, grains, and dried beans to reduce microbial contamination. As you can see, there are many ways to use ozone to reduce microbial levels on food commodities.

ULTRAVIOLET LIGHT

UV light has been used to disinfect water and process foods. UV-C-type light at a wavelength of 254 nm is used specifically for its disinfection properties (Fig. 28.4). UV light can be used to treat liquid and solid foods. UV light is fairly easy to use with foods, as it does not leave any chemical residues in the food. UV light has limited penetration. It can penetrate through water, but even turbid liquid solutions have limited penetration. It is best for surface disinfection or use with clear liquids. To get around this, some systems have been designed to treat flowing liquids in a thin layer. Many juice processors have used UV light to successfully fulfill juice HACCP requirements for the 5-log reduction of *E. coli* O157:H7 in fresh-pressed unpasteurized apple cider. UV treatment is an attractive alternative to pasteurization for its ability to retain the fresh-pressed flavor.

For successful inactivation of microorganisms, it is crucial to have a good lamp and to monitor intensity. Absorbed UV light energy is efficient at inactivating microorganisms through photophysical and photochemical means. The pyrimidine base pairs of nucleic acids (thymine in DNA, cytosine in DNA and RNA, and uracil in RNA) absorb UV light due to their aromatic ring structure. In this reaction, the production of photoproducts results in microbial inactivation. UV light readily induces thymine dimers in the DNA. This inhibits the unzipping mechanism necessary for DNA replication. While this is the main photoproduct, there are others, including cyclobutyl pyrimidine

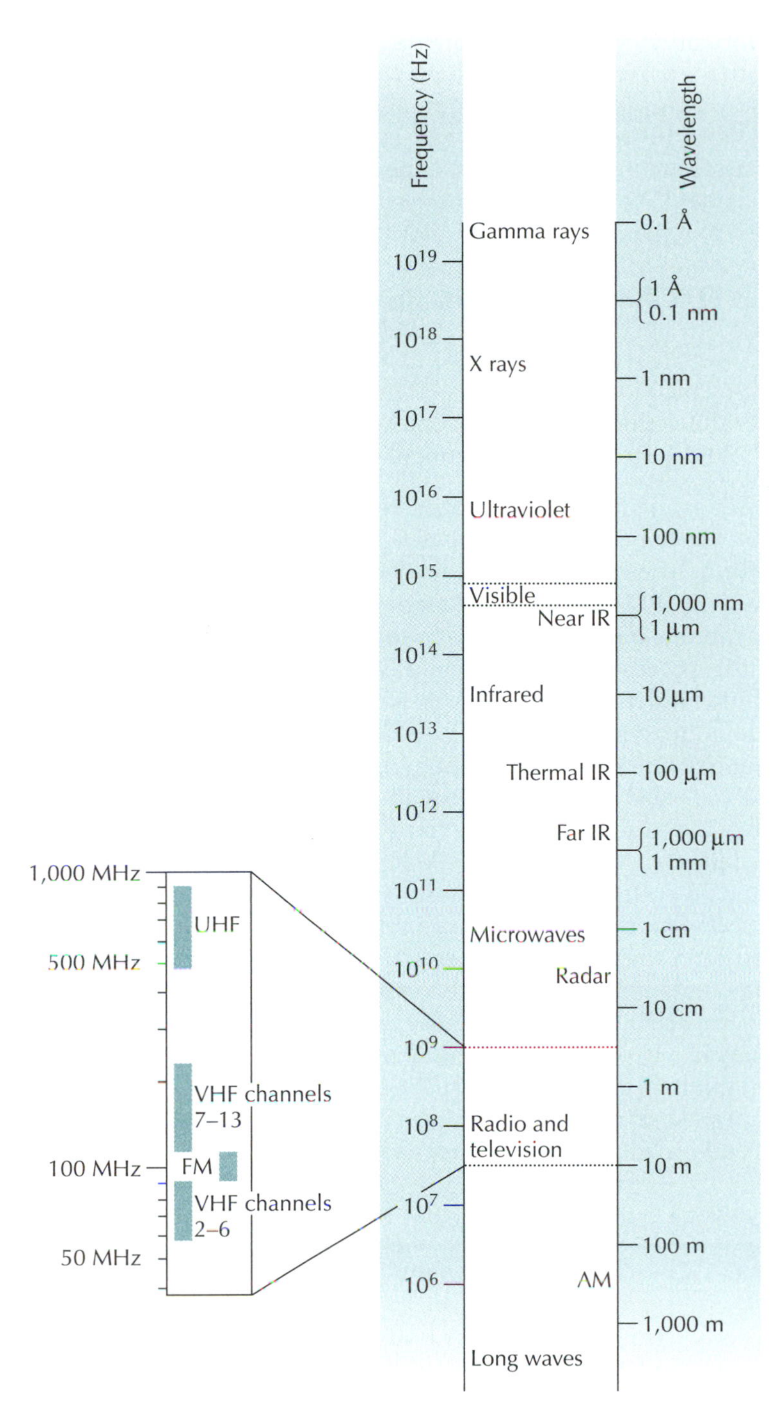

Figure 28.4 The electromagnetic spectrum is the range of all possible frequencies of electromagnetic radiation. Here you can see where visible light and UV light fit into the spectrum, as well as other familiar types of energy, including X rays. The visible-light spectrum is the color of the rainbow, with red at the low-frequency end, moving up to indigo and violet at the highest frequencies. Abbreviations: IR, infrared; UHF, ultrahigh frequency; VHF, very high frequency; FM, frequency modulation; AM, amplitude modulation. doi:10.1128/9781555817206.ch28.f28.04

dimer formation, single-strand breaks, and DNA-protein cross-links. These vary with the wavelength of the light and the sequence of the DNA. Photoreactivation occurs in the wavelength range of 330 to 480 nm due to activation of the DNA photolyase enzyme that can split thymine dimers. DNA repair mechanisms are more efficient in some bacteria than in others. Interestingly, the protozoan *Cryptosporidium parvum* appears to have this enzyme but does not use it to repair DNA damage after UV light treatment.

PULSED ELECTRIC FIELDS AND PULSED LIGHT

Interest in the utilization of electric fields to treat grape must, wine, and milk began in the 1930s. However, interest waned with the high energy costs during the 1940s. The increasing interest of consumers in fresh-like food products and the search for an environmentally friendly disinfection process brought PEFs back in favor. In PEFs, high-intensity electric fields are varied between 10 and 100 kilovolts (kV) per centimeter and the food passes through these currents. The temperature is increased minimally, about 3°C per 10 kV. This temperature change is often considered negligible. Microorganisms are destroyed by PEFs through electropermeabilization of their membranes. This membrane destruction leads to the organisms' death. Bacteria explode during this process, which inactivates between 5 and 9 log units of bacteria per treatment, depending on the food product. The potential exists now for continuous treatment systems. The PEF niche is for the application of pasteurization to liquid foods, including fruit juices and eggs. In terms of energy use, PEFs use about 80% less energy than thermal treatments.

What is the difference between PEFs and pulsed light? The latter is an irradiation system that, like UV light, does not have good penetration into the food system. It can surface disinfect a product. It is certainly effective at reducing microbial counts, including those of mold and fungi, on bread surfaces and packaging materials. The pulsed mechanism appears to reduce potential negative impacts on the foods while allowing destruction of microbial cells. Typically the pulse rate is 1 to 20 pulses per second, and a pulse may last from 300 ns to 1 ms. Compared to a continuous light system, pulsed light treatment is more effective at inactivating microorganisms due to the increase in energy in a shorter amount of time (Table 28.1). Pulsed light uses a broad spectrum of radiation.

Table 28.1 Comparative processing parameters for nonthermal inactivation of *E. coli*

Process	Parameters	Microorganism	Substrate	Log reduction
HPP	500 MPa, 15 min, 20°C	*E. coli*	Skim milk	6.5
	400 MPa, <1 min, 25°C	*E. coli*	Apple juice	8.0
Ozone	0.23–0.25 mg/liter, 1.67 min	*E. coli*	Water	4.0
	21–25 mg/liter, 3 min	*E. coli* O157:H7	Apple surface	~3.7
UV light	14 mJ/cm^2	*E. coli* O157:H7	Apple cider	5.0
Pulsed UV light	3 mJ/cm^2	*E. coli*	Liquid culture	4.3
PEFs	45 kV/cm, 64 pulses, 1.8–6 μs, 15°C	*E. coli*	Skim milk	3.0
	30 kV/cm, 43 pulses, 4 μs, 25°C	*E. coli*	Apple juice	5.0
OMFs	0.15 T (field strength), 0.05 Hz (pulse frequency)	*E. coli*	Liquid culture	2.0
Ultrasound	600 W, 28- to 100-kHz oscillations, 30 min	*E. coli*	Apple cider	<1.0
Cold plasma	0.44 W/cm^2, 25 s	*E. coli*	Liquid culture on polypropylene	4.0

Almost identical levels of inactivation were achieved by pulsed UV light using half the energy of continuous UV light. Pulsed UV light has potential for use in food processing. It is cost-effective due to the high-intensity short pulses, which can inactivate both vegetative cells and spores. The pulsed treatments have negligible effects on food quality compared to continuous UV. To better utilize this processing technology, more information is needed on pathogen inactivation in conjunction with pulsed-UV light lamp and treatment chamber design.

OSCILLATING MAGNETIC FIELDS

OMFs preserve food quality and destroy microorganisms in a short treatment time without a significant rise in temperature. A magnetic field is a force field that surrounds electric current circuits and arises from the movement of electrical charges. The potential usefulness of this technology is from the force on a stationary charged particle, the food product in this case, when the magnetic field is changing. Magnetic fields are generated when an electric current flows through a coiled wire. The magnetic field is more concentrated near the coil. The more loops of wire, the larger the cross section of each loop, so the current passing through the wire is increased and the field is stronger. Field intensities that are capable of inactivating microorganisms range from 5 to 50 teslas.

Authors' note

A tesla is a unit of measurement for magnetic fields, known as magnetic flux density.

Fields are made more intense with the inclusion of iron (or other paramagnetic or ferromagnetic materials) inside the coils. Iron cores can achieve fields with three times the intensity of air-filled coils. The coils are made of tiny superconductive filaments contained in a copper matrix which adds mechanical and thermal stability. The entire unit is placed in liquid helium, which acts as a coolant to keep the critical temperature low (around 4.2 K, which is equivalent to −269°C). Researchers are investigating different materials to use for the magnets.

Authors' note

The kelvin (K) is the unit of measurement of absolute temperature. Unlike the degree Fahrenheit and degree Celsius, the kelvin is not referred to as a degree or written with a degree sign. It is the primary unit of temperature measurement in the physical sciences.

Scientists first became interested in the effects of magnetic fields on biological processes in the early 1900s, when they observed an increase in protoplasmic streaming when a magnetic field was applied in parallel to the seaweed *Chara braunii*. OMFs cause disruptions in the enzymatic activity that is necessary for metabolic processes of microorganisms. Scientists have observed that microorganisms or tumor cells placed in an area of high-intensity magnetic fields are unable to reproduce. The theory is that the fields disrupt metabolism within the cells and thereby halt the growth of the microorganisms. This may be achieved as magnetic fields loosen the bonds between specific ions. The scientific effects and the reasoning behind them are still not well understood. Overall, magnetic fields have one of three different effects on microbial cells (yeasts and bacteria). These are (i) inhibitory, (ii) stimulatory, or (iii) no observed effect. For a food preservation treatment, the first effect of these three is the only beneficial one. Before this technology is considered for food preservation purposes, these methods must show consistent beneficial inhibitory results.

It is believed that OMFs can be used to preserve solid or liquid foods. Solid foods could be treated in sealed pouches and subjected to OMFs. Liquid foods could be pumped through a pipe in a continuous flow through the fields. OMF treatments are conducted at atmospheric pressure and with slight heat treatment to 2 to 5°C. In tests using potatoes, the bacterial microflora was inactivated more readily than fungi. Preliminary data suggest that OMFs can be useful as a processing methodology, but much more information is needed to fully understand the mechanisms involved, kinetics of inactivation, resistant pathogens, critical process factors, and effects of microbial inactivation.

ULTRASOUND

The use of high-frequency sound waves dates back to 1927, when it was observed that ultrasound accelerated the rate of conventional chemical reactions. There are two types of ultrasound: (i) high-frequency, low-energy, diagnostic ultrasound in the megahertz range and (ii) low-frequency, high-power ultrasound in the kilohertz range. The forces generated by ultrasound act on molecules, like microorganisms, in solution. The energy generated is not enough to break chemical bonds, but it results in an indirect phenomenon known as cavitation. Cavitation is the generation of bubbles in the liquid as the radiating wave energy propagates (Fig. 28.5). These bubbles go through two stages. They pulsate rapidly or expand to a critical size and collapse violently, sometimes called implosion. The activity of the bubbles amplifies the energy by more than 11 orders of magnitude. This level of energy is sufficient to break chemical bonds. Thermal conductivity effects play a role in the development of smaller bubbles. More energy is created as the bubbles collapse. In fact, collapsing cavities are like tiny microreactors, filled with heat and oxygen radical species, like hydrogen peroxide and hydroxyl radicals that can react with microbial cells.

Authors' note

It is not easy to imagine the physical processes behind ultrasound. Imagine the force of air bubbles dispersing through a jet in a whirlpool. Think about cavitation like the movement of bubbles, but these bubbles are on the near-microscopic level in size. As the bubbles implode, they release energy that can lyse microorganisms. Imagine how water through a whirlpool jet sloughs off dead skin cells.

These reactions cause cell disruption and lead to the inactivation of bacteria, molds, yeasts, and viruses. Microbial inactivation by ultrasound increases with increased temperature, called thermosonication. As described above, microbial inactivation occurs through cavitation effects on microbial cell walls through the interaction with oxygen radicals. Bacterial spores and gram-positive bacteria are more resistant to inactivation by sonication than are vegetative cells and gram-negative bacteria. Like HPP, ultrasound can reduce the activity of enzymes in foods. This is due to the depolymerization of macromolecules. Ultrasound is best known for its use in cleaning and surface decontamination. The cavitation effects are stronger than the adhesion forces (van der Waals attraction) holding particles to the surface. Surface disinfection using ultrasound may be useful in produce and poultry washing or in the removal of biofilms.

Ultrasonic bubbles can be generated by small or larger wands or rails that are submerged as part of a larger system. Ultrasound may be useful in the processing of liquids, like fruit juices or liquid eggs. Ultrasonic interactions with food components like proteins and carbohydrates are complex and depend on the ultrasonic intensity, duration, and environmental conditions. For example, ultrasound may affect protein denaturation by causing changes in the tertiary protein structure. This may be advantageous or negative depending on the food product. Ultrasound may be used to assist extraction of food components. Ultrasound applications in the wine and beverage industry are growing. High-power ultrasound equipment has large continuous-flow treatment chambers, with a flow of between 5 and 15 liters per minute. Larger systems can be generated by using a sequence of flow cells. Current ultrasound systems have energy efficiencies of 85%, which shows that the energy generated by the transducer (responsible for creating an alternating electrical field) is nearly all delivered into the medium. From this brief description, you can see the potential use of ultrasound and understand a bit about how it works. The physical processes of

Figure 28.5 Cavitation after water drop impact. Image courtesy of J. J. Harrison. doi:10.1128/9781555817206.ch28.f28.05

cavitation as ultrasound waves transmit through liquids and the energy generated by the implosion of tiny gas bubbles are more complex than one would first imagine!

CONCLUSIONS

The nonthermal processing methods discussed here all have the potential to enhance microbial food safety by inactivating microorganisms while protecting food quality. In general, microbial inactivation varies depending on the food substrate. There are a myriad of examples where food producers have benefited from the inclusion of the nonthermal technologies discussed within this chapter. No doubt there will be many more. A few examples include the following:

- use of ozone in surface wash water at meat-processing plants replacing organic acids at a cost savings of >$100,000 annually
- healthier broilers in poultry houses due to treatment of drinking water with ozone to reduce microbial contamination
- long-lasting high-quality fresh-like guacamole treated with HPP to inactivate the polyphenol oxidase that causes enzymatic browning
- sliced deli meats treated with HPP and free of nitrites and other preservatives
- PEFs used to produce fruit juices with enhanced fresh-like sensory properties
- family farms continuing to be able to sell apple cider; after treatment with ozone, the cider is comparable in safety to pasteurized juice but retains its "farm fresh" characteristics.

While food microbiologists are typically vested in food safety issues, food scientists should also focus on issues of world hunger, food security, and sustainability. It has been speculated by the American Dietetic Association that current agricultural production worldwide could be enough to feed everyone on Earth, but issues of access to food are critical. Some scientists feel that nonthermal technologies like high-pressure and pressure-assisted thermal sterilization may be used for the production of humanitarian daily rations that have long shelf lives. These could be similar to the ready-to-eat meals provided to U.S. military personnel.

Summary

- Foods made using nonthermal processing technologies retain the fresh-like characteristics absent in thermally processed foods.
- Under the appropriate conditions, these nonthermal processing methods can inactivate microorganisms (Table 28.1) and sometimes enzymes as well depending on the process and dose. This is certainly a great benefit for the use of HPP at specific doses.
- Ozone (O_3) is a highly reactive and useful disinfectant. Ozone is applicable for use in water disinfection and in food production, including produce wash water.
- Surface disinfection is an important area of use and of development for ozone, UV, and ultrasound.

- Novel treatment methods like cold plasma, OMFs, and ultrasound may be useful in the future once we understand how to better implement them in food production.
- For each treatment process, there are foods that are good candidates and those that are not. This may depend on moisture content, air, pH, fluidity, turbidity, shape, and other physical and chemical parameters.
- Scientists continue to work to improve food safety by enabling the food industry to make better decisions about how to utilize nonthermal processing methods to reduce the risk of foodborne illness. Vast research and development into novel nonthermal processing methods continue.

Suggested reading

Zhang, H. Q., G. V. Barbosa-Cánovas, V. M. Balasubramaniam, C. P. Dunne, D. F. Farkas, and J. T. C. Yuan (ed.). 2011. *Nonthermal Processing Technologies for Food.* IFT Press, Wiley-Blackwell, Ames, IA.

Questions for critical thought

1. Why do scientists want to explore the use of nonthermal processing for food production?
2. Describe the physical and chemical reactions that are involved in nonthermal processing methods.
3. The packaging used in HPP is important. While this was not described within this text, can you discuss important parameters for efficient packaging? (These may be flexibility, moisture resistance, size, or shape. Would metal cans work well in HPP?)
4. There are several parameters that can make ozone a more effective disinfectant for food processing. You are a processor in the fresh-cut industry. You want to use ozone and get the most disinfectant properties from the gaseous ozone you are mixing in your wash water. How might you engineer your production environment to be successful in using ozone?
5. Describe which novel nonthermal processing technology you think will be the most useful treatment method in 25 years. Or do you think another method that is not described here will fit this description?
6. How do pulses and oscillations enhance treatment efficiency and microbial disinfection over continuous-type treatments (light, electric fields, and magnetic fields)?
7. You are interested in the preservation of salad ingredients, including salad dressing, so that you can ship them across the country. To increase the shelf life and retain fresh-like characteristics, you want to use novel nonthermal technologies. How will you choose to process your product? Is there a benefit to using specific technologies for specific foods?
8. Compare and contrast the advantages and disadvantages of HPP, UV light, ozone, PEFs, OMFs, cold plasma, and ultrasound.

29

Sanitation and Related Practices

LEARNING OBJECTIVES

The information in this chapter will enable the student to:

- recognize the link of Hazard Analysis and Critical Control Points (HACCP) and food safety objectives (FSOs)
- discuss the benefits of sanitation, good manufacturing practices (GMPs), and the HACCP system with respect to human safety and food quality
- outline the process for development of a HACCP program
- recognize the limitations of HACCP
- outline the basic concepts of GMPs
- discuss the proper application of sanitizing agents
- integrate sanitation, GMPs, and HACCP
- identify practices that can result in the introduction of chemical, biological, or physical hazards into food
- differentiate FSOs from performance objectives
- identify segments of the food industry for which HACCP is mandatory

INTRODUCTION

Most people in the food industry are aware of Hazard Analysis and Critical Control Point (HACCP) systems. Indeed, few students majoring in food science have not heard of HACCP by the end of their first year of college. This chapter focuses primarily on HACCP; however, before a HACCP program can be introduced or developed, basic building blocks must be in place. Specifically, in an industrial setting, implementation of good manufacturing practices (GMPs) and a sound sanitation program are key elements. Food safety objectives (FSOs) were introduced by the International Commission on Microbiological Specifications for Foods in 2002. Programs including HACCP will provide the evidence that FSOs are met. The implementation of GMPs and of sanitation and HACCP systems is perhaps even more important now to address public concern that food supplies are safe and secure from intentional contamination by terrorists.

FOOD SAFETY OBJECTIVES

Consumers expect that food processors employ measures to ensure the safety of the food they consume. FSOs are the maximum frequency and/or concentration of a microbiological hazard in a food at time of consumption that is considered acceptable contributing to the appropriate level of

doi:10.1128/9781555817206.ch29

protection. FSOs can be set for any food hazard, chemical or biological; in this chapter, only hazards of microbial origin are considered. FSOs and performance objectives (POs) can be used by a government agency (e.g., the Food and Drug Administration [FDA] or U.S. Department of Agriculture [USDA]) or other authority to communicate food safety levels to industry and to other governments. Basically, the FSO functions as a tool to translate a public health goal (e.g., level of aflatoxin in peanuts) to measurable attributes.

The ability of a food processor to use one processing technique over another is not set by the FSO. Indeed, the goal is to ensure that the maximum hazard level specified at consumption is not exceeded. Different countries may use different technologies or adopt new technologies more rapidly for processing a given food, for example, milk. Milk is generally pasteurized, a thermal process, to render it safe; other technologies that do not involve thermal processing, including irradiation and high hydrostatic pressure, may also be used. Evaluation of techniques to determine equivalence in reaching a particular level of safety is required to ensure consumer protection and to facilitate fair trade.

FSOs must include a number of attributes: they should be quantitative, verifiable, and technically feasible. FSOs can then be incorporated into principles of both HACCP and GMPs. Hypothetical FSOs include aflatoxin in peanuts not exceeding 15 μg/ml, *Listeria monocytogenes* in ready-to-eat foods not exceeding 100 colony-forming units (CFU)/g, and *Salmonella enterica* serovar Enteritidis in eggs not exceeding 1 egg per 100,000. The potential microbiological hazards associated with a food may represent such a low risk that an FSO is not needed. Examples of such foods include carbonated beverages, most breads, granulated sugar, and pineapple. FSOs are typically very low, since they represent the maximum level of a hazard at the point of consumption. FSOs are not the same as microbiological criteria (see chapter 6). HACCP systems provide the most reliable evidence that FSOs (and POs) are being achieved, since measuring this level is impossible in most cases.

Authors' note

Governments use FSOs and POs to communicate public health goals to be met by food processors by HACCP.

GOOD MANUFACTURING PRACTICES

In 1969, the FDA published regulations on GMPs that provide basic rules for food plant sanitation. In 1986, the FDA published a revised version that expanded the original regulations. The GMPs can be found in the *Code of Federal Regulations* under Title 21, part 110, as "Current Good Manufacturing Practices in Manufacturing, Packing, or Holding Human Food" (referred to hereafter as CGMPs). To clarify what the *Code of Federal Regulations* contains, the following explanation may help. Proposed regulations are first published in a daily publication, the *Federal Register*, setting forth rulings by the executive branch and agencies of the federal government. These are then codified and published as the *Code of Federal Regulations*. The *Code of Federal Regulations* is divided into 50 titles that represent areas that are subject to federal regulation.

The CGMPs are divided into seven subparts, each detailing requirements related to various operations or areas in food-processing facilities. The requirements are established to prevent contamination of a product from direct and indirect sources. Subparts D and F are reserved; therefore, only subparts A to C, E, and G are discussed here.

General Provisions (Subpart A)

Subpart A contains a list of definitions, the underlying reason for CGMPs linked to the Food, Drug and Cosmetic Act (referred to hereafter as the Act); responsibilities imposed on plant management regarding personnel; and exclusions. Definitions are provided to maintain consistency; however, they are often vague and open to interpretation. The criteria and definitions used to determine whether a food is *adulterated* (made unfit for human consumption by the addition of impurities) are covered under sections 402(a)(3) and 402(a)(4) of the Act. A food is considered adulterated if it contains any filth or putrid and/or decomposed material or is otherwise unfit for use as food. The Act goes on to state that food is adulterated if it has been prepared, packed, or held under unsanitary conditions; may become contaminated with filth; or has been rendered injurious to human health.

Criteria for disease control, cleanliness, and the education and training of personnel are included. These requirements are designed to prevent the spread of disease-causing microorganisms from workers to food contact surfaces and the food itself. Workers in direct contact with food or areas that indirectly contact food must conform to hygienic practices. These include maintaining personal cleanliness, wearing clean clothing, removing jewelry, washing hands, and wearing appropriate protective garments (gloves, hairnets, and beard covers). Personnel responsible for plant sanitation should receive education and exhibit a level of competency so that they can pass the information on to workers in a facility.

The exclusions section holds exempt food establishments that are engaged solely in the harvesting, storage, or distribution of raw agricultural commodities [section 201(r) of the Act defines this as any food in its raw or natural state, including all fruits that are washed, colored, or otherwise treated in their unpeeled natural form prior to being marketed] that are ordinarily cleaned, prepared, treated, or otherwise processed before being marketed.

Buildings and Facilities (Subpart B)

Subpart B covers regulations for plant and grounds, which state that the grounds of a food plant should be maintained in a condition that will protect against the contamination of food. The plant design and construction must be such that operations within a plant can be separated and that materials used for walls, floors, etc., facilitate cleaning and sanitizing. The sanitary operations section covers general requirements related to general maintenance, cleaning and sanitizing compounds, toxic-material storage, pest control, sanitation of food contact surfaces, and storage and handling of cleaned equipment and utensils. The last section covers sanitary facilities and controls and provides minimum requirements for sanitary facilities and accommodations. Facilities and accommodations include, but are not limited to, the water supply, plumbing, sewage disposal, toilet facilities, hand-washing facilities, and disposal of rubbish and offal (animal parts that are considered waste, generally applied to meat- and poultry-processing plants).

Equipment (Subpart C)

The equipment and utensils section specifies that the materials used and the design of plant equipment and utensils must allow adequate cleaning and proper maintenance. Emphasis is placed on preventing the adulteration of food by microbial contamination or by other contaminants, such as lubricants or metal fragments. Moreover, equipment used in the manufacturing, holding, or conveying of food must be designed and constructed for easy cleaning and maintenance. Cooling equipment (freezers and cold storage) and control instruments (e.g., pH meters and thermometers) must be accurate and adequately maintained. Finally, compressed gases that are introduced into food or used to clean food contact surfaces must be free of contaminants that are considered unlawful indirect food additives.

Production and Process Controls (Subpart E)

Subpart E is the most detailed subpart and is broken down into several sections and subsections that contain rules to ensure the fitness of raw materials and ingredients, to maintain manufacturing operations, and to protect finished products from deterioration. The processes and controls section stresses that all operations must be conducted in accordance with adequate sanitation principles. Quality control programs must be in place to ensure that food is suitable for human consumption and that packaging materials are safe and suitable. Quality control sanitation programs are required to ensure compliance with the Act.

The subsection on raw materials and ingredients contains descriptions for how those materials should be inspected, handled, washed, and segregated to minimize contamination and deterioration. Materials must not contain microorganisms or must be treated during manufacture to ensure that the levels of microorganisms that produce food poisoning or other human illness no longer present a problem and do not render the food adulterated. Compliance with regulations for natural toxins and other deleterious (harmful) substances is of the utmost importance and should be verifiable. Holding of raw materials must be adequate to protect against contamination and must be consistent with requirements for that material. Therefore, frozen materials must be kept frozen, dry materials must be protected against moisture and temperature extremes, and material scheduled for rework must be identified as such.

The handling of food in an appropriate way to prevent contamination is covered in the subsection on manufacturing operations. Required conditions are listed for the handling of foods that support the rapid growth of undesirable microbes, particularly those that cause foodborne illness. Requirements to protect several types of foods, for example, batters, sauces, and dehydrated foods, from contamination or adulteration are stated. Various operations, such as mechanical manufacturing (washing, cutting, and mashing), heat blanching, and filling, assembling, and packaging, must be performed to protect against contamination. To facilitate the tracking process, companies should code individual consumer packages with appropriate information (most companies routinely code each package). This is also important in the voluntary recall of products that are in violation of food laws and regulations.

The final section of subpart E, warehousing and distribution, is brief but nonetheless extremely important. The section states that the storage and transport of finished food must be conducted under conditions that prevent physical, chemical, and microbial contamination. The conditions must also prevent deterioration of the food and the container.

DALs (Subpart G)

The section of subpart G covering natural or unavoidable defects in food for human use that present no health hazards was established because it is not possible, and never has been possible, to grow (in open fields), harvest,

In the absence of good agricultural practices, field crops can be contaminated by bacteria in manure and then, when eaten, can make people sick.
Reprinted with permission from CartoonStock.

and process crops that are totally free of natural defects. Certain defects that present no human health hazard, particularly in raw agricultural commodities, may be carried through to the finished product. The defect action level (DAL) represents the limits at or above which the FDA may take legal action to remove a product from the consumer market. The fact that the FDA has established DALs does not mean that a manufacturer need only stay below that level. Manufacturers cannot mix or blend a product containing any amount of defective food at or above the current DAL with another lot of the same or another product. This is not permitted by the FDA and renders the food unlawful regardless of the DAL of the finished food. Companies cannot use DALs to mask poor manufacturing practices. The FDA clearly states that the failure to operate under GMPs will result

Quantitative Microbial Risk Assessment of Foods

DONALD W. SCHAFFNER, Department of Food Science, Rutgers University, New Brunswick, New Jersey

Microbiological risk assessment is an emerging tool for evaluating the safety of food. This text box only introduces the topic; if you want to learn more, there are entire books on the subject (see reference). Risk assessment is often described as being one part of risk analysis. Risk analysis is described as a process composed of three components. Those components are risk assessment (the systematic, scientific evaluation of known or potential risks), risk management (the inherently nonscientific process of weighting and then selecting various policy alternatives), and risk communication (the process by which information is exchanged about risk, among interested parties).

Risk assessment is often further broken down into four components: hazard identification, exposure assessment, hazard characterization, and risk characterization. The relationship among all these parts can be seen in the figure. The original definition of risk assessment dates back to a seminal 1983 document published by the National Academy of Sciences (NAS). The NAS document reviewed the state of management of the risk assessment process in the federal government. The NAS four-component definition (hazard identification, exposure assessment, hazard characterization, and risk characterization) has since been adapted to a wide variety of risks (pesticides, air pollution, hazardous waste, cellular telephones, nanomaterials, etc.).

The Codex Alimentarius Commission develops food standards and guidelines to protect the health of consumers and ensure fair food trade practices. It adapted this risk assessment definition to specifically address microbial hazards in food. The Codex Alimentarius Commission defines the different components of microbial risk assessment as follows: hazard identification is a qualitative process to identify hazards in a food or group of foods; exposure assessment is a qualitative or quantitative evaluation of the chance that a microbial hazard is ingested and causes an adverse health effect; hazard characterization describes the adverse effects of microbial ingestion, using a dose-response function if available; and finally, risk characterization integrates the three previous steps to obtain an overall risk estimate.

In the last decade, academic researchers, national regulatory agencies, and international expert bodies have developed a number of microbial risk assessments for various pathogen-product combinations (e.g., *Escherichia coli* O157:H7 in ground beef and *Vibrio parahaemolyticus* in raw oysters). A microbial risk assessment may consider a single stage (e.g., retail) or multiple stages (farm to fork) in the supply chain. The primary outcome of a microbial risk assessment is an estimate of the chance of illness from consumption of the food product in question. Once the risk model is developed, scenario analysis can be used to see the impact of control measures and mitigation strategies on the overall risk to consumers. A sensitivity analysis can also be performed to identify factors that have the most significant influence on the risk. Inputs and outputs in quantitative microbial risk assessment are usually represented by statistical distributions instead of single point values. This means that a typical input might be the average concentration of a pathogen on a raw material, e.g., 1 CFU or cell per 100 g, but that this concentration is known to vary by ±1 log CFU. The use of such distributions provides a more realistic and accurate assessment of risk than one that is based on a single point value input such as the mean or the worst-case value.

The very first national government and peer-reviewed microbial risk assessments for food were published in 1998. In 12 short years an entire scientific subdiscipline has developed. The communities of quantitative food microbiologists used the framework developed for chemical risk assessment to develop dozens of federal and international microbial risk assessments. These are continuously being revised as additional data are collected and new modeling approaches are developed. Microbial risk assessment not only is a useful tool for global food safety but also is used by the food industry as a means to understand and manage risk.

Schaffner, D. W. (ed.). 2008. *Microbial Risk Analysis of Foods*. ASM Press, Washington, DC.

The relationship among the parts of the risk assessment process, risk communication, and risk management. doi:10.1128/9781555817206.ch29.fGuestBox

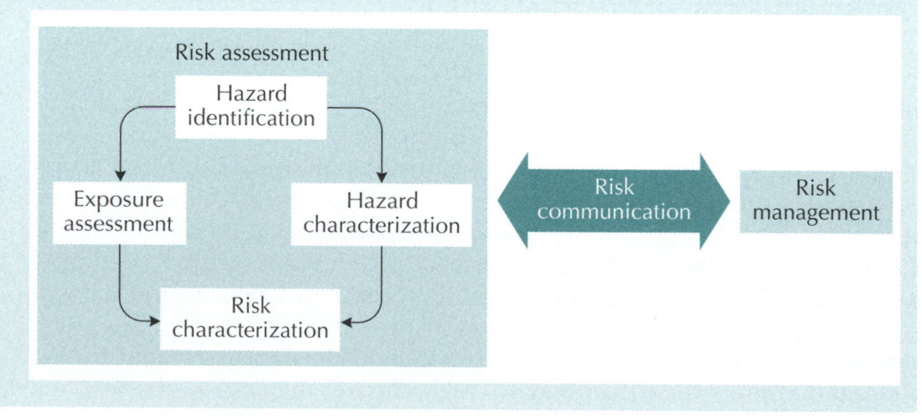

in legal action even though products might contain natural or unavoidable defects at levels lower than the currently established DALs.

An extensive list of DALs is available for products from allspice to wheat flour. The defect may be mold, rodent filth, insects, rot, or mammalian excreta. Below are a few examples of DALs (listed as product, defect, and action level).

- Apple butter: rodent filth; average of four or more rodent hairs per 100 g of apple butter
- Cocoa beans: mammalian excreta; mammalian excreta at 10 mg or more per pound
- Fig paste: insects; >13 insect heads per 100 g of fig paste in each of two or more subsamples
- Peaches, canned and frozen: moldy or wormy; average of 3% or more of fruit by count is wormy or moldy
- Potato chips: rot; average of 6% or more of pieces by weight contain rot

Although not specified or required in the CGMPs, maintaining accurate records to document compliance with laws and regulations is implied. Records from all phases of the food operation should be maintained to verify that raw materials, food-packaging materials, and the finished product comply with the provisions of the Act. Records related to processes used to destroy or prevent the growth of microorganisms of public health concern are also important. To facilitate tracking of finished products and recalls, records related to distribution must be maintained. Record keeping also aids in the development of a HACCP program.

SANITATION

Many students, and for that matter food plant workers, incorrectly consider sanitation to be only the removal of trash and the cleaning of equipment. However, sanitation includes a host of activities that are ultimately designed to prevent the adulteration of a product during processing. The true goal of any sanitation program is to protect the health of the consumer. In addition, a sound sanitation program increases the shelf life of a product, thereby minimizing economic loss, and prevents contamination of a product with materials that may offend the consumer (e.g., pieces of wood, rocks, or machinery). During the past 50 years, significant advances have been made in productivity and the ways that foods are processed. These advances have enhanced the safety and shelf life of products, but they require strict sanitation programs to ensure that contamination does not occur.

Before sanitation can be discussed, it must be defined with respect to the food industry. *Sanitation* is the creation of a comprehensive program designed to control microorganisms in a food-processing plant. A sanitation program must be based on sound science to ensure that healthy workers handle food in a clean environment. The goal of the program is to prevent contamination of food with pathogens that cause human illness and to limit the introduction and growth of microorganisms that result in spoilage. To achieve this goal, personal and environmental hygiene must be practiced. Sanitation is considered an applied science because of the human health issues that could arise should a breakdown in sanitation practices occur.

Within the sanitary-operations section of the CGMPs, general requirements are covered for the maintenance of buildings, fixtures, and other physical facilities; pest control; sanitation of food contact surfaces; and storage and handling of cleaned equipment and utensils. Depending on the product, the USDA may have jurisdiction over food-processing facilities, specifically those associated with meat, poultry, and eggs (according to the Federal Meat Inspection Act, Poultry Products Inspection Act, and Egg Products Inspection Act). Other agencies may also become involved. For instance, the Environmental Protection Agency may be involved if issues arise over water pollution (Federal Water Pollution Control Act) or air pollution (Clean Air Act) or through the use of pesticides (Federal Insecticide, Fungicide, and Rodenticide Act).

The benefits of sanitation can be numerous, but first and foremost sanitation aids in protecting the consumer from illness. An effective sanitation program facilitates the development of an effective HACCP program. Inspections of food-processing plants, not only by federal inspectors but also by other companies (e.g., if company A buys a product from company B for use in a company A product), are becoming more stringent. An effective sanitation program can limit the number of negative consumer contacts related to food spoilage (a product has an off-odor or -flavor) and foodborne illness. These types of problems weaken consumer confidence, causing decreased sales and increased claims. A stringent sanitation program can increase the storage life and shelf life of food, minimizing economic loss. Regular cleaning and maintenance of processing equipment and heating and cooling systems permit more efficient operation and reduce the microbial load in the facility and the product. In fact, the spread of microbes from cooling systems has been linked to the contamination of food and outbreaks of foodborne illness. Overall, a sound sanitation program can increase the quality of the product, improve customer relations, and increase the trust of inspectors associated with regulatory agencies.

Before sanitation standard operating procedures (SSOPs) can be developed, a sanitation audit must be conducted. A review of the facilities aids in identifying areas of noncompliance and establishing corrective actions. A systematic approach should be used when conducting an audit. Starting at the receiving area, check food materials (ingredients) for temperature, odor, color, and appearance. Next, move to the general plant area and inspect walls, floors, drains, and footbaths. Processing equipment, tanks, pumps, pasteurization charts, and retort charts can also be inspected. While moving through the plant, observe employee practices and determine compliance. Check for employees who appear ill, have open lesions, are wearing jewelry, or are walking in restricted areas; check for the use of protective devices (hairnets, beard nets, and gloves) and footbaths. The building and the area around the outside of the building must also be checked for signs of pests, broken windows, a damaged roof, open doors and windows, standing water, and trash. Recall that sanitation is more than trash removal.

Microbes can be brought into processing facilities on raw materials or picked up by food during processing as it passes through or over handling equipment. Employees can spread microbes to equipment and to food. Microbes from process waters during washing, conveying, or preparation can contribute to the microbial load of the product. Microorganisms can also be present in ingredients that may be added to the product. This is particularly important if ingredients are added after a process designed to reduce or eliminate microbes has been completed.

Sanitization (using germicidal or sanitizing agents) of food contact surfaces is the key to maintaining a sanitary food-processing environment, thereby reducing the number of microorganisms associated with a food. Sanitization involves the treatment of a food contact surface so that the process effectively destroys vegetative cells of microorganisms capable of causing human illness and adequately reduces populations of other microbes. The process does not need to reduce populations of bacterial spores. The treatment should not have an adverse impact on the food (impart an off-odor or -color) or the safety of the consumer.

The main criterion for selecting a chemical sanitizing agent is the ability of the agent to kill microorganisms of concern within a specified time. While chlorine or other chemical sanitizing agents can be extremely effective in reducing the microbial load, they must be used appropriately. In food-processing facilities, chlorine compounds are the most widely used sanitizing agents. A number of factors can reduce or effectively inactivate sanitizers. Sanitizers can react with organic materials (soil, food, bacteria, oils, etc.) that have not been removed from equipment surfaces, reducing the effectiveness of the sanitizer. Impurities in water, including iron, manganese, nitrites, and sulfides, can react with a sanitizer and decrease its effectiveness. Although temperature is generally not a factor, temperatures below 10°C and above 90°C can dramatically influence the availability and action of some sanitizers. Generally, increasing the concentration of a sanitizer increases the rate of destruction of the microorganisms. However, in the case of hypochlorites (e.g., sodium hypochlorite), this is not the case. For example, the addition of hypochlorite to water to give 1,000 parts per million (ppm) of chlorine produces a solution that requires three times as long to kill the same number of bacteria as a solution containing 25 ppm. This is because hypochlorites are produced by combining chlorine with hydroxides. The addition of hypochlorite results in the addition of a proportional amount of alkali that raises the pH of the solution. A 1,000-ppm solution of chlorine has a pH of ~11.0, whereas a 25-ppm solution has a pH of 8.0. In general, the effectiveness of chlorine and iodine compounds decreases as the pH increases.

Chlorine sanitizers may be the most popular sanitizers, but a number of other chemical sanitizers are available that are more useful than chlorine compounds for certain applications. Iodophors are good as skin disinfectants, since they do not irritate the skin at recommended levels of use. They can be used to clean equipment but may stain epoxy and polyvinyl chloride and may cause discoloration of starchy food. Hydrogen peroxide is effective against biofilms and may be used on all types of surfaces (equipment, floors, and walls). Ozone, glutaraldehyde, and quaternary ammonium are also effective. The efficacies of these compounds against certain microbes differ from those of chlorine compounds. For example, quaternary ammonium sanitizers are effective against molds and gram-positive bacteria but are ineffective against most gram-negative bacteria.

Sanitizing can be accomplished without the use of chemical agents. Ultraviolet (UV) irradiation is used to destroy microorganisms in hospitals. In Europe, UV light units are commonly used to disinfect drinking water and food-processing waters. Steam can be used, but it is costly and not particularly effective. In fact, condensation from the application of steam may complicate the cleaning process and provide moisture for surviving microorganisms. Hot water may be useful for sanitizing small components (eating utensils and small containers), but like steam, the process is expensive because of high energy costs.

A sound sanitation program results in the manufacture of foods that are safe and wholesome and that meet legal obligations. Sanitation is the responsibility of every person in a food-processing plant. Sanitation must be practiced every day, a policy the food company must reinforce through proper training and encouragement of employees. If every person understands his or her responsibility in a sanitation program, individual efforts will ensure that a consistent product is produced and consumer confidence is maintained. If executed properly, sanitation should reduce (perhaps eliminate) concern about the spreading of human diseases and food poisoning. Moreover, if sanitation is maintained, a product with a reduced microbial population will be produced and waste and spoilage will be minimized.

SSOPs

SSOPs are guided by CGMPs detailing a specific sequence of events necessary to perform a task to ensure sanitary conditions. To prevent direct product contamination, SSOPs should contain a description of procedures that a food-processing facility will follow to address the elements of preoperational and postoperational sanitation. A set of written SSOPs must be developed and maintained by federally inspected and state-inspected meat and poultry plants.

The USDA Food Safety and Inspection Service (FSIS) felt that SSOPs would permit each company to consistently follow effective sanitation procedures, thereby reducing the likelihood of direct product contamination or adulteration. The SSOPs must include information related to daily preoperational and operational procedures for the plant personnel responsible for monitoring daily plant sanitation activities. Companies must determine whether the SSOPs are effective and, if not, take corrective action. Corrective actions include procedures to ensure the appropriate disposition of contaminated products, restoring sanitary conditions, and preventing the recurrence of direct contamination or product adulteration. SSOPs written for meat and poultry plants must be signed and dated by the individual with overall authority over the establishment to ensure that the company will implement the SSOPs.

Although written SSOPs are not required for all food production establishments, they are the cornerstones of HACCP. SSOPs go beyond specific processes. HACCP plans are developed as a proactive approach to ensure the safety of a product. Interfacing CGMPs and sanitation (SSOPs) permits the development of an effective HACCP program.

HACCP

The development of HACCP programs by companies has become routine since the mid-1990s. In some instances, the FDA and the USDA FSIS have published final rules mandating the development and implementation of HACCP programs to enhance the safety of domestic and imported seafoods, juices, and meat and poultry. In fact, HACCP can be used by all segments of the food industry to effectively and efficiently ensure farm-to-table food safety in the United States (or, for that matter, in any country). HACCP is a proactive, prevention-oriented program that addresses food safety through the analysis and control of biological (predominantly pathogenic microorganisms), chemical (including those naturally produced by microorganisms), and physical (metal, wood, and plastic) hazards. The hazards may come from raw material production or the handling, manufacturing, distribution, and consumption of the finished product.

A food processor must identify potential sources of food safety hazards for each product and production line (process). Preventative steps can then be identified to reduce or eliminate potential hazards. Since this is a proactive approach, the manufacturer determines in advance what to do with a product or process should a preventative step fail. This requires continuous monitoring and documentation to verify that the appropriate corrective action was taken. Records can be reviewed to aid in improvement of the plan. HACCP is a continuous system, and HACCP plans require updating as ingredients and processes change and as new technologies emerge.

Authors' note

A firm foundation for development of a solid HACCP system is strict adherence to GMPs and SSOPs. Without implementation of GMPs and SSOPs, an effective HACCP system cannot be achieved.

Background

The HACCP concept was developed nearly 50 years ago by the National Aeronautics and Space Administration (NASA) and Natick Laboratories as the Failure Mode Effect Analysis system for use in aerospace manufacturing. Through collaborative work by the Pillsbury Company, NASA, and Natick Laboratories, the process was applied to foods produced for the U.S. space program. This was done to ensure that the food used would be 100% free of bacterial pathogens. Could you imagine being in space and developing a case of projectile vomiting and severe diarrhea? This initial work led to what is now HACCP.

Although HACCP was introduced to food-manufacturing companies in the early 1970s, the concept was met with limited interest. However, in 1985, a subcommittee of the Food Protection Committee of the National Academy of Sciences made a strong endorsement of HACCP. The National Advisory Committee on Microbiological Criteria for Foods was formed in 1988 and given the mission of developing material to promote understanding of HACCP. HACCP is now practiced by food manufacturers and food service establishments, and even in the home.

HACCP Basics

The HACCP system is based on seven principles, or steps, which are discussed in more detail below. They are as follows: (i) conduct a hazard analysis, (ii) determine the critical control points (CCPs), (iii) establish critical limits, (iv) establish monitoring procedures, (v) establish corrective actions, (vi) establish verification procedures, and (vii) keep records and documentation. There are a number of approaches that can be taken in the development of a HACCP program; however, they all stress a commonsense approach to food safety management. Template HACCP plans can be obtained through the FDA and USDA FSIS for specific products for which HACCP have been mandated.

Before a HACCP plan can be developed, there are a number of activities that must be completed. Obviously, management must support the development and implementation of HACCP. Without such support, efforts to develop and implement HACCP will likely prove futile. A HACCP team should be assembled and should include line workers and personnel involved in sanitation and in production and quality assurance. Including personnel involved in marketing and communication may also be appropriate. To balance the team and bring a fresh perspective, an outside expert may be beneficial. In small companies, it may be difficult to establish an effective HACCP team; this is when an outside expert(s) is especially useful.

The HACCP team must recognize that a generic HACCP plan cannot be developed for all food products produced in a manufacturing plant. A HACCP plan should be specific for each product. However, once a HACCP plan is developed, it may be appropriate to use the plan as a template for the development of subsequent plans. A description of the product should include the name, formulation, method of distribution, and storage requirements. The intended use and anticipated customers for the food should be established. The product may not be intended for use by the general public but, rather, by infants, the immunocompromised, or some other high-risk group.

The development and verification of a flow diagram describing the production process are important steps that are often overlooked. The flow diagram is essential for hazard analysis and establishing CCPs (Fig. 29.1).

Figure 29.1 Representative flow diagram and CCPs for a nonthermally processed product. The flow diagram with CCPs is for peeled refrigerated minicarrots. Note that GAPs, GMPs, and a sanitation program are key elements in the production of a safe, wholesome product. doi:10.1128/9781555817206.ch29.f29.01

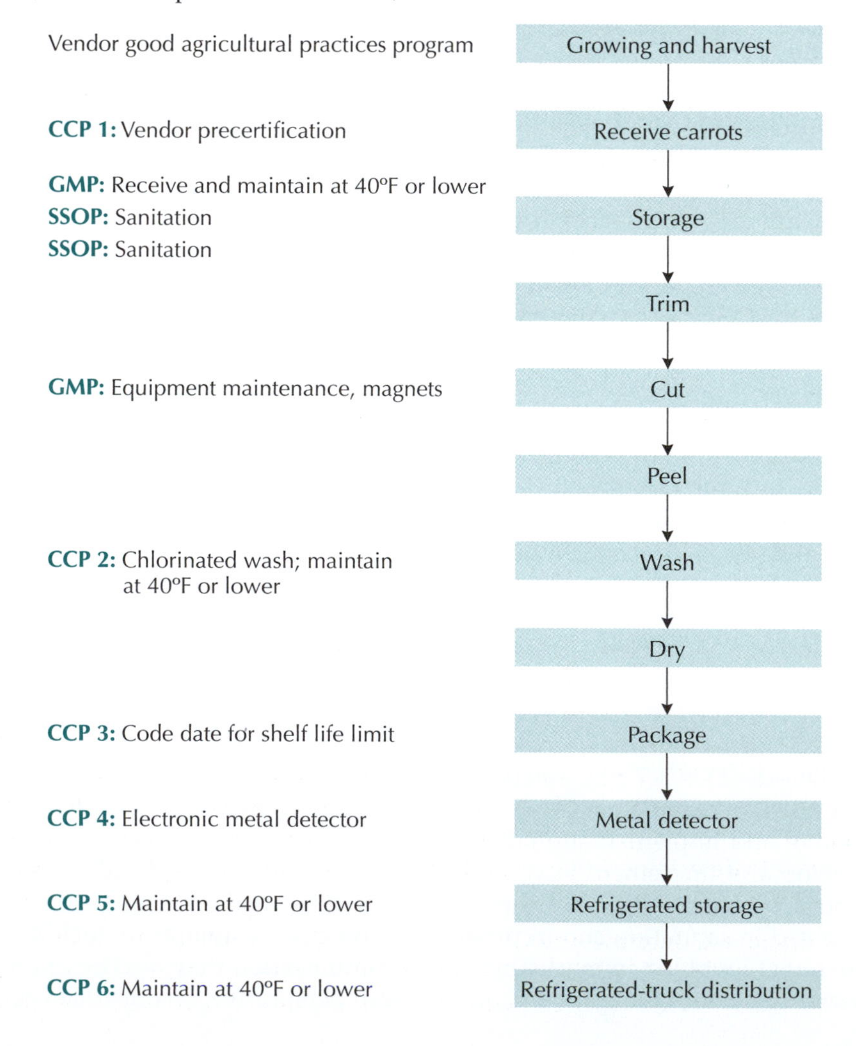

Moreover, the diagram becomes a guide and is essential in the verification process. Even though a flow diagram has been constructed, a "walk-through" of the production facility must be done to verify that the diagram is accurate. For example, new equipment, such as metal detectors, may have been added or the route for rework may not have been accounted for. The flow diagram should be modified if and when necessary.

The following paragraphs discuss the seven principles of HACCP, providing a brief description of each.

Principle 1: hazard analysis. Through hazard analysis, a list of hazards that are likely to cause injury or illness if not controlled is developed. This should be done systematically, starting with raw materials and following through each step of the food production flow diagram. Hazards that present little or no risk to the consumer should not be considered in a HACCP plan. Indeed, those hazards are generally addressed within GMP programs, a building block for the HACCP system. A hazard as set forth by the National Advisory Committee on Microbiological Criteria for Foods for HACCP is "a biological, chemical, or physical agent that is reasonably likely to cause illness or injury in the absence of its control." Even though a hazard may not be considered for inclusion in a HACCP plan, this does not mean that food processors can ignore the hazard, especially if the hazard fits the definition of adulteration.

Once the hazards have been identified through the review of ingredients, the entire production process, equipment, storage, and distribution, each hazard must be evaluated. During hazard evaluation, the HACCP team decides which hazards present a significant risk to consumers and must be addressed through the HACCP plan. Generally, this involves determining the severity of the hazard in terms of the illness or injury it could produce and the likelihood of occurrence. For example, the likelihood that an undercooked beef patty might be contaminated with *Escherichia coli* O157:H7 might be low, but the illness that could result is severe and the potential for subsequent illness (e.g., disease involving kidney failure) is high. Therefore, these hazards should be addressed through HACCP. Alternatively, the hazard may be small pieces of cardboard that enter when ingredient packages are opened during mixing. The likelihood of this occurring may be high, but the severity of the hazard is low (the cardboard may not be very tasty, but it presents little, if any, health hazard). This type of hazard can be addressed through GMPs. The evaluation of the severity of hazards can be based on epidemiological information, scientific research, company records, and information provided by government agencies.

During this process, the HACCP team must identify appropriate control measures. Since not all hazards can be prevented or eliminated, the term "control" is more appropriate. Control measures may vary significantly from one food process to another, even for the same hazard. To control a hazard, a change in processing, such as the addition of metal detectors, or a certificate of analysis from a supplier indicating the absence of a given hazard may be required. Additionally, existing plans, including GMPs and sanitation, can be reviewed to determine whether the hazard is controlled through those plans.

Principle 2: determining CCPs. Once all significant hazards and control measures have been identified, the CCPs to control each hazard must be identified. A CCP can be any point, step, or procedure at which control can be applied and a food safety hazard (chemical, biological, or physical) can be controlled (prevented, eliminated, or reduced to an acceptable level). A CCP may be a cooler, a metal detector, or a package label indicating storage and cooking instructions. Many processes may be redundant, for example, passing dry materials through multiple sieves to control entry of physical hazards into the final product or the use of multiple metal detectors on a production line. In either case, the last sieve or metal detector is considered the CCP.

The temptation in establishing CCPs in a process is to assign each step as a CCP. This is particularly true for raw products. The number of CCPs for any process should be kept to a minimum to facilitate monitoring and documentation (Fig. 29.1). Otherwise, records are generated that have little meaning with respect to the effectiveness of the HACCP plan. Keep in mind that a CCP may control a hazard that has occurred upstream in the process. For example, the hazard might be enteric pathogens in ground beef being used for cooked beef patties. The cooking process is the CCP, since this step will kill pathogens of concern. This does not mean that the ground beef can be left at room temperature or otherwise abused prior to or after being cooked. Indeed, keeping the ground beef at ≤4°C may also be a CCP.

Principle 3: critical limits. Critical limits must be set for each CCP. A critical limit is based on predetermined tolerances that must be met at a CCP to prevent, eliminate, or reduce a biological, chemical, or physical hazard to an acceptable level. Critical limits for preventative measures may be set for such factors as time, temperature, pH, water activity, salt concentration, moisture level, and even the presence or concentration of preservatives. Each food company is responsible for ensuring that the critical limits indicated in its HACCP plan will control the identified hazard. Critical limits may be based on in-house experimental studies or derived from regulatory standards and guidelines or review of the scientific literature. Validation (see principle 6 below) that the critical limits are effective and being met is the key to HACCP.

Principle 4: monitoring procedures. Monitoring is conducted to ensure that a CCP and its limits are effective. The results of monitoring are documented and provide a record for future use in verification. Monitoring may be done automatically (e.g., by computer) or manually. Monitoring activities include measurement of the pH, temperature, and moisture level or water activity. Monitoring must be a planned sequence of observations or measurements to assess whether a CCP is under control. If it is not conducted as a routine, the HACCP plan becomes compromised and contaminated products may reach the consumer. Ideally, monitoring should be continuous and allow advance warning of a problem before violation of a critical limit occurs. However, when continuous monitoring is impractical, a monitoring interval must be established, procedures for the collection of data and equipment to be used must be defined, and the person responsible for monitoring must be identified. The frequency of monitoring will depend in large part on the product and the type of measurement made.

The monitoring procedure(s) for a given CCP should be rapid, which rules out the use of time-consuming analytical tests. Therefore, microbial testing is rarely used for monitoring CCPs. Personnel involved in monitoring must be properly trained, understand the importance of monitoring, and know the proper action to take should deviation from critical limits occur. For example, if ground beef patties were to be cooked for 60 s but the belt increased in speed and the patties were cooked for only 30 s, what should be done? To address this deviation, corrective actions must be established.

Principle 5: corrective actions. HACCP is a proactive system to control hazards that may be associated with a food product. Therefore, corrective actions designed to address deviations from established critical limits should be written into the HACCP plan. Corrective actions should address how the company will fix or correct the cause of deviation to ensure that the process is brought under control and that critical limits are achieved, indicate what is to be done with a product in which deviation from the critical limit occurred, and facilitate the process for determining whether adjustments to the HACCP plan are required.

Corrective action for the beef patties that were not cooked adequately would include correcting the equipment problem, i.e., adjusting the conveyor belt speed. Moreover, the product involved in the deviation must not be released. The product could be sent through the oven again to receive the proper time-temperature treatment, disposed of as waste, or perhaps used in some other product.

Principle 6: verification procedures. Verification within and of the HACCP plan is conducted at several levels. The verification process determines whether the HACCP plan is being followed and whether records of monitoring activities are accurate. The entire HACCP plan should be revalidated through the use of outside auditors. Verification should be done to ensure that the HACCP plan is functioning satisfactorily and in compliance, if required, with government regulations. In the case of cooking ground beef patties, the time-temperature parameters must be verified to kill, for example, *E. coli* O157:H7.

Verification at some level, a given critical limit or the entire HACCP plan, should take place on a regular and predetermined schedule. This may be daily, weekly, monthly, or yearly. The frequency is determined by the operation being verified.

Principle 7: record keeping. Record keeping is the heart of a HACCP plan. All records generated in association with a HACCP plan must be on file at the food establishment, particularly those relating to CCPs and any action on critical deviations and product disposition. Records provide a means for implementation of changes by the HACCP team to assess and document the safety of products. HACCP plan records should include the names of team members, a product description and intended use, flow diagram with CCPs indicated, types of hazards, critical limits, and a description of monitoring, corrective actions, and types of records.

HACCP: beyond the Food Processor

Whether dining at a fine restaurant or enjoying a meal at a local burger establishment, customers expect that the food served will be wholesome and safe. Operators of food service and retail establishments are not required to utilize HACCP. However, many such establishments have voluntarily implemented a food safety management system based on HACCP principles. This proactive approach is encouraged by the FDA to ensure that the food served or sold at a given establishment is safe. In fact, utilizing HACCP-based systems can help food service and retail establishments meet regulations associated with the Food Code.

The flow of food through a food service operation is critical to limiting the contamination of the finished product. Of particular concern is cross-contamination, whereby a cooked or ready-to-eat food comes into contact with an uncooked or raw product. The holding temperature of cold or hot foods is also critical to preventing or limiting the growth of microbes that may be present on a food. In order to ensure that the food is handled appropriately, a flow diagram must be developed (Fig. 29.2). The process of developing a HACCP plan for a food service facility is clearly very similar

Figure 29.2 The flow and handling of product in a food service facility are of the utmost importance to ensuring that the product served to the consumer is wholesome and safe. The flow diagram outlines the flow of foods intended to be served hot or cold. doi:10.1128/9781555817206.ch29.f29.02

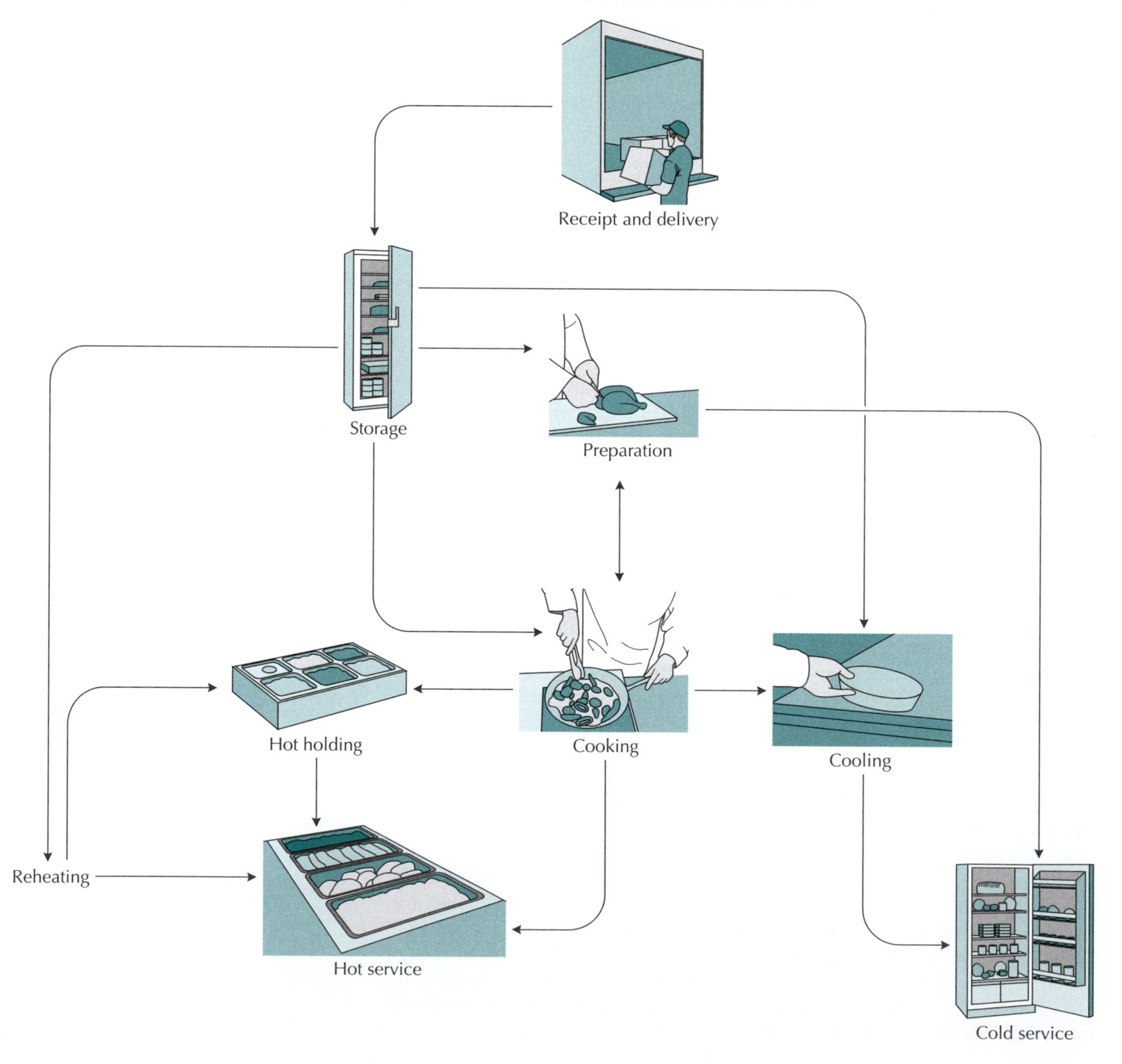

to that used by the food processor. In fact, HACCP plans can be developed for the home by using the same approach.

HACCP Established for Specific Food Industries

In 1996, the USDA FSIS established the Pathogen Reduction and HACCP system program for meat- and poultry-processing plants. The Pathogen Reduction and HACCP rule encompasses more than HACCP. The plan requires all meat and poultry plants to develop and implement HACCP to improve the safety of their products, sets pathogen reduction performance standards for *Salmonella*, requires all meat and poultry plants to develop and implement written SSOPs, and requires animal- and poultry-slaughtering plants to conduct testing for generic *E. coli* to verify the adequacy of their process controls for the prevention of fecal contamination. The plan was precipitated by a large outbreak of foodborne illness caused by the consumption of undercooked hamburgers that were contaminated with *E. coli* O157:H7.

In 1996, the FDA published the *Fish and Fishery Products Hazards and Control Guide* to assist processors in the development of HACCP plans. The plans must address the various hazards to which seafood can be exposed from the fishery to the table, including viruses, bacteria, parasites, natural toxins, and chemical contaminants. Under the regulation, seafood companies must also write SSOPs. The FDA will periodically inspect seafood processors and review HACCP records to determine how well a company is complying over time. The action by the FDA to implement a mandatory HACCP system for the seafood industry is in part linked to the knowledge that more than one-half of the seafood eaten in the United States is imported from almost 135 countries.

The FDA Center for Food Safety and Applied Nutrition proposed the HACCP fruit and vegetable juice regulation in 1998, which applies to juice products in both interstate and intrastate commerce. As in other HACCP regulations, juice processors are required to evaluate their manufacturing processes to determine whether any chemical, biological, or physical hazards could contaminate their products. The real concern is microbiological hazards. The most notable aspect of this regulation is that processors must implement control measures that achieve a 5-log-unit reduction in the numbers of the most resistant pathogens in their finished products compared with levels that may be present in untreated juice. A combination of methods can be used to achieve this goal. Citrus processors can use surface treatments that achieve the 5-log-unit pathogen reduction. Processors are exempt from the microbiological-hazard requirements of the HACCP regulation if they are making shelf-stable juices or concentrates by using a single thermal-processing step. Finally, retail establishments that package and sell juice directly to consumers are not required to comply with the regulation. A driving force behind this HACCP regulation was a large outbreak of foodborne illness linked to the consumption of unpasteurized apple juice that was contaminated with *E. coli* O157:H7.

CONCLUSIONS

Implementation of GMPs and sanitation and HACCP programs prevents the sale of adulterated food products to the consumer. These measures also have a positive impact by decreasing the number of cases of foodborne illness. In today's global marketplace, measures need to be in place to prevent intentional contamination of food or to prevent food that is potentially contaminated from reaching the consumer. Governments are establishing FSOs to translate a public health goal to measurable attributes that allow a company to set control measures for processes. GMPs and HACCP can be used to ensure that FSOs are met. The programs discussed above have mainly concentrated on foods postharvest. On the farm, measures such as good agricultural practices (GAPs) are being used to decrease biological, chemical, and physical hazards that may be associated with raw agricultural products. Indeed, many fruit and vegetable processors are now requiring producers (farmers) to implement GAPs. Reducing the microbiological load of a food at the farm level should aid in keeping the microbiological load of the finished product in an acceptable range.

Summary

- HACCP can be used as evidence that FSOs are being met.
- FSOs are not the same as microbiological criteria.
- GMPs and sanitation are the building blocks of a HACCP program.
- GMPs are designed to prevent the manufacture of food under unsanitary conditions and the sale of adulterated food.
- Sanitation includes personal and environmental hygiene.
- HACCP is a proactive approach to effectively and efficiently ensure farm-to-table food safety.
- HACCP can be effectively applied to food service facilities to ensure safety of prepared food.
- HACCP is mandatory for meat- and poultry-processing plants, fruit and vegetable juice producers, and seafood processors.

Suggested reading

Gould, W. A. (ed.). 1994. *CGMP's/Food Plant Sanitation.* CTI Publications, Inc., Baltimore, MD.

Marriott, N. G. (ed.). 2006. *Principles of Food Sanitation*, 5th ed. Springer Publishing, New York, NY.

National Advisory Committee on Microbiological Criteria for Foods. 1998. Hazard Analysis and Critical Control Point principles and application guidelines. *J. Food Prot.* **61:**762–775.

U.S. Department of Health and Human Services, Food and Drug Administration, Center for Food Safety and Applied Nutrition. 2006. *Managing Food Safety: a Manual for the Voluntary Use of HACCP Principles for Operators of Food Service and Retail Establishments.* http://www.fda.gov/downloads/Food/FoodSafety/RetailFoodProtection/ManagingFoodSafetyHACCPPrinciples/Operators/UCM077957.pdf.

Questions for critical thought

1. Why is it necessary to have DALs for raw agricultural products?
2. What are the seven HACCP principles? Before a HACCP plan can be developed, what needs to be done?
3. Explain why GMPs and sanitation are considered building blocks (prerequisites) for the development of HACCP.
4. Why would testing to determine whether an FSO has been achieved be impractical?

5. Select a product, such as bagged chopped lettuce or precooked hamburger patties, and develop a HACCP plan for it.
6. Using this book and other resources, provide a definition for FSO, PO, and performance criterion. Now try to develop an example for each.
7. When sanitizing agents are used, what factors can influence their efficacy? Explain why.
8. The FDA and USDA FSIS made HACCP mandatory for meat and poultry processors, fruit and vegetable juice producers, and the seafood industry. Write one page for each, detailing events that necessitated mandatory HACCP for those industries.
9. Define adulteration of a food product. Provide a scenario that could lead to a product being considered adulterated.
10. Explain why HACCP can be used to demonstrate that FSOs are being achieved.
11. You have been hired by a small company to develop and direct the company's HACCP program. The owner of the company was against hiring you, since he believes that GMPs and sanitation already cover all aspects of HACCP. You must explain to him what a HACCP program does that GMPs and sanitation do not. He requests that you provide him with a two-page report.
12. Establishing a HACCP program for a restaurant can be difficult since food may be served cooked (grilled steak) or raw (house salad). If possible, visit the kitchen of one of your school's dining halls and record the flow of food through the kitchen. Now determine CCPs for hot and cold foods. Consider whether the flow of food needs to be changed to ensure that raw foods do not cross-contaminate cooked foods. This exercise could be conducted in teams, and then the teams could compare the HACCP programs that they have developed.

Critical Thinking Skills

We are bombarded with hundreds of pieces of (often conflicting) information each day. What should we believe? When should we be skeptical? The guidelines below will help you to critically evaluate scientific information.

Consider the source of the information.

What is the medium of publication?

Is it the Internet, a flier, a book, a magazine, TV, or a peer-reviewed journal? (Papers published in peer-reviewed journals must pass the review of other experts in the field. Peer reviews ensure that the experiments are properly designed, that the data are correctly interpreted, and that the conclusions follow from the results. Peer-reviewed journals are the "gold standard" for scientific publication.) Websites can be made by anyone at little cost, making it easy to champion any cause. Books are expensive to publish and take more time and effort to produce. They are subject to more scrutiny, but their information can go out-of-date faster.

Who is the source of the information?

Is it a "scientist," a "doctor" (in what discipline and from where?), an "expert" (with what credentials?), "a nutritional counselor" (for which there are no recognized professional credentials), a "registered dietician" (whose training is quite rigorous), "a consumer activist," or someone else? Do they have a degree in their field of expertise? Have they published papers on the topic in peer-reviewed journals?

Which organizations do the authors represent?

Why do these organizations exist? Which other issues are the organizations involved with? Are they for profit or nonprofit? Who funds the organization? Dig down one or two layers.

If the information is on a web page, what type of web page is it?

There is a difference between "web pages" where anyone with access to a computer can say anything they want and "pages on the Web" (electronic renderings of highly respected peer-reviewed print journals such as *Science*, *Proceedings of the National Academy of Sciences*, or, for that matter, government publications such as the *Federal Register*). There are also distinctions among web pages. Generally, ".edu" pages have the most credibility since they are associated with educational institutions. ".com" pages vary in credibility from the opinion of "some guy named Joe" to the *opinion* of a giant corporation. Pages ending with ".org" can be at either end of the spectrum. Be aware of these issues when visiting a website.

Be quantitative.

Numbers can be used to influence your opinion. Do not be afraid of simple math.

Put numbers into terms you understand. It is better to be approximately right than clueless. For example, parts per million (ppm) = micrograms per kilogram (μg/kg) = "one in a million." (This is accurate.)

Use approximations to convert between units. Approximations are admittedly wrong, but they get one close to the right answer.

Two kilometers are about a mile.

One meter is about a yard.

A liter is about a quart.

A kilogram is about 2 pounds.

There are about 30 grams in an ounce. (Actually, a more accurate value is 28.3, but 30 allows us to use mental math.)

One centimeter is about 2×10^{-18} light-years.

Understand what the numbers mean.

Be especially aware of percentages. Is the report of "a 100% increase in the incidence of protrusive umbilical disambiguation" derived from "1 in a million" changing to "2 in a million"? (Beware also of the use of big words to confuse and confound. A protrusive umbilical disambiguation is an "outie" belly button.) Are the numbers "estimated," "probable," "related," "confirmed," or "as many as"? Examine the claim, "Experts estimate that there may be as much as a 20% increase in *x*-itis among some people who drink as little as three cups of coffee per day." In addition to all of the qualifiers, the 20% increase may represent a change from 1 per 100,000 to 1.2 per 100,000.

Compare the number with a reference value you know and understand.

Is a given number a lot or a little?

The population of United States is about 300,000,000. Remember that when considering the numbers in the examples below.

The U.S. Environmental Protection Agency considers a hazard actionable if it causes more than one in a million excess deaths. So, is one in a million a lot?

In the United States every year, about 300 people are killed by lightning, about 50,000 people are killed in auto accidents, and about 3,000 people are killed by foodborne disease.

Here is a simple example. At the time of writing, the U.S. national debt was about $14 trillion. That probably means nothing to you. But compare it with something you understand. Take that $14 trillion and divide it by 300,000,000, the number of people in the United States. The answer is that every man, woman, and child in the United States carries a debt of about $47,000. Now the number is not so abstract.

Compare information with data from other sources.
The World Wide Web is great for comparing data. If the same number comes up again and again, it is probably right. Alternatively, these numbers may have originated from the same, but incorrect, source. For many years 9,000 was a widely reported number for annual deaths in the United States from foodborne bacteria. However, this number was traced back to the inaccurate estimate of a single individual. Five thousand deaths then became the accepted figure, but that estimate has recently been revised down to 3,000 to reflect the actual data. Note also that by the nature of the beast, this number can *only* be *estimated*.

Follow the money.
How is the website funded? Is the "Society for Sane Snacking" funded exclusively by snack companies?

When in doubt, doubt.
This is the close corollary of "If it sounds too good to be true, it probably is."

Useful Websites for Food Safety Information

Website[a]	Comments[b]
www.usa.gov	This is the grand portal for any website related to the federal government. If in doubt, start here.
www.fda.gov	The home page for the FDA provides information about everything the FDA does. This includes items related to food, drugs, and medical devices.
www.fda.gov/Food/FoodSafety/default.htm	The subset of the main FDA portal that provides information on everything related to food safety, from allergens to illness to product-specific information for food manufacturers.
www.foodsafety.gov	The gateway to federal food safety information is presented in a consumer-friendly fashion.
www.usda.gov	This is the main USDA homepage and covers everything under the jurisdiction of the USDA. It is easy to follow the links to food safety issues.
www.fsis.usda.gov	The Food Safety Inspection Service is the regulatory arm of the USDA. This site contains information ranging from recalls to the correct cooking temperature for pork.
www.epa.gov	The Environmental Protection Agency has regulatory oversight of pesticides, but the site also has information on microbial food safety, food irradiation, and current issues related to food microbiology.
www.fda.gov/RegulatoryInformation/Legislation/FederalFoodDrugandCosmeticActFDCAct/default.htm	This site contains the entire text of the Food Drug and Cosmetic Act. "The Act" is the foundational document for all food laws and regulations.
www.accessdata.fda.gov/scripts/fcn/fcnNavigation.cfm?rpt=eafusListing	All food additives are listed on this site.
www.cdc.gov/foodsafety	The Centers for Disease Control and Prevention is responsible investigating the epidemiology of foodborne disease. The site also contains consumer information. The explanation of differences between the 5,000 estimated deaths in 1999 and the 2011 estimates of 3,000 deaths from foodborne illness is especially informative.
www.fightbac.org	This site is devoted to informing the public of safe food handling practices.[c]
www.ift.org	The Institute of Food Technologists is the professional society for food science and technology.

Website[a]	Comments[b]
foodsafe.ucdavis.edu/index.html#	Carl Winter, a professor of food microbiology at the University of California, puts food safety to song. This is a very clever site and lots of fun.
www.acsh.org	The ACSH is an independent organization dedicated to bringing sound peer-reviewed science on controversial issues to the public and to policy makers. A wide variety of topics are covered on this site.
www.cast-science.org	The Council for Agricultural Science and Technology plays a role similar to that of the ACSH, but it is more narrowly focused and generates its own expert reports.
www.foodinsight.org	The site of the International Food Information Council provides resources in the area of food safety and nutrition. The site is very up-to-date, for example, providing information on energy drinks, high-fructose corn syrup, and functional foods.
www.cspinet.org	The Center for Science in the Public Interest is the premier consumer advocacy group focusing on food and nutrition. Sometimes it sides with government and industry; more often it does not.
www.truthinlabeling.org	This site is mainly devoted to the fight against monosodium glutamate and is easy to debunk. See its claims that flavor enhancers are made from aborted human embryonic kidney cells.
www.pure-food.com	Automatically redirects to www.organicconsumers.org/irradlink.cfm, a largely anti-food irradiation site.
www.notmilk.com	This site consists of a few hundred links to other sites that the authors have not evaluated.
www.raw-milk-facts.com	Webmaster Randolph Jonsson extols the benefits of drinking raw milk. The Centers for Disease Control and Prevention includes milk pasteurization among the last century's top 10 advances in public health. Whom do *you* trust?
www.foodprotection.org	The International Association for Food Protection provides food safety professionals a forum to exchange information in its peer-reviewed journals and meetings.

[a]These links were active as of 1 June 2011.

[b]FDA, Food and Drug Administration; USDA, U.S. Department of Agriculture; ACSH, American Council on Science and Health.

[c]It is estimated that as much 70% of foodborne disease originates in the home.

Glossary

Acid tolerance response The response of cells to an initial acid treatment which allows them to survive more severe acid treatments. Abbreviated ATR.

Adenosine triphosphate A compound that serves as cellular energy currency. Abbreviated ATP.

Aerobe An organism that requires oxygen to grow.

Aerobic plate count A count that provides an estimation of the number of aerobic microorganisms in a food. Abbreviated APC.

Anaerobe An organism that cannot grow in the presence of air.

Antibody Immunoglobulin protein, produced by B cells (or plasma cells derived from B cells), that binds with a specific antigen.

Antibody titer Measure of the amount of antibody present, usually given in units per milliliter of serum or determined in an end point by dilution.

Antigenic Capable of eliciting an immune (antibody) response. Small molecules may be nonantigenic. Large molecules may have several antigenic sites.

AOAC International An organization involved in validation of testing methods. Formerly the Association of Official Analytical Chemists.

APC *See* Aerobic plate count.

Ascospore A heat-resistant reproductive fungal spore.

ATP *See* Adenosine triphosphate.

ATR *See* Acid tolerance response.

Attaching and effacing lesions Lesions that occur when a bacterium (e.g., *Escherichia coli* O157:H7) adheres to the surface of an intestinal epithelial cell, resulting in loss of microvilli (effacement).

Autoxidation The spontaneous oxidation of a substance.

a_w *See* Water activity.

Bacteremia The presence of bacteria in the blood.

Bioprocess A process that has a biological basis, e.g., fermentation, an enzyme-catalyzed reaction such as the production of high-fructose corn syrup, etc.

Bradyzoite Small stage in various protozoa; it develops into a cyst.

Bureau of Alcohol, Tobacco, Firearms and Explosives An agency of the U.S. Department of Justice that has regulatory authority over beer, wine, and liquor (as well as firearms, explosives, and tobacco). Abbreviated BATF.

Capsid The protective protein structure or coat that encases viral nucleic acid (RNA or DNA).

Catabolic pathway A pathway that leads to the breakdown or utilization of a compound. All pathways for the utilization of sugars are catabolic.

Cell-mediated immunity Immunity in which an antigen is bound to receptor sites on the surface of sensitized T lymphocytes that have been produced in response to prior immunizing experience with that antigen and which manifests through the macrophage response with no intervention of antibody.

Centers for Disease Control and Prevention The agency within the Department of Health and Human Services that tracks foodborne illness and helps solve outbreaks. It has no regulatory authority. Abbreviated CDC.

CFU *See* Colony-forming unit.

Chelate To bind ions.

Chlamydospores Asexually produced resting spores of certain fungi.

-cidal A suffix indicating the ability to kill.

Colony-forming unit A colony on an agar plate that in theory arises from a single bacterial cell. Abbreviated CFU.

Commercial sterility A level of sterility that indicates that an item is free of organisms that can cause illness.

Compatible solutes "Harmless" compounds which cells accumulate to equilibrate their internal water activity with the water activity of the environment.

Cortex The spore structure responsible for resistance properties, presumably through dehydration of the core.

Cucurbits A group of plants that includes squashes, cucumbers, and pumpkins.

Cytokines Substances or compounds that are produced and secreted by cells of the immune system. These substances play a significant role in the immune response.

Cytotoxic Lethal to cells.

Definitive host The host in which a parasite achieves sexual maturity. If there is no sexual reproduction in the life of the parasite, the host most important to humans is the definitive host.

Differential media Media that allow specific bacteria to be visualized (e.g., through a color reaction) among a population of other bacteria.

Disulfide linkage A covalent bond formed between two cysteine molecules in different parts of a protein; serves to stabilize the protein shape.

E_h *See* oxidation-reduction potential.

Emetic toxin A toxin that causes vomiting.

Empirical Determined by trial and error without regard to theory or a hypothesis.

Endemic disease A disease that is continually prevalent in a specific population or area.

Endophthalmitis A type of infection of the eye.

Endospore A bacterial spore formed in the body of the mother cell.

Endotoxin A toxin structurally associated with the cell.

Enterocolitis Inflammation of the lining of the intestine.

Enterotoxin A toxin that acts in the gastrointestinal tract.

Environmental Protection Agency A federal agency whose regulatory authority includes some food-related issues, such as sanitizer efficacy and pesticides. Abbreviated EPA.

Enzyme-linked immunosorbent assay An immunodiagnostic test designed to detect the presence of fixed antibody through linkage with an enzymatic reaction. Abbreviated ELISA.

Epidemiology The study of epidemics; it is used to determine the factors that lead to an outbreak of foodborne illness.

Eukaryote Any of the single-celled or multicellular organisms whose cell contains a distinct, membrane-bound nucleus.

Exopolysaccharide A sugar polymer that is excreted by, and exterior to, the cell. "Slime" is made of exopolysaccharides.

Exotoxin An excreted toxin.

Extrinsic factor An external factor, such as temperature or atmosphere, that influences the ability of microbes to grow in a food.

Facultative Able to do something that is not in the preferred mode. For example, a facultative anaerobe can grow in the absence of oxygen but grows better in its presence.

FDA *See* Food and Drug Administration.

Fecal-oral route A route of disease transmission from fecal matter to the body via the oral cavity.

Fecal-oral transmission The transmission of pathogens from one person's feces to another person's mouth, usually due to poor hand washing.

50% lethal dose The concentration of a substance that will kill 50% of a population. Abbreviated LD_{50}.

Flaccid paralysis A paralysis characterized by limp or "floppy" muscles, as opposed to rigid paralysis.

Food and Drug Administration The U.S. government agency that has legal authority over all foods except meat, poultry, eggs, and alcohol. It is part of the Department of Health and Human Services. Abbreviated FDA.

Food Safety and Inspection Service The food safety regulatory arm of the U.S. Department of Agriculture. It inspects all meat- and poultry-processing plants. Abbreviated FSIS.

Food safety objective A quantitative goal for the frequency of a particular foodborne illness. Abbreviated FSO.

FSIS *See* Food Safety and Inspection Service.

FSO *See* Food safety objective.

Fungicide A compound that kills fungi (i.e., yeasts and molds).

GAPs *See* Good agricultural practices.

Gastroenteritis Broadly speaking, a disease or illness that acts in the gut.

Generally recognized as safe A legal classification of food additives in use before 1958; it includes additives affirmed as safe since that time. Abbreviated GRAS.

Genetic fingerprinting A nucleic acid-based technique that provides specific identification (a "fingerprint") of a microorganism.

Genotype A trait that is characterized genetically (as opposed to phenotype).

Germinant A compound that induces spore germination.

Germination The first irreversible step in the process by which a spore becomes a vegetative cell.

Glyco- A prefix meaning "containing a sugar."

GMPs *See* Good manufacturing practices.

Good agricultural practices Prescribed practices, such as the use of potable water for rinses, prohibition against fertilizing with human manure, and good worker hygiene, that help ensure the microbial safety of food at the farm level. Abbreviated GAPs.

Good manufacturing practices Prescribed practices, such as rodent control programs, use of hairnets and gloves, and prohibition of jewelry, that help ensure the safety of food in food processing plants. Abbreviated GMPs.

GRAS *See* Generally recognized as safe.

Grocery Manufacturers' Association A trade association that provides technical support and lobbies on behalf of member food companies. Abbreviated GMA.

Guillain-Barré syndrome A disease in which paralysis starts at the hands and feet and progresses toward the trunk. Because of this symptomatic similarity to botulism, botulism is often misdiagnosed as Guillain-Barré syndrome.

HACCP *See* Hazard Analysis and Critical Control Points.

Halotolerant Able to tolerate high salt concentrations.

Hazard Analysis and Critical Control Points A proactive, prevention-oriented program that addresses food safety through the analysis and control of biological, chemical, and physical hazards. Abbreviated HACCP.

Hemolysin A compound that causes lysis of red blood cells.

Hemolytic Able to break open red blood cells.

Hemorrhagic colitis A disease characterized by bloody diarrhea.

Hepatomegaly Enlargement of the liver.

Hermetically sealed Sealed under a vacuum.

Heterofermentative Forming lactic acid, acetic acid, and ethanol as fermentation products. Also called heterolactic.

Homeostasis An attempt by a microorganism to maintain a constant intracellular state, e.g., maintenance of pH.

Homolactic Forming only lactic acid as a fermentation product. Also called homofermentative.

Homology A measure of similarity, usually when comparing protein or genetic sequences.

Host An organism that provides an environment in which a second organism lives. In the case of foodborne pathogens, the host is the victim.

Humoral immune response Binding of an antigen to a soluble antibody in blood serum. Also the entire process by which the body responds to an antigen by producing antibody to that antigen.

Hydrophilic Orienting toward water (literally, water loving).

Hydrophobic Orienting away from water (literally, afraid of water).

Immunity State in which a host is more or less resistant to an infective agent; preferably used in reference to resistance arising from tissues that are capable of recognizing and protecting the animal against "nonself."

Immunocompromised Having an immune system that is unable to combat normal disease processes.

Inoculum (pl., inocula) The population of organisms initially present that initiate growth.

Intervention An action taken to prevent an adverse effect. For example, the use of an acid wash on slaughtered beef is an intervention that reduces the probability of bacteria being carried over to the food.

Intrinsic factor A property, such as pH or water activity, inherent in a food.

Isoelectric point The pH at which a protein has no net charge.

Isolate A strain of a bacterium obtained ("isolated") from a specific source. Strains of the same species that are environmental or clinical isolates can be quite different.

Kilogray A unit of absorbed radiation equal to 1 joule of energy. Abbreviated kGy.

Larva Progeny of any organism that is markedly different in body form from the adult.

Lavage A wash.

LD_{50} *See* 50% lethal dose.

Leukocyte A white blood cell.

Low-acid food A food with a pH of >4.6 and a water activity of >0.85.

Lyse To break open.

Lysis The breakage of cells.

Lysozyme An enzyme that degrades cell walls.

MAP *See* Modified-atmosphere packaging.

Meningitis Inflammation of the tissue (meninges) surrounding the brain and spinal cord.

Mesophile An organism with an optimal growth range of 20 to 45°C.

4-Methylumbelliferyl-β-D-glucuronide A substrate used to determine production of β-D-glucuronidase by *Escherichia coli* O157:H7. Abbreviated MUG.

Modified-atmosphere packaging Storage or packing of foods under elevated levels of carbon dioxide. Abbreviated MAP.

Monoclonal antibody An antibody that detects a specific cell target.

Most probable number A statistical method for estimating small populations of bacteria. Abbreviated MPN.

MPN *See* Most probable number.

MUG *See* 4-Methylumbelliferyl-β-D-glucuronide.

Mutagenesis The process of creating a mutant by altering an organism's DNA.

Mycotoxins A generic term for the chemical toxins formed by fungi.

Necrotic Dead (in reference to tissue).

Neurotoxin A toxin that acts on nerves.

New-variant Creutzfeldt-Jakob disease A human neurological disease caused by prions. Abbreviated nvCJD.

Nosocomial infections Infections that are acquired in a health care facility. (Hospitals are very dirty places.)

nvCJD *See* New variant Creutzfeldt-Jakob disease.

Oocyst The cystic form resulting from sporogony in the Apicomplexa protozoa; the oocyst may be covered by a hard, resistant membrane.

Opportunistic pathogen An organism that causes illness only in people with some preexisting medical condition.

Organic acids Acids formed from an organic (carbon-containing) compound, e.g., acetic, lactic, and citric acids. There are also mineral acids such as phosphoric and nitric acids, but these are not as important in foods.

Osmotolerant Tolerant of conditions (i.e., high salt or sugar concentrations) in which there is little available water (low water activity).

Outgrowth The process by which a germinated spore becomes a vegetative cell.

Oxidation-reduction potential The ability to accept or donate electrons. Negative oxidation-reduction potentials are associated with anaerobic conditions. Abbreviated ORP or E_h.

Pandemic Epidemic over an especially wide geographic area.

Paracrystalline Materials having only short- or long-range order in their structures. True crystals have long-range order. Paracrystalline structures have a certain degree of fixed structure, but not enough to be characterized as true crystals.

Parasite An organism that obtains nourishment and shelter from another organism.

Parasitophorous vacuole A vacuole within a host cell that contains a parasite.

Pasteurization A heat process designed to kill pathogens but not necessarily spoilage organisms.

Phagosome A membrane structure formed around a cell absorbed by phagocytosis.

Phenotype An observable characteristic such as the ability to ferment a specific sugar, motility, etc.

Planktonic cells Single cells that are not physically associated with other cells.

Plasmid A piece of circular DNA, separate from the chromosome, that contains genes for traits such as virulence, toxin production, bacteriocin production, and various biochemical traits. Plasmids can be transmitted among different species of bacteria.

Plasmolysis Shrinkage of a cell due to water loss.

Poliomyelitis A degenerative disease of the muscles that causes paralysis. Also called polio.

Polyclonal antibody An antibody that detects many cellular targets.

Polymerase chain reaction A technique in molecular biology that amplifies a few copies of DNA to quantities (thousands to millions) sufficient for further analysis. It is used in food microbiology for identification, detection, and quantification of target bacteria. Abbreviated PCR.

Pomaceous fruits A group of fruits that includes apples and pears.

Premunition Resistance to reinfection or superinfection, conferred by a still-existing infection that does not destroy the organisms of the infection already present.

Prion An unusual type of protein which, by changing shape, causes transmissible encephalopathies.

Prokaryote An organism primarily characterized by the lack of a true nucleus and other membrane-bound cell compartments such as mitochondria and chloroplasts.

Prostration An extreme state of exhaustion.

Proteolytic Producing enzymes that degrade proteins.

Pseudoappendicular syndrome A syndrome in which, in broad terms, an individual experiences symptoms commonly associated with appendicitis.

Psychrophile An organism that "loves" to grow in the cold, has an optimum growth temperature of 15°C, and cannot grow at 30°C.

Psychrotroph An organism that can grow in the cold but has an optimum growth temperature of >20°C and can grow at 30°C.

Pulsed-field gel electrophoresis A modification of standard electrophoretic methods that, through the use of alternating voltages, allows large pieces of DNA to be separated. Abbreviated PFGE.

Quorum The number of organisms that must be present before action can be taken.

Ready-to-eat food A food that can be eaten without further cooking, such as a delicatessen meat.

Reiter's syndrome An autoimmune disease characterized by arthritis and inflammation around the eye.

Resuscitated Brought back to life.

Reverse transcription-polymerase chain reaction The process of making a double-stranded DNA molecule from a single-stranded RNA template through the enzyme reverse transcriptase. Abbreviated RT-PCR.

***rpoS* genes** Genes that encode DNA-dependent RNA polymerase.

Saprophyte An organism that survives by living off dead or decaying plant material.

SASP *See* Small acid-soluble proteins.

Scolex The head or holdfast organ of a tapeworm.

SEA *See* Staphylococcal enterotoxin A.

Selective media Media that select for the growth of specific bacteria by inhibiting the growth of other bacteria that may be present.

Semilogarithmic Pertaining to a relationship or graph where the plot of one variable on a logarithmic scale and the other on a linear scale yields a straight line. Graphs of heating lethality for bacteria are semilogarithmic when the log of the remaining viable cells is plotted against the duration of heating.

Septicemia A gross, whole-body infection.

Serological Able to cause an antibody response.

Serotypes Varieties of a bacterial species that respond to different antibodies.

Sigma factor A protein that binds to a DNA-dependent RNA polymerase.

Small acid-soluble proteins Spore proteins that confer resistance properties. Abbreviated SASP.

Sporadic Occurring randomly.

Sporulation The process by which a vegetative cell forms and releases a spore.

Staphylococcal enterotoxin A One serological type of staphylococcal enterotoxin. Abbreviated SEA.

-static A suffix indicating the ability to inhibit or stop something.

Sublimation The process by which a compound goes from the solid to the gas phase without becoming a liquid.

T cells A type of lymphocyte with a vital regulatory role in immune response; so called because they are processed through the thymus. Subsets of T cells may be stimulatory or inhibitory. They communicate with other cells involved in an immune response by protein hormones called cytokines.

Temperature abuse The holding of food at temperatures that permit microbial growth, i.e., 40 to 140°F.

Tenesmus The sensation of an urgent need to defecate while being unable to do so.

Thermophile An organism that grows at high temperatures.

Titer The concentration of a substance in a solution as determined by titration.

Toxicoinfection An enterotoxin that, as a result of an infection, causes fluid loss.

Transmissible spongiform encephalopathy A disease, such as scrapie, kuru, or "mad cow disease," that is caused by prions. Abbreviated TSE.

Transposable elements Pieces of DNA that can move from one chromosomal location to another.

TSE *See* Transmissible spongiform encephalopathy.

Turkey "X" disease A disease of turkeys for which the causal agent was not initially known.

USDA *See* U.S. Department of Agriculture.

U.S. Department of Agriculture A cabinet-level government department that has legal authority over meat, poultry, and eggs. Abbreviated USDA.

Vascular Related to blood vessels.

Vehicle A source or carrier.

Viable but nonculturable A term applied to cells that cannot be cultured by conventional methods but that still cause illness if ingested. Abbreviated VNC.

Virulent Causing illness.

VNC *See* Viable but nonculturable.

Water activity The measure of water available for microbial growth and chemical reactions, defined as the equilibrium relative humidity of a product. Abbreviated a_w.

Xerotolerant Capable of tolerating dry conditions.

Zoonosis A disease of animals that is transmissible to humans.

Answers to Crossword Puzzles

Chapter 3

Across
2. cortex
5. dehydration
7. low acid
8. botulinum
9. commercial

Down
1. resistance
3. germination
4. sporulation
5. dipicolinic
6. core

Chapter 8

Across
3. pasta
4. dairy
8. tenesmus
9. cyclic
10. hydrophobic
11. *Bacillus*

Down
1. spore
2. *Bacillus cereus*
5. emetic
6. *anthracis*
7. diarrheal

Chapter 10

Across
3. inactivated
5. Botox
8. anaerobic
9. psychrophilic
10. soil
11. infant

Down
1. neurotoxin
2. wound
4. neurological
6. nitrite
7. diaphragm

Chapter 16

Across
1. antigenicity
4. minimal
7. microslide
9. summer
12. sanitation

Down
1. AGR
2. temperatures
3. toxin
5. vomiting
6. people
8. lower
10. shock
11. rapid
13. abuse

Chapter 20

Across
5. malt
6. vinegar
7. bacteriophage
10. bread
11. yeast
12. amylases
14. Pasteur
15. unleavened

Down
1. malolactic
2. sourdough
3. homolactic
4. pasteurization
7. breweries
8. LAB
9. expand
13. hops

Chapter 25

Across
3. dissociate
10. essential
13. acetic
14. bacteriocidal
15. lysozyme
16. extrinsic
17. propionic

Down
1. antimicrobial
2. nitrites
4. sulfites
5. hurdle
6. preservatives
7. plants
8. chemical
9. parabenzoic
11. above
12. adaptation

Chapter 26

Across
2. probiotic
4. bacteriocin
6. pediocin
7. market
10. nisin
11. gram positive
13. bile
14. fermentation
15. indigenous

Down
1. plasmid
3. biopreservation
5. refrigeration
8. urogenital
9. lantibiotic
12. operon

Chapter 27

Across
1. proportional
5. hermetically
6. freezing
8. pasteurization
9. lethal
10. application
13. denaturation
15. drying
17. raisins
18. nonthermal
19. physical

Down
2. pasteurization
3. logarithmic
4. spores
7. radioactive
8. psychrotrophs
11. ionizing
12. penetration
14. Napoleon
15. decimal
16. DNA

Answers to Selected Questions for Critical Thought

Chapter 1

1. Spores in Pasteur's flask would have germinated, grown, and "disproved" his theory.

2. Spontaneous generation could also be demonstrated using flasks that had cotton plugs. If spontaneous generation were true, we would be in big trouble. No process that kills bacteria (such as heat) would be useful because new bacteria would spontaneously arise to take their place.

3. Who knows what the future holds? We will undoubtedly "discover" "new" pathogens. Perhaps there will be an immunization for noroviruses, the leading cause of foodborne disease. Acceptance of food irradiation might allow produce to be "pasteurized." Can you think of more?

4. These are matters of personal opinion.

Chapter 2

1. These assumptions are always false for streptococci which exist in chains. They are also false when cells clump. The statement also assumes that the proper medium is used and that there are no "viable but nonculturable" cells. Incubating cells at the wrong temperature or atmosphere keeps bacteria from forming colonies.

2. There are 24 colonies on the plate. The 24 colonies come from the 10^{-1} (or 10-fold dilution), so at this point there are 240 colony-forming units (CFU) in the tenth of a milliliter that was plated. To get number of CFU per milliliter, multiply by 10. The answer is 2,400 CFU/ml, usually expressed as 2.4×10^3 CFU/ml.

3. The smaller population of pathogens would be obscured by the larger population of total bacteria on the plate.

4. In this example of 3, 3, 2, 1 positives, one starts from the highest dilution that has all positives. Thus, the pattern becomes 3, 2, 1. The most-probable-number (MPN) table yields a numerical value of 150, which is multiplied by the 10-fold dilution for the second "3" to give a bacterial level of 1,150 MPN/g in the sample. Note that the table is constructed to factor in the 0.1-g inocula for the first three tubes.

5. Intrinsic factors such as pH, salt, inhibitors, etc., are inherent to the food. Extrinsic factors such as temperature and atmosphere are external to the food. pH is important because it provides the dividing line for the processing of "high-acid" and "low-acid" food. Temperature is important because it controls microbial growth rates. Cold temperatures can be very inhibitory.

6. The "homeostatic" response is always "on" and increases the activity of proton pumps. The "acid tolerance response" is triggered by moderate acid and then allows the cell to survive under strong acid. "Acid shock proteins" are turned on by strongly acidic conditions. The homeostatic response in the most important since it is the first line of defense for growth at low pH.

7. The amount of sucrose required to lower the water activity (a_w) further would be beyond its solubility limit.

8. The a_w is calculated from *moles* of the solute. One mole of sucrose yields two moles (mol) of monosaccharides. The a_w would decrease.

9. The calculations are based solely on the *addition* of salt. In reality, there are other ions that depress the a_w so that less salt is needed. (This is much like the freezing point of water. In theory, it is 0°C, but in food systems, other solutes make it lower.) One could test this explanation by making different concentrations of some other solute (like glucose) and then determining how much salt is needed to reduce the a_w. If the hypothesis is correct, the amount of salt needed would decrease with increasing glucose concentrations.

10. Use ratios in your calculations. This approach is not very sophisticated, but it always works.

a. The molecular mass of NaCl is 58.4 g/mol.

b. 5 g × (1 mol/58.4 g) = 0.0856 mol (follow the units).

c. When salt disassociates in water, 1 mol of salt gives rise to 2 mol of ions. Thus, for 5% salt one gets 0.17 mol. (For 3% salt, just multiply these by 3/5 to get 0.102 mol.)

d. Turn to the equation $a_w = 55.5/(55.5 + \text{moles of solute})$. For 5% salt, the a_w is 55.5/55.67 = 0.996. For 3% salt, the a_w is 0.998. The 5% salt exerts no microbial significant depression on the a_w. Reformulating to 3% salt would have no effect on microbiological safety.

Chapter 3

1. The required F_o is equal to a $D_{250°F}$ of 0.2 min times the desired 12*D* reduction of *C. botulinum* spores. This gives the legally mandated F_o of 2.4.

This can be expressed as follows:

$$2.4 \text{ min} / 1\text{-log reduction } (0.2 \text{ min})^{-1} = 12\text{-log reduction} \quad \textbf{(1)}$$

For the case of a shorter time, substitute the appropriate values into equation 1.

$$2.16 \text{ min}/1\text{-log reduction } (0.2 \text{ min})^{-1} = 10.8\text{-log reduction} \quad \textbf{(2)}$$

For the case of a lower temperature, substitute the appropriate values into equation 1. In this case, however, because the temperature changed, the *D* value also changes. You have to calculate the new *D* value. The temperature is conveniently 232°F, exactly 18°F lower than the required 250°F.

The z value is 18°F, so the $D_{232°F}$ must be 2 min. Substitute the appropriate values into equation 1:

$$2.4 \text{ min}/1\text{-log reduction } (2.0 \text{ min})^{-1} = 1.2\text{-log reduction} \quad \textbf{(3)}$$

The reduced temperature is much worse than the reduced time.

2. The z value is the number of (Celsius or Fahrenheit) degrees required to change the D value by a factor of 10. The z value for the problem is 15°C.

3. Either answer is acceptable if it is supported by some facts. For example, I would rather be a spore because they just sit around and wait for the right conditions to appear. Then they spring into action.

4. The exosporangium protects the spore from lytic enzymes. The cortex helps maintain the core in the dehydrated state.

5. It is permitted because *C. botulinum* cannot grow at a pH below 4.6. (This is the reasoning behind the different process requirements for low-acid and high-acid foods.)

6. "Gene expression" means that the genes are "turned on" to do whatever they do, i.e., code for proteins, regulatory signals, etc. So genes can be turned on or off at different times and in different places.

Chapter 4

1. The first part of the question you must address on your own. A differential medium will allow for the selection of a specific microbe, for example, *E. coli* O157:H7.

2. The method cannot distinguish live from dead cells.

3. The pour plate method permits greater separation of cells, resulting in a more accurate determination of cell numbers in a sample.

4. Bacteria often adhere to lipids that would be found in hot dogs, preventing dispersion. Also, clumping of cells can occur on large pieces of hot dog. Both would result in underestimating the microbial load of the hot dogs.

5. Bacteria could multiply in the dilution medium prior to being plated.

6. Swab test: swab area can be adjusted by the individual conducting the test; works on all surfaces; diluents can be used for standard plate count (SPC) or directly in commercial test kits. Sticky-tape test: predetermined sample area; transfer only to solid media; no buffers or additional tools required (i.e., swab, diluents, or test tubes).

Chapter 5

1. Bacteria may be clumped or adhered to certain components in a food matrix. Processing a sample will decrease the likelihood that microbial numbers are underestimated.

2. Samples containing known populations of bacteria would need to be tested using the new method and the Food and Drug Administration (FDA)-approved method. Validation of results by a third party should be sought.

3. Student must answer independently.

4. Student must answer independently.

Chapter 6

1. Student must answer independently.

2. Standards—part of a law, ordinance, or administrative regulation.

3. The presence of generic *Escherichia coli* in a sample indicates possible fecal contamination. These criteria might be used to address existing product quality or to predict the shelf life of the food. Coliforms are recommended over *E. coli* and aerobic plate counts (APCs), because coliforms are often present in higher numbers than *E. coli* and the levels of coliforms do not increase over time when the product is stored properly. Presence of coliforms may suggest inadequate sanitation.

4. Student must answer independently.

5. Fecal coliforms include *Klebsiella*, *Enterobacter*, and *Citrobacter* species, but these organisms may be considered false-positive indicators of fecal contamination since they can grow in nonfecal niches, including water, food, and waste. The remaining portion of the question must be answered independently by the student.

6. $n = 5$, number of samples tested; $c = 0$, number samples that can be positive; $m = M = 0$, upper limit for target organism in a given sample. If one sample had 10 colony-forming units (CFU), it would be rejected.

7. *S. aureus* must achieve a critical number such that the toxin associated with foodborne illness achieves a concentration capable of causing a human health risk. There is zero tolerance for *Salmonella* in ready-to-eat foods.

8. Differences in growth of bacteria could be associated with properties of the food, including pH and water activity. The temperature at which the food is held can also impact microbial growth. Perhaps one product is more heavily contaminated with a microbe that will grow at refrigeration temperatures.

Chapter 7

1. Food safety and quality were certainly important before the regulations that became reality in 1906 and 1938 from the U.S. Department of Agriculture (USDA) and Food and Drug Administration (FDA). The FDA actually has its roots in the USDA. As you can see by reading about the creation and development of these agencies, some aspects were certainly dictated by history and some aspects by chance. In reviewing the various acts that were passed in history, you should see how the agencies were becoming more like the ones we have today.

2. *The Jungle* provided the motivation needed to secure an important role for the USDA. The novel is still used today to showcase historical problems as well as contemporary social issues surrounding industrialized animal agriculture. *The Jungle* exposed the unsanitary conditions in the Chicago meat packing industry. This ignited public outrage, which gave President Theodore Roosevelt support and reason to investigate the tale told by Sinclair. The Federal Meat Inspection Act was passed by Congress in June 1906 in response. This act established the sanitary requirements for the meat packing industry and is as important today as it was over 100 years ago. These include mandatory inspection of livestock before slaughter, mandatory postmortem inspection of every carcass, and explicit sanitary standards for abattoirs. In addition to these, the act granted the agency the right to conduct inspections at slaughter and processing operations and enforce food safety regulatory requirements. As a result, the agency grew in size, hiring more than 2,000 inspectors to carry out inspection activities at 700 establishments by 1907.

3. Student must answer independently.

4. Student must answer independently.

5. Epidemiological surveillance is the ongoing systematic collection, recording, analysis, interpretation, and dissemination of data reflecting the current health status of a community or population. The Centers for Disease Control and Prevention (CDC) works with state public health agencies to monitor foodborne illness. Both active and passive surveillance has helped identify changes in outbreak scenarios within the United States. Passive surveillance occurs when health agencies are contacted by physicians or laboratories, which report illnesses or laboratory results to them. In active surveillance, the health agencies regularly contact physicians and laboratories to make sure that reportable diseases have been reported and required clinical specimens or isolates have been forwarded to state laboratories for further analysis. FoodNet is an important example of active surveillance organized by the CDC.

Chapter 8

1. *B. cereus* may be more prevalent in Europe than in the United States because food consumption patterns are different. For example, the British use lots of cream. Sauces are more heavily used than in the United States. There may be greater consumption of rice and other foods associated with *B. cereus* spores.

2. "Rectal tenesmus" is analogous to the "dry heaves," but at the other end.

3. Elements of a case study should include heating followed by temperature abuse, with the appropriate time to illness and symptoms.

Chapter 9

1. In general, the organism merely needs to survive on a given food. In a survival state, the metabolic activity of the cell is extremely low.

2. Student must answer independently.

3. The pathogenesis of Guillain-Barré syndrome (GBS) induced by *Campylobacter jejuni* is not clear. The bacterial lipopolysaccharide causes an immune response. The antibodies produced recognize not only the lipopolysaccharide but also peripheral nerve tissue. This is likely a major mechanism of *Campylobacter*-induced GBS.

4. Initial treatment may not be effective, prolonging the illness, or in the case of a multiresistant strain, the infection may not be treatable. Although ciprofloxacin is effective in treating *Campylobacter* infections, resistant strains have emerged. The resistance may be linked to the use of similar antibiotics (fluoroquinolones) in poultry.

5. Under unfavorable conditions, the organism can essentially remain dormant and cannot be easily recovered on growth media. In the viable-but-nonculturable (VBNC) state, the organism enhances its ability to survive until favorable growth conditions occur. The role of these forms as a source of infection for humans is not clear.

6. (i) In the United States, children are not regularly exposed to *Campylobacter*, so they do not develop immunity as they get older. (ii) The flagella.

7. Thermal processing; pasteurization of milk.

Chapter 10

1. Vichyssoise soup is served cold. Heating it (or the cheddar cheese soup) would inactivate the toxin.

2. Consider a case where a large mass of food (such as mushrooms) is cooked, left in a big bowl on the side of the stove, and then added to the main dish just prior to serving. Cooking the food kills the competing vegetative cells but allows the spores to live in a large warm anaerobic mass of food. The spores germinate and grow. The toxin they make is not inactivated since the food is not reheated before serving.

3. Boiling inactivates botulinal toxin.

4. High-acid foods are allowed to contain botulinal spores because *C. botulinum* cannot grow at a pH of <4.6.

5. The foods could be contaminated by other bacteria that elevate pH to a range that allows growth. Botulism could also be caused by "low acid" varieties of tomatoes or relishes that contain some tomatoes but not enough to make the product acidic.

6. I would tell her that botulinal toxin type C does not kill humans. Her concerns should be strictly aesthetic.

7. This problem is not that hard. The amount of toxin required to kill 50% of the people (LD_{50}) when injected under the skin is 10^{-9} ounce. Let's assume that the oral dose is 10-fold higher and that the study was done with pure toxin. Assume that unpurified toxin is 100-fold less toxic. This would increase the LD_{50} to 10^{-6} ounce. Also assume that none of the toxin is inactivated during dispersal and that it is (somehow) dispersed in single 10^{-6}-ounce doses. So, how many people would be killed? Fifty percent of one million would die.

Chapter 11

1. Your imaginary outbreak should include an appropriate food, temperature abuse, and a long interval between preparation of the food and its consumption. Your outbreak should also contain an element of humor.

2. This is up to the student but should be well reasoned and show an understanding of the difference between a cell and a spore.

3. Steam tables are for keeping hot foods hot, not for heating up cold foods. With a temperature of only 212°F (compared to an oven's 375°F), the steam tables would take too long to heat the food. The food would be in the abuse zone long enough to make foodborne disease a real possibility. So what do I do? The Swedish meatballs are the most microbiologically sensitive food. They have low acidity and lots of nutrients. *Clostridium perfringens* would love these. They go into my oven. The chicken parmesan is less likely to cause a problem, so I go to my neighbor, return her hedge clippers, and ask if she would put the chicken parmesan in her oven. The lasagna has lots of acidic sauce, so I am not too worried. It goes on the grill. Everything goes onto the steam tables, but only when it is good and hot.

4. A case is one person getting sick. An outbreak is a large number of cases in a relatively small geographical area. The text suggests 2,772 cases in 40 outbreaks. Dividing outbreaks into cases = 69 cases per outbreak.

5. Temperature abuse is the biggest cause of *C. perfringens* toxicoinfection.

6. *C. perfringens* has the advantages of being anaerobic, being a sporeformer, and having an optimum temperature slightly higher than that of other pathogens.

Chapter 12

1. Characteristics of *E. coli* O104:H4 (student must find additional strain characteristics): hybrid of enterohemorrhagic *E. coli* (EHEC) + enteroaggregative *E. coli* (EAEC), ferment sorbitol within 24 h, lactose positive, β-glucuronidase positive, missing Stx1 and *eae* genes, has only Stx2, can adhere to intestine cells, has incubation period of 7 to 9 days (O157:H7, 3 to 4 days).

2. Possession of an attaching-and-effacing (AE) gene (*eae*) and Stx production.

3. Stxs act by inhibiting protein synthesis. Transfer of the toxin to the Golgi apparatus is essential for intoxication; however, the mechanism of entry of the A subunit and particularly the role of the B subunit remain unclear. Although the entire toxin is necessary for its toxic effect on whole cells, the A1 subunit can cleave the *N*-glycoside bond in one adenosine position of the 28S ribosomal RNA (rRNA) that comprises 60S ribosomal subunits. This elimination of a single adenine nucleotide inhibits the elongation factor-dependent binding to ribosomes of aminoacyl-bound transfer RNA (tRNA) molecules. Peptide chain elongation is stopped, and overall protein synthesis is suppressed, resulting in cell death.

4. Let's use enterotoxigenic *E. coli* (ETEC) as an example. Humans are the principal reservoir for ETEC. Therefore, improper hand washing practices could result in cross-contamination of food. This scenario can occur even after foods have been properly cooked or thermally processed.

5. There have been outbreaks of *E. coli* O157:H7 linked to the consumption of high-acid foods, including apple cider. The pathogen, although not capable of growth at such extremely low pH, is capable of surviving and subsequently causing human illness.

6. (i) No. (ii) ETEC is generally not associated with outbreaks in the United States. (iii) True. (iv) True.

7. The diet of animals prior to slaughter could be changed in an effort to reduce the shedding of *E. coli* O157:H7 in the feces. This may have an impact on reducing contamination of the carcass during processing.

8. There is no one single or simple answer to this question. Implementation of good agricultural practices, good manufacturing practices, Hazard Analysis and Critical Control Points, and test-and-hold programs and greater consumer awareness on proper handling and processing of food all would be beneficial.

Chapter 13

1. There are many reasons that *Listeria monocytogenes* was not recognized before 1980. Listeriosis appears as geographically distant sporadic outbreaks. Only with the advent of computer networks and centralized identification could these outbreaks be recognized. Moreover, the medical profession did not know about listeriosis. If physicians did not recognize it, they could not order the right tests and listeriosis remained unreported. The increased sales of refrigerated foods might contribute to increased cases and therefore increased recognition. It is also possible that as sanitation improved, there were fewer competitors against *L. monocytogenes* in foods, allowing it to become more prominent.

2. The pros of having a numerical tolerance are as follows. (i) With a zero tolerance, companies are reluctant to look for *L. monocytogenes* (and often look for *Listeria* species instead). This "don't ask, don't tell" policy can lead to increased cases of listeriosis. (ii) Testing specifically for *L. monocytogenes* would allow companies to identify it and nip the problem in the bud. (iii) Countries that have numerical tolerances have similar rates of listeriosis as do countries having a zero tolerance. This suggests that the zero tolerance provides no public health benefit. Given this last argument, it appears that neither criterion has a public health impact. Decisions in the United States will be made on other grounds. (Consider the political fallout of asking your congresswoman, "How many of those deadly bacteria do you want to allow in my food?")

3. Sorry, but you have to make up your own outbreak. I have written mine at the beginning of the chapter.

4. See Fig. 13.6.

5. Several characteristics make *L. monocytogenes* a successful foodborne pathogen. (i) It grows at refrigeration temperatures. (ii) It can cross membrane barriers, including the placental and brain barriers. (iii) It survives in food processing environments for long periods. (iv) It is ubiquitous. (v) There is a long period between ingestion and the onset of symptoms. (vi) It multiplies within the host cells.

Chapter 14

1. *Salmonella* requires a water activity (a_w) of >0.90 to grow. The organism would be expected to survive.

2. The isolate would react with the C_1 group antiserum, indicating that antigens 6 and 7 are present. Flagellar antigen agglutination in poly-C flagellar antiserum and subsequent reaction of the isolate with single-group H antisera would confirm the presence of the r antigen (phase 1). The empirical antigenic formula of the isolate would then be 6,7:r. Phase reversal in semisolid agar supplemented with r antiserum would immobilize phase 1 salmonellae at or near the point of inoculation, thereby facilitating the recovery of phase 2 cells from the edge of the zone of migration. Serological testing of phase 2 cells with poly-E and 1-complex antisera would confirm the presence of the flagellar 1 factor. Confirmation of the flagellar 5 antigen with single-factor antiserum would yield the final antigenic formula 6,7:r:1,5, which corresponds to serovar Infantis.

3. *Salmonella enterica* subsp. *enterica, S. enterica* subsp. *salamae, S. enterica* subsp. *arizonae, S. enterica* subsp. *diarizonae, S. enterica* subsp. *houtenae, S. enterica* subsp. *indica,* and *S. enterica* subsp. *bongori.* The most common serovars in the United States are Enteritidis and Typhimurium.

4. The pathogen colonizes the ovaries of laying hens, making it difficult to eradicate.

5. The resistance genes are located on the chromosome; therefore, all progeny will carry the genes.

6. In the host, the capsule can block antigen recognition and protect against destruction by phagocytic cells. In a processing environment, capsule can facilitate biofilm formation by the organism, protecting it from the action of sanitizing agents.

7. Induction of the acid tolerance response (ATR) would permit survival of the pathogen in low-acid foods such as orange juice. The ATR would allow the organism to survive passage through the low pH of the stomach and acidic environments in phagocytic cells.

Chapter 15

1. There are four species in the genus *Shigella: Shigella dysenteriae* (group A), *S. flexneri* (group B), *S. boydii* (group C), and *S. sonnei* (group D). Shigellae are zoonotic.

2. Shigellae are shed in the feces of symptomatic and asymptomatic carriers. In day care centers, the pathogen is spread by the fecal-oral route (children placing dirty hands in each other's mouths).

3. Two important characteristics: (i) the production of bloody diarrhea or dysentery and (ii) the low infectious dose.

4. Treatment with antibiotics may result in lysis of the bacterial cell releasing Shiga toxin, which is associated with hemolytic-uremic syndrome (HUS).

5. Complications arising from the disease include severe dehydration, intestinal perforation, septicemia, seizures, HUS, and Reiter's syndrome.

6. Infectious dose, in simple terms, is the number of cells required to cause an infection. *Shigella* has an infectious dose of ≤100 cells, which is low compared to those of other pathogens.

7. Virulent strains of *Shigella* are invasive when grown at 37°C but noninvasive when grown at 30°C. This strategy ensures that the organism conserves energy by synthesizing virulence products only when it is in the host.

Chapter 16

1. Many things contributed to this outbreak. The source of the *Staphylococcus aureus* was the boil on the chef's hand. If the *S. aureus* had not been there, none of the other errors would have mattered. The chef should not have been handling food, or at the very least his hand should have been bandaged and gloved. Note that gloves are not an absolute barrier for bacteria; 0.01% of bacteria on hands can be transmitted to food (R. Montville et al., *J. Food Prot.* **64:**845–849, 2001). Secondary causes of the outbreak include preparing the food too far in advance and holding the food at temperatures that allow its growth.

2. The passengers got sick even though the food was ultimately cooked because the enterotoxin is extremely heat resistant.

3. Do *I* practice what I preach? Giving a short microbiology lesson to the cabin attendant never helps. Whether or not I eat the undercooked chicken depends on many factors. How hungry am I? What other food is available? How undercooked is it? True confession: I was in this situation once and ate the chicken, saving a small piece for my son, a lawyer, in case any illness followed. (It did not.)

4. Enterotoxins act in the stomach. Exotoxins are released from the bacteria. Botulinal toxin is an exotoxin that is not an endotoxin. (It is a neuroparalytic agent.)

5. You need two buckets since staphylococcal toxins cause vomiting and diarrhea. (An airplane is a good place to contract staphylococcal food poisoning because the lavatory is so small that one can vomit in the sink without leaving the toilet.)

6. Signal transduction is like a telegraph that transmits information from outside the cell to the inside of the cell. This signals the cell to respond in a metabolically appropriate way. For example, when the cells sense that water activity is decreasing, that signal may be transduced across the cell membrane to increase the synthesis of compatible solutes (see chapter 2) that help cells maintain their viability.

Chapter 17

1. *V. vulnificus* causes wound infections.

2. *V. vulnificus* causes wound infections.

3. Infection associated with the consumption of raw or undercooked seafood.

4. Cold water temperatures are thought to result in a cold-induced viable-but-nonculturable (VBNC) state, in which the cells remain viable but are no longer culturable on the routine media normally employed for their isolation.

5. The pathogen can be killed through exposure to horseradish-based sauces. Application of these sauces to raw oysters will not make the oysters safe to eat, since they do not kill bacteria within the oyster.

6. Most infections are associated with the consumption of raw or undercooked shellfish. Shellfish are bivalves, and so the organism would be concentrated in the stomach of the shellfish and not influenced by sanitation.

Chapter 18

1. Yes. The pathogen still causes serious cases of foodborne illness and even death.

2. *Listeria monocytogenes* is of greater concern to pregnant women and has a higher mortality rate.

3. Yersiniae may have acquired a number of human genes that enable them to undermine key aspects of the physiological response to infection.

4. The pig. Pigs are the only animal species from which *Yersinia enterocolitica* of biovar 4, serogroup O:3 (the variety most commonly associated with human disease), has been isolated with any degree of frequency.

5. In children older than 5 years and adolescents, acute yersiniosis often causes pain in the abdomen that is mistaken for appendicitis.

6. Human volunteers cannot be used to study yersiniosis, since they may develop other diseases. Infections with *Y. enterocolitica* are noteworthy for the large variety of immunological complications, such as reactive arthritis, carditis, and thyroiditis, which follow acute infection. Of these, reactive arthritis is the most widely recognized.

7. Ability to enter mammalian cells is associated with invasion, a heat-stable enterotoxin, pYV plasmid, and hemolysin production.

8. Immunological complications, such as reactive arthritis, carditis, and thyroiditis. Other autoimmune complications of yersiniosis, including Reiter's syndrome, uveitis, acute proliferative glomerulonephritis, collagenous colitis, and rheumatic-like carditis, have been reported, mostly from Scandinavian countries.

Chapter 19

1. Heterofermentative bacteria can only make one adenosine triphosphate (ATP). Homofermentative bacteria make two ATPs. *But*, heterofermentative bacteria can generate that one ATP from pentoses, while heterofermentative organisms cannot use pentoses at all. Hence, "In the land of the blind, the one-eyed man is king." Another folk saying might be "An ATP in the hand is worth two in the bush."

2. Use the hexose: it yields twice as much energy as the pentose.

3. "Fermentation" can be used to mean (i) any bioprocess, such as an amino acid fermentation; (ii) the process by which microorganisms make energy in the absence of oxygen; or (iii) food fermentations where the microbes make changes in the food to alter properties important to consumers.

4. Acidity controls the degree of ionization of the protein's amino acids. Changes in protonation can cause changes in conformation. Consider what happens when an egg is dropped into acid. The protonation of its proteins changes and causes the conformation to change to one that is insoluble in water (i.e., coagulate). Clearly, coagulation affects the binding of water.

Chapter 20

1. Yeasts that favor the production of carbon dioxide (as a leavening agent) are used to make bread (although you can smell a bit of alcohol in freshly baked bread). Yeasts that make more ethanol (and carbon dioxide for carbonation) are more prevalent in beer.

2. Barley contains the fermentable carbohydrate used in beer fermentation. Unfortunately, the carbohydrate is starch, which yeast cannot ferment.

Mashing converts the starch to glucose, which the yeast can use. Grape juice contains sugars that the yeast can ferment directly, so no mashing is required.

3. An adjunct is a fermentable carbohydrate that is less expensive than the malt it replaces. Corn (which is very cheap in the United States) is the most common adjunct, although other cereals and grains can be used. Purity laws in countries such as Germany forbid the addition of adjuncts in the brewing of beer. In the United States, inexpensive (usually temporary) adjunct professors are hired to teach courses that might otherwise be taught by expensive permanent faculty members.

4. A certain amount of "genetic drift" occurs when the inocula are used repeatedly. The fermentation needs to be "reset" periodically using defined inocula.

5. Draft beer contains live yeasts, which could spoil the product. It must be refrigerated to keep them from growing. Bottled beer does not require refrigeration because it is pasteurized.

6. If I gave you the answer, you wouldn't answer the question.

7. William Franklin signed the charter of Queen's College, which eventually became Rutgers University (the academic home of two of the authors of this book, Drs. Matthews and Montville).

Chapter 21

1. The limited shelf life of comminuted (ground and blended) muscle foods has been attributed to (i) a higher initial microbial load due to use of a poorer-quality product for grinding, (ii) contamination during processing, and (iii) the effects of the comminution of the muscle tissue.

2. Temperate waters: *Vibrio*, *Moraxella*, and *Pseudomonas*. Tropical waters: *Bacillus*, coryneforms, and *Micrococcus*. Other factors: method of harvest, method of storage, type of fish.

3. The major microbial inhibitors in raw milk are lactoferrin and the lactoperoxidase system. Natural inhibitors of less importance include lysozyme, specific immunoglobulins, folate, and vitamin B_{12}-binding systems. Lactoferrin, a glycoprotein, acts as an antimicrobial agent by binding iron. Lactoperoxidase catalyzes the oxidation of thiocyanate and the simultaneous reduction of hydrogen peroxide, resulting in the accumulation of hypothiocyanite.

4. Generally, a chlorinated water wash reduces the microbial load of a product by 1 to 2 log colony-forming units (CFU)/g.

5. Lactic acid bacteria, many streptococci, and certain sporeformers. The milk would start to brown as the milk sugar starts to caramelize.

6. Most bacteria do not grow when the water activity (a_w) is ≤ 0.85.

7. Generally, bacteria start to produce proteases at lower populations than lipases. Most bacteria preferentially utilize protein prior to degrading lipids.

8. The pH of most fruits and vegetables is not conducive to the growth of bacteria.

Chapter 22

1. Most molds will not produce mycotoxins at temperatures of <15°C. Most molds will not grow at a water activity (a_w) of ≤0.60. In developing countries, the key would be to store grains that have an a_w of ≤0.60; this would prevent mold growth and thereby mycotoxin production.

2. Patulin is a lactone, and it produces teratogenic effects (that is, it causes fetal malformations [birth defects]) in rodents, as well as neurological and gastrointestinal effects. *Penicillium expansum* produces patulin as it rots apples and pears. Consequently, it is also important as an indicator of the use of poor raw materials in juice manufacture.

3. Aflatoxins are among the few mycotoxins covered by legislation. Statutory limits are imposed by some countries on the amounts of aflatoxin that can be present in particular foods.

4. *Aspergillus* species are conveniently "color coded," and the color of the conidia (nonmotile spores) serves as a very useful starting point in identification. The microscopic morphology is also important in identification. The most effective medium for rapid detection of aflatoxigenic molds is *Aspergillus flavus* and *Aspergillus parasiticus* agar. Under incubation at 30°C for 42 to 48 h, *A. flavus* and *A. parasiticus* produce a bright orange-yellow colony reverse, which is diagnostic and readily recognized. A selective medium, dichloran-rose bengal-yeast extract-sucrose agar, is used for the enumeration of *Penicillium verrucosum* and *Penicillium viridicatum*. *P. verrucosum* produces a violet-brown reverse coloration on this agar.

5. Aflatoxin B_1 is toxic in its native form. Aflatoxins inhibit oxygen uptake in the tissues by acting on the electron transport chain and inhibiting various enzymes, resulting in decreased production of adenosine triphosphate (ATP).

6. The presence of *Fusarium verticillioides* in corn is a major concern because of the possible widespread contamination of corn and corn-based foods with its toxic metabolites, especially the fumonisins. The main human disease associated with *F. verticillioides* is esophageal cancer. Mycotoxins produced by *F. verticillioides* include fumonisins, fusaric acid, fusarins, and fusariocins.

Chapter 23

1. Taxonomy is the science of classifying, identifying, and naming species and arranging them into a classification system. Members of the kingdom Protista may be the most diverse of all organisms. Protozoa are classified based on cellular structures; for example, *Giardia* has flagella but *Cryptosporidium* does not. After classifying protozoa on cellular structures and other physical or physiological classifications, genus and species differences will be based on genetic similarities.

2. Both oocysts and helminth eggs may be spread in the environment via contaminated soil or water. The initial contamination event may occur from human or animal feces from sewage sludge or from raw or undercomposted manure or biosolids.

3. *Cryptosporidium* oocysts are immediately infectious upon being excreted in feces, while *Cyclospora* and *Toxoplasma* oocysts are shed unsporulated and become infectious and sporulate in the environment after about 6 days of exposure to oxygen and the environment.

4. *Toxoplasma* is an extraintestinal parasite. *Toxoplasma* oocysts may be transmitted in three very different ways: oocysts may be ingested as described in answer 2 above, or oocysts may be ingested by an animal and lead to the formation of tissue cysts that can be ingested in undercooked meats and cause disease, or *Toxoplasma* may cause congenital infection in an unborn fetus when a pregnant woman ingests *Toxoplasma* oocysts or tissue cysts.

5. Sometimes, the lining of the small intestine is damaged by the infection. This may lead to digestive difficulties even after the infection is gone. About half of those who are infected with *Giardia* develop problems with lactose (milk sugar) intolerance. This may last up to 6 months after the infection is treated. Symptoms of lactose intolerance may include diarrhea, gas, cramps, and bloating after the ingestion of milk products. For additional information, see T. B. Garder and D. R. Hill, *Clin. Microbiol. Rev.* **14:**114–128, 2001.

6. There are many ways to respond to this question. Body comparisons are below. While all helminths have complex life cycles, generally speaking the trematodes have a more complicated life cycle. The trematodes have at least two hosts, with one often a mollusk.

Feature	Roundworm	Tapeworm	Fluke
Classification (taxonomy)	Nematodes	Cestodes	Trematodes
Body shape	Elongated cylindrical body, tapered at both ends	Distinct body form with a head called a scolex and a segmented body composed of proglottids; each segment contains a complete set of organs	Fairly simple body shaped like a diamond or a leaf, with a small sucker on the anterior end

Chapter 24

1. The commonly accepted cause of bovine spongiform encephalopathy (BSE) is a misshapen prion protein. This misshapen protein causes other normal proteins to become misshapen, and these proteins clump together to form sheets. The presence and detection of this altered or misshapen protein are regarded as a marker for the disease. Current tests work by detecting the infectious agent in a sample of central nervous system tissue, either brain or spinal cord. Other laboratory methods for detection include histopathological examination and staining for the misshapen protein.

2. There are many ways to differentiate viruses, protozoa, and prions based on size, life cycle, and complexity. One important difference is the fact that out of this group, bacteria may replicate on their own outside of a host. This difference has vast implications for public health importance and food safety issues.

3. There are many good answers to this question, including heat, freezing, irradiation, chemical disinfection, etc.; however, it is important to note that prions are resistant to heat, even that at temperatures during rendering.

4. Check out food safety and public health or medical diagnostics. One suggestion for answering this question is to look up various tests that are marketed for food safety, designed by the myriad of food testing companies.

5. The change in prion tertiary structure influences the action of the protein. The normal prion (PrP^c) is composed of α-helices, while the misshapen PrP^{sc} is composed of runs of β-pleated sheets. This structural change is reflected in the protein's activity and response to denaturing enzymes and chemicals. The conversion of PrP^c into PrP^{sc} appears to be modulated by PrP^{sc} consumed in the diet.

6. Good sanitation and attention to good manufacturing practices can reduce the likelihood of viruses in foods. While viruses may be present in foods, viruses never grow in foods, as suggested in the answer above.

Chapter 25

1. Even if bacteria did not grow, foods would not last forever. There are many quality issues involving endogenous enzymes, loss of nutrients, color changes, and other reactions (such as the Maillard reaction and lipid oxidation) that occur over time and would render the food unacceptable.

2. This is a tough nut to crack. One might freeze the food since microbes do not grow at freezing conditions. But there could still be issues with lipid oxidation, freezer burn, etc. One could also irradiate the foods, but if the marketing department does not like preservatives, it will like irradiation even less.

3. See Fig. 25.1 In low-acid (high-pH) foods, the acid dissociates (the proton comes off, leaving an ionized acid) and cannot pass through the membrane to do any harm. High-acid foods keep the acid protonated, so it can cross the membrane. In the neutral cytosolic environment, the acid dissociates and acidifies the interior of the cell.

4. The *ideal* antimicrobial would have to be safe, effective against all species of bacteria, yeasts, and molds, have no odor or taste, function at all pH levels, be natural and multifunctional (i.e., do other things, such as prevent lipid oxidation), and be very inexpensive. (There are no ideal antimicrobials.)

5. The pK_a of acetic acid is 4.5.

a. To solve at pH 5.5,

$$pH = pK_a + \log [A^-]/[HA]$$

$$5.5 = 4.5 + \log [A^-]/[HA]$$

where A^- is the anion form of the acid (A), H is a proton, and AH is the protonated form of the acid. Let's try solving this in word math: 1 (subtract 4.5 from both sides of the equation) is equal to the logarithm of $[A^-]/[HA]$. If the log of the number is equal to 1, that number is 10. So at pH 5.5, there will be 10 times more anion than acid. In other works, only 10% of the acid being used is in a form that is effective.

b. At pH 3.5, solve in symbolic math:

$$3.5 = 4.5 + \log [A^-]/[HA]$$

$$-1 = \log [A^-]/[HA]$$

There will be 1/10 as much A^- as HA. Now you have 10-fold more of the effective acid.

c. If someone knew that acetic acid is not very effective at pH 5.5 but decided to just add more acetic acid, they would have to add 100-fold more to have the same effective concentration as there would be at pH 3.5. (It would taste like vinegar.)

The pK_a concept can be very important to product developers. If you are math phobic, just remember that at the pK_a, the amounts of protonated and unprotonated acids are equal. Every pH unit above the pK_a decreases the amount of protonated acid by a factor of 10. Organic acids are not very useful above their pK_a values but work fine below them.

Chapter 26

1. Only you can check the temperature in your refrigerator. (Is it between 40 and 140°F?)

2. The pH drops because when temperature abuse occurs, the lactic acid bacteria are able to grow and produce acid, thus inhibiting *C. botulinum* growth.

3. In the early days of bacteriocin research, for an inhibitor to be characterized as a bacteriocin, it had to be a (i) plasmid-mediated (ii) ribosomally made (iii) small protein (iv) that was bactericidal to a closely related group of bacteria (v) which have specific binding sites for it, but it (vi) did not inhibit itself and (vii) was made in the stationary phase of growth. (Yes, we know, that's seven criteria.)

4. If the bacteriocin acted against the bacterium that produced it, that bacterium would effectively be committing suicide.

5. Since the genes needed for bacteriocins to work are on the same operon, they can be transcriptionally regulated as a group.

6. There is no fixed answer to this question.

Chapter 27

1. The z value is a measure of how much the reaction rate (D) changes with time and thus would be analogous to the activation energy (E_a) in chemical kinetics.

2. The equation to use in both cases is

$$W \text{ (log reduction)} \times D \text{ (minutes/log reduction)} = \text{process time (minutes)}$$

(Note how the units cancel out.)

In the case of a 10% reduction in time,

$$W D \times 0.2 \text{ min}/D = 2.16 \text{ min}$$

Solved for W, $2.16/0.2 = 10.8$-log reduction.

In the case of a 10% reduction in temperature, the equation above can be used again if you know the $D_{225°F}$. The z value is the number of degrees required to change the D value 10-fold. In this case, the z value is 25°F, serendipitously the same as the difference in temperature. So there is a 1-log change in the D value; $D_{250°F}$ becomes a $D_{225°F}$ of 2.0 min. Now you can use the equation

$$W D \times 2.0 \text{ min}/D = 2.4 \text{ min}$$

Solved for W, there is only a 1.2-log reduction. (Question within an answer: Why is the influence of temperature so much greater than that of time?)

3. The z value is the number of degrees required to change the D value 10-fold. In this case the z value is 15°C. (Every increase of 15°C changes the D value by a factor of 10.)

Chapter 28

1. Nonthermal processing methods are desirable in that they can maintain the fresh-like characteristics of food products while inactivating enzymes and microorganisms for increased shelf life and may also inactivate pathogenic microorganisms.

2. Students should research independent responses to this question.

3. Students should research independent responses to this question.

4. Students should research independent responses to this question.

Chapter 29

1. It is not possible, and never has been possible, to grow (in open fields), harvest, and process crops that are totally free of natural defects. Certain defects that present no human health hazard, particularly in raw agricultural commodities, may be carried through to the finished product.

2. The Hazard Analysis and Critical Control Points (HACCP) system is based on seven principles, or steps, which are discussed in more detail below. They are as follows: (i) conduct a hazard analysis, (ii) determine the critical control points (CCPs), (iii) establish critical limits, (iv) establish monitoring procedures, (v) establish corrective actions, (vi) establish verification procedures, and (vii) keep records and documentation. A flow diagram of the process must be developed.

3. The "Current Good Manufacturing Practices in Manufacturing, Packing, or Holding Human Food" (CGMPs) detail requirements related to various operations or areas in food processing facilities. The requirements are established to prevent contamination of a product from direct and indirect sources. Sanitation includes a host of activities that are ultimately designed to prevent the adulteration of a product during processing. The true goal of any sanitation program is to protect the health of the consumer. In addition, a sound sanitation program increases the shelf life of a product, thereby minimizing economic loss, and prevents contamination of a product with materials that may offend the consumer (e.g., pieces of wood, rocks, or machinery). Implementation of CGMPs and sanitation standard operating procedures (SSOPs) ensures that HACCP only focuses on true human health hazards.

4. Food safety objectives (FSOs) are typically very low, since they represent the maximum level of a hazard at the point of consumption. Programs including HACCP will provide the evidence that FSOs are met.

5. Student must answer independently.

6. FSOs are the maximum frequency and/or concentration of a microbiological hazard in a food at time of consumption that is considered acceptable. Performance objectives (POs) can be used by a government agency (e.g., the Food and Drug Administration [FDA] or U.S. Department of Agriculture [USDA]) or other authority to communicate food safety levels to industry and to other governments. Basically, the FSO functions as a tool to translate a public health goal (e.g., level of aflatoxin in peanuts) to measurable attributes. The performance criterion is the frequency or concentration of a hazard in a food that must be achieved by the application of one or more control measures to provide or contribute to a PO or FSO.

Index

D

E

F

G

H

I

M

P

Q

R

W

X

Y

Z